Low-Abundance Proteome Discovery

Low-Abundance Proteome Discovery

State of the Art and Protocols

Pier Giorgio Righetti

Miles Gloriosus Academy, Via Archimede 114, Milano 20129, Italy

Egisto Boschetti

11, rue de la Cote a Belier, 78290 Croissy-sur-Seine, France

AMSTERDAM • BOSTON • HEIDELBERG • LONDON
NEW YORK • OXFORD • PARIS • SAN DIEGO
SAN FRANCISCO • SINGAPORE • SYDNEY • TOKYO

ELSEVIER

Elsevier
225 Wyman Street, Waltham, MA 02451, USA
The Boulevard, Langford Lane, Kidlington, Oxford, OX5 1GB, UK
Radarweg 29, PO Box 211, 1000 AE Amsterdam, The Netherlands

Library of Congress Cataloging-in-Publication Data
Righetti, P. G.
  Low-abundance proteome discovery : state of the art and protocols / Pier Giorgio Righetti and Egisto Boschetti.
    pages cm
  Includes index.
  ISBN 978-0-12-401734-4
1. Proteomics. 2. Proteomics–Methodology. 3. Proteins–Analysis. I. Boschetti, Egisto. II. Title.
  QP551.R524 2013
  572'.6–dc23

                        2013007406

British Library Cataloguing in Publication Data
A catalogue record for this book is available from the British Library

ISBN: 978-0-12-401734-4

CONTENTS

ACKNOWLEDGMENTS

We thank all the colleagues, who sent us their original figures for reproduction, and all scientists and post-doctoral researchers, who have collaborated with us during the years for the development of the ligand libraries and their applications, in particular, Drs. Luc Guerrier, Lee Lomas, Vanitha Thulasiraman, David Hammond, Frederic Fortis, Alexander V. Kravchuk, Annalisa Castagna, Bernard Monsarrat, François Berger, Paolo Antonioli, Elisa Fasoli, Alfonsina D'Amato, Angela Bachi, Giovanni Candiano, Olivier Meilhac, Damien Lavigne, Clara Esteve, Francesco Di Girolamo, Vincenzo Cunsolo, Angelo Cereda, Alessia Farinazzo, Martha Mendieta, Carolina Simó, and Lau Sennels.

Achtung, achtung, and again achtung! If you are looking at a book that:

1. will offer a thorough description of mass spectrometry in proteomics and related protein and peptide identifications via databases;
2. will give a detailed account of two-dimensional electrophoresis, perhaps the oldest proteomics tool;
3. or provide an in-depth illustration of two-dimensional chromatography (or if you like MudPIT),

then we are afraid you are looking at the wrong book. There is indeed a plethora of such books dealing with this three-pronged description of proteomics tools, most of them telling similar stories over and over again. We felt there is definitely no need for another comparable book in the modern proteomics panorama.

The present book stands out in this panorama because it does not touch at all on any of these three fields, although it deals extensively with proteomics. We maintain that the point of view presented here is unique in that it covers in depth the field of low- to very-low-abundance protein discovery by means of combinatorial peptide ligand libraries (CPLLs). It describes how this technique can help in digging protein compositions to an incredible degree of any biological fluid and in fact in any tissue from any origin (be it mammalian, plant, bacterial, virus, etc., except perhaps for extraterrestrial life) where most biomarkers or not-yet-discovered components of important pathways still lay undisclosed. The technique was launched in 2005, with a paper in *Electrophoresis*, and after a couple of years of slow growth, began taking on momentum and being applied to numerous biological problems as the CPLL beads became commercially available. Not a single negative report came from users, with perhaps the exception of a couple of papers that reported questionable results *only* because the technique had not been understood and thus had not been used properly. All scientists who adopted it reported excellent data in hundreds of publications, and all were enthusiastic about it.

In an attempt to popularize it and reveal all the intricacies of this unique methodology, this book deals with CPLLs in depth and shows how their proper use could make you a winner in your field. To that aim, for instance, Chapter 4, which is at the core of this book, forming its cornerstone, deals with all aspects of CPLLs, including the secrets of combinatorial peptide synthesis and how to use CPLLs for reducing the dynamic concentration range of any proteome, and thus it highlights those rare species that are normally hidden to view. The theory of dynamic protein concentration range as well as the analysis of numerous thermodynamic and kinetics parameters are discussed. In addition, Chapter 8 introduces step by step any possible methodology pertaining to the proper use of CPLLs driving the practitioner to technical success.

The book is bound between two hard covers, renaissance type (engraved leather), called Antiphon and Polyphony. You are encouraged to read these sections: They tell some cute and unbelievable stories. Or, if you prefer, you can call them "Book Ends," like the famous album by Simon and Garfunkel. From one of this album's songs, "We'll Never Say Goodbye," we have extracted that part of their lyric that seems to have been especially written for us scientists:

> We walk the halls of learning
>
> And serve a proud tradition
>
> The flame of truth is burning

To clarify our vision

Look at how the future gleams

Gold against the sky!

We hope that your future in science will flash golden sparks against the sky. Enjoy it!

Egisto Boschetti

Pier Giorgio Righetti

Paris and Milano, December 31, 2012

PIER GIORGIO RIGHETTI

Prof. Pier Giorgio Righetti is a Ph. D. in Organic Chemistry from the University of Pavia in 1965, post-doctoral fellow from MIT and Harvard (Cambridge, Mass, USA), and former Professor at the State Universities of Milan, of Calabria, of Verona and at Milan's Polytechnic (Italy). Currently, he is the President of the *Miles Gloriosus Academy* in Milan.

He is a member of the Editorial Board of international journals such as Electrophoresis, Journal of Chromatography A, Journal of Proteomics, Journal of BioTechniques, Proteomics, and Journal of Proteomics - Clinical Applications. He is the author of the book Proteomics Today (Hamdan M., Righetti P.G.), published by Wiley-VCH, Hoboken, 2005. He has published more than 650 scientific articles in the international literature.

Prof. Righetti has developed and patented isoelectric focusing in immobilized pH gradients, multicompartment electrolyzers with isoelectric membranes, membrane-trapped enzyme reactors operating in an electric field and temperature-programmed capillary electrophoresis.

On 650 articles listed by Web of Knowledge (Thomson Reuters), Righetti scores 18,000 citations and H-index of 59. In the past five years (2007–2011), he has received citations ranging from 1000 to 1200 per year.

Awards:
- CaSSS (California Separation Science Society) award (October 2006, San Francisco, USA).
- Csaba Horvath Medal, presented on April 15, 2008 by the Connecticut Separation Science. Council (Yale University) in Innsbruck, Austria.
- In 2011 (February), he has been nominated as the honorary member of the Spanish Proteomics Society in Segovia, Spain.
- Beckman Award and Medal, presented on February 13, 2012 at the MSB2012 meeting in Genève, Switzerland.

EGISTO BOSCHETTI

Internationally recognized as an expert on various aspects of protein separation by liquid chromatography, Dr. Boschetti in his last decade served Ciphergen Biosystems Inc. as Vice President R&D between 2002 and 2006, and then Bio-Rad Laboratories Inc. as R&D Director between 2006 and 2009 within CEA organization. Previously he had a position as Research Director at BioSepra and its predecessors (now belonging to Pall Corporation) as far back as the inception of the chromatography business in the beginning of '80s.

With a degree in biochemistry from the University of Bologna (Italy) and an MBA, (Paris), Dr. Boschetti is the founder in 1994 of Biosphere Medical Inc in Boston (now belonging to Merit Medical Systems Inc.), where he served as President at the early stage of the Company development. While with this company he files a dozen of patents in the domain of therapeutic embolization for the development of medical devices presently largely used worldwide.

In 2007 he also co-founded Senova Systems Inc. in Sunnyvale (California) where from the beginning 2007 he actively serves as a Chairman of the Scientific Board. Within this company new patented pH solid-state sensing devices have been developed.

More recently Egisto Boschetti is established as an Independent International Consultant within his domain of expertise, supporting and advising French and foreign companies especially in the USA.

His extensive experience in the domain of proteins is witnessed by over 240 scientific publications in international journals and books, and the application of more than 60 patents. In the last few years, he has developed biochips for protein analysis by mass spectrometry.

Among other important accomplishments in the bio-purification domain are the design of a variety of composite solid-phase sorbents today extensively used in downstream processing. Coupling chemistry, ligand design and immobilization, impurity tracking and removal, protein discovery and identification are related fields of expertise.

In proteomics field a major achievement is the development of the technology of dynamic concentration range reduction with peptide ligand libraries and concomitant discovery of many low-abundance species (e.g. new allergens) as well as the development of new concepts in isoelectric fractionation involving so-called Solid-State Buffers.

1D	mono-dimensional
2D	two-dimensional
AD	Alzheimer's disease
ADH	alcohol dehydrogenase
AF	amniotic fluid
AKAP	A-kinase anchor proteins
BAL	bronchoalveolar lavage
BALF	bronchoalveolar lavage fluid
BiCPIT	bidimensional chromatography protein identification technology
CA	carrier ampholytes for IEF
CDAII	congenital dyserythropoietic anemia type II
CEX	cation exchanger
CF-IEF	continuous-flow isoelectric focusing
CHAPS	3-[(3-cholamidopropyl)dimethylammonio]-1-propanesulfonate
CHO	Chinese hamster ovary
cIEF	capillary isoelectric focusing
CM	carboxymethyl
CNS	central nervous system
COH	ovarian hyper stimulation
ConA	concanavalin A
CPLL	combinatorial peptide ligand library
CSF	cerebrospinal fluid
Ctrl or CTRL	control
Cy2, Cy3, Cy5	cyanine dyes
CZE	capillary zone electrophoresis
Da	Dalton
De-Streak	bis-(2-hydroxyethyl)disulfide
DIGE	differential in gel electrophoresis
DMF	dimethyl formamide
DOC	deoxycholate
DRC	dynamic range compression
DTT	di-thiothreitol
EC	European Community
EDTA	ethylenediaminetetraacetic acid
ELISA	enzymatic-linked immunosorbent assay
ESI-IT-MS	electrospray ionization ion-trap mass spectrometry
ESI-MS	electrospray ionization-mass spectrometry
FAB	fast atom bombardment
FFE	free-flow electrophoresis
FF-IEF	free flow-isoelectric focusing
Fmoc	fluorenylmethyloxycarbonyl
FT	flowthrough

FTICR	Fourier transform ion cyclotron resonance
GMGC	Global *Musa* Genomics Consortium
GO	gene ontology
HAP	high-abundance proteins
HAPs	hazardous air pollutants
Hb	hemoglobin
HCP	host-cell proteins
HEPES	hydroxyethyl piperazineethanesulfonicacid
hFF	human follicular fluid
HIV	human immunodeficiency virus
HPDP	N-[6-(biotinamido)hexyl]-3′-(2′-pyridyldithio)propionamide
HPLC	high-performance liquid chromatography
HuPO	human proteome organization
ID	internal diameter or identification (according to the context)
IEF	isoelectric focusing
IgG	immunoglobulin G
IMAC	immobilized metal affinity chromatography
INPPO	International Plant Proteomics Organization
IPG	immobilized pH gradient
IVF	*in vitro* fertilization therapy
LAP	low-abundance proteins
LC-MS/MS	liquid chromatography-tandem mass spectrometry
LC-MS	liquid chromatography-mass spectrometry
LDL	low density lipoprotein
LPE	liquid phase electrophoresis
LTP	lipid transfer protein
MALDI TOF	matrix assisted laser desorption ionization—time of flight
MASCP	Multinational Arabidopsis Steering Committee Proteomics
MCE	multicompartment electrophoresis
M-LAC	multilectin affinity chromatography
Mr	molecular mass
MS	mass spectrometry
MudPIT	multidimensional protein identification technology
NAF	nipple aspirate fluid
NCBI	National Center for Biotechnology Information
NCI	National Cancer Institute
NMP	dimethyl pyridine
NP or NP40	nonidet P40
OHS	ovarian hyperstimulatin syndrome
OS	organic solvent mixture
PAGE	polyacrylamide gel electrophoresis
PBS	phosphate buffered saline
PCR	polymerase chain reaction
PEG	polyethylene glycol
pI	isoelectric point
PKA	phospho kinase A
PMSF	phenylmethylsulfonylfluoride
PNGase	peptide-N-glycosidase
PPP	plasma proteome project

PSA	prostate specific antigen
PTM	post-translational modifications
PVP	polyvinylpyrrolidone
RBC	red blood cell
RP	reverse phase
RPC	reverse phase chromatography
RSA	recurrent spontaneous abortion
RuBisCO	ribulose-1,5-diphosphate carboxylase/oxygenase
SCX	strong cation exchanger
SDS	sodium dodecyl sulfate
SF	synovial fluid
SPE	solid-phase extraction
SSB	solid-state buffers
SVMP	snake venom metalloproteinase
SVOCs	semivolatile organic compounds
TCA	trichloroacetic acid
TCEP	Tris(2-carboxyethyl)phosphine hydrochloride
TFA	trifluoroacetic acid
Tris	Tris(hydroxymethyl)aminomethane
TUBE	tandem-repeated ubiquitin-binding entities
TUC	2 M thiourea-7 M urea-2% CHAPS
UCA	9 M urea-50 mM citric acid, pH 3
VOCs	volatile organic compounds
WGA	wheat germ agglutinin
ZE	zonal electrophoresis

Trade Names:

- Affibodies: Trade name of Affibody AB, Sweden.
- Ampholine, Resource, HiTrap, Sepharose, Sephadex: Trade names of GE Healthcare.
- Atto: Trade name of Atto-Tec GmbH.
- Centricon: Trade name of Millipore Corporation MA, USA.
- Cleanascite: Trade name of Biotech Support Group Inc. NJ, USA.
- Coomassie: Trade name of Imperial Chemical Industries PLC.
- COPAS: Trade name of Union Biometrica Inc MA, USA.
- Cy2, Cy3, Cy5 dyes: Trade names of Biological Detection Systems Inc.
- De-Streak, DryStrips: Trade names of Amersham Biosciences.
- iTRAQ: Trade name of AB Sciex Inc.
- LTQ-Orbitrap: Trade name of Thermo Fisher Scientific Inc.
- Mars, Off-Gel: Trade names of Agilent Technologies Inc.
- Octopus: Trade name of Dr Wber GmbH, Germany.
- Progenesis SameSpots: Nonlinear Dynamics, Ltd.
- ProteoMiner, Rotofor SELDI, ProteinChip, Bio-Plex, Protean, PDQuest, Versa-doc: Trade names of Bio-Rad Inc.
- ProteomLab: Trade name of Beckman Coulter Inc.
- ProteoPrep, Seppro: Trade names of Sigma Aldrich Inc.
- Proteosep: Trade name of Eprogen Inc.
- QconCAT: Trade name of PolyQuant GmbH.
- Quantum Dot/COPAS: Trade names of Union Biometrica Ltd.
- QuickGO: Trade name of European Bioinformatics Institute.
- Rapigest: Trade name of Waters Inc.
- Sequest: Trade name of Proteome Software Inc.
- Texas Red: Trade name of Life Technologies Corp.
- Triton X: Trade name of Union Carbide Corp.

ProteoMiner (combinatorial peptide ligand libraries in general) is a product manufactured and distributed by Bio-Rad Inc. Its use for the reduction of dynamic concentration range and removal of protein impurities is covered by patents belonging to Bio-Rad.

Segovia, Convent de la Santa Cruz, in the year of our Lord 1495, Tomàs de Torquemada, nominated Grand Inquisitor of the Kingdoms of Castille, Leòn, Catalogna, Aragon, and Valencia, by King Ferdinand of Aragon and the Queen Isabel of Castille, has set up his tribunal in the Lonja de la Seda, a flamboyant Gothic building annexed to the convent. As the building is full of light and built on clear stone, he has transformed it into a dark room, veiling most of the large gothic windows with drapes and paneling the walls with dark-brown chestnut wood. He has set up a high desk, from which he can dominate the benches in which the suspected heretics are seated in handcuffs and chains. Two Dominican judges help him during the trials, a senior scribe, Girine-Isag and a younger one, Enki-Mansum. The terror he has created in the country and the fierce sentences and auto-da-fè (resulting in the burning of the condemned at the stake in public plazas) generated so much hate toward him that at one point he was forced to travel with a bodyguard of 50 mounted guards and 250 armed men. He is now poring over a thick pile of papers, absorbed in the study of the documents.

From a hidden door on the right side of the cathedra, enters the elder friar.

Girine-Isag: Magister, magister….

Tomàs (*to himself*): *Holy Nail of the Holy Cross, these two scribes always pop up when you least expect them.* Loud: Yes, what is up, Girine?

Girine-Isag (*breathing heavily from the scuttle*): Grand Inquisitor, I have an important book delivered by diplomatic route from the Vatican. It seems that the matter is quite urgent.

Tomàs (*to himself*): *those Vatican chaps always ready to give orders and interfere with other people's work …* Loud: A book? What am I supposed to do with it, Girine? I deal with souls and bodies, I bring to trial those heretics that hide everywhere in Spain and all over the Catholic world, always ready to ravage the word of the Bible. I can torture them, break their bones, reduce them to pulp, if needed, but a book? Shall I burn it at the stake? Not a big show to offer to the populace in the Plaza Major!

Girine: Magister, to start with, this book is written in an obscure language and contains all sorts of magic formulas and demonic sentences. Listen to this: They describe a devilish machine, called Orbitrap, by which they claim they can grind some invisible spirits called proteins and, via their identification, trace back the origin of life …

Tomàs (*to himself*): *Holy Cross…* Loud: But this is sheer heresy, only God is at the origin of life and death in earth. Yet, I could make this Orbitrap machine to a profitable use if it could grind not just invisible spirits but also bodies made of flesh and bones! Tell me more about this book.

From a hidden door on the left side of the cathedra, enters the younger friar, with a sardonic smile. He also holds a book in his hands.

Enki-Mansum: Great Inquisitor, here in these paragraphs the two authors claim that they have holy beads, blessed by the Virgin Mary, with which they capture these invisible spirits called proteins and make them visible to humans.

Girine (*to himself, sighing*): *Those arrogant young scribes, they always have to put up a show in front of the big boss.* Loud: what do you know about these exoteric formulae, Enki? Young scribes are supposed to keep their mouth shut and learn from the elderly and learned ones.

Tomàs (to himself): Here we go again, these two scribes always quarreling and trying to have the upper hand on each other. I wish I could accuse them of heresy and maybe burn them at the stake as well. Imagine, what a spectacle, the Church going against the Church! Loud: But this is terrible too. The only holy grains I know of are the grains of the rosary, not those magic beads … let me read this sentence (*he starts leafing through the book pages*). Holy Nails, look here, they claim that these beads are coated with some magic little spirits that will interlace with these invisible proteins and bring them to life and visibility. But only God, the Seraphims and Cherubims can see the invisible and protect life. This is anathema. Girine, Enki, send immediately messengers with orders to capture the two heretics, to be brought here to Segovia in chains. I want to put such a torque on them that they will repent, cry for mercy, abjure and embrace faith again. Otherwise … the flames of hell will envelop them in an eternal shroud. By the way, where do they come from?

Girine: Magister, one is from France …

Enki (popping in brusquely): and the other from Italy.

Tomàs: Hum, in France there is King Charles VIII, who is now at war against Spain and the Pope, so extraditing the French chap might require some secret maneuvering. What about the other one? Italy is a mosaic of dukedoms, fiefs, and republics.

Girine and Enki (simultaneously): He is in the dukedom of Ludovico il Moro, in Milano …

Girine (to himself): One of these days I will force this young scribe Enki to wear a cilice, not just around his waist but around his mouth as well!

Tomàs: Hum, Ludovico is also an ally of the French King, but surely he will bend to the will of the Pope. It might very well be that the Italian heretic has been to school with Leonardo da Vinci; I understand he has moved to Milano from Florence, leaving the court of Lorenzo il Magnifico. Certainly a big loss for Lorenzo. They say that Leonardo is a real genius and this Italian heretic might have learned these strange theories from him. And I suspect that Leonardo himself is not a devout Catholic; on the contrary, his brain might be full of heresies as well! Well, send out these messages with orders to capture the two heretics and bring them in chains to Segovia, by any ways and means, bribes and other "secular tools." (*He hands them two warrants of arrest that he had meanwhile prepared and signed.*)

Segovia, six months later. Tomàs is again sitting in his high desk, on a Savonarola armchair, poring as usual over piles of papers and edicts. From the two lateral doors both Girine and Enki break in simultaneously.

Girine and Enki (in unison): Magister, magister ….

Tomàs (to himself): Here we go again; now these two fools of scribes play the game of echoing each other! Loud: What's up, fellows? Such important news as to distract me from my Christian duties in cleansing up God's Church from bad weeds and antichrists?

Girine (enraged): Will you shut up, Enki, and let the elders speak? Grand Inquisitor, mission impossible! As the two heretics got wind of the arrest warrants, they both fled to Basel, a town well known to protect all agitators, anarchists, antichrists, rebels, free-thinkers, you name it! It is a stronghold, with a powerful army ready to fight any invasion and the kind of place definitely out of reach of the *longa manus* of the Vatican!

Tomàs (sighing): Yeah, I know that quite well. It seems that all free-thinkers are taking refuge there. There are rumors that the Town Council has also offered asylum to Erasmus von Rotterdam, a clergyman with heretic ideas as well. The same rumors have it that he is planning to write a book by the title of *Stulticiae Laus*. Instead of chanting the eulogy of God, he sings a hymn to folly! Well, the times they are a-changin'. Next thing you know, one of these days those astronomers who are popping up in all the schools and universities will one day declare that the earth is not any longer at the center of the universe; it will be reduced to a filthy slipper running after some big star, you bet. (*To himself*): *While I spend my life chasing marranos, moriscos, any sort of heretics from without, I wonder if the fall of the Church will come from within. Perhaps Savonarola was right in denouncing the filth and corruption in the Church and certainly Pope Alessandro VIII the Borgia is not the proper Saint Pastor to be the guide of Christianity. I would not be surprised if Savonarola would soon be burned at the stake, and this will not be by my own hand or will.* Loud: Well, if we cannot punish the authors, at least we can incriminate their book. By the way, what is the name of the two heretics?

Girine and Enki (in unison): Boschetti, the French, and Righetti, the Italian.

Tomàs (with determination): Funny names for two antichrists. Well, we have two books, hand me the first one (*he grabs the first book from Enki's hands*). Here is the hot branding tool, let us mark its damnation (*he presses a hot iron on the book cover, which is branded with a purple P*). *As the two scribes look at him with stupor, Tomàs fulminates: libri qui signati sunt litera P. rubra indicantur esse prohibiti.* Now Girine, hand me the other copy and you two follow me.

The trio moves out from the Lonja and, mumbling prayers, enter into the refectory, where grand logs are burning in the fireplace. Tomàs takes the book and, with the help of the two scribes, starts tearing off the pages and throwing them into the high flames, where the book is quickly consumed.

Tomàs: Thus shall perish all enemies of the Holy Church. Let the ashes be dispersed in the countryside and may the wind blow away their sinful deeds! Heresy shall never prevail.

CHAPTER **1**

Introducing Low-Abundance Species in Proteome Analysis

1.1   THE PROTEOMIC ARENA

Proteomics, defined as a technical method to study the protein composition of a biological sample by a global approach, was proposed more than a decade ago and is now largely used to rapidly decipher the protein composition of cell extracts, tissues, and biological fluids. A complete knowledge of complex proteome compositions is considered to be a starting point for a better understanding of (i) the control of subtle metabolic pathways, (ii) the interaction of some protein components with nucleic acids to control the biosynthesis and the cell cycle, (iii) the communication among cells and tissues to the formation of organs and organisms, and (iv) pathological processes and their correction when deciphered at early stages.

Reaching all these goals requires resolving several obstacles that frequently prevent progress in proteome analysis. This is the case for (i) reducing the suppression effect by species that are extremely concentrated, (ii) catching transient species that appear only under certain circumstances, (iii) solubilizing species that are very hard to get unless deeply denatured, (iv) improving the dynamic range of analytical instrumentation and methods, (v) finding effective technologies to analyze isoforms, resulting from post-translational modifications and (vi) showing species that are present at very low levels of concentration. All these issues should show that the preferred method is usually to get native proteins, which is one of the biggest challenges in proteomics.

The quest for traces of gene products is challenging because analytical instrumentation extends "only" between 3 and 5 orders of magnitudes, while the dynamic concentration range extends well beyond this number. To circumvent this difficulty, sample treatments dealing with enrichment processes have been developed.

The crude reality of proteomics is that in spite of the relatively modest number of coding genes between 20 and 25,000—the count continues to fluctuate (Pertea et al., 2010)—the number of proteins that constitute a proteome is much larger. Actually, if it is assumed that a cell contains 1000 genuine proteins, each of which is present in twenty different glycosylated forms and ten different cleaved variants, one would end up with 200,000 protein forms. This would be the figure without counting the numerous other post-translational modifications in addition to several splicing variants that currently contribute to manufacture a very large variety of proteins from the original gene coded protein. A good example is given by antibodies that might contain more than 1 million different epitope sequences that add to the complexity of blood proteomes, for example. Therefore proteomics is largely more complex than genomics. Whereas genomics is "linear" (a given number of coding genes where the question of dynamic concentration range is absent), proteomics is further complicated by a dramatically large expression difference generating extreme variations between the lowest and highest expressions of proteins.

The coding genes are not all necessarily expressed in a single cell or tissue at a given time, thereby producing differences from cell extracts of organs or between biological fluids from the same organism while naturally the genome is the same. It is within this context that the notion of proteomes related to specific cell functions was born, and hence functional proteomics came to be seen as the basis of protein presence deciphering along with the differential protein expression significance. Permanent and transient functions are present in the same cell with different protein expressions over time or over different stimulations; this renders proteome investigations quite difficult, especially for low-abundance species that are expressed either for a very limited time or at a very low number of copies or both. Neither fact means that their behavior is of minor importance for cell life; on the contrary, it might be essential for the maintenance of cell integrity or as a proper response to external stimulations. The understanding of these phenomena is of utmost importance for the global knowledge of cell functioning. All these facts point to the necessity of paying a great deal of attention (much more than was initially anticipated) to low- and very low-abundance proteins; these proteins are generally ignored because the technologies and methods currently available are not sufficiently sensitive.

Hundreds of scientific publications underline the major limitations of current approaches for proteome investigations. Taken separately, each method could easily be seen as ineffective in obtaining a global detailed vision of the proteome on a structural level, not even considering its functional aspects. There are methods available that can resolve only high-abundance species, and other methods that can resolve only relatively small polypeptides. Among analytical approaches are those defined as top-down and bottom-up, capable, respectively, of accessing a relatively modest number of proteins in their native form and getting a whole picture indirectly through analysis of breakdown peptides generating confusion between native proteins and their truncated forms. For instance, electrophoresis-based methods (which are still the most commonly analytical methods used to date) are not appropriate for polypeptides of masses lower than 5 kDa; on the contrary, mass spectrometry contributes to the analysis of small-size polypeptides. Strongly alkaline proteins are poorly resolved by classical two-dimensional electrophoresis (Bae et al., 2003). Polypeptides comprising a large number of hydrophobic amino acids folded so as to repel water cannot be properly solubilized and consequently cannot be analyzed using nondenaturing methods.

All these technical limitations make a quite modest estimation of the analytical coverage for proteomics composition: probably no more than 30% of expressed proteins are detectable using standard analytical methods.

With gene expression differences, large effects can be determined: They are not only limited to the pathological situations but importantly also to the normal cell life cycle. Figure 1.1 depicts one of the spectacular macroscopic effects detectable in protein expression differences without the help of instruments.

It is within this general situation that proteins expressed at low-abundance gain in importance while constituting a real technical difficulty even for specialized or experienced investigators.

FIGURE 1.1
Two phenotypes of the same genome. A genome can differently express proteins, giving rise to distinct organism phenotypes. For instance, at one life stage the coding genes of *Parnassius apollo* are expressed as caterpillar (image on the left), a larval form of Lepidoptera insects. At another life stage of the same animal (image on the right), coding genes are expressed in the adult insect. *Pictures adapted from Wikipedia.*

1.2 LOW-ABUNDANCE PROTEIN DEFINITION AND THEIR TRACKING INTEREST

The concentration difference among highly expressed proteins (e.g., albumin in serum) and proteins that are expressed only in a few copies can span over 12 orders of magnitudes. Under these circumstances and considering that the most concentrated proteins could reach several mg/mL, there are many species that are directly undetectable. The presence of high-abundance proteins is not a characteristic of blood serum, but is rather a relatively common situation in numerous biological fluids, cell extracts (e.g., hemoglobin in human red blood cells where locally the concentration is few hundreds of mg per mL; actins in other eukaryotic cells), and different species (e.g., ribulose-1,5-bis-phosphate carboxylase/oxygenase in plant leafs). Alongside high-abundance proteins there are other polypeptides whose concentration is extremely low. Nevertheless their biological importance is not marginal; on the contrary, their physiological effect can be huge even at an extremely low concentration. For instance, transcription factors and other regulatory proteins are present permanently or transiently at concentrations that are extremely difficult to assess, and as a consequence they are very hard to isolate in amounts compatible with complete characterization studies. In general, heterologous expressions are necessary to get enough material for in-depth investigations. This situation represents all the challenges and difficulties involved in reaching a final solution for the full deep deciphering of proteomes. A typical example of a quite well-known proteome is human serum where high-abundance species co-exist with low- and very low-abundance ones. The first comprehensive situation of the dynamic concentration range was elegantly demonstrated by Anderson et al. (2002, 2004) with the anticipation that, when using two-dimensional electrophoresis only, resolving the large difference between the concentrated and the most diluted proteins would be highly problematic.

Close to the major protein—albumin—many others are present: This group is composed of quite familiar species, but others are poorly known or even completely unknown. Figure 1.2 represents schematically categories of proteins with their relative abundance in human plasma. Naturally there are large overlaps since a single protein with its post-translational modification differently concentrated could belong to both medium-abundance and low-abundance proteins. The border between intermediate abundance and low abundance is relatively subjective; however, it could be placed below ng/mL. Generally, below this concentration it is very difficult to detect proteins in bulk, directly from an untreated sample with current instrumentations and methods. In serum the first dozen of most abundant proteins represents about 90% of the proteome mass, and from the remaining 10%, medium- and low-abundance proteins represent 1% of the global number of proteins or less.

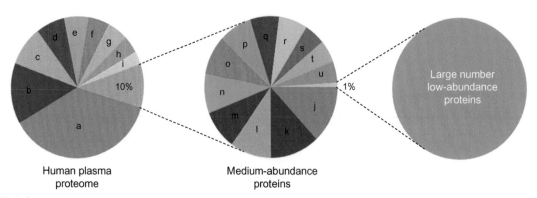

FIGURE 1.2

Schematic representation of the proportion of low-abundance proteins in plasma. While a dozen proteins represent more than 90% of the protein content, less than 0.1% is accounted, comprising many proteins of low and very low-abundance. The first "cake" on the left comprises the following protein: a: albumin, b: immunoglobulins G; c: transferrin; d: haptoglobin; e: transthyretin; f: alpha-2-macroglobulin; g: alpha-1-antitrypsin; h: fibrinogen; i: hemopexin.

The central "cake" represents several other proteins of intermediate abundance: j: apoliproprotein-1; k: apolipoprotein-B; l: alpha-1-acidic glycoprotein; m: lipoproteins-A; n: coagulation factor H; o: ceruloplasmin; p: C4 complement; q: Complement factor B; r: prealbumin; s: C9 complement factor; t: C1q complement factor; u: C-reactive protein. The circle on the right represents the mass of all other plasma proteins, most of which are of low abundance summing over several hundred thousand proteins (including isoforms but not antibodies).

Many proteins can be classified in the low bundance category. There are naturally all proteins that are involved in transcription and translation mechanisms (and most of those that are involved in cell regulation) that are permanently present or not. Membrane proteins, especially receptors, are generally of low concentration and are also quite difficult to solubilize without denaturation; hence their detection is challenging. Allergens, whatever their origin, can also be of very low concentration, with yet an extreme effect on immunology systems. A lot remains to be discovered in this domain where only relatively high-abundance allergens are known as part of international classifications (see Section 6.8 for more details). Early-stage metabolic misregulations can also be due to the expression of polypeptides. If normally present in very low amounts, their concentration can change; this constitutes the basis of the quest for biomarkers using proteomic technologies. Detecting very low-abundance misexpressed proteins at an early stage of a disease might be challenging, but it is largely rewarding (see the discussions in Sections 6.2, 6.3, and 6.4). Post translational modifications generating a number of possible protein isoforms can also be classified as low-abundance species. The ratio among isoforms could in addition be taken as a signal of potential cell disorder; hence detecting low-abundance isoforms can be an interesting route to explore. Under the category of low-abundance proteins are also impurities that are generally present in purified biologicals intended for therapeutic applications. Their detection, identification, quantitation, and elimination pose major challenges for the biopharma industry (see Sections 7.2.2 and 7.2.3 for more details).

In proteomic investigations, access to very low-abundance species is in certain circumstances similar to searching for a "fossil" species. As a typical example of what has been found when a large amount of red blood cell lysate was treated with a small volume of peptide library beads was the observation of the presence of certain gene products that were supposed to be absent. Actually they are present only as early expression events during the cell maturation cycle. This was the substitute presence with case of ε and ζ embryonic haemoglobin chains, both of which are normally absent in mature cells (Roux-Dalvai et al., 2008). This finding suggests other possible discoveries from other cell maturation processes, such as in lymphocytes. In another domain, embryogenesis processes could be investigated with the detection of early-stage and/or transient profound expressions explaining phenotyping effects. Transient events are quite frequent in the life cycle and are probably associated with the expression or the repression of the biosynthesis of specific proteins (e.g., certain plasma protein in labor; Yuan et al., 2012).

It is within this concept that it can also be postulated that in biological disorders differences in protein expressions produce markers generated early on at preliminary stages of biological function alterations, at a stage when they cannot normally be detected without the intervention of an external enrichment or amplification process. What is described as "looking for a needle in a haystack" (Veenstra, 2007; Tichy et al., 2011) could thus be largely simplified with the benefit of evidencing an "early" protein expression (Qian et al., 2008), an essential aspect of detecting a disease much earlier. One particularly useful aspect of this approach is shown by the early diagnosis of endometriosis, undetectable by ultrasonography, with a noninvasive proteomics analysis of plasma samples after proper enrichment of low-abundance species instead of laparoscopic procedures (Fassbender et al., 2012). Other examples of the interest of early access to protein markers for diagnostic applications are also reported by Lorkova et al. (2012) for ovarian cancer detection, by Marrocco et al. (2010) when detecting candidate biomarkers for liver cirrhosis-associated hepatitis B, and by Meng et al. (2011) when describing a way to discover protein marker candidates for breast cancer. With affordably easy methods for protein enrichment and for masking the suppression effect of high-abundance proteins, the detection of low-abundance species for diagnosis is no longer a real problem; however, what is more questionable is the selection of the most pertinent ones capable of discriminating for the right disease since many common misexpressions have been evidenced in different pathological cases. Actually, all newly discovered misexpressed proteins must resist the confrontation of already found markers from other diseases.

1.3 CATEGORIES OF LOW-ABUNDANCE PROTEINS

Very dilute proteins are considered to be a function of the type of biological extract. Low-abundance proteins in serum (e.g., leakage proteins from tissues) are not necessarily very dilute in tissue extracts; conversely, albumin that is largely present in blood may not be part of some tissue extracts when they are not contaminated by blood.

Various categories of low-abundance proteins can be defined. A cell proteome contains proteins that are present only in trace form or are even absent except at a given cell cycle or cell stage of maturation. Thus regulation proteins are rare proteins and difficult to detect when using a global approach such as proteomics analysis.

The most relevant example of proteins that are difficult to detect are those that derive from signaling. Here phosphorylated species are the clear target. With estimated thousands phosphorylated motifs (Thingholm et al., 2009), this group of proteins is one of the most abundant in terms of diversity, which contrasts greatly with the least abundant character of each component. They are considered to be of low abundance, and with their function of binary switches that control many cell functions, they represent a group of very important proteins for cell life, but their transient situation makes their detection and identification challenging. Their importance is witnessed by the fact that deregulations of phosphorylation-based signaling pathways are at the basis of a large number of pathological situations (Ding et al., 2007) and constitute a great target for the development of drugs. This protein group is generally considered to be of low abundance, and most related investigations involve enrichment methods (see Section 3.3.4). When phosphorylated species are part of membrane proteins, the challenges are even harder (Orsburn et al., 2011); the first step is thus the solubilization of membrane proteins and then the selection of the phosphorylated forms prior to an in-depth analysis.

Other post-translational modification effects (e.g., formation of isoforms) can also be undetectable using current analytical methods and hence are of low abundance. Nevertheless, the important point is to know the relative amount of isoforms of the same protein. When analyzing a two-dimensional electrophoresis gel, it is quite common to find several isoforms aligned along the axis of pH with different isoelectric points. This is evidenced by discrete spots, with intensities decreasing over the pH scale (in one direction or another). Among these distinguishable isoforms are others that are difficult to detect because of their too low concentration. This information could be important for the normal course of post-translational protein modification that should normally follow a predetermined rule as far as the proportionality of single isoforms is concerned. In pathological situations, this proportionality can easily be modified and in this case would be considered important data for properly understanding possible biological disorders. The modification of isoform proportionality in pathological conditions was recently discovered thanks to the compression of dynamic concentration range of proteins from human carotid artery atherosclerotic plaque (Malaud et al. 2012). These expression differences (isoforms present at trace levels) would not have been reachable without a preliminary enrichment treatment of the sample.

Antibodies present in blood are considered as belonging to the category of high-abundance proteins and are treated as such. They are in fact globally removed by immuno-subtraction (see Section 3.2.2) because they mask part of the signal of protein traces. However, if one considers that antibodies are very diverse even if they have similar physicochemical characteristics such as similar molecular mass and similar shape and are macroscopically indistinguishable, they have a very different composition at their hypervariable domain. If one considers that the concentration of antibodies of IgG class is about 10 mg/mL and that perhaps 1 or more million diverse antibody species coexist, their individual average concentration is around 1–10 ng/mL, which is below the sensitivity of current proteomics methods and instruments and would not be detected individually. Thus each or many specific antibodies present in blood could be qualified as being of low abundance. This is a sort of paradox in proteomics, but this reality opens up an opportunity for developing immuno-proteomics, a still underinvestigated domain that potentially holds rich information about infections and inflammation without even considering the large interest in autoimmune diseases. The presence of a given antibody or the combination of certain antibodies could signal the advent or the presence of specific physiological or pathological situations.

In another domain, proteins interacting with nucleic acids, such as transcription factors, including the large number of post-translationally modified species, are another group of high-impact and very low-abundance expressed proteins. They represent the second largest group of human genes (about 6% of total or about 1500 genes) where not much more than 5% has been separated and characterized (Gadgil et al., 2001). Their very low-abundance expression is the reason this group is not yet well known, while it appears essential for understanding one of the most important pathways in living cells.

Key roles in a large number of biological function regulations are played by the so-called post-translational protein modifications. Although these modifications are made in a large majority of proteins, they are extremely different, several of them coexisting in the same protein. Moreover, a quite large dynamic complexity is present. All of this contributes not only to enriching the already large diversity, but indicates that most of these modifications are present in very low amounts, creating difficulties for identification processes. Post-transaltionally modified proteins are thus considered to be of low abundance, requiring specific methodologies for their identification based on the integration of both top-down and bottom-up approaches (Seo et al., 2008).

Membrane proteins are also difficult to detect and thus to identify/characterize because of both their poor solubility in nondenaturing aqueous media and their low abundance. For these studies specific methods have been proposed (Vertommen et al., 2011), especially for improved solubilization, but the enrichment methods are not well developed.

Among a number of other low-abundance protein categories are genetically modified organisms that are suspected to express differently some proteins or express other proteins at very low concentration

(Widjaja et al., 2009; Agrawal et al., 2011). Some allergens are also categorized as being of low abundance, especially those that are not well known and discovered thanks to adapted proteomics investigations (Shahali et al., 2012). Finally, protein additives can be classified as being of low abundance—for instance, those in wines (see Section 3.5) and impurities present in biopharma products as depicted in Section 7.2.2.

The above groups of proteins qualified to be of low abundance probably do not constitute an exhaustive list, but the list indicates unambiguously that the field to cover is extremely large, which with no doubt requires the development of dedicated methods.

1.4 THE PROPOSAL FOR LOW-ABUNDANCE PROTEIN ACCESS

It is the aim of the present book to give details, application results, and future possibilities of a technology that today appears to be extremely promising in detecting low-abundance species with the capability to evidence small expression differences among the same species.

Therefore, throughout the different chapters the reader will become aware of various aspects of the approach with the description of the principle and of the molecular mechanisms of the enhancement of very dilute species, while concomitantly reducing the high-abundance proteins, as well as applications in various domains either in proteomics investigations or even in domains such as protein purification and molecular interactions.

While making statistical calculations based on low-abundance protein investigations compared to current proteomic studies, it appears that at present the former domain represents a marginal contribution, with about 10–50 times fewer studies. However, most papers mentioning low-abundance proteins highlight the importance of this category of proteins in various domains. For instance, the domains where low-abundance proteins are mentioned are, according to published reports, around cell proteins, biomarker discovery, and serum proteins.

Knowing the importance of low-abundance proteins in both the understanding of protein functions within the cells and especially their interactions, on the one hand, and the interest of detecting biomarkers at an early stage related to metabolic disorders, on the other hand, the study of this category of proteins becomes of outmost importance.

This situation justifies why, in keeping with the main aim of this book which is to teach the use of peptide ligand libraries, other recent technologies are mentioned and referenced. Together with comparative data that follows, the different chapters should give an objective picture of the current capabilities useful in digging deep in proteome compositions.

In spite of its young age—it was introduced in late 2007—the number of publications dealing with peptide libraries as a tool in proteomic studies is growing at a relatively high rate. From a short statistical review of 280 published works, it appears that 36% of papers deal with the technology, which is normal for a novel technology where the knowledge is to be built by a variety of experiences. Interestingly enough, 34% of the publications describe its applicability for low-abundance protein tracking and 27% for biomarker discovery (see Figure 1.3). An almost exhaustive list of scientific publications on the technology is given at the end of the book covering the period from 2005 to 2012.

A more in-depth analysis of published papers shows that the technology-oriented papers were mostly devoted to comparative studies to position the technology among the best known proteomics methods (16% of total) and 11% focused on reproducibility and quantitation. Combinations of the peptide library technology with other well-established methods have also been largely explored. Surprisingly, a number of reviews have been published indicating the interest of editors and scientists in this young technology and its potential applications. The applications have been split into two

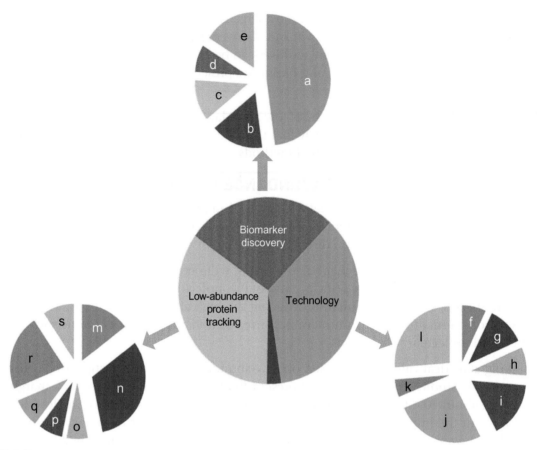

FIGURE 1.3

Statistical review of publications involving combinatorial peptide ligand libraries.

The central "cake represents the major general applications sorted out from all screened papers.

The broken cake at the top is the subgroup of biomarker discovery from various biological extracts. a: serum or plasma; b: other biological fluids; c: secretomes, cell and tissue extracts; d: infection detections; e: general documents on biomarkers.

The subset broken cake on the right represents the collection of papers dealing with technological characterizations: f: technology basics; g: reproducibility and quantitation; h: elution methods; i: comparative studies; j: general reviews; k: technology combinations; l:various other items.

The subset broken cake on the left represents the collection of papers within the domain of proteome deciphering of various biological extracts. m: serum and plasma; n:other biological fluids; o: plants proteins; p: cell extracts and secretome; q: allergen discovery; r: agrifood; s: various other biological material.

main groups: (i) detection of low-abundance proteins as a way to demonstrate the promise of the technology to track species that are normally undetectable and (ii) discovery of biomarkers, particularly those that are inaccessible with current methods.

As far as low-abundance protein tracking is concerned, published papers cover a large spectrum of extracts from biological fluids (see, for example, Smith et al., 2011), from tissues/cells (Rho et al., 2008), and from agrifood proteins and plants (Baracat-Pereira et al., 2012). In addition, allergens have also been the object of specific studies, with the discovery of interesting, novel, low-abundance unexpected species (see Section 6.8). Among the biological fluids are blood serum or plasma, urine, saliva, cerebrospinal fluid, snake venom, milk, bile, coelomic fluid, cervical lavage, follicular fluid, and haemolymph. Cell extracts also revealed a large number of unexpected proteins. This is the case, for instance, for red blood cells and platelets where hundreds of novel proteins never described before having been identified. In plant proteomics, combinatorial peptide ligand libraries largely contributed to the identification of low- and very low-abundance species such as in spinach leaves, Arabidopsis leaves, Hevea exudates, avocado, olive and banana pulps, and others (for details see Section 5.2 and the bibliography at the end of the book). Contribution in trace protein tracking in food products and particularly beverages is also outstanding. For instance, intrinsic protein traces as well as protein additives have been found in a number of drinks such as beer, wines (red and white), nonalcoholic and alcoholic aperitifs, cola drinks, and others. For the first time the possibility of tracking frauds was described as a possible target for this technological application.

A major application of this technology is of course the discovery of protein potential markers that are of diagnostic interest. The possibility of finding traces of proteins opens up the likelihood of discovering early-stage markers when a metabolic disorder can be treated more easily.

Many different protein expression differences have been found in a large number of biological extracts and for a number of possible targets. The most frequent biological material from where expression modifications were investigated is serum and plasma. This is a readily available material, and its collection is not really considered as invasive and does not require multiple manipulations before analysis. No general or normalized rules have been described for the use of plasma for discovering biomarkers of diagnostic interest except a procedure that allowed reaching low-abundance proteins using peptide libraries under a peculiar scheme (Leger et al., 2011) that could be generalized for large preliminary studies. Nevertheless, already a simple treatment of serum ends up in a very different protein pattern, as is evidenced by two-dimensional electrophoresis (see Figure 1.4).

The most important diseases analyzed were first cancer with a number of examples as, for instance, breast cancer, prostate cancer, epithelial ovarian cancer, pancreas cancer, and lung cancer. Diabetes was also intensively investigated using peptide libraries, with interesting technical results. Blood serum or plasma was also used to track diseases involving lipoproteins or cardiovascular issues.

Protein expression modifications have been evaluated using other biological fluids. Urine, for instance, as a totally noninvasive material (Decramer et al., 2008; Kentsis et al., 2011), has been explored in the case of disorders related to pregnancy and ovarian cancer. Cerebrospinal fluid, saliva, and bile have also been extensively explored.

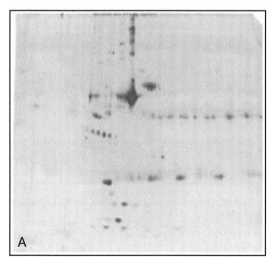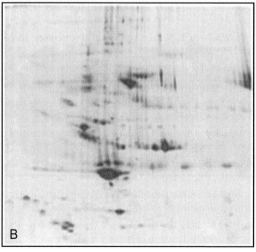

FIGURE 1.4
Two-dimensional comparative electrophoresis maps of human serum before (A) and after (B) treatment with a combinatorial peptide ligand library. The volume of serum treated was 10 mL on 1 mL of peptide beads under physiological conditions of ionic strength and of pH. Captured proteins were then desorbed using a solution of 9 M urea containing 50 mM citric acid. First migration dimension: IPG 3-10 nonlinear; second dimension: 8–18% T polyacrylamide gel in sodium dodecyl sulfate. The spot count of the control (A) was 195, while it increased to 487 after peptide library treatment (B).

Other types of biological material necessitating biopsies followed by tissue extraction prior to low-abundance protein enrichment have only marginally been explored with the described technology. However, important findings have been shown concerning carotid atherosclerotic plaques (Malaud et al., 2012).

Since the biomarker discovery is probably one of the most important future applications of low-abundance species in proteomics, another type of classification was also determined using the type of disease as a discriminating factor (see Figure 6.3; Chapter 6).

It appears quite clear then that combinatorial peptide ligand libraries are effective in evidencing low-abundance proteins from a large spectrum of biological material. This propensity that has been taken by a number of scientists around the world justifies an extended review like the one presented in this book, not only with the objective of disseminating the technology applications, but also of offering the most appropriate explanations of the theory and the practical involvement of the entire process extending from the construction of libraries to protein capture and recovery in order to cover proteome compositions as exhaustively as possible. The multiple technological approaches should in addition open the mind of creative scientists to explore proteomes by introducing variations in either the composition of peptide libraries or in the mode of using them alone or in association with synergistic technologies.

A chronological list of scientific publications involving combinatorial peptide ligand libraries for proteomics investigations is given at the end of this book.

1.5 References

Agrawal GK, Rakwal R. Rice proteomics: A move toward expanded proteome coverage to comparative and functional proteomics uncovers the mysteries of rice and plant biology. *Proteomics*. 2011;11:1630–1649.

Anderson LN, Polanski M, Pieper R, et al. Human plasma proteome. *Mol Cell Proteomics*. 2004;3:311–326.

Anderson NL, Anderson NG. The human plasma proteome: History, character, and diagnostic prospects. *Mol Cell Proteomics*. 2002;2:845–867.

Bae SH, Harris AG, Hains PG, et al. Strategies for the enrichment and identification of basic proteins in proteome projects. *Proteomics*. 2003;3:569–579.

Baracat-Pereira MC, de Oliveira Barbosa M, Magalhães MJ, et al. Separomics applied to the proteomics and peptidomics of low-abundance proteins: Choice of methods and challenges—A review. *Genet Mol Biol*. 2012;35:283–291.

Decramer S, Gonzalez de Peredo A, Breuil B, et al. Urine in clinical proteomics. *Mol Cell Proteomics*. 2008;7:1850–1862.

Ding SJ, Qian WJ, Smith RD. Quantitative proteomic approaches for studying phosphotyrosine signaling. *Expert Rev Proteomics*. 2007;4:13–23.

Fassbender A, Waelkens E, Verbeeck N, et al. Proteomics analysis of plasma for early diagnosis of endometriosis. *Obst & Gynecol*. 2012;119:276–285.

Gadgil H, Jurado LA, Jarrett HW. DNA affinity chromatography of transcription factors. *Anal Biochem*. 2001;290:147–178.

Kentsis A. Challenges and opportunities for discovery of disease biomarkers using urine proteomics. *Pediatr Int*. 2011;53:1–6.

Leger T, Lavigne D, Le Caër JP, et al. Solid-phase hexapeptide ligand libraries open up new perspectives in the discovery of biomarkers in human plasma. *Clin Chim Acta*. 2011;412:740–747.

Lorkova L, Pospisilova J, Lacheta J, et al. Decreased concentrations of retinol-binding protein 4 in sera of epithelial ovarian cancer patients: a potential biomarker identified by proteomics. *Oncol Rep*. 2012;27:318–324.

Malaud E, Piquer D, Merle D, et al. Carotid atherosclerotic plaques: Proteomics study after a low-abundance protein enrichment step. *Electrophoresis*. 2012;33:470–482.

Marrocco C, Rinalducci S, Mohamadkhani A, D'Amici GM, Zolla L. Plasma gelsolin protein: A candidate biomarker for hepatitis B-associated liver cirrhosis identified by proteomic approach. *Blood Transfus*. 2010;8:s105–s112.

Meng R, Gormley M, Bhat VB, Rosenberg A, Quong AA. Low abundance protein enrichment for discovery of candidate plasma protein biomarkers for early detection of breast cancer. *J Proteomics*. 2011;75:366–374.

Orsburn BC, Stockwin LH, Newton DL. Challenges in plasma membrane phosphoproteomics. *Expert Rev Proteomics*. 2011;8:483–494.

Pertea M, Salzberg L. Between a chicken and a grape: Estimating the number of human genes. *Genome Biol*. 2010;11:206–213.

Qian WJ, Kaleta DT, Petritis BO, et al. Enhanced detection of low abundance human plasma proteins using a tandem IgY12-SuperMix immunoaffinity separation strategy. *Mol Cell Proteomics*. 2008;7:1963–1973.

Rho JH, Qin S, Wang JY, Roehrl MHA. Proteomic expression analysis of surgicalhuman colorectal cancer tissues: Up-regulation of PSB7, PRDX1, and SRP9 and hypoxic adaptation in cancer. *J Proteome Res.* 2008;7:2959–2972.

Roux-Dalvai F, Gonzalez de Peredo A, Simó C, et al. Extensive analysis of the cytoplasmic proteome of human erythrocytes using the peptide ligand library technology and advanced spectrometry. *Mol Cell Proteomics.* 2008;7:2254–2269.

Seo J, Jeong J, Kim YM, Hwang N, Paek E, Lee KJ. Strategy for comprehensive identification of post-translational modifications in cellular proteins, including low abundant modifications: Application to glyceraldehyde-3-phosphate dehydrogenase. *J Proteome Res.* 2008;7:587–602.

Shahali Y, Sutra JP, Fasoli E, et al. Allergomic study of cypress pollen via combinatorial peptide ligand libraries *J Proteomics.* 2012;77:101–110.

Smith MP, Wood SL, Zougman A, et al. A systematic analysis of the effects of increasing degrees of serum immunodepletion in terms of depth of coverage and other key aspects in top-down and bottom-up proteomic analyses. *J Proteomics.* 2011;11:2222–2235.

Thingholm TE, Jensen ON, Larsen MR. Analytical strategies for phosphoproteomics. *Proteomics.* 2009;9:1451–1468.

Tichy A, Salovska B, Rehulka P, et al. Phosphoproteomics: Searching for a needle in a haystack. *J Proteomics.* 2011;74:2786–2797.

Veenstra TD. Global and targeted quantitative proteomics for biomarker discovery. *J Chromatogr B.* 2007;847:3–11.

Vertommen A, Panis B, Swennen R, Carpentier SC. Challenges and solutions for the identification of membrane proteins in non-model plants. *J Proteomics.* 2011;74:1165–1181.

Widjaja I, Naumann K, Roth U, et al. Combining subproteome enrichment and Rubisco depletion enables identification of low abundance proteins differentially regulated during plant defense. *Proteomics.* 2009;9:138–147.

Yuan W, Heesom K, Phillips RJ, et al. Low abundance plasma proteins in labour. *Reproduction.* 2012;144:505–518.

CHAPTER 2

Chromatographic and Electrophoretic Prefractionation Tools in Proteome Analysis

2.1 INTRODUCTION

It was only in 2002 that Anderson and Anderson rang the bell on the extreme complexity of biological fluids, with particular regard to human serum/plasma, whose dynamic range was thought to be as high as 12 orders of magnitude, and whose set of most abundant proteins (twenty of them) would colonize as much as 99% of the territory, leaving little space for sampling the low- to very low-abundance proteins. This was timely to some extent, as discussed below. It had two immediate beneficial effects. On the one side, it convinced the HuPO (Human Proteome Organization) to start a pilot project called PPP (Plasma Proteome Project), to which thirty-five laboratories in thirteen countries adhered with the following aims: (i) evaluate the advantages and limitations of depletions, fractionations, and mass spectrometry technology platforms; (ii) compare reference specimens of human serum and EDTA (ethylenediamine tetraacetic acid), heparin, and citrate-coagulated plasma; (iii) create a publicly available knowledge database (www.bioinformatics.med.umich.edu/hupo/ppp; www.ebi.ac.uk/pride). On the other side, it set in motion the first deep prefractionation tool, that is, immuno-depletion, an idea that originated in fact in Leigh Anderson's lab, who in 2003 published extensive investigations on the human serum proteome that had been depleted by immunosubtraction of the first six major-abundance species (Peiper et al., 2003a,b).

As stated above, though, this warning was not so timely, since scientists working on proteomics had fully realized their very high complexity not just in mammalian systems but in general in every living organism. In fact, it could be stated that the first (and most powerful) prefractionation tool was the two-dimensional (2D) gel map analysis, as epitomized by O'Farrell's classic paper (1975). Although we all assume that the 2D electrophoresis technique is just an analytical method, it is indeed also a two-dimensional fractionation method: The large surface area of the final 2D gel plate can be thought of as a fraction collector, where thousands of polypeptide chains are distributed in the charge/mass territory ready to be eluted for any desired analytical method. This is so much the case that, already in 1977/1978, Anderson and Anderson (1977, 1978a,b) started thinking of "large-scale biology" and building instrumentation (called the ISO-DALT system) for preparing and running a large number of O'Farrell-type gels together. This approach greatly enhanced reproducibility and comparison among the resulting protein maps while enabling a very large number of samples to be handled in a short time. A variation (termed BASO-DALT), giving enhanced resolution of basic proteins by employing nonequilibrium focusing in the first dimension, was also described (Willard et al., 1979). These approaches made practical the Molecular Anatomy Program at the Argonne National Laboratory in the United States, the object of which was to be able to fractionate human cells and tissue with the ultimate aim of being able to describe completely the products of human genes and how these vary between individuals and in disease (Anderson and Anderson, 1982, 1984). It is also thanks to the Herculean efforts of the two Andersons that present-day 2D gel maps have reached such a high stage of evolution; it is worth a perusal through two special issues of *Clinical Chemistry* (Young and Anderson, 1982; King, 1984) to see the flurry of papers published and the incredible level of development that had already been reached in the years 1982–1984! For that matter, the interest in 2D gel mapping grew so rapidly that the journal *Electrophoresis* was forced to publish a large number of special issues devoted to this topic [of which the Stakhanovist guest editor was certainly Julio Celis, who alone was the curator of no less than 10 special issues during 1989–1999) (Celis, 1989, 1990 a&b, 1991, 1992, 1994a&b, 1995, 1996, 1999)], followed by Dunn (1991, 1995, 1997, 1999, 2000), and by Williams (1998) which burdened the journal so greatly that the publisher John Wiley and Sons was forced to stop the deluge and fund a new journal, *Proteomics*. We dare say that with the birth of this journal in January 2001 proteomic science came of age.

It is not the aim of this chapter to review in detail the 2D gel electrophoresis methodology that has had a prima donna role in launching modern-day proteomics. Here we will only recall the major prefractionation tools developed in those days to enhance the visibility of the low-abundance proteome, with the proviso that such tools will be limited to chromatographic and electrophoretic approaches. As such, except for a few cases, these approaches are not per se based on bio-affinity recognition, but separate proteins on the basis of the physicochemical approaches, namely, size, shape, surface charge, and overall hydrophobicity. What this book will deal with at length are bio-affinity-based separations, of which, of course, the tool of combinatorial peptide ligand libraries will play the most prominent role.

2.2 PREFRACTIONATION TOOLS IN PROTEOME ANALYSIS

As stated above, we will survey the major approaches to prefractionation, with the aim of enhancing the visibility of minor proteinaceous species in any proteome (so-called low-abundance proteins) and give them figures of merit. We will start with chromatographic tools, followed by electrokinetic methodologies. Many of the methods surveyed here are still much in vogue in proteome analysis, often in combination, either in the same class (e.g., two-dimensional chromatography or 2D electrophoresis) or across classes (combining chromatographic with electrophoretic tools). Some of these tools have also been surveyed recently by Ly and Wasinger (2011).

2.2.1   Sample Prefractionation via One-dimensional Chromatographic Approaches

Fountoulakis's group has developed this approach extensively, as presented below. In a first procedure, Fountoulakis et al. (1997) and Fountoulakis and Takàcs (1998) adopted affinity chromatography on heparin gels as a prefractionation step for enriching certain protein fractions in the bacterium *Haemophilus influenzae*. This Gram-negative bacterium is of pharmaceutical interest and has already been sequenced; its complete genome has been found to comprise ca. 1740 open reading frames, although not more than 100 proteins had been characterized by 2D gel map analysis. Heparin is a highly sulphated glucosaminoglycan with affinity for a broad range of proteins, such as nucleic acid-binding proteins and growth and protein synthesis factors. Because of its sulphate groups, heparin also functions as a high-capacity cation exchanger. Thus, prefractionation on heparin-affinity gel beads was deemed suitable for enriching low-copy number gene products. In fact, about 160 cytosolic proteins bound with different affinities to the heparin matrix and were thus highly enriched prior to 2D PAGE (polyacrylamide gel electrophoresis) separation. As a result, more than 110 new protein spots, detected in the heparin fraction, were identified, thus increasing the total identified proteins of *H. influenzae* to more than 230. In a second approach (Fountoulakis et al., 1998), the same lysate of *H. influenzae* was prefractionated by chromatofocusing. In the eluate, two proteins, major ferric iron-binding protein (HI0097) and 5'-nucleotidase (HI0206), were obtained in a pure form, with another hypothetical protein (HI0052) purified to near homogeneity. Four other proteins, aspartate ammonia lyase (HI0534), peptidase D (HI0675), elongation factor Ts (HI0914), and 5-methyltetrahydropteroyltriglutamate methyltrasferase (HI1702), were strongly enriched by the chromatofocusing process. Approximately 125 proteins were identified in the eluate collected from the column. Seventy of these proteins were identified for the first time after chromatofocusing, the majority of them being low-abundance enzymes with various functions. Thus, with this additional step, a total of 300 proteins could be identified in *H. influenzae* by 2D gel map analysis, out of a total of ca. 600 spots visualized on such maps from the soluble fraction of this microorganism. In yet another approach, the cytosolic soluble proteins of *H. influenzae* were prefractionated by Fountoulakis et al. (1999a) by hydrophobic interaction chromatography (HIC) on a phenyl column. The eluate was subsequently analyzed by 2D electrophoresis mapping, followed by spot characterization by MALDI-TOF MS. Approximately 150 proteins, bound to the column, were identified, but only 30 for the first time. In addition, most of the proteins enriched by HIC were represented by major spots, so that enrichment of low-copy-number gene products was only modest. In total, with all the various chromatographic steps adopted, the number of proteins so far identified could be increased to 350.

The same heparin chromatography procedure was subsequently applied by Karlsson et al. (1999) to the prefractionation of human foetal brain soluble proteins. Approximately 300 proteins were analyzed, representing 70 different polypeptides, 50 of which were bound to the heparin matrix. Eighteen brain proteins were identified for the first time. The polypeptides enriched by heparin chromatography included both minor and major components of the brain extract. The enriched proteins belonged to several classes, including proteasome components, dihydropyrimidinase-related proteins, T-complex protein 1 components, and enzymes with various catalytic activities.

In yet another variant, Fountoulakis et al. (1999b) reported enrichment of low-abundance proteins of *E. coli* by hydroxyapatite chromatography. The complete genome of *E. coli* has been sequenced, and its proteome analyzed by 2D electrophoresis mapping. To date, 223 unique loci have been identified, and 201 protein entries were found in the SWISS-PROT 2-D PAGE on ExPASy server using the Sequence retrieval system query tool (http://www.expasy.ch/www/sitemap.html). Of the 4289 possible gene products of *E. coli*, about 1200 spots could be counted on a typical 2D gel map when ca. 2-mg total protein was applied. Possibly, most of the remaining proteins were not expressed in sufficient amounts to be visualized following staining with Coomassie Blue. Thus, to enhance low-abundance species, it was felt necessary to perform a prefractionation step, this time on hydroxyapatite beads.

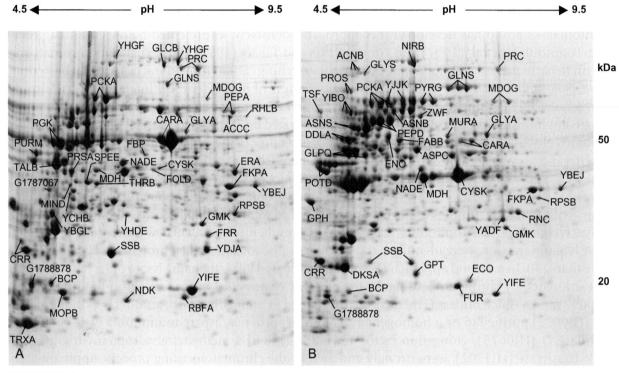

FIGURE 2.1

2D gel analysis of pools 19 (A) and 21 (B) eluted from a hydroxyapatite prefractionation of a total *E. coli* lysate. The proteins were eluted with an ascending gradient of 5.5M NaCl. The samples were analyzed on a pH 3-10 nonlinear IPG strip, followed by a 9-16%T gradient SDS-PAGE gel. The gels were stained with colloidal Coomassie Blue, destained with water and scanned in an Agfa DUOSCAN machine. Protein identification was by MALDI-TOF MS. The abbreviated names/numbers next to the protein spots are those of the *E. coli*, database (ftp://ncbi.nlm.nih.gov/genbank/genomes/bacteria(Ecoli/ecoli.ptt). *(From Fountoulakis et al., 1999b, by permission)*

By this procedure, approximately 800 spots, corresponding to 296 different proteins, were identified in the hydroxyapatite eluate. About 130 new proteins that had not been detected in 2D gel plates of the total extract were identified for the first time. This chromatographic step, though, enriched both low-abundance and major components of the *E. coli* extract. In particular, it enriched many low-Mr proteins, such as cold-shock proteins. An example of their results can be seen in Figure 2.1, displaying 2D electrophoresis maps of fractions from pools 19 (panel A) and 21 (panel B) collected from the hydroxyapatite prefractionation of *E. coli* total lysate.

In another variant, Krapfenbauer et al. (2003) adopted a double prefractionation scheme: ultracentrifugation for collection of separate cytosolic, mitochondrial, and microsomal fractions, followed by additional fractionation of the cytosolic fraction via ion-exchange chromatography. They could thus detect and identify some 437 rat brain proteins, about double those that had been previously catalogued.

Although the work presented by Fountoulakis's group is impressive and truly innovative in prefractionating samples for proteome analysis, some notes of caution should be offered:

- In general, huge amounts of salts are needed for elution from chromatographic columns, up to 2.5 M NaCl.
- As a consequence, loss of entire groups of proteins could ensue during the various manipulations (dialysis of large salt amounts, concentration of highly diluted pools of eluted fractions).
- The eluted fractions do not represent narrow pI cuts, but in general are constituted by proteins having pIs within a large pH range. Thus, one should still use wide pH gradients for their analysis in a 2D gel mapping protocol.

In particular, it is known that high salt amounts can induce irreversible adsorption of proteins to the resin used for fractionation, resulting in loss of spots, especially those of low abundance, during the subsequent 2D electrophoresis mapping. When eluting from ion-exchange columns, moreover, not only the collected fractions will be quite dilute, necessitating a concentration step, but they will also contain huge amounts of salt, rendering them incompatible with the IEF/IPG (isoelectric focusing/ immobilized pH gradients) first-dimension step. Thus, although the approach by Fountoulakis group could be quite attractive for analysis of some protein fractions in a total cell lysate, it might not be the best for exploring the entire proteome, due to all potential losses described above.

A very peculiar aspect of chromatographic prefractionation is the concomitant use of several superimposed columns, all of them used simultaneously with one single buffer for adsorption and one single buffer for elution of captured proteins, as proposed by Guerrier et al. (2005). The basic principle, as shown in Figure 2.2, is based on a set of solid-phase chemistries serially connected in a stack mode (in this case, an assembly of seven blocks), equilibrated in the same binding physiological buffer. The protein sample is applied at the top and it crosses all the chemistries, each of them capable of capturing a group of proteins, so that when the sample reaches the bottom-positioned block, most if not all proteins are removed. Once the protein adsorption phase is over and all the sequences of stacked beds are washed with a physiological saline to clean each solid phase, the sectional columns are dissociated and singularly used to strip out captured proteins by groups. This operation is performed by using a buffer that is compatible with the following analytical methods, such as mass spectrometry and two-dimensional electrophoresis. There is only one flow-through fraction for the whole experiment. The entire operation has to be done under nonoverloading conditions. By the choice of orthogonal chemistries, their sequence, and the loading volume amount, and by proper selection of the binding buffer, it is expected that the bound proteins will be quite different in each of the capturing blocks. With this chromatographic separation approach the redundancies (the same protein found in different fractions) are virtually eliminated. When N chemistries are used, N+1 fractions are typically obtained. Figure 2.2B shows SDS-PAGE profiles of the seven fractions recovered from a seven-bed column separation of human serum (the FT fraction is not represented). It has been reported that the number of mass spectrometry species observed was more than doubled compared to a classical unimodal/multistep elution chromatography.

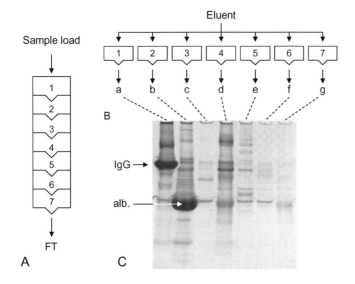

FIGURE 2.2

Scheme of fractionation of human serum using stacked columns. A: Assembled stack of seven capture columns (from 1 to 7) filled with different sorbents. B: Disassembled construct ready for elution of the captured proteins. C: SDS-PAGE analysis of the seven eluates from single disassembled columns (FT, fraction not shown). Adsorption buffer was Tris-phosphate (for all columns); elution was operated by means of either a mixture of trifluoroacetic acid-water-acetonitrile-isopropanol or of ammonia-water-acetonitrile-isopropanol. The positions of IgG (fraction 1) and of HSA (fraction 2) are highlighted. IgGs appear as a single band at MrB155 kDa since disulfide bonds have not been reduced (Guerrier et al., 2005).

As a variant of this approach, Guerrier and Boschetti (2007) and Guerrier et al. (2007) have developed a chromatographic scheme exploiting mixed-bed columns for separation of a target protein from a complex mixture, via a simpler, three-step procedure, as follows: (1) the selection of resins to capture, under a complementary manner, most proteins but not the target one, (2) the selection of at least one resin capable of capturing the target protein with the least number of other co-adsorbed species, and (3) the definition of the resin blend for the elimination of the maximum number of impurities. The entire selection needs to be operated under the same physicochemical conditions of buffer, salinity, and pH. This is illustrated in Figure 2.3. The mixed-bed sorbent is positioned as a first column and directly connected to a second column packed with the most selective sorbent for the target protein itself issued from a previous screening method (upper panel A). When the sample is loaded on the top of the first column, most impurities are adsorbed by the mixed-bed resin, while the target species is captured by the second sectional column with a possible co-adsorption of few other species. Panel B (lower) illustrates the remaining steps: After uncoupling, the eluate from the first column will contain the vast majority of impurities, whereas the eluate from the second sectional column, aided by a judicious and optimized gradient elution, yields the purified target protein. The procedure was successfully applied to several real cases to identify proteins from mass spectrometry signals as, for instance, YAP-1, a yeast protein expressed in *E. coli*, and a protein from *Helicobacter pylori* (Guerrier et al., 2008).

In yet another variant, an elegant way to simplify proteome prefractionation while increasing the detectability of low-abundance species via anion and cation exchangers, for isoelectric group separations, has been proposed by Fortis et al. (2008) by exploiting the principle of mixed-bed chromatography with several columns aligned in a sequence. The concept is based on the modification of protein net charge by surrounding pH values. The very unique difference from regular ion-exchange

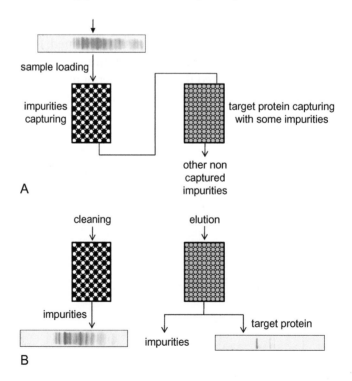

FIGURE 2.3

Setup of two columns in a cascade fashion for the separation process of a target protein from accompanying impurities (upper panel A). The first column where the initial crude sample is injected is made by a blend of selected sorbents able to capture impurities; the second column is filled with a sorbent able to capture the target protein. The noncaptured protein impurities are found in the final flow-through (FT). Lower panel B: after uncoupling, the first column eluate will comprise the vast majority of impurities; the second column, in turns, can be desorbed by sequential solutions, allowing the separation of additional co-adsorbed impurity traces from the target, purified protein. The three horizontal, rectangular boxes represent SDS-PAGE profiling of the various protein solutions before and after purification. *(From Boschetti and Righetti, 2011, by permission)* chromatography is that here the pH is produced by solid-state buffers (SSBs, prepared via the Immobiline chemicals, as adopted for generating IPGs) instead of liquid buffers. This difference leads to the preparation of beds of solid-state buffers admixed with ion exchangers. Basically, a protein is positively charged when the environmental pH is below the pI and is negatively charged when the pH is above the pI. When a protein mixture is loaded into a column of ion exchange resin mixed with beads of solid buffers, the latter material, by conditioning the surrounding pH, regulates the ionization of individual proteins. Species that have a net charge complementary to the ion-exchange resin are captured, while others remain in solution and are found in the flow-through. The flow-through is then directly injected into a second mixed-bed column where the same ion-exchange resin is mixed with a second solid buffer conferring to the environment a different pH value. Thus the proteins present are differently ionized, and the same mechanism as described for the first column takes place. The second column could then be followed by a third column and so on.

At the end of this process, the columns are disconnected from each other and the proteins captured by the ion exchanger eluted and collected independently. They contain proteins of different isoelectric point ranges depending on the type of solid buffer used. A scheme of the sequence of operations is shown in Figure 2.4, in which this principle is applied to a cation exchanger (CEX) buffered with different SSBs at pH 6.2, 5.4, and 4.6. To obtain successful results, specific rules must be followed.

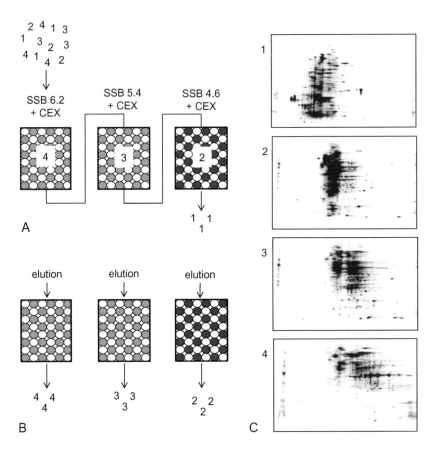

FIGURE 2.4
Schematic representation of protein separation (*E. coli* protein extract) by isoelectric point groups using a sequence of mixed beds constituted of solid-state buffers (SSBs) and ion exchangers. The mixed beds are here constituted of a cation exchanger (CEX), and the pH of solid-state buffers aligned are, as indicated, between 6.2 and 4.6. The protein crude solution (protein group "1": pI<4.6; protein group "2": pI 5.4-4.6; protein group "3": pI 6.2-5.4; protein group "4": pI>6.2) is injected in the column sequence (upper panel A); proteins of pI below 4.6 (group "1") are found in the effluent of the last column since they do not acquire the right ionic charge to be captured (right side of panel A). Lower panel B: the columns are uncoupled and the proteins above 6.2, between 5.4 and 6.2 and between 4.6 and 5.4 are desorbed separately from each sectional column. Right panel C: Two-dimensional electrophoresis of collected groups (pI range 3-10). Silver staining. (*Adapted from from Boschetti and Righetti, 2012*)

If a cation exchanger is used throughout the sequence, the top sectional column contains the solid buffer of the highest pH value and the bottom sectional column contains the one with the lowest pH value. On the contrary when an anion exchange resin is used, the solid buffers are aligned under a reversed pH sequence. The ionic strength should be very low to prevent salt competition with the ion-exchange resin; in turn, protein–protein interactions can be limited by the presence of non-ionic chaotropic dissociating agents such as low concentrations of urea. Moreover, the loading should not exceed the binding capacity of the ion-exchange resin that is part of the bed blend. Elution from each sectional column individually is achieved classically by either increasing the ionic strength or modifying the pH so that analytes and resin repulse to each other.

This chromatographic technology has been used successfully for separation of a large number of protein mixtures, among them *E. coli* extracts, *Pichia pastoris* cell lysate extracts (Fortis et al., 2008) and human serum proteins after having removed high-abundance proteins (Restuccia et al., 2009). Interestingly, compared to pI fractionation techniques, it does not require amphoteric molecules such as carrier ampholytes and the volume of the loading solution (or its dilution) is not critical because the capture of proteins by each sectional column induces a concentration effect. The separation process itself lasts between thirty minutes and a couple of hours depending on the column size. A noxious phenomenon occurring when separating proteins by isoelectric focusing under an electrical field is precipitation at their isoelectric point values. In the present process the protein aggregation is prevented first by the presence of a certain level of ionic strength and second by the action of dissociating agents. A more extensive description of this methodology can be found in Boschetti and Righetti (2012).

2.2.2 Two-dimensional Chromatography Coupled to Mass Spectrometry

Although this, too, is an analytical tool, aimed at exploring to the maximum possible depth any proteome, it can be regarded as a prefractionation tool as well, since a large number of fractions can indeed be obtained, all amenable to protein identification via MS equipment. In fact, 2D chromatography was born out of frustrations over some shortcomings inherent to the standard 2D electrophoretic mapping. High-throughput analysis of proteomes is challenging because each spot from 2D PAGE must be individually extracted, digested, and analyzed, a time-consuming process. In addition, owing to the limited loading capacity of 2D PAGE gels and the detection limit of staining methods, 2D PAGE has an insufficient dynamic range for complete proteome analysis. These shortcomings have encouraged development of alternative methods, which could be 1D or 2D chromatographic/electrophoretic methods, such as high-performance liquid chromatography (HPLC), capillary isoelectric focusing (cIEF), capillary zone electrophoresis (CZE), or micro-capillary chromatography (Banks et al., 1994). Just as with 2D PAGE, MS would be the method of choice for identifying proteins resolved by liquid separation protocols. This would have the advantage that the coupling with MS would be online, thus eliminating the lengthy steps for transferring proteins to MS equipment in 2D PAGE. We will first describe 1D methodologies, since these were the first ones developed early on. Most of these methodologies belong to the so-called bottom-up approaches since first a digestion of the protein mixture is made and produced peptides are fractionated contrary to the so-called top-down approaches where the proteins are fractionated first and each fraction is digested and analyzed then by mass spectrometry.

Among the 1D chromatography steps, in the early 1990s a flurry of papers appeared hyphenating CZE to MS instrumentation. Wahl et al. (1994), for instance, have claimed attomole level sensitivity by coupling CZE to ESI-MS in narrow-bore (10 μm ID) capillaries. Moseley et al. (1991) and Deterding et al. (1991), by coupling CZE to fast atom bombardment (FAB)-MS, reported sensitivities of the order of low femtomoles, under high separation efficiencies in the CZE step (410,000 theoretical plates) for bioactive peptides. Weinmann et al. (1993, 1994) described coupling of CZE to ^{252}Cf plasma desorption mass spectrometry, both off- and online, with sensitivities in the sub-picomolar range. These authors were able to analyze not only hydrophilic, but also hydrophobic peptides, the latter dissolved into aqueous acetic acid buffers containing up to 20% 2-propanol or 25% acetonitrile. In another approach, a number of other reports have appeared, in general coupling the eluate of a CZE run to an ESI-MS system (Banks and Whitehouse, 1996; Rosnack et al., 1994; Hsieh et al., 1994). In a review (Figeys and Aebersold, 1998), many of these developments have been listed and evaluated; in addition, the possibility of coupling microfabricated devices (CZE on a chip) to ESI-MS has been discussed.

Nevertheless, these early reports were not concerned at all with proteome analysis, but simply with development of such hyphenated techniques; thus, very simple samples were analyzed and only a handful of peptides separated. More recently CZE was reassessed as a potential tool in the proteome field. Among the different CZE variants, cIEF (capillary Isoelectric Focusing) appeared promising, since it would combine a high-resolution step, based on surface charge (pI) with an MS step, the latter providing a highly accurate Mr value. Tang et al. (1997) coupled cIEF to ESI-MS, whereas Yang et al. (1998) and Jensen et al. (1999) hyphenated cIEF to FTICR-MS (Fourier-transform in cyclotron resonance-mass spectrometry). With both approaches, however, very few proteins were electrophoretically resolved. When an *E. coli* lysate was analyzed by cIEF coupled to FTCIR-MS, a few hundred proteins were resolved, but only a few of them were identified, presenting an inadequate situation in proteomic studies. An alternative approach to separation would be to digest the entire protein mixture into a peptide mixture and use tandem MS (MS-MS) for generating amino-acid-sequence-specific data. It is in fact agreed that knowledge of the number of peptides generated by proteolysis, coupled with assessment of their relative masses, allows superior protein identification. McCormack et al. (1997) identified proteins from complex mixtures by first digesting them into peptides and subsequently loading the generated peptides onto a reverse-phase (RP) column. Upon elution, the isolated peptides were fed into an ESI-MS instrument.

However, in order to match the unique resolving power of 2D PAGE, 1D chromatography is not the best approach; two-dimensional chromatography would definitely be a better choice. Nilsson et al. (1999) have devised such an approach, consisting of preparative 2D liquid-phase electrophoresis (LPE) coupled to MALDI-TOF MS for final identification of the separated analytes. The samples under analysis were cerebrospinal fluid, in the first case, and human pleural eluate in the case of the last authors. Since 2D LPE has a high loading capacity, low-abundance proteins could be detected and identified. Another interesting approach consists in using a variety of 2D HPLC approaches, coupled to MS. In one instance, Raida et al. (1999) coupled two chromatographic columns, first a cation exchanger followed by RP-HPLC. The effluent of this last column was fed into an ESI MS machine, and the peptide masses assessed. Opiteck et al. (1997, 1998) described two different chromatographic approaches for separation of complex protein mixtures. In one approach, an *E. coli* lysate was injected onto a cation-exchange column and the eluates were fed stepwise into an RP-HPLC column. The eluate after this last step was sprayed into an ESI mass spectrometer. In a second system, size-exclusion chromatography was coupled to an RP-HPLC column, again for analysis of *E. coli* cells. In both instances, however, too few proteins could be identified.

In yet another approach, Link et al. (1999) devised a discontinuous 2D chromatography methodology, utilizing strong-anion exchange HPLC for a first-dimension separation. Portions of the eluate were digested with trypsin and analyzed by RP-microcolumn HPLC. The eluted peptides were finally characterized by tandem mass spectrometry. Whereas this is an offline method, an online approach, by which a peptide mixture from a digested *S. cerevisiae* was fed into a biphasic 2D capillary column packed with a strong cation exchanger (SCX) juxtaposed to RP beads was also described (Link et al., 1999). Peptides were displaced iteratively by salt from the SCX resin into the RP pearls; this was followed by a classic eluant for RP columns, feeding the eluted peptides into an ESI MS-MS. After re-equilibrating the RP beads, raising the salt concentration displaced more peptides from the SCX resin onto the RP pearls, thus reiterating the process. In a single experiment, this method could resolve and identify 189 unique proteins from a *S. cerevisiae* whole-cell lysate via SEQUEST interrogation.

Ultimately, in fact, two-dimensional methodologies became the established technique, as reported by Wolters et al. (2001) and Washburn et al. (2002, 2003) and reviewed by Link et al. (2004), Paoletti et al. (2004), Florens and Washburn (2006), and Fournier et al. (2007). This 2D chromatography method has become the standard one in proteome analysis, and it has gone down into history with the unfortunate acronym of MudPIT (multidimensional protein identification technology). In reality this is a pompous acronym, since the multidimension is simply two-dimensional as in 2D gel mapping, as it couples a strong cation exchanger to a reversed-phase resin. Additionally, calling one's own brainchild MudPIT (hole full of mud) does not seem a happy choice; personally we would have chosen BiCPIT (bidimensional chromatography protein identification technology), which is precisely what it is, not at all multidimensional!

2.3 SAMPLE PREFRACTIONATION VIA DIFFERENT ELECTROPHORETIC TECHNIQUES

Unlike its chromatographic counterpart, preparative electrophoretic techniques are scarcely used today (Righetti et al., 1992), due to the cumbersome instrumental setup and the limited load ability. This limitation stems from the very principle of electrophoresis: Electrokinetic methodologies are quite powerful, in fact much too powerful. This comes with a steep price: This extraordinary power (call it wattage) has to be dissipated from the separation chamber, thus seriously limiting any attempt at scaling-up, such as increasing the thickness of a gel slab or the diameter of a column. However, not all electrophoretic techniques suffer from these limitations: In conventional IEF in soluble, amphoteric buffers (CA), the typical conductivity at the steady state is about 2 orders of magnitude lower than in ordinary buffers (Righetti, 1980); in turn, immobilized pH gradients offer a further decrement of conductivity of at least another order of magnitude as compared to CA-IEF (Mosher et al., 1986). Considering, in addition, that proteins fractionated by any IEF/IPG preparative protocol are both isoelectric and isoionic (in fact, fully desalted), their interfacing with the first dimension of a 2D electrophoresis map is a straightforward procedure.

2.3.1 Rotationally Stabilized Focusing Apparatus

There is a long history behind this remarkable invention by M. Bier, going back to the doctoral thesis of S. Hjertén (1967): As he was trained in astrophysics, his apparatus was a "Copernican revolution" in electrokinetic methodologies. Hjertén was, in fact, the first one to propose electrophoretic separations in free zone (i.e., in the absence of anticonvective, capillary media, such as polyacrylamide and agarose gels), but he had to fight the noxious phenomenon of electrodecantation, induced by gravity. He thus devised rotation of the narrow-bore tubes used as electrophoretic chambers around a horizontal axis, mimicking celestial planet motions! This must have spurred the fantasy of Svensson-Rilbe, in those days a colleague at the Uppsala University, who finally described a mammoth multicompartment electrolyzer, capable of fractionating proteins in the gram range (Jonsson and Rilbe, 1980). The cell was assembled from forty-six compartments, accommodating a total sample volume of 7.6 L, having a total length of 1 m—hardly user friendly! Cooling and stirring were affected by slow rotation of the whole apparatus in a tank filled with cold water.

Bier's fifty-year-long love affair with preparative electrophoresis in free solution produced, as a last evolutionary step, a remarkable gadget, dubbed the Rotofor (Bier, 1998): It was happiness for all parties involved on both sides of the counter, users and producers alike. This preparative-scale device is capable of being loaded with up to 1 g of protein, in a total volume of up to 55 mL. A mini-version of it, with a reduced volume of about 18 mL, as well as a micro-Rotofor of only 2.5 mL volume comprising just ten compartments, is available. A typical instrumental setup is shown in Figure 2.5: It is assembled from twenty sample chambers, separated by liquid-permeable nylon screens, except at the extremities, where cation- and anion-exchange membranes are placed against the anodic and

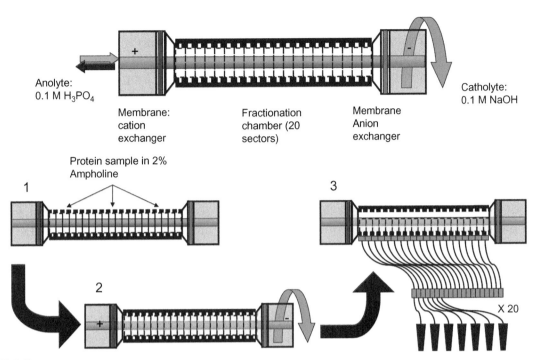

FIGURE 2.5

Scheme of a rotationally stabilized focusing apparatus operation. The upper drawing shows the assembled apparatus with anolyte and catholyte reservoirs and the ability of rotating on its axis (green arrow). The blue horizontal central cylinder with inlet and outlet arrows represents the central cooling finger. 1: sample loading under static conditions; 2: focusing operation under voltage and rotation; 3: simultaneous elution of the content of the twenty chambers via a manifold of 20 needles connected to 20 collection tubes.

cathodic compartments, respectively, so as to prevent diffusion within the sample chambers of noxious electrodic products. At the end of the preparative run, the twenty focused fractions are collected simultaneously by piercing a septum at the chambers' bottom via twenty needles connected to a vacuum source. The narrow-pI range fractions can then be used to generate conventional 2D electrophoresis maps. This is the original approach described by Hochstrasser et al. (1991).

In a further development, this methodology has taken another, unexpected turn: The device is used directly as the first dimension of a peculiar 2D gel methodology, in which each fraction is further analyzed by hydrophobic interaction chromatography, using nonporous reversed-phase HPLC. Each peak collected from the HPLC column is then digested with trypsin, and subjected to MALDI-TOF MS analysis and MSFit database searching. By this approach, Wall et al. (2000) have been able to resolve a total of about 700 bands from a human erythroleukemia cell line. It should be stated, though, that the pI accuracy of this methodology (which is still based on conventional CA-IEF) is quite poor: It ranges from ±0.65 to ±1.73 pI units, a large error indeed. On a similar line of thinking, Davidsson et al. (2001) have subfractionated human cerebrospinal fluid and brain tissue, whereas Wang et al. (2002) have mapped the proteome of ovarian carcinoma cells. This instrument is quite popular in proteome prefractionation and has been used, together with the miniaturized version, for a variety of applications, including the exploration of the proteome of grape berry cell cultures (Sharathchandra et al., 2011). All these attempts were made with the objective of discovering the maximum number of protein components, most of them being of low abundance and not detectable prior to fractionation.

2.3.2 Continuous Free-flow Isoelectric Focusing

This liquid-based IEF technique, free-flow isoelectric focusing (FF-IEF), relies on the principle of free-flow electrophoresis, described long ago by Hannig (1982). In FF-IEF, samples are continuously injected into a carrier ampholyte solution flowing as a thin film (0.4 mm thick) between two parallel plates and, by introducing an electric field perpendicular to the flow direction, proteins are separated by IEF according to their different pI values and finally collected into up to 96 fractions. The key advantages of the method are improved sample recovery (due to absence of gel media or membranaceous material) and sample loading capacity (due to continuous sample feeding; this step is not rate limiting). The fundamental principles of FFE, including commercial instrumentation, have been described in a recent review (Krivankova and Bocek, 1998).

It is of interest to recall here that perhaps one of the first applications of continuous-flow IEF for preparative protein fractionation came from Fawcett (1973), who, however, devised a vertical, continuous-flow chamber filled with a Sephadex bed as an anticonvective medium. In one version, Gianazza et al. (1975) fractionated synthetic, wide-range carrier ampholytes in a chamber with just twelve outlets at the bottom as sample collection ports. The system could be run for weeks almost unattended by arranging a continuous sample input fed by constant hydrostatic pressure via a Mariotte flask. An example of a modern FF-IEF apparatus is shown in Figure 2.6. It can be appreciated that the three colored protein species, continuously injected in the lower right part of the vertical cell, become focused on the upward migration, and thus their stream-lines remain parallel to the electrode rods and stop their electrophoretic migration. Instrumental in the success of the method, especially when run for long periods of times, is the constancy of the elution profile at the collection port, so that the same protein species is always collected into the same test tube. Thus, electroendoosmotic flow should be suppressed via a number of ways, including glass-wall silanol deactivation and addition of polymers (e.g., 0.1% hydroxypropylmethyl cellulose), providing dynamic wall coating and proper liquid viscosity.

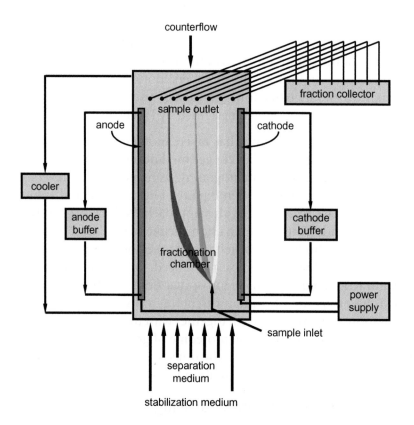

FIGURE 2.6
Schematic drawing of the modern Hannig apparatus used in the free-flow isoelectric focusing mode. A stable pH gradient is generated by focused carrier ampholytes as the liquid curtain moves upstream. The three colored proteins halfway toward the chamber top become focused, and thus they move upward as filaments parallel to the electrode wires, ceasing their electrophoretic migration upon attaining their respective pI values.

The first report on the use of CF-IEF for prefractionation of total cell lysates (HeLa and Ht1080 cell lines), in view of a subsequent 2D electrophoresis map, is perhaps the one of Burggraf et al. (1995), who collected individual or pooled fractions for further 2D gel analysis. Theoretical modeling of the process was later presented by Soulet et al. (1998). In a variant of the above approach, Hoffman et al. (2001) have proposed CF-IEF as the first dimension of a 2D electrophoresis map, the eluted fractions being directly analyzed by orthogonal SDS-PAGE. In turn, individual bands in the second SDS dimension were eluted and analyzed by ESI-IT-MS. By this approach, they could identify a number of cytosolic proteins of a human colon carcinoma cell line. One advantage of CF-IEF is immediately evident from their data: Large proteins (e.g., vinculin, Mr 116.6 kDa) could be well recovered and easily identified; on the contrary, recovery of large Mr species has always been problematic in IPG gels.

2.3.3 Continuous Free-flow Electrophoresis

The original Hannig (1982) apparatus was, in fact, designed for zonal electrophoresis (ZE) and for separation of intact cells and organelles, which would have very low-diffusion coefficients in the ZE mode. This apparatus went through successive designs and improvements, from an original liquid descending curtain to the present commercial version, dubbed Octopus, exploiting an upward liquid stream (Kuhn and Wagner, 1989), as shown in Figure 2.7. This approach has been recently applied to improved proteome analysis via continuous free-flow purification of mitochondria from *S. cerevisiae* (Zischk et al., 2003) and of cytosolic proteins in human colon carcinoma (Hoffmann et al., 2001). These authors reported a considerable increment in purity of the mitochondria via ZE-FFE as compared to their isolation via differential centrifugation. Moreover, there was

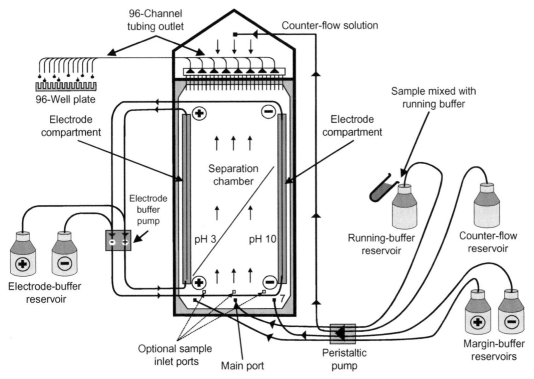

FIGURE 2.7
Schematic representation of the continuous FFE apparatus (Octopus). Separation chamber dimensions: 500x100x0.4 mm); electrode length: 500 mm (electrical contact with the separation area is made via a polyacetate PP60 membrane); distance between electrodes: 100 mm; chamber depth: 0.4 mm, created by insertion of polyvinyl chloride papers between the separation plates. The separated protein samples are collected into 96-well plates via an in-line multichannel outlet. The volume of each fraction is typically ca. 2 mL. *(Courtesy of Dr. H. Wagner)*

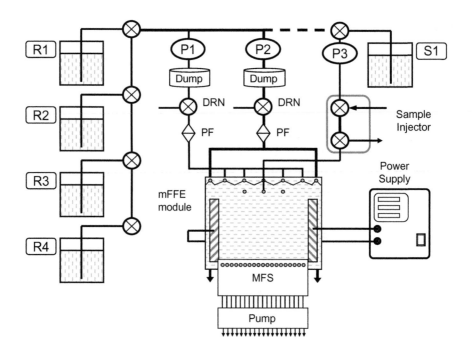

FIGURE 2.8

Scheme of the miniaturized ZE-FFE apparatus. Reservoir R1 is used for the electrophoresis buffer, whereas R2 (0.1 M aqueous NaOH-ethanol, 50:50, v/v), R3 (0.01 M HCl) and R4 (80% aqueous ethanol) are used for washing cycles. S1 is the sample reservoir, P1-P3 are peristaltic pumps for pouring the solutions into the separation chamber, the electrode reservoirs and the sample port, respectively. D is a dumper, DRN a drain duct, PF is a fuse unit for excessive pressure. The proteins separated in the micro-FFE chamber subsequently enter the micromodule fraction separator (MFS) and are collected into individual fractions. *(From Kobayashi et al., 2003, by permission)*

an extra benefit: 2D electrophoresis maps of proteins extracted from ZE-FFE purified mitochondria exhibited a much higher level of undegraded proteins. On the contrary, about 49% of the proteins from crude mitochondria, as isolated via centrifugation, were found to be degraded or truncated. The ZE-FFE instrument has been miniaturized by Kobayashi et al. (2003). Their separation chamber is barely 66 x 70 mm in size, with a gap between the two Pyrex glass plates of only 30 μm (Figure 2.8). The liquid curtain and the sample are continuously injected from five and one holes, respectively, at the top. At the bottom of the chamber, a micromodule fraction collector, consisting of nineteen stainless steel tubes, is connected perpendicular to the chamber and liquid stream.

2.3.4 The Gradiflow

The Gradiflow is a multifunctional electrokinetic membrane apparatus that can process and purify protein solutions based on differences of mobility, pI, and size (Margolis et al., 1995; Horvàth et al., 1996). Its interfacing with 2D gel map analysis was demonstrated by Corthals et al. (1997), who adapted this instrument for prefractionation of native human serum and enrichment of protein fractions. In a more recent report (Locke et al., 2002), this device was also shown to be compatible, in the prefractionation of Baker's yeast and Chinese snow pea seeds total cellular extracts, with the classical denaturing/solubilizing solutions of 2D PAGE maps, comprising urea/thiourea and surfactants. Whereas, in the case of size fractionation, this can be achieved with polyacrylamide- coated membranes at different %T and %C for sieving of macromolecules in given Mr ranges, its use for separating approximate pI fractions is more complex and can in general only provide two fractions in two different pI ranges. Instrumental to this setup is the use of low-conductivity buffers, such as those devised by Bier et al. (1984), so as to allow reasonably high-voltage gradients and relatively short separation times. This technique, though, has been described only at the methodological level and has not proven its worth, as yet, in "mining below the tip of the iceberg" for finding low-abundance and membrane proteins (see below).

2.3.5 Sample Prefractionation via Multicompartment Electrolyzers with Isoelectric Membranes

This kind of equipment represents perhaps the most advanced evolutionary step in all preparative electrokinetic fractionation processes (Righetti, 1990). The miniaturized version of the original apparatus devised by Righetti et al. (1989, 1990, 1992) is shown schematically in Figure 2.9: It consists of a block of chambers sandwiched between an anodic and a cathodic reservoir. The apparatus is modular and can accommodate up to eight flow chambers, six for sample collection and the two extreme ones as electrodic reservoirs. In the miniaturized version (which is the best option in proteome analysis, since often the amount of sample is rather minute), external reservoirs and pumps for large sample volumes, present in the original 1990 version, have been abolished. This system is shown for the following reason: Two isoelectric membranes, facing each flow chamber, act by continuously titrating the protein of interest to its isoelectric point. They can be envisaged as highly selective membranes, which retain any proteins having pIs in between their limiting values, and which will allow transmigration of any nonamphoteric, nonisoelectric species. For that, the only condition required is that $pI_{cm} > pI_p > pI_{am}$, where the subscripts cm and am denote cathodic and anodic membranes, respectively, and p is the protein having a given isoelectric point between the two membranes. For this mechanism to be operative, it is necessary that the two isoelectric membranes possess good buffering capacity (β), so as to be able to effectively titrate the protein present in the flow chamber to its pI, while ensuring good current flow through the system. Wenger et al. (1987) had in fact synthesized amphoteric, isoelectric Immobiline membranes and demonstrated that indeed they are good buffers at their pI. These membranes are, of course, made with the same technology used to make IPG strips; that is, they exploit the Immobiline buffers, but not for creating a pH gradients, but simply for defining any single pH value along the pH scale, such value being then the pI of a given membrane.

The concept of isoelectric Immobiline membranes is indeed revolutionary and deserves further comments. Such membranes can be envisaged as pH-dictating assemblies in an IEF separation, much like a pH-stat unit is set up for controlling, for example, the pH during a biochemical reaction or during in vitro tissue growth. Each species that is tangent or crosses such isoelectric membranes is titrated to the pH of the membrane (provided it does not overcome its intrinsic β value). For amphoteric compounds, this results in a drastic change in mobility, which could reach zero if the two membranes

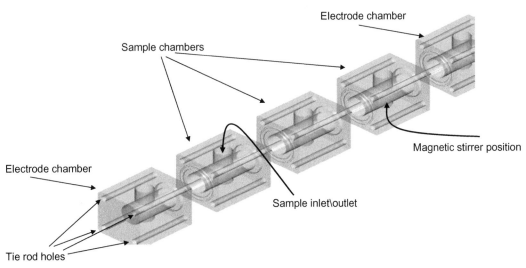

FIGURE 2.9
Exploded view of the miniaturized multicompartment electrolyzer operating with isoelectric membranes. An assembly with only five chambers is shown, the central three for sample focusing and collection into isoelectric traps of any desired pI interval, the ones at the two ends functioning as electrodic chambers connected to the power supply.

delimiting a single-flow chamber have pIs encompassing the protein of interest to be trapped in said chamber. Any other amphoteric species with lower or higher pI value will be forced to exit from such a chamber toward either the anode or the cathode. Thus it is clear that, with a proper set of membranes, it is possible to define in a given chamber isoelectric conditions for just a single component of a protein mixture, which ultimately will be arrested as the sole isoelectric species in such a chamber. Moreover, if the protein concentration is high enough, the macroion present in the liquid stream will possess enough buffering power to control the pH, in the absence of exogenous ions migrating through the system.

Although this system was used originally for extreme purification of a single protein form, in preparation for crystallization experiments or other biochemical assays requiring absolute purity, it was easy to envisage its application to proteome analysis. In this last case, instead of using very narrow pI gaps in between two membranes delimiting a chamber, one could have used large pI gaps, from one up to several pH units, for trapping groups of proteins in a given chamber. Such narrow-pI range families could then be analyzed on an IPG strip encompassing just the pI values of such a family. The advantages of such a procedure are immediately apparent:

- It offers a method that is fully compatible with the subsequent first-dimension separation, a focusing step based on Immobiline technology. Such a prefractionation protocol is precisely based on the same concept of immobilized pH gradients; thus protein mixtures harvested from the various chambers of this apparatus can be loaded onto IPG strips without any need for further treatment, in that they are isoelectric and devoid of any nonamphoteric ionic contaminant.
- It permits harvesting a population of proteins having pI values precisely matching the pH gradient of any narrow (or wider) IPG strip.
- As a corollary of the above point, much reduced chances of protein precipitation will occur. In fact, when an entire cell lysate is analyzed in a wide gradient, there are fewer risks of protein precipitation. On the contrary, when the same mixture is analyzed in a narrow gradient, massive precipitation of all nonisoelectric proteins could occur, with a strong risk of co-precipitation of proteins that would otherwise focus in the narrow pH interval.
- Due to the fact that only the proteins co-focusing in the same IPG interval will be present, much higher sample loads can be operative, permitting detection of low-abundance proteins.
- Finally, in samples containing extreme ranges in protein concentrations (such as human serum, where a single protein, albumin, represents>60% of the total species), one could assemble an isoelectric trap narrow enough to just eliminate the unwanted protein from the entire complex, this too permitting much higher sample loads without interference from the most abundant species.

Herbert and Righetti (2000) have applied precisely this technology to sample prefractionation, in preparation for 2D PAGE analysis, by miniaturizing the original apparatus as shown in Figure 2.9. Ideally, such an instrument would hold in each chamber just the total liquid volume that could be adsorbed in the IPG strip for the first-dimension run. In properly made strips, such reswelling sample volume could be as much 1 mL. However, total volumes of 3-4 mL per chamber could be a valid alternative, since they would permit running a few IPG strips for duplicate or triplicate analysis. It might be asked if this procedure could lead to losses of proteins, either by isoelectric precipitation or by adsorption onto the isoelectric membranes. Herbert and Righetti (2000) performed a protein assay on the starting and ending products and could confirm a 95% protein recovery. Additionally, a similar assay performed on the ground and extracted isoelectric membranes failed to reveal any appreciable amount of proteinaceous material bound or adsorbed onto those surfaces.

Since this prefractionation process is quite time consuming, also due to the fact that all proteins have to cross the isoelectric membranes which, being composed by polyacrylamide, would exert some sieving, Cretich et al. (2003) and Fortis et al. (2005 a & b) have produced them in the form

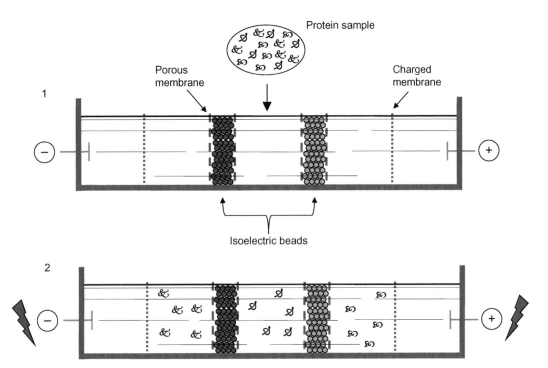

FIGURE 2.10
Scheme of a multicompartment electrolyzer operating with isoelectric beads instead of isoelectric membranes. Upper drawing (1): initial loading stage; lower scheme: attainment of the final focusing stage (2).

of isoelectric beads and proven that they indeed exert the required buffering power and can be used in the multicompartment electrolyzer in lieu of membranes, thus considerably reducing focusing and separation time (see Figure 2.10). The present prefractionation method is proving a formidable tool in proteome analysis (see, e.g., Herbert et al., 2003, 2004, 2007; Pedersen et al., 2003; Righetti et al. 2001, 2003, 2004, 2005a&b, 2007 for original data and reviews) not only because it provides the much needed improvement in resolution, but also because of the highly desirable increment in sensitivity, due to the possibility of loading a much higher sample amount in any desired narrow pH interval. The philosophy of Herbert and Righetti's protocol (2000) is to allow for such a high-protein loading that only Coomassie stain will be needed for detection; silver staining is in fact not very suitable for MS procedures, since the aldehydes used in most protocols generally cross-link proteins and render them unavailable for further analysis. In addition, the glutaraldehyde-free silver stains commonly used for MS analysis are less sensitive and are often not better than colloidal Coomassie.

2.3.6 Off-Gel IEF in Multicompartment Devices

This is the latest evolution in preparative techniques based on contact with IPG matrices (or membranes) (Bjellqvist et al., 1982). Just like the MCE technique, Off-Gel electrophoresis has been devised for separation of peptide and proteins according to their pI and for their direct recovery in solution without adding buffers or ampholytes (Ros et al., 2002). The principle is to place a sample in a liquid chamber that is positioned on top of an IPG gel. The gel buffers a thin layer of the solution in the liquid chamber, and the proteins are charged according to their pI values and to the pH imposed by the gel. Theoretical calculations and modeling have shown that the protonation of an ampholyte occurs in the thin layer of solvation closed to the IPG gel/solution interface (Arnaud et al., 2002). Upon application of a voltage gradient, perpendicularly to the liquid chamber, the electric field penetrates into the channel and extracts all charged species (those having pI values above and below the pH of the IPG gel), thus vacating them from the sample cup. After separation, only the globally "neutral" species (pI=pH of the IPG gel) remain in solution. This technique offers high separation efficiency and allows easy recovery of the purified compounds directly in the liquid phase. However, due to the diffusion limitation of large molecules from the IPG gel to the well, the device is mostly used for peptide fractionation after trypsin hydrolysis of protein mixtures.

In a further extension of this initial work, the Off-Gel electrophoresis format was improved and adapted to a multi-well device, composed of a series of compartments of small volume (100 to 300 μL) and compatible with the instruments used in modern separation science (Michel et al., 2003). In a most recent version, a multi-electrode setup that significantly improves protein separation efficiency has been developed. Here, the electric field is applied by segments between seven electrodes connected in series to six independent power supplies. The aim of this strategy is to distribute evenly the electric field along the multi-well system, and as a consequence to enhance electrophoresis in terms of separation time, resolution, and protein collection efficiency, while minimizing the overall potential difference and therefore the Joule heating (Tobolkina et al., 2012).

Examples of such fractionation cells are shown in Figure 2.11, which displays two such assemblies, one with eight, the other with twenty chambers (the remaining two being for anolyte and catholyte, respectively). It can be seen that, underlying such chambers is an IPG gradient gel, spanning any desired pH gradient, rather than a fixed-pH gel (an isoelectric membrane, in fact) as in previous, single-cup equipment. Thus, at the end of the electrophoretic run, each chamber should, in principle, contain only that peculiar pI cut delimited by the pH value of the gel segment underlying a given cup. As the underlying gel contains a continuous pH gradient, each cup will not contain a single population of pI values, but rather a narrow (or wider) pI interval according to the pH span of the gel segment forming the floor of the given chamber. How well this device would work is illustrated in Figure 2.12, which shows the prefractionation of an entire *E. coli* lysate, run in the 10-cup assembly

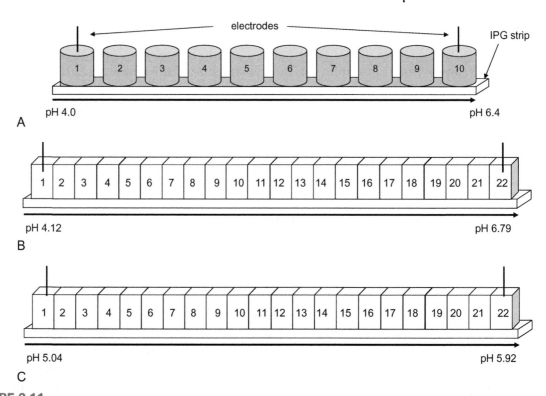

FIGURE 2.11

Experimental setup used to perform Off-Gel separations with multi-cup devices, composed of either ten (A) or twenty-two (B, C) wells. The pH gradients used in each case are presented under the gels. *(From Michel et al., 2003, by permission)*

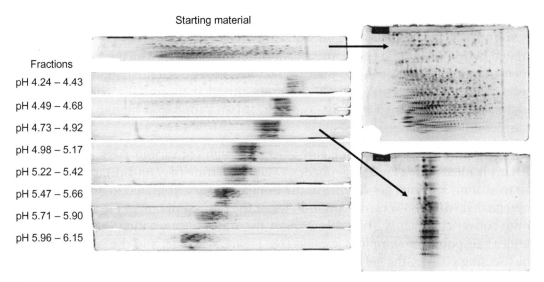

FIGURE 2.12

Immobilzsed pH gradient analyses of an *E. coli* protein extract fractionated by Off-Gel electrophoresis in the ten-cup device of Figure 7A. Each strip was used to analyze the fraction present in one given well. *(From Michel et al., 2003, by permission)*

shown in Figure 2.11, delimited at the bottom by an IPG pH 4.0–6.4 interval. One would thus expect that each cup would, at the end, contain a population of proteins spanning approximately a pI interval of 0.2 pH units; this fact is fully confirmed by the IPG strips showing the pI interval of each eluted fraction and by the corresponding 2D electrophoresis maps. In addition, the same authors have demonstrated that this fractionation device, when operated in rather narrow ranges, can ensure a resolution of ca. 0.1 pH units between two neighboring protein species, since it can resolve β-lactoglobulins A and B, known to have pIs differing, precisely, by this ΔpH value. More recently, Off-Gel electrophoresis has been applied, with remarkable success, not just to intact proteins but to peptides obtained by tryptic digestion (Hörth et al., 2006; Chenau et al., 2008; Geiser et al., 2011). After such peptide separations, the number of identified proteins increases substantially.

2.3.7 Prefractionation via Isoelectric Focusing in Sephadex Beds

In 2002 Görg et al. developed a simple, cheap, and rapid sample prefractionation procedure based on flatbed isoelectric focusing (IEF) in granulated gels. Complex sample mixtures were prefractionated in Sephadex gels containing urea, zwitterionic detergents, dithiothreitol, and carrier ampholytes. After IEF, up to ten gel fractions alongside the pH gradient were removed with a spatula and directly applied to the surface of the corresponding narrow pH range immobilized pH gradient (IPG) strips as the first dimension of 2D gel electrophoresis. The major advantages of this technology are the highly efficient electrophoretic transfer of the prefractionated proteins from the Sephadex IEF fraction into the IPG strip without any sample dilution and the full compatibility with subsequent IPG-IEF, since the prefractionated samples are not eluted, concentrated, or desalted, nor does the amount of the carrier ampholytes in the Sephadex fraction interfere with subsequent IPG-IEF. This prefractionation protocol allows loading higher protein amounts within the separation range applied to 2D gel plates and facilitates the detection of less abundant species. This system is also highly flexible, since it allows both small-scale and large-scale runs, and separation of different samples at the same time.

Although these data look impressive, in reality this is a rediscovery of the famous Radola technique, described in the 1970s (Radola, 1973 a&b, 1975). The technique became extremely popular in those days, to the point that the producer of the technique released on the market a trough, a sample applicator, and a fractionation grid that would "guillotine" the focused gel slurry into twenty fractions (for a more thorough description of the principle and the setup, see Righetti,

1983, pp. 129–136). There is also a cute story behind it, as told to one of us (PGR) by the producer scientists: The Sephadex G-200 superfine beads used to form the granulated bed was contaminated by some material (perhaps salts or other chemicals), which severely interfered with proper focusing, generating wavy and distorted bands. So, the scientists bought large quantities of Sephadex from the competing company, simply rinsed it thoroughly with distilled water to remove the contaminants, dried it again, and sold it (we suppose at higher prices) as their own material under another trade name!

How the technique works is shown in Figure 2.13: A trough (of various sizes, typically 40 x 20 cm or 20 x 20 cm) is filled with a layer of Sephadex G-200 superfine (4 g/100 mL), with a gel layer thickness of up to 10 mm. The total gel volume in the trough can vary from 200 to 800 mL. The Sephadex layer (impregnated with 2–3% carrier ampholytes of the desired pH range) is allowed to lose water in air to the correct consistency (typically 25% of the water impregnating the beads is allowed to evaporate). At this point the sample is applied (see Figure 2.13, *left*) and the voltage applied. At the end of the focusing process, the gel layer is divided into twenty segments by a fractionation grid (see Figure 2.13, *right*), and the granulated bed is scooped up with a spatula. The proteins can be easily recovered simply by a gel filtration step or else (as suggested by Görg (2002), in the event small gel layers and sample loads are adopted), the recovered Sephadex gel can be directly applied to the surface of an IPG gel of the desired pH interval if 2D gel mapping is planned after this prefractionation step. The Radola technique offers the advantage of combining high-resolution, high sample loads and easy recovery of sample components. The load ability of this technique is quite amazing: Radola (1975) has fractionated 10 g pronase E in an 800 mL of gel layer and obtained excellent band resolution even at this remarkable load capacity. This method would thus appear to be ideal for prefractionation of large volumes of sera in search for low-abundance species and for biomarker discovery. Why this facile and powerful method is not in use today remains a mystery to us: Perhaps scientists shun simple methods requiring essentially inexpensive and fancy instrumentation. Perhaps they think that if they were to use such an unsophisticated method (yet, we repeat, extremely powerful), they would look like Clark Kent, whereas everyone thinks of themselves as Superman, of course! It is nice that Görg et al. (2002, 2008) have resurrected this brilliant idea of Radola: Let us drink to that! Just to paraphrase a famous song by Woody Guthrie: "Good ideas never die, they just fade away."

2.3.8 Two-dimensional Mapping

As stated in the introduction, we do not want to deal with this technique, first since, per se, this is not a prefractionation tool and second because, after 1975, with the classical O'Farrell paper, hundreds of reviews and book chapters appeared dealing extensively with this methodology to the

Sample loading   Sample harvesting

FIGURE 2.13
Experimental setup for proteome prefractionation by the "Radola technique" in granulated layers of Sephadex G200. Left drawing: sample application, by collecting a bed segment, mixing it with the sample and pouring it back into the separation tray. Right drawing: sample fractionation and collection upon steady-state focusing. The Sephadex gel layer is segmented into twenty fractions by a grid and each zone scooped up with a spatula.

point that yet another description would be redundant. It is sufficient to refer the reader to a series of recent reviews by Rabilloud's group (Rabilloud 2009a,b; Rabilloud et al., 2010; Rabilloud and Lelong, 2011; Rabilloud 2012a,b) which will supply answers even to questions they were afraid to ask. We cannot possibly compete with the "oracle" Rabilloud, who could easily challenge even the Cumana Sybil! The other source of information on any possible aspect of 2D electrophoresis mapping is, of course, the classical manual of Westermeier (2006), which is so popular that it has gone through four editions.

Yet, since indeed 2D PAGE can also be considered as a prefractionation tool, in which hundreds to thousands of polypeptide chains are deposited in the charge/mass 2D space, it is worthwhile to make some final comments here. Although both the first (IEF/IPG) and second (SDS-PAGE) separation dimensions appear to be performing at their best, there remains the fact that, due to the vast body of polypeptide chains present in a tissue lysate, especially at acidic pH values (pH 4-6), resolution is still not enough. Some papers have thus proposed that this problem could be overcome by running a series of narrow-range IPG strips (covering no more than 1 pH unit) on large size gels (18 or longer cm in the first dimension, large-format slabs in the second dimension), which would dramatically increase the resolution (Corthals et al., 2000). The entire, wide-ranging two-dimensional map would then be electronically reconstructed by stitching together the narrow-range maps. This might turn out to be a Fata Morgana, though: It is still true that, even when using very narrow IPG strips, they have to be loaded with the entire cell lysate, containing proteins focusing all along the pH scale. Massive precipitation will then ensue, due to aggregation among unlike proteins, with the additional drawback that the proteins that should focus on the chosen narrow-range IPG interval will be strongly underrepresented, since they will represent only a small fraction of the entire sample loaded.

That this is a serious problem has been debated by Gygi et al. (2000). These authors analyzed a yeast lysate by loading 0.5 mg total protein on a narrow-range (nr) IPG pH 4.9–5.7. Although they could visualize by silvering ca. 1500 spots, they were disappointed that a large number of polypeptides simply did not appear in such a two-dimensional electrophoresis. In particular, proteins from genes with codon bias values of<0.1 (low-abundance proteins) were not found, even though fully one-half of all yeast genes fall into that range. The codon bias value for a gene is its propensity to use only one of several codons to incorporate a specific amino acid into the polypeptide chain. It is known that highly expressed proteins have large codon bias values (> 0.2). Thus, these authors conclude that, in reality, when analyzing protein spots from 2D PAGE maps by mass spectrometry, only generally abundant proteins (codon bias>0.2) can be properly identified. Thus, the number of spots is not representative of the overall number or classes of expressed genes that can be analyzed.

Gygi et al. (2000) have calculated that, when loading only 40 μg total yeast lysate, as done in the early days of 2D gel mapping, only polypeptides with an abundance of at least 51,000 copies/cell could be detected. With 0.5 mg of starting protein, proteins present at 1000 copies/cell could now be visualized by silvering, but those present at 100 and 10 copies/cell could not be detected. These authors thus concluded that the large range of protein expression levels limits the ability of the 2D gel-MS approach to analyze proteins of medium to low abundance, and thus the potential of this technique for proteome analysis is likewise limited. This is a severe limitation inasmuch as it is quite likely that the portion of proteome we are presently missing is the most interesting one from the point of view of understanding cellular and regulatory proteins—considering that such low-abundance polypeptide chains will typically be regulatory proteins. A robust mitigation of the problems listed above appears to be the adoption of a number of electrophoretic prefraction tools described in this chapter, especially the multicompartment electrolyzers with Immobiline membranes and, of course, Off-Gel electrophoresis. These two methodologies indeed can provide narrow pI cuts of prefractionated proteomes encompassing the width of any narrow IPG gradient adopted in the subsequent two-dimensional gel map analyses.

Before closing this chapter, though, we would like to report something more on two-dimensional gel electrophoresis maps which has not been dealt with in previous reviews. Although there would appear to be no more space for improvement of 2D map analysis (a standard detailed method is described in Section 8.26), in reality there have been a couple of recent modifications worth discussing. In one, Millioni et al. (2010a) have described a novel second-dimension setup by which this dimension is performed not in a conventional square- or rectangular-size gel, but in a radial surface. This has the advantage of permitting resolution of closely adjacent bands, representing strings of isoforms of similar or identical mass but of closely spaced isoelectric points. When used in a monodimensional, SDS-PAGE format, this system allows the simultaneous running of sixty-two sample tracks. Radial electrophoresis is performed by using a horizontal apparatus, where the lines of force of the electric field are determined by circular and concentric electrodes. A special software then allows conversion of the gel slab image from radial to Cartesian shape. It should also be considered that the modern trend is to run larger and larger 2D gels, since it has been theoretically predicted, and verified in practice, that even the largest-size 2D gels available (18 x 20 cm in size) would still not permit optimal resolution of polypeptide spots in the two-dimensional plane: Singlets would constitute the minority, whereas doublets and even triplets and quadruplets would form a substantial part of the spot population (Campostrini et al., 2005).

Aware of these problems, recently some companies introduced even longer IPG strips, up to 24-cm length. For complex proteomes, this appears to be an intelligent solution, although it does not, per se, represent a novelty in the field. Perhaps unknown to present-day users, the problem of spot resolution in 2D gel map analyses was acutely felt since the inception of the technique. Thus, already in 1984, Young deployed large-size gels and named them, humbly, "giant" gels, since they had a napkin-size: 39 x 37 cm! He developed them out of a sense of frustration, after 2½ years of research for proteins induced by adrenal glucorticoid administration, of which he could detect none. Yet, when the gel size was increased by roughly six-fold in area, and the amount of protein loaded increased up to 100-fold, he could see a constellation of protein changes induced by the hormone. An even more bountiful harvest was made by Klose and Zeindl (1984): When adopting slightly larger gels (42 x 33 cm, 0.85 mm in thickness) for analyzing an epithelial-like human larynx carcinoma cell line, uniformly labeled with ^{14}C-amino acids: >10,000 different polypeptide spots could be revealed in a single 2D map, a remarkable achievement indeed. It would thus appear that, in modern times, we are slowly inching toward the achievements of the 1980s—so much so that, in 2002, Oguri et al. (2002) presented "cybergels" of 70 x 67 cm dimensions, able to resolve and detect some 6677 polypeptide spots from rat hippocampal neurons.

Aware of this trend, Millioni et al. (2010b) have reported novel instrumentation for performing large-size (> 25 cm) two-dimensional gels. For running the first dimension, they developed a power supply that could deliver a voltage of up to 15,000 V while allowing regulation of current (up to 200 µA) onto each individual focusing (IPG) strip. The IEF strip tray can accommodate as many as twelve IPG strips up to 45 cm in length. The electrodes are sliding on a ruler, thus permitting running, besides 45 cm strips, any IPG strip length. For the second dimension, this apparatus includes a second power supply that allows performing electrophoresis at high amperage (400 mA) and a Peltier system that allows a 10–80°C temperature control. In another setup, Westermeier and Görg (2011) have described a horizontal flatbed apparatus able to run a large- size 2D map (up to 30 x 40 cm) in a chest-of-drawers format accommodating four gel slab units (see Figure 2.14). According to these authors, such a horizontal setup allows running two-dimensional gel electrophoresis maps with considerably less horizontal and vertical streaking as compared to standard vertical set-up systems. Additionally, this system allows seeing twice as many spots as the comparable vertical system (see Figure 2.15).

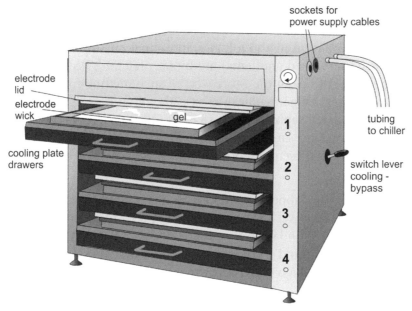

FIGURE 2.14
"Chest of drawers" for running 2D maps in a horizontal setup. This tower allows running four 2D gels simultaneously. *(Courtesy of Dr. Reiner Westermeier, Heidelberg, Germany)*

12.5 % Flatbed NF pH 5-8 24 cm   12.5 % Vertical pH 5-8 24 cm

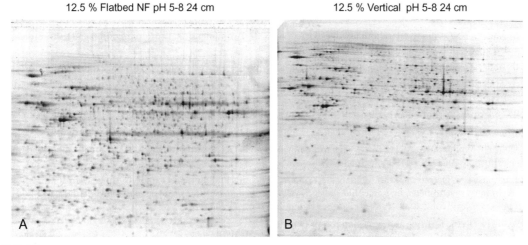

FIGURE 2.15
Comparison of 2D electrophoresis maps of a cell lysate run in the horizontal flatbed system (panel A) and in the vertical, conventional setup (panel B), both gels in the IPG pH 5-8 interval and in the same 24-cm size. About twice as many spots can be visualized in the horizontal gel assembly as compared to the vertical one. *(Courtesy of Dr. Reiner Westermeier, Heidelberg, Germany)*

2.4 CONCLUSIONS

We hope that the present chapter will provide some guidance to the many methods available for prefractionating complex proteomes in view of detecting low-abundance proteins that are otherwise difficult to find. Among the chromatographic methods, particular attention needs to be paid to coupling cation and anion exchangers (or other types of chromatographic beds) intimately admixed with granulated, solid-state buffers (the latter made with Immobiline technology). This method permits to maintain a constant local pH within serially-connected columns depending on the selected solid buffers even if they are different from column to column. This would not have been possible, of course, if liquid percolating buffers were used. This technique seems to be quite powerful, yet thus far it has only rarely been adopted. The reason could be that the solid-state buffers are not commercially available. Although their synthesis is quite facile, modern scientists do not seem to want to be bothered by extra synthetic work: They would rather buy everything off the producers' shelves. Among the electrophoretic prefractionation methodologies, those that rely on focusing techniques (if using immobilized buffer or immobilized gradients) are by far preferred, both because of their high resolving power and their ease of operation. Moreover, all such methods are directly compatible with subsequent 2D PAGE analyses. Of particular interest is the recently rediscovered focusing technique in horizontal layers of granulated Sephadex beds (a technique much in vogue in the 1970s): It is very powerful, capable of handling large sample volumes and very large protein loads, and absolutely user friendly. It is hoped that potential users will not shun away from it only because it is not a "glamour" methodology!

All in all, the latest techniques appear to represent incremental development rather than real breakthroughs. Actually finding novel species within complex proteomes is a question not only of size or performance but also of abundance, along with a question of signal suppression of very high-abundance species. With this in mind the following chapter describes approaches devised to access very low concentrated species through current treatments of initial crude samples prior to fractionation and/or analysis.

2.5 References

Anderson NG, Anderson NL. Analytical techniques for cell fractions. XXI. Two-dimensional analysis of serum and tissue proteins: Multiple isoelectric focusing. *Anal Biochem.* 1978a;85:331–340.

Anderson NL, Anderson NG. High resolution two-dimensional electrophoresis of human plasma proteins. *Proc Natl Acad Sci U S A.* 1977;74:5421–5425.

Anderson NL, Anderson NG. Analytical techniques for cell fractions. XXII. Two-dimensional analysis of serum and tissue proteins: Multiple gradient-slab gel electrophoresis. *Anal Biochem.* 1978b;85:341–354.

Anderson NL, Anderson NG. The human protein index. *Clin Chem.* (Special issue: Two-dimensional gel electrophoresis)1982;28:739–748.

Anderson NL, Anderson NG. Some perspectives on two-dimensional protein mapping. (1984) *Clin Chem.* (Special issue: two-dimensional electorphoresis) 1984;30:1898–1905.

Anderson NL, Anderson NG. The human plasma proteome: History, character, and diagnostic prospects. *Mol Cell Proteomics.* 2002;1:845–867.

Arnaud IL, Josserand J, Rossier JS, Girault HH. Finite element simulation of Off-Gel trade mark buffe ring. *Electrophoresis.* 2002;23:3253–3261.

Banks Jr JF, Quinn JP, Whitehouse CM. LC/ESI-MS determination of proteins using conventional liquid chromatography and ultrasonically assisted electrospray. *Anal Chem.* 1994;66:3688–3695.

Banks JF, Whitehouse CM. Detection of fast capillary electrophoresis peptide and protein separations using electrospray ionization with a time-of-flight mass spectrometer. In: Karger BL, Hancock WS, eds. *High Resolution Separation and Analysis of Biological Macromolecules. Methods in Enzymology;* vol. 270. Part. A San Diego: Academic Press; 1996:486–518.

Bier M. Recycling isoelectric focusing and isotachophoresis (a review). *Electrophoresis.* 1998;19:1057–1063.

Bier M, Mosher RA, Thormann W, Graham A. Isoelectric focusing in stable preformed buffer pH gradients. In: Hirai H, ed. *Electrophoresis'83.* Berlin: de Gruyter; 1984:99–107.

Bjellqvist B, Ek K, Righetti PG, et al. *J Biochem Biophys Methods.* 1982;6:317–339.

Boschetti E, Righetti PG. Mixed beds: Beyond the frontiers of classical chromatography of proteins. In: Grushka E, Grinberg N, eds. *Advances in Chromatography*, vol. 50. Boca Raton: CRC Press; 2012:1–46.

Boschetti E, Righetti PG. Mixed-bed chromatography as a way to resolve peculiar protein fractionation situations. *J Chromatogr.* 2011;879:827–835.

Burggraf D, Weber G, Lottspeich F. Free flow-isoelectric focusing of human cellular lysates as sample preparation for protein analysis. *Electrophoresis.* 1995;16:1010–1015.

Campostrini N, Areces LB, Rappsilber J, et al. Spot overlapping in two-dimensional maps: A serious problem ignored for much too long. *Proteomics.* 2005;5:2385–2395.

Celis JE (Guest Ed.). Paper Symposium: Protein Databases in Two Dimensional Electrophoresis. *Electrophoresis.* 1989;10:71–164.

Celis JE (Guest Ed.). Paper Symposium: Cell Biology. *Electrophoresis.* 1990a;11:189–280.

Celis JE (Guest Ed.). Paper Symposium: Two dimensional gel protein databases. *Electrophoresis.* 1990b;11:987–1168.

Celis JE (Guest Ed.). Paper Symposium: Two dimensional gel protein databases. *Electrophoresis.* 1991;12:763–996.

Celis JE (Guest Ed.). Paper Symposium: Two dimensional gel protein databases. *Electrophoresis.* 1992;13:891–1062.

Celis JE (Guest Ed.). Paper Symposium: Electrophoresis in cancer research. *Electrophoresis.* 1994a;15:307–556.

Celis JE (Guest Ed.). Paper Symposium: Two dimensional gel protein databases. *Electrophoresis.* 1994b;15:1347–1492.

Celis JE (Guest Ed.). Paper Symposium: Two dimensional gel Protein databases. *Electrophoresis.* 1995;16:2175–2264.

Celis JE (Guest Ed.). Paper Symposium: Two dimensional gel protein databases. *Electrophoresis.* 1996;17:1653–1798.

Celis JE (Guest Ed.). Genomics and proteomics of cancer. *Electrophoresis.* 1999;20:223–428.

Chenau J, Michelland S, Sidibe J, Seve M. Peptides OFFGEL electrophoresis: A suitable pre-analytical step for complex eukaryotic samples fractionation compatible with quantitative iTRAQ labeling. *Proteome Sci.* 2008;6:9–19.

Corthals GL, Molloy MP, Herbert BR, Williams KL, Gooley AA. Prefractionation of protein samples prior to two-dimensional electrophoresis. *Electrophoresis.* 1997;18:317–323.

Corthals GL, Wasinger CV, Hochstrasser DF, Sanchez JC. The dynamic range of protein expression: a challenge for proteomic research. *Electrophoresis.* 2000;21:1104–1115.

Cretich M, Pirri G, Carrea G, Chiari M. Separation of proteins in a multicompartment electrolyzer with chambers defined by a bed of gel beads. *Electrophoresis.* 2003;24:577–581.

Davidsson P, Paulson L, Hesse C, Blennow K, Nilsson CL. Proteome studies of human cerebrospinal fluid and brain tissue using a preparative two-dimensional electrophoresis approach prior to mass spectrometry. *Proteomics.* 2001;1:444–452.

Deterding LJ, Moseley MA, Tomer KB, Jorgenson JW. Nanoscale separations combined with tandem mass spectrometry. *J Chromatogr.* 1991;554:73–82.

Dunn MJ (Guest Ed.). Paper Symposium: Biomedical applications of two-dimensional gel electrophoresis. *Electrophoresis.* 1991;12:459–606.

Dunn MJ (Guest Ed.). 2D Electrophoresis: From protein maps to genomes. *Electrophoresis.* 1995;16:1077–1326.

Dunn MJ (Guest Ed.). From protein maps to genomes: Proceedings of the Second Siena Two-Dimensional Electrophoresis Meeting. *Electrophoresis.* 1997;18:305–662.

Dunn MJ (Guest Ed.). From genome to proteome: Proceedings of the Third Siena Two-Dimensional Electrophoresis Meeting. *Electrophoresis.* 1999;20:643–1122.

Dunn MJ (Guest Ed.). Proteomic reviews. *Electrophoresis.* 2000;21:1037–1234.

Fawcett JS. Continuous-flow isoelectric focusing and isotachophoresis. *Ann N Y Acad Sci.* 1973;209:112–126.

Figeys D, Aebersold R. High sensitivity analysis of proteins and peptides by capillary electrophoresis-tandem mass spectrometry: Recent developments in technology and applications. *Electrophoresis.* 1998;19:885–892.

Florens L, Washburn MP. Proteomic analysis by multidimensional protein identification technology. *Methods Mol Biol.* 2006;328:159–175.

Fortis F, Girot P, Brieau O, et al. Amphoteric, Buffering Chromatographic Beads for Proteome Pre-Fractionation. I: Theoretical Model. *Proteomics.* 2005a;5:620–628.

Fortis F, Girot P, Brieau O, et al. Isoelectric beads for proteome pre-fractionation. II: Experimental evaluation in a multicompartment electrolyzer. *Proteomics.* 2005b;5:629–638.

Fortis F, Guerrier L, Girot P, et al. A pI-based protein fractionation method using solid-state buffers. *J Proteomics.* 2008;71:379–389.

Fountoulakis M, Langen H, Evers S, Gray C, Takacs B. Two-dimensional map of Haemophilus influenzae following protein enrichment by heparin chromatography. *Electrophoresis.* 1997;18:1193–1202.

Fountoulakis M, Langen H, Gray C, Takacs B. Enrichment and purification of proteins of Haemophilus influenzae by chromatofocusing. *J Chromatogr A.* 1998a;806:279–291.

Fountoulakis M, Takacs B. Design of protein purification pathways: Application to the proteome of Haemophilus influenzae using heparin chromatography. *Protein Expr Purif.* 1998b;14:113–119.

Fountoulakis M, Takacs MF, Takacs B. Enrichment of low-copy-number gene products by hydrophobic interaction chromatography. *J Chromatogr A.* 1999a;833:157–168.

Fountoulakis M, Takacs MF, Berndt P, Langen H, Takacs B. Enrichment of low abundance proteins of Escherichia coli by hydroxyapatite chromatography. *Electrophoresis.* 1999b;20:2181–2195.

Fournier ML, Gilmore JM, Martin-Brown SA, Washburn MP. Multidimensional separations-based shotgun proteomics. *Chem Rev.* 2007;107:3654–3686.

Geiser L, Dayon L, Vaezzadeh AR, Hochstrasser DF. Shotgun proteomics: A relative quantitative approach using Off-Gel electrophoresis and LC-MS/MS. *Methods Mol Biol.* 2011;681:459–472.

Gianazza E, Pagani M, Luzzana M, Righetti PG. Fractionation of carrier ampholytes for isoelectric focusing. *J Chromatogr.* 1975;109:357–364.

Görg A, Boguth G, Köpf A, et al. Sample prefractionation with Sephadex isoelectric focusing prior to narrow pH range two-dimensional gels. *Proteomics.* 2002;2:1652–1657.

Görg A, Lück C, Weiss W. Sample prefractionation in granulated sephadex IEF gels. *Methods Mol Biol.* 2008;424:277–286.

Guerrier L, Boschetti E. Protocol for the purification of proteins from biological extracts for identification by mass spectrometry. *Nature Protocols.* 2007;2:831–837.

Guerrier L, D'Autreaux B, Atanassov C, Khoder G, Boschetti E. Evaluation of a standardized method of protein purification and identification after discovery by mass spectrometry. *J Proteomics.* 2008;71:368–378.

Guerrier L, Lomas L, Boschetti E. A simplified monobuffer multidimensional chromatography for high-throughput proteome fractionation. *J Chromatogr.* 2005;1073:25–33.

Guerrier L, Lomas L, Boschetti E. A new general approach to purify proteins from complex mixtures. *J Chromatogr A.* 2007;1156:188–195.

Gygi SP, Corthals GL, Zhang Y, Rochon Y, Aebersold R. Evaluation of two-dimensional gel electrophoresis-based proteome analysis technology. *Proc Natl Acad Sci U S A.* 2000;97:9390–9395.

Hannig K. New aspects in preparative and analytical continuous free-flow cell electrophoresis (a review). *Electrophoresis.* 1982;3:235–243.

Herbert BR, Righetti PG. A turning point in proteome analysis: Sample pre-fractionation via multicompartment electrolyzers with isoelectric membranes. *Electrophoresis.* 2000;21:3639–3648.

Herbert B, Pedersen SK, Harry JL, et al. Mastering proteome complexity using two-dimensional gel electrophoresis. *PharmaGenomics.* 2003;3:22–36.

Herbert B, Righetti PG, McCarthy J, et al. Sample preparation for high-resolution two-dimensional electrophoresis by isoelectric fractionation in an MCE. In: Simpson RJ, ed. *Purifying Proteins for Proteomics.* Cold Spring Harbor: Cold Spring Harbor Lab. Press; 2004:431–442.

Herbert BR, Righetti PG, Citterio A, Boschetti E. Sample Preparation and Pre-fractionation Techniques for Electrophoresis-based Proteomics. In: Wilkins MR, Appel RD, Willams KL, Hochstrasser DH, eds. *Proteome research: Concepts, technology and applications.* 2nd ed. Berlin: Springer; 2007:15–40.

Hjertén S. Free zone electrophoresis. *Chromatogr Rev.* 1967;9:122–219.

Hochstrasser AC, James R, Pometta D, Hochstrasser D. Preparative isoelectrofocusing and high resolution 2-dimensional gel electrophoresis for concentration and purification of proteins. *Appl Theor Electrophor.* 1991;1:333–337.

Hoffmann P, Ji H, Moritz RL, et al. Continuous free-flow electrophoresis separation of cytosolic proteins from the human colon carcinoma cell line LIM 1215: A non two-dimensional gel electrophoresis-based proteome analysis strategy. *Proteomics.* 2001;1:807–818.

Hörth P, Miller CA, Preckel T, Wenz C. Efficient fractionation and improved protein identification by peptide OFFGEL electrophoresis. *Mol Cell Proteomics.* 2006;5:1968–1974.

Horvàth ZS, Gooley AA, Wrigley CW, Margolis J, Williams KL. Preparative affinity membrane electrophoresis. *Electrophoresis.* 1996;17:224–226.

Hsieh FY, Cai J, Henion J. Determination of trace impurities of peptides and alkaloids by capillary electrophoresis-ion spray mass spectrometry. *J Chromatogr A.* 1994;679:206–211.

Jensen PK, Pasa-Tolić L, Anderson GA, et al. Probing proteomes using capillary isoelectric focusing-electrospray ionization Fourier transform ion cyclotron resonance mass spectrometry. *Anal Chem.* 1999;71:2076–2084.

Jonsson M, Rilbe H. Large-scale fractionation of proteins by isoelectric focusing in a multi-compartment electrolysis apparatus. *Electrophoresis.* 1980;1:3–14.

Karlsson K, Cairns N, Lubec G, Fountoulakis M. Enrichment of human brain proteins by heparin chromatography. *Electrophoresis.* 1999;20:2970–2976.

King JS (Guest Ed.). Special Issue: Two-Dimensional Gel Electrophoresis. *Clin Chem.* 1984;12:1897–2195.

Klose J, Zeindl E. An attempt to resolve all the various proteins in a single human cell type by two-dimensional electrophoresis: I. Extraction of all cell proteins. *Clin Chem.* 1984;30:2014–2020.

Kobayashi H, Shimamura K, Akaida T, et al. Free-flow electrophoresis in a microfabricated chamber with a micromodule fraction separator. Continuous separation of proteins. *J Chromatogr A.* 2003;990:169–178.

Krapfenbauer K, Fountoulakis M, Lubec G. A rat brain protein expression map including cytosolic and enriched mitochondrial and microsomal fractions. *Electrophoresis.* 2003;24:1847–1870.

Krivánková L, Bocek P. Continuous free-flow electrophoresis (a review). *Electrophoresis.* 1998;19:1064–1074.

Kuhn R, Wagner H. Application of free flow electrophoresis to the preparative purification of basic proteins from an E. coli cell extract. *J Chromatogr.* 1989;481:343–350.

Link AJ, Eng J, Schieltz DM, et al. Direct analysis of protein complexes using mass spectrometry. *Nat Biotechnol.* 1999;17:676–682.

Link AJ, Jennings JL, Washburn MP. Analysis of protein composition using multidimensional chromatography and mass spectrometry. *Curr Protoc Protein Sci.* 2004;Chapter 23:Unit 23.1.

Locke VL, Gibson TS, Thomas TM, Corthals GL, Rylatt DB. Gradiflow as a prefractionation tool for two-dimensional electrophoresis. *Proteomics.* 2002;2:1254–1260.

Ly L, Wasinger VC. Protein and peptide fractionation, enrichment and depletion: Tools for the complex proteome. *Proteomics.* 2011;11:513–534.

Margolis J, Corthals G, Horvàth ZS. Preparative reflux electrophoresis. *Electrophoresis.* 1995;16:98–100.

McCormack AL, Schieltz DM, Goode B, et al. Direct analysis and identification of proteins in mixtures by LC/MS/MS and database searching at the low-femtomole level. *Anal Chem.* 1997;69:767–776.

Michel PE, Reymond P, Arnaud IL, et al. Protein fractionation in a multicompartment device using Off-Gel isoelectric focusing. *Eletrophoresis.* 2003;24:3–11.

Millioni R, Miuzzo M, Puricelli L, et al. Improved instrumentation for large-size two-dimensional protein maps. *Electrophoresis.* 2010a;31:3863–3866.

Millioni R, Miuzzo M, Antonioli P, et al. SDS-PAGE and two-dimensional maps in a radial gel format. *Electrophoresis.* 2010b;31:465–470.

Moseley MA, Deterding LJ, Tomer KB, Jorgenson JW. Determination of bioactive peptides using capillary zone electrophoresis/mass spectrometry. *Anal Chem.* 1991;63:109–114.

Mosher RA, Bier M, Righetti PG. Computer simulation of immobilized pH gradients at acidic and alkaline extremes: A quest for extended pH intervals. *Electrophoresis.* 1986;7:59–66.

Nilsson CL, Puchades M, Westman A, Blennow K, Davidsson P. Identification of proteins in a human pleural exudate using two-dimensional preparative liquid-phase electrophoresis and matrix-assisted laser desorption/ionization mass spectrometry. *Electrophoresis.* 1999;20:860–865.

O'Farrell PH. High resolution two-dimensional electrophoresis of proteins. *J Biol Chem.* 1975;250:4007–4021.

Oguri T, Takahata I, Katsuta K, et al. Proteome analysis of rat hippocampal neurons by multiple large gel two-dimensional electrophoresis. *Proteomics.* 2002;2:666–672.

Opiteck GJ, Lewis KC, Jorgenson JW, Anderegg RJ. Comprehensive on-line LC/LC/MS of proteins. *Anal Chem.* 1997;69:1518–1524.

Opiteck GJ, Ramirez SM, Jorgenson JW, Moseley 3rd MA. Comprehensive two-dimensional high-performance liquid chromatography for the isolation of overexpressed proteins and proteome mapping. *Anal Biochem.* 1998;258:349–361.

Paoletti AC, Zybailov B, Washburn MP. Principles and applications of multidimensional protein identification technology. *Expert Rev Proteomics.* 2004;1:275–282.

Pedersen SK, Harry JL, Sebastian L, et al. Unseen proteome: Mining below the tip of the iceberg to find low abundance and membrane proteins. *J Proteome Res.* 2003;2:303–312.

Pieper R, Su Q, Gatlin CL, et al. Multi-component immunoaffinity subtraction chromatography: An innovative step towards a comprehensive survey of the human plasma proteome. *Proteomics.* 2003a;3:422–432.

Pieper R, Gatlin CL, Makusky AJ, et al. The human serum proteome: Display of nearly 3700 chromatographically separated protein spots on two-dimensional electrophoresis gels and identification of 325 distinct proteins. *Proteomics.* 2003b;3:1345–1364.

Rabilloud T. Solubilization of proteins in 2DE: an outline. *Methods Mol Biol.* 2009a;519:19–30.

Rabilloud T. Detergents and chaotropes for protein solubilization before two-dimensional electrophoresis. *Methods Mol Biol.* 2009b;528:259–267.

Rabilloud T, Chevallet M, Luche S, Lelong C. Two-dimensional gel electrophoresis in proteomics: Past, present and future. *J Proteomics.* 2010;73:2064–2077.

Rabilloud T, Lelong C. Two-dimensional gel electrophoresis in proteomics: a tutorial. *J Proteomics.* 2011;74:1829–1841.

Rabilloud T. The whereabouts of 2D gels in quantitative proteomics. *Methods Mol Biol.* 2012a;893:25–35.

Rabilloud T. Silver staining of 2D electrophoresis gels. *Methods Mol Biol.* 2012b;893:61–73.

Radola BJ. Isoelectric focusing in layers of granulated gels. I. Thin-layer isoelectric focusing of proteins. *Biochim Biophys Acta.* 1973a;295:412–428.

Radola BJ. Analytical and preparative isoelectric focusing in gel-stabilized layers. *Ann N Y Acad Sci.* 1973b;209:127–143.

Radola BJ. Some aspects of preparative isoelectric focusing in layers of granulated gels. In: Arbuthnott JP, Beeley JA, eds. *Isoelectric Focusing.* London: Butterworths; 1975:182–197.

Raida M, Schulz-Knappe P, Heine G, Forssmann WG. Liquid chromatography and electrospray mass spectrometric mapping of peptides from human plasma filtrate. *J Am Soc Mass Spectrom.* 1999;10:45–54.

Restuccia U, Boschetti E, Fasoli E, et al. pI-based fractionation of serum proteomes versus anion exchange after enhancement of low abundance proteins by means of peptide libraries. *J Proteomics.* 2009;72:1061–1070.

Righetti PG. Molarity and ionic strength of focused carrier ampholytes in isoelectric focusing. *J Chromatogr.* 1980;190:275–282.

Righetti PG. *Isoelectric Focusing: Theory, Methodology and Applications.* Amsterdam: Elsevier; 1983.

Righetti PG, Wenisch E, Faupel M. Preparative protein purification in a multi-compartment electrolyzer with Immobiline membranes. *J Chromatogr.* 1989;475:293–309.

Righetti PG. *Immobilized pH Gradients: Theory and Methodology.* Amsterdam: Elsevier; 1990.

Righetti PG, Wenisch E, Jungbauer A, Katinger H, Faupel M. Preparative purification of human monoclonal antibody isoforms in a multi-compartment electrolyzer with Immobiline membranes. *J Chromatogr.* 1990;500:681–696.

Righetti PG, Wenisch E, Faupel M. Preparative electrophoresis with and without immobilized pH gradients. In: Chrambach A, Dunn MJ, Radola BJ, eds. *Advances in Electrophoresis.* vol. 5. Weinheim: VCH; 1992:159–200.

Righetti PG, Castagna A, Herbert B. Prefractionation techniques in proteome analysis. *Anal Chem.* 2001;73:320A–326A.

Righetti PG, Castagna A, Herbert B, Reymond F, Rossier JS. Prefractionation techniques in proteome analysis. *Proteomics.* 2003;3:1397–1407.

Righetti PG. Bioanalysis: Its past, present and some future. *Electrophoresis.* 2004;25:2111–2127.

Righetti PG, Castagna A, Antonioli P, Boschetti E. Prefractionation techniques in proteome analysis: the mining tools of the third millennium. *Electrophoresis.* 2005a;26:297–319.

Righetti PG, Castagna A, Herbert B, Candiano G. How to bring the "unseen" proteome to the limelight via electrophoretic pre-fractionation techniques. *Biosci Rep.* 2005b;25:3–17.

Righetti PG, Antonioli P, Cecconi D, Campostrini N. Pre-fractionation of the hidden Proteome. In: Poole CF, Wilson ID, eds. *Encyclopedia of Separation Science.* (Online Edition). 2007:1–11.

Ros A, Faupel M, Mees H, et al. Protein purification by Off-Gel electrophoresis. *Proteomics.* 2002;2:151–156.

Rosnack KJ, Stroh JG, Singleton DH, Guarino BC, Andrews GC. Use of capillary electrophoresis-electrospray ionization mass spectrometry in the analysis of synthetic peptides. *J Chromatogr A.* 1994;675:219–225.

Sharathchandra RG, Stander C, Jacobson D, Ndimba B, Vivier MA. Proteomic Analysis of Grape Berry Cell Cultures Reveals that Developmentally Regulated Ripening Related Processes Can Be Studied Using Cultured Cells. *PLoS One.* 2011;6:e14708.

Soulet N, Roux-de-Balmann H, Sanchez V. Continuous flow isoelectric focusing for purification of proteins. *Electrophoresis.* 1998;19:1294–1299.

Tang W, Harrata AK, Lee CS. Two-dimensional analysis of recombinant *E. coli* proteins using capillary isoelectric focusing electrospray ionization mass spectrometry. *Anal Chem.* 1997;69:3177–3182.

Tobolkina E, Cortés-Salazar F, Momotenko D, Maillard J, Girault HH. Segmented field Off-Gel electrophoresis. *Electrophoresis* 2012;in press.

Wahl JH, Gale DC, Smith RD. Sheathless capillary electrophoresis-electrospray ionization mass spectrometry using 10 μm I.D. capillaries: analyses of tryptic digests of cytochrome c. *J Chromatogr A.* 1994;659:217–222.

Wall DB, Kachman MT, Gong S, et al. Isoelectric focusing nonporous RP HPLC: A two-dimensional liquid-phase separation method for mapping of cellular proteins with identification using MALDI-TOF mass spectrometry. *Anal Chem.* 2000;72:1099–1111.

Wang H, Kachman MT, Schwartz DR, Cho KR, Lubman DM. A protein molecular weight map of ES2 clear cell ovarian carcinoma cells using a two-dimensional liquid separations/mass mapping technique. *Electrophoresis.* 2002;23:3168–3181.

Washburn MP, Ulaszek R, Deciu C, Schieltz DM, Yates 3rd JR. Analysis of quantitative proteomic data generated via multidimensional protein identification technology. *Anal Chem.* 2002;74:1650–1657.

Washburn MP, Ulaszek RR, Yates 3rd JR. Reproducibility of quantitative proteomic analyses of complex biological mixtures by multidimensional protein identification technology. *Anal Chem.* 2003;75:5054–5061.

Weinmann W, Baumeister K, Kaufmann I, Przybylski M. Structural characterization of polypeptides and proteins by combination of capillary electrophoresis and (252)Cf plasma desorption mass spectrometry. *J Chromatogr.* 1993;628:111–121.

Weinmann W, Parker CE, Baumeister K, et al. Capillary electrophoresis combined with 252Cf plasma desorption and electrospray mass spectrometry for the structural characterization of hydrophobic polypeptides using organic solvents. *Electrophoresis.* 1994;15:228–233.

Wenger P, de Zuanni M, Javet P, Gelfi C, Righetti PG. Amphoteric, isoelectric Immobiline membranes for preparative isoelectric focusing. *J Biochem Biophys Methods.* 1987;14:29–43.

Westermeier R. *Electrophoresis in Practice, A Guide to Theory and Practice.* Weinheim: VCH; 2006.

Westermeier R, Görg A. Two-dimensional electrophoresis in proteomics. In: Janson JC, ed. *Protein Purification: Principles, High Resolution Methods, and Applications.* 3rd ed. New York: Wiley-VCH; 2011:411–439.

Willard KE, Giometti C, Anderson NL, O'Connor TE, Anderson NG. Analytical techniques for cell fractions. XXVI. A two-dimentional electrophoretic analysis of basic proteins using phosphatidyl choline/urea solubilization. *Anal Biochem.* 1979;100:289–298.

Williams KL (Guest Ed.). Strategies in proteome research. *Electrophoresis.* 1988;19:1853–2050.

Wolters DA, Washburn MP, Yates 3rd JR. An automated multidimensional protein identification technology for shotgun proteomics. *Anal Chem.* 2001;73:5683–5690.

Yang L, Lee CS, Hofstadler SA, Pasa-Tolic L, Smith RD. Capillary isoelectric focusing-electrospray ionization Fourier transform ion cyclotron resonance mass spectrometry for protein characterization. *Anal Chem.* 1998;70:3235–3241.

Young DA. Advantages of separations on "giant" two-dimensional gels for detection of physiologically relevant changes in the expression of protein gene-products. *Clin Chem.* 1984;30:2104–2108.

Young DS, Anderson NG (Guest Ed.). Special Issue: Two-Dimensional Gel Electrophoresis. *Clin Chem.* 1982;28:737–1092.

Zischk H, Weber G, Weber PJA, et al. Improved proteome analysis of Saccharomyces cerevisiae mitochondria by free-flow electrophoresis. *Proteomics.* 2003;3:906–916.

Current Low-Abundance Protein Access

3.1 GENERAL SITUATION

The global analysis of protein extracts intrinsically suffers from very heterogeneous compositions where, in many cases, a limited number of proteins represent more than 80–90% of the total protein amount. This is the case of (i) several mammalian fluids such as serum, cerebrospinal fluid, follicular fluid and urine with albumin, immunoglobulins, and a few other proteins; (ii) bird eggs (white and yolk) with, for example, ovalbumin; (iii) milk from mammals with caseins and lactoglobulins; (iv) plant extracts such as leaf proteins with the large dominance of ribulose-1,5-biphosphate carboxylase/oxygenase; and (v) cell extracts such as red blood cells with hemoglobin and carbonic anhydrase. In all of these well-known examples minor proteins are undetectable even when using the most sensitive available methods because dominant proteins hide the least concentrated species of similar physicochemical properties as well as the sunlight that prevents the observation of stars in a clear sky during the day.

It is from these examples that the idea of removing high-abundance proteins became a reality.

In this respect, chromatographic techniques and especially affinity chromatography represented the right model to start with. Most of the tools were already available even if they were not completely adapted for proteomic study applications.

Synthetic and natural ligands for the separation of albumin (a major protein in blood plasma) have been known for years. Ligands for the separation of glycoprotein families are also available, such as lectins described in the 1970s or immobilized organic boronic acid which is well known to reversibly esterify carbohydrates containing 1,2-*cis*-diol structures.

In the early days of the concept, the major question was to select ligand technologies minimizing the phenomenon of nonspecific binding that would co-subtract other proteins, including those that could be considered of biological or diagnostic interest. This is extremely important because, if present, the nonspecific binding issue would partially cancel the interest of subtraction or depletion technology.

As described in previous chapters, this approach should also be considered as part of the panoply of other methods capable to enhance the signal of low-abundance species such as fractionation and narrow range focusing.

Among other important general considerations concerning the treatment of biological extracts, the limited availability of samples is one of several difficulties. This is an important issue because, after having removed the abundant species present, all other proteins will remain dilute (and even more dilute than in the initial sample), with almost no possibility of getting concentrated, whatever the means, without a large risk of losing many species.

The removal of high-abundance proteins also has an impact on the reduction of the dynamic concentration range whose extent depends on the capability of specifically targeting the most representative proteins. This being said, low-abundance species are not concomitantly concentrated. Thus, in spite of some limited reduction of dynamic range, the treated sample is dilute, and the detection of medium- to low-abundance protein is achievable only using very sensitive analytical methods.

3.2 HIGH-ABUNDANCE PROTEIN SUBTRACTION

Several solid-phase extraction approaches have been designed for the removal of undesired proteins. This is the case with immobilized dyes, metal ion-chelating agents, protein ligands, and antibodies.

For clarity, in this section protein subtraction or depletion will be described under two distinct aspects that were more or less developed chronologically with the aim of increasing the selectivity of removing undesired material from biological samples. In both parts, the removal of high-abundance proteins was a common goal, but the solid-phase adsorbents used were different. Nevertheless, they both borrow not only the principles but also the sorbent matrix from affinity chromatography technologies that have been known for decades. A novelty was introduced, however, when sorbents were used together to form a mixed bed of solid-phase extraction, thereby eliminating the use of multiple steps, each one addressing the elimination of one single protein. This was possible when selected sorbents were compatible and operated under the same conditions. Since the desorption was not a major issue owing to the fact that only the depleted fraction was kept for proteomic analysis, the use of a mixed bed simplified the subtraction operation. Basically, and regardless of the type of sorbent used singularly or as a mixed bed, the solid-phase adsorbent is equilibrated first with the appropriate buffer (generally the same physiological saline where the sample proteins are dissolved). Then it is mixed with the biological sample for the necessary time to capture undesired proteins, and the supernatant containing in principle all other proteins is collected and used for proteomic analysis.

Disregarding a few exceptions where the goal is to attain perfect understanding of the composition of captured proteins, the solid phase with undesired removed proteins is ignored and most frequently wasted.

Although quite simple, this process must comply with some basic rules. For instance, the amount of sample mixed with the solid-phase sorbent must not exceed the binding capacity for targeted proteins; otherwise the supposed removed proteins will still be present in the treated sample. This is probably the most important and riskiest part of the process since the binding capacity of sorbents is very low, thus limiting the volume of biological sample to treat. As a direct consequence of such a situation, the amount of protein recovered is so low that only few analytical determinations are possible. However, the binding capacity is not the same for different sorbents. For instance, for the depletion of albumin from 100 µL of human serum, the volume of Cibacron Blue-agarose beads should be at least 150 µL; conversely, for the same 100 µL of serum the volume of anti-albumin immunosorbent should be at least 5 mL !! The difference of the sample to bead volume ratio directly influences the dilution factor of the depleted sample; therefore, the larger the sorbent volume for a given volume of sample, the larger the dilution factor. Outside immunosorbents where the specificity is very high, the amount of nonspecific binding is always relatively large, especially for proteins of interest that are present in very low amounts. This phenomenon—which is not a problem for regular chromatography where the objective is not to collect a sample where some proteins are removed but rather to capture the protein of interest regardless of whether other minor species are co-adsorbed (because they are eliminated in a further fractionation)—becomes a huge issue in proteomic analysis.

In proteomic solid-phase extraction, the material should not be reused because, even after having cleaned the sorbent, traces of still present proteins might contaminate the following sample. Nevertheless, because of the high cost of some adsorbents such as immobilized antibodies, suppliers indicate that the sorbent could be used dozens of times. This is so without considering that it is a common rule that upon reuse sorbents progressively lose their binding capacity, while the nonspecific binding concomitantly increases. This is due to progressive deterioration of the sorbent or the linker, or even the attached ligand with progressive hydrolysis. These phenomena depend on the mode of regeneration and the intensity of reuse. Since this is chromatography, other physicochemical parameters such as kinetics and thermodynamic are to be considered to maintain the reproducibility at an acceptable level.

The next two paragraphs describe the two generally known categories of depletion technologies: the first, called "oligospecific," assembling various ligands that are of poor or medium specificity for the protein to remove, and the second being biologically specific since it is based on antibodies.

3.2.1 Oligospecific Subtraction Methods

The term oligospecific encompasses subtraction methods that target a well-established protein (e.g., albumin) but whose specificity is not outstanding and depends on operating conditions. This is typically the case of grafted Cibacron Blue on chromatography beads. This dye ligand was described in the 1970s for the capture of human albumin from human serum. It binds albumin under physiological conditions (but also under other conditions) as a result of multiple molecular interaction that are electrostatic (by the presence of three sulfate groups) (Zhang et al., 2001), hydrophobic, by the presence of multiple aromatic moieties (Subramanian, 1984), and hydrogen bonding interactions (Andac et al., 2007). The presence of these distinct interactions is demonstrated by the ability to dissociate albumin by, respectively, high concentrations of sodium chloride, ethylene glycol, and concentrated urea, each of them targeting a well-defined molecular interaction. Nevertheless more than that is involved: This dye also binds proteins that naturally interact with nicotinamide adenine dinucleotide due to some molecular similarity in shape and behavior as described by Jmeian et al. (2009a). When using Cibacron Blue ligand, the depletion of albumin is not complete by just loading a serum sample: There is some amount of albumin found in the flow-through, which classifies this depletion method as releasing two different fractions. The first is a serum-poor albumin, and the second is an albumin-rich fraction. The latter is of difficult manipulation in proteomic investigations because the dynamic concentration range of this sample is simply much worse than the initial serum sample. The fact that a large number of other proteins are co-adsorbed with albumin renders the technology questionable because a number of species are subtracted to the global proteomics analysis, including those that could be of interest for diagnostic applications (Zhou et al., 2004; Zolotarjova et al., 2005).

Cibacron Blue ligand can easily be used at different pH and ionic strengths, with very different results. For instance, using it at acidic pH, all proteins are adsorbed, including those that are of very low abundance (Di Girolamo and Righetti, 2011). Here the major attraction intensity is played by the three sulfonate groups, reinforced by the two other types of interactions mentioned above.

Another generic subtraction way to deplete high-abundance proteins from blood plasma is depletion by hydrophobic interaction, as reported by Mahn et al. (2010). Here high- and medium- abundance proteins are calculated as being the most hydrophobic, especially albumin and IgG (heavy chains), but also transferrin and fibrinogen (Mahn et al., 2009). Those proteins all together represent about 70% of serum proteins, and the use of hydrophobic interaction chromatography appears as an attractive generic alternative for their removal. The method was described as being quite reproducible, but, when compared to immunodepletion, it showed not only that the subtraction was incomplete, but also that several medium-abundant proteins were still present in the depleted fraction. In spite of this limitation, hydrophobic depletion allowed detecting twice the number of protein spots compared to albumin immunodepletion when samples were analyzed by two-dimensional electrophoresis. Nevertheless, even if the authors suggested that the method could be improved, it is still true that there are large probabilities that many hydrophobic low-abundance proteins are co-adsorbed escaping thus their detection. This will have a negative consequence for studies devoted to proteome mapping on the one hand, and the discovery of diagnostic signatures for early-stage biomarkers, on the other hand.

Other methods targeting serum IgG propose an affinity depletion using protein A, a *Staphylococcus aureus* protein well known for many years for its antibody purification function (Eliasson et al., 1988). This protein binds to the Fc region of antibodies from human and various other species. Due to the specific binding to the Fc region of IgG, protein A has been described as removing immunoglobulin G from human serum (Zhou et al., 2004; Vestergaard & Tamiya 2007). Clearly, this method works well; however, it also suffers from (i) the poor interaction with the IgG3 subclass that is not or scarcely captured and (ii) the co-adsorption of other low-abundance proteins. Consequently it has the same drawbacks as does albumin depletion using poorly specific binders. Use of protein G instead of protein A might be slightly better because of its propensity to target all IgG subclasses (it comprises two interaction domains) without resolving the issue of co-absorbing other proteins such as some albumin (it comprises an albumin-binding domain) (Olsson et al., 1987; Ma & Ramakrishna, 2008). The specificity for IgG can, however, be much improved by genetic engineering: The regions on the gene corresponding to the albumin-binding domains and the Fab-binding region are deleted by site-directed mutagenesis. The advantage of proteinA–proteinG affinity approaches is that they are applicable to various species, which is not the case for immunoadsorption. In spite of their pretty good specificity, protein A and protein G used alone are not capable of removing all immunoglobulin G subclasses. Positioning protein G as a first-affinity depletion followed by the protein A column is probably the best combination for removing most of these immunoglobulins. This is why a protein A/G fusion entity has been proposed for IgG depletion (Whiteaker et al., 2007), with the advantage of a larger specificity and coverage toward antibody classes.

To resolve the nonspecific binding question and to find better exploitation conditions with lower molecular mass ligands compared to antibodies, some authors have proposed the development of small proteins called affibodies against, for instance, albumin with the objective of removing specifically targeted proteins from human serum. Affibodies are a class of engineered, small polypeptides composed of fifty-eight amino acids yielding a global mass of about 6 kDa. They are derived from the Z domain of protein A, which consists of three alpha helices without disulfide bonds. The affinity for a given protein is obtained by randomizing thirteen amino acids and a library obtained by phage display from where the appropriate affibody is sorted out and then expressed in a heterologous system and finally purified (Binz et al., 2005). The description of this depletion system applied to albumin demonstrated that the removal efficiency was different in serum compared to cerebrospinal fluid. In the latter biological fluid, the use of such an affibody was significantly less performant (Grönwall et al., 2007). In another study a mixture of immobilized affibodies against albumin, IgG, transferrin and transthyretin, was successfully used to remove concomitantly four high-abundance proteins from human cerebrospinal fluid (Ramstrom et al., 2009). All four proteins were effectively removed; nevertheless, the specificity assessment probably requires a more extensive evaluation for other biological fluids and species. Comparative studies highlighting exclusive technical or economical benefits of a mixed bed of affibody against a mixed bed of antibodies were not available at the time this book was prepared.

Other tentative aspects of depleting high-abundance proteins have been reported, but due to their lower application interest, they are out of the scope of this book. Nevertheless we can cite the possibility of precipitating some high-abundance proteins as reported in a recent review by Selvaraju & El Rassi (2012) that included, among other processes, a precipitation of serum albumin and serum transferrin upon incubation with dithiothreitol or tris(2-carboxyethyl)phosphine, followed by a centrifugation step to remove a precipitate that is particularly rich in albumin. Immunoprecipation was also reported among others by Warder et al. (2009), but this is relatively similar to immunodepletion with solid-phase materials. Again these precipitation techniques can easily be "contaminated" by the presence of many other proteins that may co-precipitate or are entrapped within the obtained primary precipitate of the target protein.

3.2.2 Immunodepletion

As depicted in the previous paragraph, the first generic depletion of albumin in serum was performed by using immobilized Cibacron Blue chromatography (Ahmed et al., 2003). Although this ligand binds quite strongly to albumin, the presence of nonspecific interactions engenders the co-capture of many other proteins that are as a consequence ignored for further proteomic studies. This is equally the case for the second major serum protein immunoglobulin G when removed by protein A or protein G columns (Fu et al., 2005; Govorukhina et al., 2006). To try resolving the low-selectivity question, immunoaffinity subtraction using specific antibodies has been proposed, demonstrating improved results (Steel et al., 2003; Pieper et al., 2003).

In the last few years, a flurry of immunodepletion columns has appeared on the market: Among these columns are those targeting individual major proteins and immunodepletion columns that are mixed beds of immunosorbents targeting more than one protein. Most, if not all, of these proposed solutions address the human serum proteome as being probably the biological fluid of choice for the discovery of markers of diagnostic or pharmaceutical interest. Among mixed-bed immunosorbents, one can find solutions to remove six or twelve, or fourteen, or even several dozen high- or medium-abundance serum proteins.

At the beginning, antibodies used for immunodepletion columns had a mammalian source. Since it was recognized that some nonspecific binding could be attributed to possible cross reactions, Hinerfeld et al. (2004) suggested hens' egg yolk IgY polyclonal antibodies as representing a preferred approach. Initially, this IgY immunosorbent column was made using six antibodies against the six major serum proteins; however, quite rapidly, being considered insufficient, the depletion was extended to the twelve most abundant proteins (Huang et al., 2005) and then to the twenty high-abundance proteins (Martosella et al., 2005). To better cover the human serum proteome, another polyvalent immunosorbent column was proposed, targeting up to more than seventy top proteins (Lin et al., 2009).

Since 2005, comparative studies have been performed using different depletion technologies, with a general consensus conclusion that multiple immunoaffinity sorbents offer an effective, promising depletion approach (Bjorhall et al., 2005; Zolotarjova et al., 2005).

It was within the context of comparative studies the observation that normally undetectable species became visible for serum (Vasudev et al., 2008) and other human biological fluids (Shores & Knapp, 2007; Cellar et al., 2009).

Concomitantly to a better visibility of hidden proteins, other authors reported drawbacks that might limit the use of depletion methods. Granger et al. (2005) as well as Shen et al. (2005) reported large co-depletion of low-abundance proteins when using immunosorbents, reaching large numbers that were also concomitant with the use of more sensitive instruments.

Although depletion methods are very popular today, they are not without severe drawbacks. Once the top ten or twenty most abundant proteins in sera are removed, the recovered protein mixture is more dilute than it was prior to the treatment. The chromatographic process, especially in this case where the depleted fraction is a flow-through, adds a certain degree of dilution, rendering the detection of low-abundance species even more challenging. Another important limitation is the intrinsic biological specificity of antibodies, thus restricting their use not only to one species but also to a given extract (e.g., human serum). Also, the intrinsic properties of immunoaffinity chromatography (e.g., low binding capacity due to the unfavorable stoichiometric situation and grafting technologies) limit the treatment to small biological samples, which are typically 20 to 100 μL of serum where the amount of total proteins involved is very low. This fact also limits the number of analytical determinations, especially when using two-dimensional electrophoresis on large gel plates with the necessary replicates. The serious risk of co-depletion (nonspecific binding along with a large number of associated species) concludes the list of problems associated with this subtraction technology. In a word, the co-depletion phenomenon could be so extensive that paradoxically the number of detected proteins in the removed fraction could be larger than what is found in the depleted fraction (Shen et al., 2005).

In a recent study, Yadav et al. (2011) identified with a high degree of confidence not less than 101 low-abundance gene products co-depleted using three types of immunodepletion cartridges (Mars Hu-6, Hu-14, and Proteosep 20). The number of proteins that were nonspecifically removed along with high-abundance species was increased concomitantly to the number of proteins that were targeted for removal. In this situation, the authors suggested that for the application in biomarker discovery, both the depleted fraction and the bound fraction would have to be analyzed for an exhaustive investigation. Although this statement speaks by itself to the fact that a large number of proteins escape the analysis after depletion, it does not explain how the co-depleted fraction could be identified since the situation of the presence of high-abundance proteins is much worse in the bound fraction than what it was in the initial sample. In the present case, the process could better be categorized as fractionation instead of depletion, where the first fraction (depleted sample) is more dilute than the initial sample; therefore, the low-abundance species are even more difficult to detect. At the same time the other low-abundance species that are co-depleted and present in the second fraction are also more difficult to detect due to the massive presence of high-abundance proteins. Similarly, another comparative study was made by using immunodepletion columns where one, six, twelve, or twenty major proteins were depleted using Beckman Coulter ProteomLab IgY-HAS, Agilent Technologies Hu-6HC, Beckman Coulter ProteomLab IgY-12, and Sigma ProteoPrep 20, respectively (Roche et al., 2009). When the number of depleted proteins was increased from twelve to twenty proteins, the benefits were limited, whereas when six proteins were depleted many low-abundance proteins were detected.

In another comparative study (Lin et al., 2009), twelve high-abundance and seventy-seven moderate-abundance proteins were, respectively, removed from serum and discovery results compared. The authors reported the identification of respectively 71 and 222 proteins. This indicates that the simultaneous depletion of the high- and moderate-abundance species would increase the number of novel low-abundance proteins otherwise undetectable. To counterbalance this point, it was found that the flow-through fractions still contained some of the supposed removed proteins, indicating that either the columns were overloaded or that the efficiency of depletion was low (Qian et al., 2008).

Tu et al. (2010) reported that the use of current immunodepletion platforms cannot reasonably allow discovering targeted low-abundance, disease-specific biomarkers in plasma. From analyses of immunodepleted plasma by isoelectric focusing, followed by liquid chromatography-tandem mass spectrometry (LC-MS/MS), it appeared that the increase in identified proteins compared to non-treated plasma was only close to 25%. Even more devastating was the report by Zhi et al. (2010) who candidly stated that immunosubtraction allowed seeing 10% fewer proteins than in control sera. As a confirmation of these findings, in another study conducted with the objective of searching for possible protein markers from human urine, Afkarian et al. (2010) simply concluded that *"depletion of albumin and IgG did not increase the number of identified proteins or deepen the proteome coverage."*

Even if immunodepletion has been recognized as having serious drawbacks, its use has expanded greatly in the last few years. When considering all arguments regarding the advantages and disadvantages of immunodepletion, the most important point is the real capability of revealing additional species. Clearly, this goal is reached; however, when compared to the sample treatment with peptide libraries—which is the scope of this book—the number of additional proteins detectable after immunodepletion is considerably lower than is repeatedly reported and as illustrated in one comparative analysis illustrated in Figure 3.1.

Within this context, and in order to improve the depletion process, a recent paper describes a two-stage process for producing markers for the prediction of heart attack (Juhasz et al., 2011). A first depletion of high-abundance proteins was followed by a second depletion of several dozen medium-abundance proteins, and finally the recovered fraction was fractionated by chromatography. Here the authors reported that the general trends in protein abundance were maintained; however, the matter of accuracy for determining

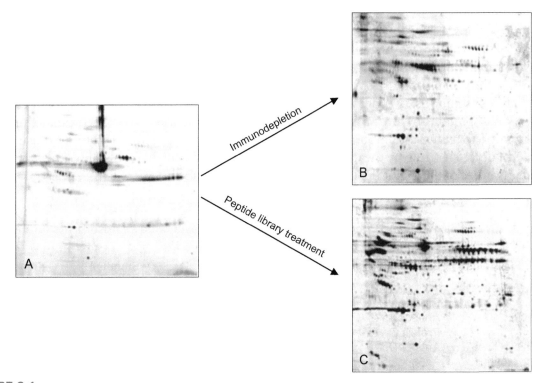

FIGURE 3.1
Two-dimensional electrophoresis analysis of human serum before (A) and after treatment, with an immunodepletion column for the removal of the six most concentrated proteins (B) and after treatment with a peptide library (C).

the expression is to be handled with care since initial differences might be compromised. In spite of these precautions, the authors demonstrated that several marker candidates emerged from the study; the counterpart was an excessive labor that rendered the approach quite questionable for extensive studies.

The reproducibility of immunodepletion seems relatively good when using the same depletion system and the same antibodies. This is an important point for studies related to clinical applications (see, for instance, Seam et al., 2007 and Tu et al., 2010).

3.2.3 Importance and Limits of Subtraction Methods

The elimination of one or more high-abundance proteins sequentially or concomitantly can clearly help to detect novel species that are otherwise ignored. This can be applied with highly specific ligands attached to a solid phase as, for instance, antibodies. As long as this applies to human serum or plasma or even other common biological fluids, the interest in selecting very specific antibodies is quite evident due to the specificity for the targets. Nevertheless, these tools are so specific that they are not usable for other biological extracts and/or other species. The cost of antibodies is relatively high; therefore, the amount of ligand grafted onto the solid phase is in all cases quite limited, and hence the binding capacity for depletion is also quite low. As a result, the sample size is small. The use of antibodies as affinity ligands is also handicapped by their high molecular mass unfavorable to large binding capacities. Upon depletion, the remaining sample ideally comprises all other proteins for further analysis. Their detectability is in principle improved by the absence of invasive proteins; however, this advantage is counterbalanced by the very small depleted sample and the fact that the treated specimen is significantly diluted, thus reducing the probability of finding many more species unless a concentration of the depleted sample is operated. Dilution is the general rule of all flow through chromatography methods as it is the depletion approach. This is without counting on the nonspecific binding and protein–protein interaction, both of which are responsible for the co-removal of a relatively large number of gene products. Thus the latter continue to escape detection even if this is for a different reason.

The dilution engendered by the depletion operation (see Table 3.1) is also detrimental to other fractionation operations, especially when the affinity for the new ligands is weak. This is the case of lectins used for the separation of glycoproteins (see Section 3.3.2). It is not recommended that lectins be used for the capture or fractionation of glycoproteins just after depletion of high-abundance proteins because the affinity constants are of the order of micromoles, and it is always preferable to use concentrated protein solutions. In order to succeed in this approach it is advisable to use an intermediate concentration of proteins by any possible tool/method, including ion-exchange chromatography but paying attention to reduce the amount of protein losses as much as possible. Figure 3.2 illustrates several results of depleted serum by the use of various commercially available tools. Table 3.2 summarizes the general features of the major depletion methods described above.

TABLE 3.1 Examples of Dilution Effects of Protein Depletion Resins Used as Adsorbents for the Capture of Proteins to Remove

Removed Proteins	Ligands	Buffer	Sample/Resin Ratio	Dilution Factor (max)
Albumin	Cibacron Blue	20 mM phosphate pH 7	0.1	20
Albumin	Antibodies	Phosphate buffered saline	0.75	180
IgG	Protein G	Phosphate buffered saline	0.05	5
Albumin & IgG	Antibodies	Phosphate buffered saline	0.02	35
Six major proteins	Antibodies	10 mM phosphate + 0.5 M NaCl	0.02	60

As a reference, the dilution factor of combinatorial peptide ligand library is at least 0.5; in other words there is a concentration effect instead of a dilution. The concentration effect increases with the increase of the sample volume.
(Adapted from Righetti et al., 2006).

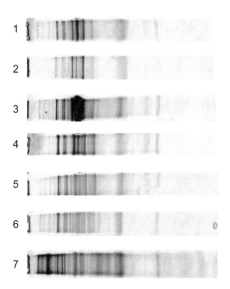

FIGURE 3.2

SDS polyacrylamide gel electrophoresis analysis of serum proteins after various treatments in view of reducing the concentration of high-abundance proteins and thus detecting more species.

1: Albumin-depleted serum by Cibacron Blue chromatography.
2: Immunodepleted serum sample by means of anti-albumin antibodies.
3: IgG-depleted serum with the use of protein G chromatography.
4: Albumin- and IgG-depleted serum proteins by means of a mixed bed of Protein A and Cibacron Blue sorbents.
5: Albumin and IgG immunodepleted serum using a mixed bed of two immunosorbents.
6: Multiple immunodepleted serum using antibodies against six major serum proteins: albumin, immunoglobulin G, Immunoglobulin A, transferrin, haptoglobin, α-1-antitrypsin.
7: Serum treated with a combinatorial peptide ligand library.

TABLE 3.2 Summary of Described Depletion Methods

Type of Ligand	Targeted Proteins	Advantages	Disadvantages
Cibacron Blue	Albumin	Easy to use Cheap High binding capacity	Nonspecific binding Many lost proteins Incomplete depletion
Hydrophobic groups	High abundance species	High binding capacity Cheap approach	Incomplete subtraction Nonspecific binding Need sample adjustment
ProteinA — ProteinG	Immunoglobulins G	Specific for Fc region Good binding capacity	Incomplete subtraction Nonspecific binding
Affibodies	High-abundance proteins	Highly target specific Low molecular mass Manageable non-specific binding	Expensive ligands
Precipitation	High-abundance proteins	Easy handling Cheap Small and large samples	Protein entrapping Nonspecific Rough method
Antibodies (immunodepletion)	High- and medium-abundance proteins	Low cross reactivity possible High specificity Easy use Well standardized Direct sample loading	Large co-depletion of LAP Dilution of the treated sample Species specific Expensive Low binding capacity

3.3 ENRICHMENT OF LOW-ABUNDANCE, POST-TRANSLATIONALLY MODIFIED PROTEIN GROUPS

The term *post-translational modifications* (PTM) signifies covalent modifications by introducing chemical groups on the polypeptides chain of proteins at one or more sites (see, for instance, the review by Walsh et al., 2005). These modifications are the result of complex enzymatic reactions that enter in action after DNA has been transcribed into RNA and then translated into proteins. Numerous enzymes are specifically dedicated to post-translational modifications, and they are at medium or low concentration. Their number is large enough to represent at least 5% of the entire proteome for most evolved eukaryotes. The modifications take place at the side chains of amino acids carrying nucleophilic groups.

Post-translational modifications are not only reactions of introducing various groups on the polypeptidic chains but also cleavages of peptides most of the time subsequent to specific protease intervention. This latter mechanism of post-translational modification is not discussed in this section.

3.3.1 Post-translational Modifications as a Means of Group Discrimination

Most proteins are post-translationally modified to serve their biological function permanently or temporarily. These modifications are very diverse and also quite complex in their own structure (e.g., glycans), complicated by a possible consensus sequence of amino acids around which the modifications take place. It is not the purpose of this section to give details on their biological role and integrated interaction with other species within a cell, but rather to give a sense of these modifications as a means of classifying protein groups and separating them for in-depth proteomics studies. Post-translational modifications basically allow grafting on the surface of a protein structure chemical or biochemical groups that can be the starting point of the family recognition. The complexity of proteomes can be better investigated and understood after fractionation of the entire proteome. This paragraph describes the principles of protein separation thanks to this peculiar diversity. It is also interesting to note that post-translational modifications can determine groups of functions; by consequence, separating them in groups, it becomes possible to restrict the investigations to a given number of protein members involved in the entire targeted function. This is the case, for instance, for phosphoproteins that are mostly involved in cell signaling.

With the use of potent modern instruments, it is possible to monitor thousands of modified sites at the same time, but post-translationally modified proteins are so numerous and diverse that it is necessary to proceed first for an enrichment of low-abundance proteins and/or a fractionation of the collected group. Although some authors indicate that the fractionation could be obtained by current methods such as cation exchange chromatography (Edelmann, 2011), it is here considered that for a proper fractionation of sub-proteomes resulting from classification of the type of modification, it is preferable to adopt more group-specific approaches.

The most important modified protein groups are analyzed below on the aspect of their separation possibilities in view of specific studies either for deciphering subproteomes or for understanding whether wrongly modified species can characterize an existing or a possible upcoming pathological situation. In spite of this simple principle, there are in reality complicated situations where the same protein could be post-translationally modified by different substituents of the same family (e.g., glycans) or of different nature (e.g., phopsphorylated glycoproteins). Moreover, for some post-translational modifications it is difficult to identify a specific ligand for separations, and finally one should remember that, in spite of enormous intellectual and technical efforts, there is a large number of nonspecific bindings making these separations quite challenging. The following paragraphs describe the main methodological approaches with reference to key published reports. It is hoped that the elements described will stimulate scientists to develop their imaginations for further technical developments with the aim of gaining better specificity.

3.3.2 Glycoprotein Global Capture and Selective Fractionation

Glycosylation of gene products is probably the most common post-translational modification in eukaryotic cells. It is so generalized that more than half of proteins are processed involving a large number of enzymatic reactions. The variety of sugars and of sugar derivatives involved creates a large diversity since the glycosylation is not limited to a single sugar grafted on the polypeptide chain, but forms complex structures with several sugars that can be also different in their linear or branched sequences (see Figure 3.3). The number of glycosylation sites could also be numerous, thus creating a very large diversity with consequences for the biological and physicochemical properties of proteins. Within this picture it becomes evident that the question of separating glycoproteins from other proteins is difficult since in appearance there is not a single property that can be taken as a starting point for affinity separation.

This large family of glycoproteins comprises high-abundance species (e.g., antibodies that are largely glycosylated) and very low-abundance species undetectable without the application of enrichment processes.

Two general types of glycosylation are known: N-glycosylation, where the carbohydrate is attached on asparagines, and O-glycosylation, where the sugars are attached on serine, threonine, and tyrosine. Nevertheless other minor glycosylation types are found especially in membrane-associated proteins (Wells et al., 2001; Hofsteenge et al., 1994).

The large diversity of glycoproteins represents a dilemma for the separation of the entire family for focused investigations. No good common denominator could be found, not even with antibodies or antibody groups, due to the large number of possible structures to target. The available technologies for the separation of glycoproteins are based essentially on lectin affinity chromatography and boronic acid chromatography.

Lectins are proteins from plant and other origins that have the properties to recognize sugars or given sugar sequences. A very large number of lectins have been described, and many of them are commercially available; they associate with protein glycans with affinity dissociation constants that are around the micromolar range. In spite of the large number of lectins, those that are most used in the separation of glycoproteins are concanavalin A (ConA), wheat germ agglutinin (WGA), and jacalin. Used under a cascade sequential configuration or as sorbent mixtures (multi-lectin chromatography), they can

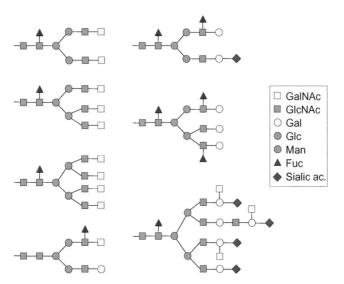

FIGURE 3.3
Typical structures of glycans found in eukaryotic cells. A number of other variations exist with complicated structures involving rarer sugars.

capture most of the N-linked glycoproteins (Yang et al., 2006). However, to have a quite exhaustive coverage, a larger number of lectins are necessary. To obtain a complete profiling, Drake et al. (2006) proposed a set of six different lectins. In spite of that, an important question remains unanswered regarding the capture of the low-abundance glycoproteins by means of lectins.

In fact, their low concentration and the weak affinity constants for their specific lectins probably leave them in the lost proteins category due to the unfavorable mass action law equilibrium. To try resolving the question, the lectin capture would consist of two different phases: (i) a reduction of the dynamic concentration range with, for instance ,the use of combinatorial peptide ligand libraries (this includes the concentration of very low-abundance species; see Chapter 4) and (ii) a multi-lectin affinity chromatography fractionation.

The affinity of lectins for the glycan moiety is not displayed on a single sugar but rather on a sequence or a structure comprising more than a single type of saccharide (see Table 3.3). Thus some lectins have an affinity for the same terminal sugar or belong to the same group of sugar affinity, but the complex glycan that interacts with the binding site has a different structure or sequence. Overlaps of sugar positioning also exist between lectin and lectin. This makes possible a computational reading integrating the similarity/diversity of glycans (Rosenfeld et al., 2007) for a given group of lectins, with benefits related to the global fingerprint that can reveal differences in the post-translational modification processes (see the schematic example in Figure 3.4). In a published study, Kuno et al. (2009) described an interesting system called differential glycan analysis using the concept of antibody-assisted lectin profiling. Briefly, the target glycoprotein present in the initial crude extract is enriched by immunoprecipitation with a specific antibody. Then, after quantitation by immunoblotting, the glycosylation profile difference against a control is analyzed by means of an antibody-overlay lectin microarray.

The question of the separation of O-glycosylated proteins is probably more challenging perhaps due to the very short glycan that further decreases the affinity constant for lectins. A general review on the use of lectins as a tool for glycoproteome deconvolution is found in Wisniewski (2010). As a possible alternative to lectin affinity chromatography, boronic acids and derivatives have been described for their affinity toward sugars. Organic boronic acids are actually known to reversibly esterify carbohydrates

TABLE 3.3 Examples of Specificity of Most Known Lectins

Lectin Name	Source	Specific Sugar Ligand
Concanavalin A	*Concanavalia ensiformis*	α-mannosidic structures (branched)
Galanthus agglutinin	*Galanthus nivalis*	(α-1,3) mannose residues
Jacalin	*Artricarpus integrifolia*	Galβ1-3GalNAcα1-Ser/Thr
Lentil lectin agglutinin	*Lens culinaris*	α -linked mannose residues
Lotus agglutinin	*Lotus tetragonolobus*	α -linked L-fucose oligosaccharides
Maackia agglutinin	*Maackia amurensis*	Neu5Ac/Gcα2,3Galβ1,3(Neu5Acα2,6)GalNac
Pea agglutinin	*Pisum sativum*	N-acetylchitobiose-linked α -fucose
Peanut agglutinin	*Arachis hypogea*	Galβ1-3GalNAcα1-Ser/Thr
Ricin agglutinin 120	*Ricinus communis*	Galβ1-4GlcNAcβ1-R
Soybean agglutinin	*Glycine max*	Terminal α- or β-linked N-acetylgalactosamine
Ulex agglutinin	*Ulex europeus*	Fucα1-2Gal-R
Wheat germ agglutinin	*Triticum vulgaris*	GlcNAcβ1-4GlcNAcβ1-4GlcNAc

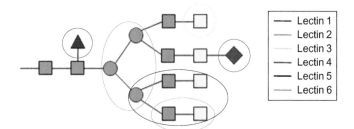

FIGURE 3.4
Portions of an N-glycan recognized by various lectins. There are lectins that recognize just a sugar and others that recognize branched regions or disaccharides. There are lectins that partially overlap to each other. The deconvolution of a combination of lectins and their sugar recognition gives information on the structure of the glycan.

containing 1,2-*cis*-diol structures. This interaction occurs regardless of the nature and position of the sugar. Immobilized boronic acid structures are therefore resins for the adsorption of glycoproteins and with a relatively large specificity (Olajos et al., 2010) to the point that they are currently used for measuring glycosylated forms of proteins (Gould et al., 1984).

Most widely used is 3-aminophenylboronic acid; however, it necessitates elution conditions (high pH and frequently associated with sugar displacers) that are not fully compatible with the integrity of the glycoprotein structure and the following proteomics studies. To lower the pKa value of boronate ligands, aminoethyl derivatives (Akparov et al., 1978) and (N-methyl)-carboxyamido groups on the phenyl ring (Singhal et al., 1991), along with catecol esters (Liu & Scouten, 1994), have been proposed.

Tracking differences in the glycan structure of certain proteins represents a diagnostic objective for certain diseases. Actually, in cancer conditions, alterations in fucosylation and sialylation, as well as changes in the glycan branching, are quite common. Their significance or their early detection may signal the beginning of a disease. Unfortunately, due to the great complexity of glycosylation and the optimization of the methodology involving more than one lectin, the approach implies a very intensive preliminary investigation.

Regardless of the specific considerations surrounding the question of glycomarker discovery, investigations on the glycoproteome itself represent a very complex approach. For instance, in elucidating some aspects, Durham & Regnier (2006) captured glycopeptides from whole serum, first using a Concanavalin A column (essentially to remove high-mannose glycans). The column effluent was then adsorbed on a Jacalin column to capture O-glycopeptide alone for further analytical determinations. This example clearly shows that the sequence of lectin columns is also critical for glycoprotein studies, with the risk of other glycans of interest escaping the affinity columns due to either displacement effects or to too low affinity.

A number of examples are reported in the literature for either elucidating the glycoproteome structure or checking glycosylation expression differences as signatures of given diseases. For references see the following reviews: An et al. (2009); Hwang et al. (2009); Tian & Zhang (2010); Butterfield & Owen (2011); and Adamczyk et al. (2012).

3.3.3 Phosphorylated Polypeptides

Protein phosphorylation is one of the most important post-translational modifications that play a key role in cell-signaling pathways (Schulze, 2010). In eukaryotic cells, phosphorylation–dephosphorylation takes place on serine, tyrosine, and threonine (Mann et al., 2002), while in bacteria, fungi, and plants it also occurs in histidine and asparagine (Besant & Attwood, 2012; Attwood, 2012) (Figure 3.5). This post-translational modification is so general that it is estimated that half of all proteins can be phosphorylated and that phosphorylation can be made at up to 100,000 sites in human proteome (Zhang et al., 2002).

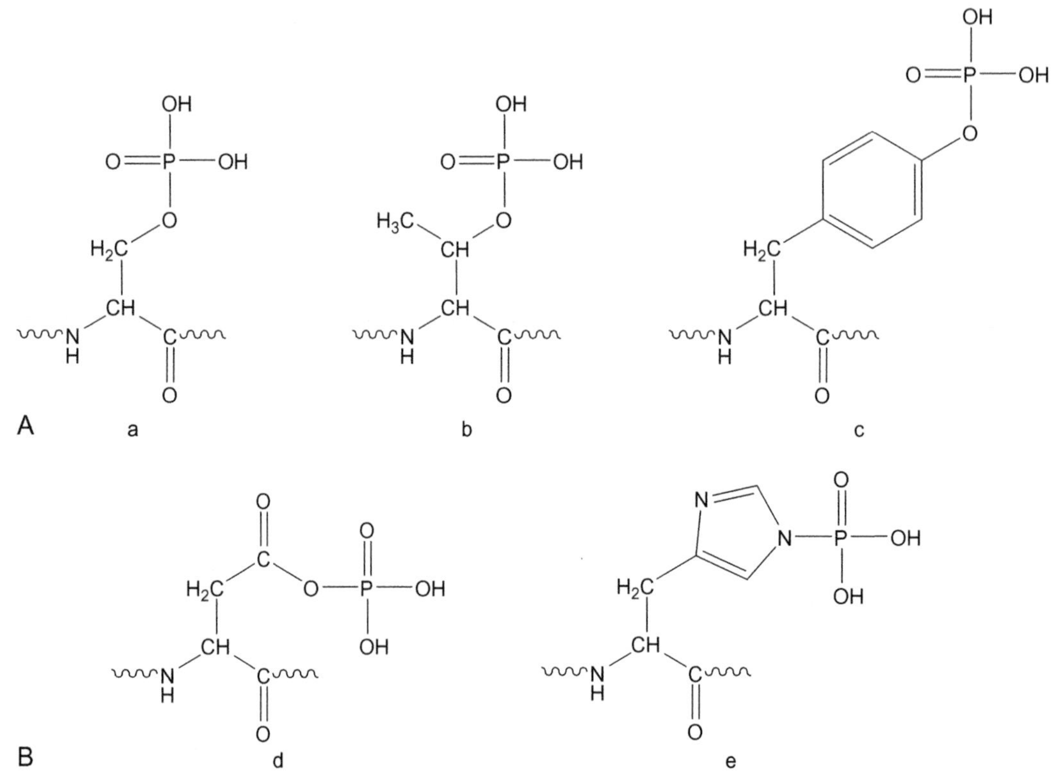

FIGURE 3.5

Most common phosphorylated structures of amino acids composing the polypeptidic chains of proteins.

A: Phosphorylated forms for eukaryotic cells. a: phospho-serine; b: phospho-threonine; c: phospho-tyrosine;
B: Phosphorylated forms of procaryotic cells. d: phospho-glutamic acid; e: phospho-histidine.

The reversible conversion of the hydroxyl side chain of a few amino acids to a phosphate ester creates complex changes in protein conformation, with major interaction effects. This is why the phosphoproteome study is considered as a very important subproteome to investigate. Its composition and functional understanding create conditions not only for novel biomarker discovery of diagnostic interest but also for the design of new pharmaceutical drugs. In spite of the large number of phosphorylated proteins, they are frequently of low abundance, thus requiring peculiar approaches for enrichment.

Various strategies for the study of phosphoproteins and phosphopeptides have been devised, as reported by, for example, Thingholm et al. (2009) and a number of other studies, are emerging. Affinity separations appear to be the most often adapted and specific. To summarize the situation, one should mention those that are the most common. The obvious and also specific way to resolve the question is to use antibodies against phosphoserine, phosphothreonine, and phosphotyrosine immobilized on classical affinity matrices. In 2002 Gronborg et al. described the antibodies antiserine and antithreonine as a means of enriching phosphorylated proteins. They first characterized the best antibodies by immunoblotting and by using an immunoprecipitation technique. Outside a variety of known phosphoproteins, they identified poly(A)-binding protein 2, a protein that was not known to be phosphorylated. In another example, Zheng et al. (2005) described the use of antibodies against phosphotyrosine. Proteins phosphorylated on tyrosine are generally of low abundance and require good antibodies for immunoprecipitation. These authors used this technique for enriching phosphotyrosine-carrying proteins; the latter were then treated with trypsin, and the resulting phosphopeptides were further enriched by metal chelate affinity chromatography (see below). With this protocol, observations were made on the presence of well-known phosphorylation sites, in addition to new phosphorylated sites in alpha-interferon signal transduction. In another study, Rush et al. (2005)

demonstrated the pivotal role of tyrosine kinase by using anti-tyrosine antibodies in oncogenic signaling pathways, reinforcing the cell proliferation and survival mechanisms. This was an interesting finding when considering that peptides containing phosphotyrosine are relatively rare and very dilute, thus precluding any possible detection without preliminary enrichment.

In most cases, once the phosphoproteins are separated, they are analyzed by mass spectrometry upon trypsin digestion. Through the use of this approach, many new proteins have been found and new phosphorylation sites have been discovered, contributing to an understanding of specific pathways. Nevertheless, it should be pointed out that immunoaffinity chromatography suffers from the availability of good quality antibodies (high specificity, low nonspecific binding, and good affinity constants) and the fact that false positives are relatively common.

A well-known method for grouping phosphorylated proteins involves immobilized metal ion chelators such as iminodiacetic acid and nitrilotriacetic acid. These chelators host selected metal ions, displaying an affinity for the phosphate groups of phosphoproteins. The very first report on the capability of chelated iron for the separation of phosphoproteins was published in 1986 by Andersson et al. Among preferred metal ions for the purpose are Fe^{+++} (Li & Dass, 1999), Ga^{+++} (Posewitz & Tempst, 1999), and other transition metal ions such as Al^{+++} and Co^{++}. This is an easy way to separate phosphopolypeptides since these sorbents are popular in liquid chromatography and they simply need to be loaded with the selected metal ion. The main issue is the co-adsorption of a number of other proteins that dock on the chelated metal ion by the affinity with their exposed histidine groups (see Section 3.4.1). One way of partially reducing these nontargeted protein bindings would be through acidification of the environmental conditions. Nonetheless, the capability of metal chelated affinity chromatography in the separation of phosphoproteins is enhanced when it is combined with other prefractionation methods (Nuhse et al., 2007), with the counterpart of increasing the working load and the number of fractions. In all cases these enrichment approaches allow increasing the capability of not only discovering novel unexpected phosphoproteins but more importantly the diversity of phosphorylation sites identified after trypsination and sophisticated mass spectrometry procedures. When justified by the numerous structures present, phosphopeptides can be fractionated to increase the potential for the discovery of novel features. Thus metal chelated affinity chromatography is combined with ion-exchange chromatography, with interesting results as described by Stone et al. (2011).

Perhaps the most effective way to target phosphorylation sites is to trypsinize the proteome and then separate all together the phosphopeptides released. This can be performed by using oxides of zirconium (ZrO_2), of titanium (TiO_2), or of hafnium (HfO_2) (Kweon & Halakansson, 2006; Zhou et al., 2007; Cantin et al., 2007; Nelson et al., 2010). These extremely dense metal oxides have been used under beaded form for years to develop fluidized beds for protein separation from crude unclarified extracts or from fermentation broths (Voute et al., 1998; Voute et al., 2000). Incidentally, one of the major drawbacks of this separation method was the interference with phosphate ions, precluding the use of phosphate buffers, and the presence of non–specific adsorption of proteins, especially those carrying phosphate groups. Thus this problem has been turned into an advantage for proteomics in the separation of phosphopeptides.

The strong interaction of these metal oxides with phosphate esters had been discussed extensively within the context of the preparation of new sorbents for uncompressible and dense material column chromatography. Zirconia surface is, for instance, populated with quite strong Lewis acids due to the presence of bridged zirconols; these structures induce interactions with Lewis bases such as phosphate groups, to the point that the mineral surface needs to be coated with inert polymers to be usable for proper separations (Dunlap & Carr, 1996). The interactions with phosphate groups have also been screened by the use of phosphonate analogs that are stable enough to protect the surface of zirconia against most nonspecific bindings (for a review see Nawrocki et al., 2004). Titania exhibit a similar behavior with a more pronounced Lewis acidity. The sorption strength of phosphopeptides on the surface of these metal oxides depends naturally on the number of phosphorylated sites present on the protein surface. Monophosphopeptides are quite easily desorbed, however, and diphosphopeptides and especially multiphosphorylated peptides are difficult to desorb with a lower yield (Thingholm et al., 2009). Globally, the specificity of metal oxide resins for phophorylated polypeptides is higher than with immobilized metal ions, even if these methods should be considered as complementary since there are phosphorylated species that are captured by one and not by the other, and vice versa.

Another mineral chromatography support that is applicable to the separation of phophorylated polypeptides is hydroxyapatite (Mamone et al., 2010), a calcium phosphate used extensively for many years for protein separations at small and large scales. Here the selectivity can be attributed to the presence of calcium ions at the surface of hydroxyapatite crystals that interact with phosphate groups. The elution of captured species is then obtained by displacement with a progressive increase of competitive phosphate ions concentration in the buffer (Fonslow et al., 2012).

Another way to isolate phosphorylated proteins or peptides is to make specific chemical derivatives at the phosphorylation site (Oda et al., 2001). Making this possible is a strong alkaline treatment of phosphorylated species (phosphoserine and phosphothreonine) that produces a beta elimination reaction with subsequent formation of a double bond. The latter can easily react with nucleophilic groups, especially sulfidryl functions attached to a tag such as biotin. This reaction cascade is in practice performed at different steps as depicted in Figure 3.6. At the end of the process, phosphopeptides or phosphoproteins are tagged and can easily be isolated by affinity chromatography. When the tag is biotin, the complementary solid phase is avidin or streptavidin to better manage elution conditions. An excellent book dealing in depth with phosphorylation in eukaryotic as well as prokaryotic organisms (and also touching on other post-translational modifications such as glycosylation and sulphation) has been recently written by Ham (2012).

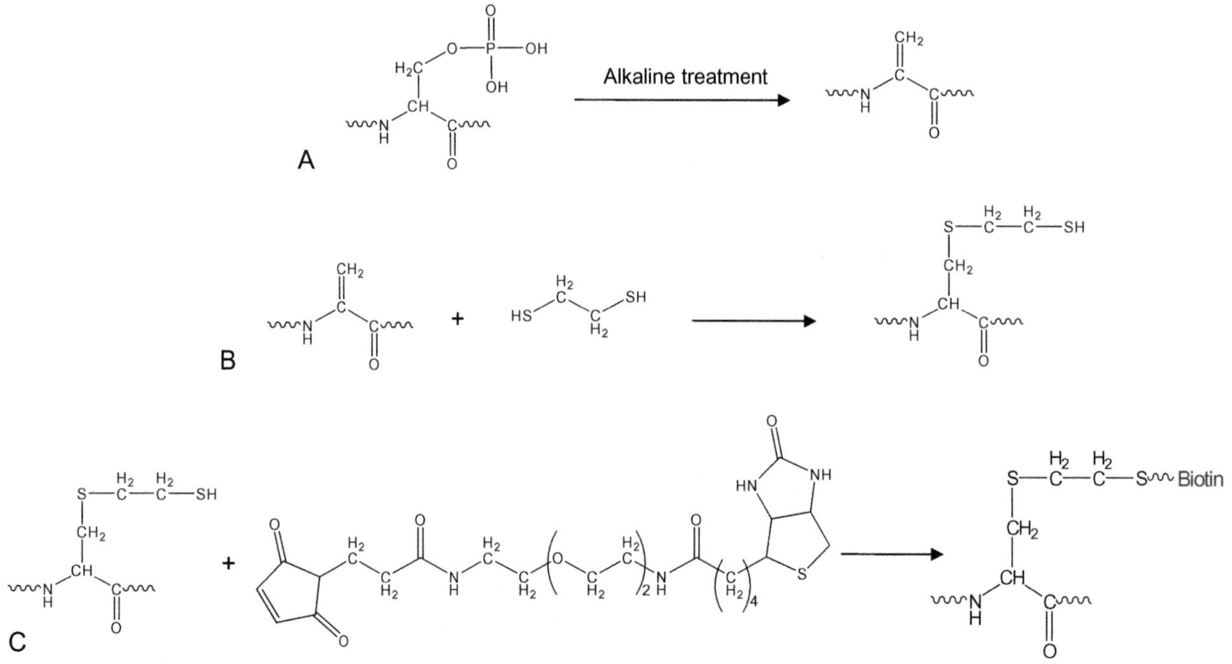

FIGURE 3.6

Chemical modifications of phoshorylated residues on serine or threonine in view of separating specifically only phosphoproteins or phosphopeptides. First, the phosphate ester is transformed under an alkaline treatment into a double bond by beta elimination (scheme "A"). Then the double bond is reacted with disulfidryl compound so as to introduce an -SH group (scheme "B"). Finally, the position is tagged by a biotin using a reactive tagged biotin (scheme "C").

At the end of these three steps the peptides or the polypeptides can easily be isolated using an avidin or a streptavidin column.

TABLE 3.4 Capturing Methods for Phosphopolypeptide Enrichment.

Type of Sorbent	Ligands Involved	Advantages	Disadvantages
Immunosorbents	- Anti-Ser antibodies - Anti-Thr antibodies - Anti-Tyr antibodies	- Highly specific - Detection of rare species	- Poor quality antibodies - Nonspecific binding - Lots of false positive
Chelated metal ions	- Fe^{+++}; Ga^{+++}; - Other transition metal ions.	- Easy approach - Available sorbents	- Docks on His-structures - Acidification required - Nonspecific binding
Heavy metal oxides	- Titanium oxide - Zirconium oxide - Hafnium oxide	- Modulation for multiple phosphorylated sites - Easy and cheap approach	- Co-adsorption of other acidic proteins - Nonspecific binding - Usable for small phosphopeptides
Hydroxyapatite	- Calcium phosphate crystals	- Easy and cheap approach - Elution by specific displacement	- Co-capture of other proteins
Streptavidin	- Biotin specific linker carrying a reactive group for the tagged phosphopolypeptide	- Targeted modification reaction of phosphate groups on Ser and Thr.	- Restricted to serine and threonine - Deglycosylation required

As an alternative to biotin, Brittain et al. (2005) proposed using a fluorous tag, thus alleviating some nonspecific binding problems as well as difficult elution of captured species. With this tag, additional benefits are also claimed related to mass spectrometry analysis.

In short, the choice of tools for the separation of phosphorylated polypeptides is relatively large (see summary Table 3.4), but there is no single ideal solution to all situations.

3.3.4 Ubiquitinylated Proteins

Ubiquitin is a 76 amino acid peptide that is largely distributed throughout the various cell compartments. Ubiquitinylation is a post-translational process used to obtain a given recognizable protein in order to enter the process of degradation (Hershko, 2005). This polypeptide is enzymatically attached to a protein; more than one ubiquitin polypeptide per protein is necessary to accomplish the process. Physiologically, this process is intended to protect the cell against the accumulation of protein aggregates subsequent to a possible process of protein ageing (see also the carbonylation process in Section 3.3.5). The amount of ubiquitinylated proteins is classified as being frequently of low abundance; therefore it is difficult to track all of them and make proper identifications (Peng, 2008). An ubiquitinylated protein can be identified based on the release of a unique peptide upon trypsinization comprising a lysine residue with an isopeptide-linked glycine-glycine sequence (Peng et al., 2003).

Various strategies have been devised for the separation/enrichment of this category of post-translational modification entities. First, antibodies against ubiquitin were used (Shimada et al., 2008) with success. This is an elegant way to resolve the situation because the peptide is relatively large with normally good recognition reactions with antibodies. The use of these tools is easy but shares the same problems described earlier for other purposes.

In addition to antibodies, various other affinity chromatography methods have been investigated. For instance Mayor & Deshaies, (2005) described the enrichment of polyubiquitinylated proteins by a two-step method. Basically, a yeast strain expressing hexahistidine-tagged ubiquitin was used, so the separation of polyubiquitinylated proteins was first performed by using a ubiquitin associated

3.4.1 Metal Ion-binding Proteins

Ion-mediated interaction with proteins is based on the ability of immobilized metal ions to interact with some well-known exposed polypeptide residues. Metal ions are part of a chelating resin on which complexing agents, such as iminodiacetic or nitrilotriacetic acid, are chemically grafted. The metal ion adsorbed on the resin is therefore available to interact with some protein epitopes. The mechanism generally involves exposed histidyl residues of the protein; however, other amino acid residues such as tryptophan, phenylalanine, and arginine can also contribute to the interaction phenomenon (Porath and Olin 1983).

Metal chelated affinity chromatography is influenced by several parameters such as the nature of the immobilized chelating group, the chelated metal ion, the environmental pH, as well as the ionic strength as described by El Rassi & Horwath. (1986).

Metal ion mediated protein separation was used not only to group proteins with a common affinity for a given metal ion but also to capture IgGs. Immunoglobulins G actually are a peculiar protein group for the presence of a highly conserved histidyl cluster at the junctures of CH2 and CH3 domains of the Fc fragment (Burton, 1985), explaining the easy adsorption on these metal chelating resins. The interaction with polyclonal and monoclonal antibodies generally occurs with either nickel ions or copper ions, and then they can be efficiently eluted and recovered from a variety of biological feed stocks (Kagedal, 1989). The affinity of antibodies for cupper ions is relatively high since they can displace competitive-wise a number of less tightly bound proteins.

To achieve high levels of antibody purity, a washing step with sodium chloride can be helpful to desorb and eliminate weakly co-adsorbed proteins (Hale & Beidle, 1994); the elution is operated by a variety of options such as reduction of pH, competition with imidazole, and free chelating agents. This general principle, which is very popular in preparative purification processes of biopharmaceuticals, has been adapted to proteomic investigations. A full review of the application of immobilized metal ion affinity chromatography in the proteomics field has been published by Sun et al. (2005), comprising a variety of technological aspects and separation properties.

The use of this affinity sorbent has been extended to a large number of metal ions displaying interesting variants in the affinity for proteins, even if the most common is Ni^{++} chelated on nitrilotriacetic acid. To demonstrate the various possibilities, a very peculiar group of proteins (uranyl-binding proteins) was separated on uranyl cations attached on a specifically designed sorbent for the purpose of studying a subproteome with respect to a radiation effect (Basset et al., 2008).

Immobilized metal ions are equally extensively used in phosphoproteomics investigations because of the propensity of phosphoric acid residues of amino acids to interact with certain metal ions such as gallium once chelated on solid supports. This has been discussed in Section 3.3.3.

As a fractionation method, metal ion affinity chromatography largely contributed to the discovery of novel proteins from crude extracts in combination with selected analytical technologies (Sun et al., 2008). This technology has been successfully applied to the characterization of metalloproteomes. For instance, it contributed to elucidation of the proteome of *Haloferax volcanii*, a microorganism isolated from the Dead Sea and used as a model for archaeal cell physiology (Kirkland et al., 2008).

Immobilized metal ions have been found useful as a prefractionation technique applied to groups of proteins from various extracts of various sources. In this respect, protein groups from a marine cyanobacterium with properties to bind cobalt, iron, manganese and nickel ions have been fractionated and identified (Barnett et al., 2012).

Finally, it should be mentioned that metal chelated affinity chromatography also allowed identifying candidate biomarker human fluids as reported by Murphy et al. (2006) and Levin et al. (2010).

3.4.2 Protein Kinase Subproteome

Phosphorylation of protein, a common regulatory signaling mechanism in cells, is intimately connected with kinases forming a group of over 500 species in eukaryotic cells (Manning et al., 2002). The general approach for the separation of this category of proteins is the use of ATP-competitive kinase inhibitor ligands, even if their specificity varies according to the type of kinase (Bain et al., 2007; Davies et al., 2000). Small organic inhibitors such as those based on pyrido[2,3-d]-pyrimidine have been grafted on chromatographic beads and used to isolate a group of about fifty proteins related to kinases from HeLa cells (Wissing et al., 2004). The main issue is that all these ligands unfortunately do not allow capturing all kinases. Moreover, nonprotein kinases are also co-captured as a sort of nonspecific binding, such as glyceraldehyde-3-phosphate dehydrogenase and lactate dehydrogenase, both interacting with the same ligands. As a possible alternative, several affinity resins have been prepared with small ligands and then used sequentially with the result of capturing different groups of kinases (Wissing et al., 2007). Due to the relative heterogeneity of these enzymes (different structures and different substrates for phosphorylation), it is quite difficult to overcome all difficulties, especially the fact that the substrates used as ligands may also serve different purposes for other proteins. This produces the capture of other proteins that are considered as contaminants because they do not belong to the same subproteome. This field remains open for the development of novel technologies, especially when the low-abundance kinases are the target of the investigation.

3.4.3 Heparin-binding Proteins

Heparin is a very common affinity chromatography natural ligand that has been extensively used for the purification of a number of proteins. Due to its biological role as well as its peculiar structure (glycosaminoglycan comprising anionic groups such as carboxylates and sulfates, both accounting for ion-exchange properties), heparin interacts with several protein groups. It interacts with coagulation factors (Clifton et al., 2010) as a result of its important anticoagulation role within this complex enzymatic cascade; it interacts with lipoproteins (O'Brien et al,. 1993); and it also interacts with growth factors (Hnasko & Ben-Jonathan, 2003; Cussenot, 1997) and transcription factors (Egly et al., 1984; Burton et al., 1991; Lionneton et al., 2001) because it mimics quite well nucleic acids to which all these factors are tightly associated.

Heparin affinity chromatography was used to reduce high-abundance proteins and also to enrich for low-abundance species. In the former case, there are examples associating heparin and protein G with a claimed improvement of depletion efficiency of high-abundance proteins from serum. In other examples, it is reported that, when compared to immunoaffinity depletion of the six most important targets of human serum, additional spots have been found (Lei et al., 2008). Nevertheless, this depletion approach is very risky because a large number of proteins are also co-depleted not only by the effect of the heparin ligand, but also because of the nonspecific binding of the second protein G column. The additional proteins found with this treatment may not necessarily be attributed to a better depletion, but most probably to the enhancement of low-abundance proteins that were concentrated by the heparin ligand itself. It is in fact in this domain that heparin affinity chromatography can do a good job as reported in several proteomic studies. One of these studies refers to the protein enrichment of soluble protein fraction of the bacterium *Haemophilus influenzae* by applying immobilized heparin (Fountoulakis et al., 1998). Another study deals with the detection of signaling low-abundance proteins, with the aim of finding biomarkers and drug targets (Krapfenbauer & Fountoulakis, 2009). Here the biological extract was enriched with heparin and the eluate digested by trypsin and analyzed by MudPIT from where it was found that about half of all identified proteins had a clear affinity for heparin.

The discovery of tissue-specific biomarkers was also described using heparin affinity chromatography upon enrichment of low-abundance proteins as a way to overcome the question of a large dynamic concentration range of proteins. By that way and using two-dimensional electrophoresis, fourteen spots were found as differentially expressed when compared to samples from normal lung and pulmonary adenocarcinoma. Among them, transgelin, transgelin-2, and cyclophilin A were identified as serious candidates (Rho et al., 2009).

3.4.4 Cyclic Nucleotide Interactome

Cyclic nucleotide binding proteins are by themselves a quite large group of proteins involved in signaling pathways. While conceptually it should be easy to separate these proteins by their natural affinity, several issues have to be overcome: (i) cyclic nucleotide ligands are not directly immobilizable and need to be chemically modified for the purpose, with consequences in the protein recognition specificity; (ii) there are proteins that have some affinity for these ligands but are not part of the targeted biological process; (iii) protein–protein interactions increase the burden of other side effects. How cyclic nucleotide-dependent signaling develops is nicely presented and discussed by Medvedev et al. (2012).

With regard to the necessity of modifying the cyclic nucleotides prior to grafting on agarose beads, Aye et al. (2009) used a combination of affinity chromatography methods for the elucidation of the PKA-AKAP (A-kinase anchor proteins) interaction specificity. This was accomplished by using three different cyclic AMP analogs attached to agarose particles at different positions to enrich for PKA (Protein Kinase A). The first was a cAMP where an aminohexylamino spacer was attached in the position 2 of the pyrimidine ring; the second was a cAMP where the same spacer was attached in the position 8 of the imidazole ring; and the third was similar to the second in which a methoxyl group replaced the hydroxyl group at the position 2' of the ribose. This investigation concluded about the large complexity of the molecular interactions where isoforms have different affinities for the investigated cAMP derivatives. The identification of more than ten AKAPs was also achieved; each AKAP may be involved in different PKA isoforms activation in various compartments of the cell. Demonstration was made on the use of natural ligand derivatives attached on a chromatographic support, thus exposing different sides of their structure for the elucidation of their biological role.

In attempting to enlarge and improve the harvest of this subproteome, four other cyclic nucleotide analogs have been made, grafted on solid supports and used to capture proteins from rat heart tissue extracts (Scholten et al., 2006). The cyclic nucleotides cAMP and cGMP were chemically attached on agarose beads by means of a spacer grafted in the 2- and 8-position of the nucleotide. Experiments showed that many protein species that naturally interact with these cyclic nucleotides were effectively captured, but other proteins outside the signaling process and more generally involved in the interaction with nucleic acids were co-captured. Known cAMP/cGMP binding proteins were identified among the captured species such as kinases and phosphodiesterases. However, other interesting proteins were also identified such as six different AKAPs, among them sphingosine kinase type1-interacting protein, another potential AKAP. These discoveries were not only related to the analogs of cyclic nucleotides used, but also to the improvement of the specificity after the protein capture of the crude extract by different washings with displacement agents such as ADP and GDP. This operation resulted in the removal of several nonspecifically adsorbed proteins. Although a number of reports attest to a creative approach for the separation of this low-abundance protein group, work remains to be done to maintain a molecular specificity for a simple and proper global proteomics analysis. A great summary of the situation is given in a recent review by Medvedev et al. (2012) where the authors underlined the importance of this interactome in the domain of characterizing targets for drug development with fewer side effects and possible new pharmacological properties.

3.4.5 Isatin-binding Proteins

Found mostly in the mammalian brain (Crumeyrolle-Arias et al., 2009), isatin exhibits a peculiar role at the crossroad of pharmacological activities around viral antidotes, bactericides, and anti-apoptotic activity. Therefore all proteins that have an affinity for this compound have a clear pharmacological interest that can easily be approached by proteomic investigations as reported by Medvedev et al. (2012). Isatine-binding proteins are separated by affinity chromatography on isatin analogs grafted on Sepharose beads (Crumeyrolle-Arias et al., 2009) with the isolation of a group of proteins from a brain tissue extract made in the presence of the non-ionic detergent Triton X-100. A proteomic analysis evidenced several dozen proteins even if the expected number was considered larger. In fact the authors hypothesized that the harvest was incomplete since known targets for isatin inhibition such as guanylate cyclase and monoamine oxidase were not found. It is postulated here that these proteins are of low abundance and that their affinity for the ligand was not high enough for their capture. This situation could have been resolved by a first operation of global enrichment using the combinatorial peptide ligand library as described in the following chapter, before the chromatographic step on isatin-Sepharose. Nevertheless, similarities and differences in proteins were found from rat and mouse brain extraction with common gene products of about one-third. According to the authors, this difference could be related to the pharmacological effects of isatin. This model could be taken for a number of other pharmacological molecules capable of interaction with a group of proteins to then investigate globally their molecular action.

3.4.6 Other Low-Abundance Protein Grouping Methods

The approaches described above are certainly not exhaustive with regard to group proteins by their similar properties toward generic natural or synthetic ligands with the objective of enhancing low-abundance species. We limited the description to the most important and those that have proven their efficacy in various situations. Nevertheless, we believe we should mention nucleotide binding proteins that have already been separated by solid-phase technologies. In a study, Xu et al. (2007) used PolyU ligands in view of detecting RNA-binding proteins from Arabidopsis leaf extracts that escape the current two-dimensional electrophoresis analysis. By that way, twenty-six novel proteins were identified by MALDI-MS. The authors reported that identified species were expressed at levels lower than 0.1% by distinct genes. In another study, Ni et al. (2010) used a PolyA affinity column to search for chloroplast proteins of a very low transcription level that cannot be detected directly without enrichment processes. More than twenty chloroplast proteins, coded by eighteen distinct genes, were identified by MALDI-MS. These studies showed the capabilities of joining specific enrichment processes for species that are related to the same type of interaction, with analytical determinations based either on mass spectrometry or two-dimensional electrophoresis.

Although all separation technologies by solid-state extraction with group ligands are continuously optimized, most of these methods make use of relatively complex mechanisms, especially when more than one protein is involved with some level of competition. When the mechanism involves protein isoforms, the complexity increases by an order of magnitude. In addition, these processes may generate several fractions that need to be analyzed separately or in parallel with quite intensive labor. Finally, it is never certain that the recovery is absolute; actually, it is lower than 100%, with some level of losses, especially for proteins that are present in trace amounts.

3.5 IMPORTANCE AND LIMITS OF CLASSICAL AFFINITY-ENRICHMENT METHODS

Group fractionation is a way to simplify the analysis of protein fractions, especially when the initial sample is treated for either removing the high-abundance species or compressing the dynamic concentration range with a special focus on the enrichment of low-abundance proteins. Although this approach appears logical and straightforward, it hurts the technicians because it produces various issues: (i) it increases the workload, (ii) it increases the amount of data to be dealt with, and (iii) it produces false positives. These three limitations should be put in perspective to sort out situations when it is advisable to associate sample treatment and subfractionation or when it is necessary to think about novel rational approaches with better specificity.

3.5.1 Staying within Current Technology Knowledge

To date, a number of technologies are known to separate proteins by groups, with the aim of detecting novel species for either improving the understanding of proteomes or for searching for early misexpression signatures. Most of these methods derive from affinity chromatography of proteins used for years in the field of preparative protein purification with few or extensive modifications, depending on the purpose. Even if many efforts have been deployed to perfect these methods, they all suffer from a number of disadvantages; the main one is the nonspecific binding, where the risk of false positives is always present even if to a limited extent. To maintain a workload within reasonable limits, compromises need to be accepted. For instance, a double fractionation in view of eliminating false positives from the first fractionation approach is very rare because it multiplies the number of fractions and hence the analysis time. In addition, the amount of data accumulated by an extended number of analytical determinations with replicates can become so huge that it even discourages scientists who claim to have time. The day when analytical determinations and data analysis can be performed on a large scale within a short time, the question of extended fractionation may be reconsidered. In fact, this could eventually be performed under high-throughput conditions using automatic systems.

TABLE 3.6 Technical Summary of Protein Grouping Methods Based Either on their Similar Post-translational Modifications or their Similar Property to Bind a Category of Ligands

Groups of Proteins	Main Separation Methods	Advantages	Disadvantages
Glycosylated proteins	Lectins and boronic acids	Good specificity Possible solutions for all glycoproteins	Lectins are weak binders
Phosphorylated proteins	See Table 3.4	See Table 3.4	See Table 3.4
Carbonylated proteins	Tagging carbonyls with hydrazine derivatives (e.g. biotin)	Specific approach for carbonyl-containing polypeptides	Laborious method Possible false positives
Ubiquitinylated proteins	Antibodies	Specific to ubiquitin	Low binding capacity Cost
Metal ion binding proteins	Chelated metal ion chromatography	Easy to perform Specificity modulation by metal ion selection	Metal ion leakage Low specificity
Protein kinases	Small organic inhibitors	Selection of specific binders	Not all kinases are captured Possible nonspecific binding
Heparin-binding proteins	Immobilized heparin	Good for coagulation factors	Capture of disparate groups of proteins
Cyclic nucleotide-binding proteins	cAMP derivatives	Could be very specific for a difficult category of proteins to separate	May require chemical modification of cAMP to be effective
Isatin-binding proteins	Grafted isatin analogues	Good for biopharma studies	May require different analogs to resolve all situations
Nucleotide-binding proteins	PolyU or PolyA grafted resins	Good specificity	Solid-phase synthesis not easy to implement

At present, known methods for protein grouping are already quite diverse, and combinations have not been used to cover all the situations; yet the field is not completely exploited. Far from suggesting alternatives, it is advised that one pays attention to disadvantages so that, when possible, one can try circumventing them or integrating them under a synergistic manner. Table 3.6 summarizes the situation of the main protein group's fractionation, giving both advantages and disadvantages.

3.5.2 Pushing the Imagination beyond Validated Fields

All current methods suffer from a certain lack of specificity and hence a certain degree of nonspecific binding. This situation stimulates thoughts for better methods of separation for protein groups by means of improved affinity-based methods.

Several avenues are open for the development of novel or improved approaches. One of them is the design of antibodies, thus enlarging a method that has already proven its efficiency for many applications. Their application in post-translational modifications is still limited because of a number of issues related to the large diversity of the considered antigens. In the case of glycans, the question is further complicated by the poor antigenicity of the targets. To circumvent this obstacle, perhaps one could consider the consensus structures around such glycans and try building specific antibodies. Their specificity will then be restricted to single glycans, therefore the approach would necessitate creating many antibodies, each of them specific for a single structure. Although this represents a certain technological investment, at the end the investigations on glycoconjugates for fundamental studies and/or diagnostic applications could be accelerated and rewarding. These antibodies could be used singularly for the segregation of groups of glycans, or they could be used as mixed beds for a full collection of these structures. They could be used not only for the separation of glycoproteins, but also possibly for glycopeptides. The affinity for the various structures present in a glycopeptide or glycoprotein mixture could be exploited as a means of enriching rare species if the solid phases are used under large overloading conditions, as is the case for peptide libraries (see Chapter 4).

Similarly, the phosphorylation diversity could be investigated using antibodies constructed against exposed consensus diversity, not just against phosphoserine, phosphothreonine, and phosphotyrosine. Despite the fact that the structure diversity seems very large in phosphoproteomics, as mentioned in Section 3.3.3, the advantage could be worth the technological effort.

Ligand design based on biomimetic principles as initiated by Lowe and co-workers (Roque & Love 2005a) is another very interesting way to develop group-specific synthetic ligands. Few examples have been published in the last few years. For instance, artificial receptors for glycoproteins have been reported (Gupta & Lowe, 2004), with the preparation of several dozen structures from where the most effective ones have been retrieved. The interaction with saccharides is formed within an organized apolar cavity where hydrogen bonding from the presence of donor and acceptor structures creates the specific interaction. Specificity has been proven by different analytical ways, and the elution is operated by borate buffers as displacing agents (see boronic acid specificity for 1,2-cis-diol structures in Section 3.3.2). It is here hypothesized that the approach could be re-elaborated for even better specificity and extension to biomimetic-targeted structures for well-determined sugars. These structures could be used sequentially, in parallel or as mixed beds, depending on the objective of the application. Biomimetic ligands against phosphorylated structures could be designed by the same technology, taking as a model the phosphotyrosine-, phosphoserine-, or phosphothreonine-binding domains as described by Roque & Love (2005b). Similar approaches have successfully been taken for the design of biomimetic ligands for IgG antibody separations (El Khoury et al., 2012).

Another approach that has been experienced in the last few years for some applications could be generalized. This is the application of the biotin–avidin system. Such an approach involves a preliminary operation at the level of sample treatment. The post-translational modifications will first be taken as a target for chemical modification or physicochemical specific interaction with molecules that are tagged with a biotin terminal structure (see the scheme in Figure 3.6 for phosphoprotein separation). Glycanes could interact with a biotinylated structure carrying, for instance, a boronic acid molecule or another biomimetic structure; alkylated peptides could be captured with mixed-mode ligands; glutationylated post-translational modifications could be targeted with reversible structures involving mercury or other reactive species for thiol groups. Carbonylated and phosphorylated amino acids have already been processed using this technology, but the current known method could be refined for better specificity. The biotin tag itself on the bifunctional molecule could be attached by many possible variations, using cleavable or positively charged or negatively charged groups so as to simplify the elution from avidin columns or to enhance the docking on the protein structure. Biotin itself could be replaced by other structures such as fluorous tags to improve the level on nonspecific binding on the one hand and to facilitate the elution from the solid phase on the other hand. Such an approach could also simplify the interpretation of mass spectrometry analysis.

3.6 INTEGRATION OF LOW-ABUNDANCE PROTEIN ACCESS METHODS

Reaching low-abundance proteins by the use of methods described in this chapter might be challenging, especially if a single technical approach is chosen. There are issues related to high-abundance protein depletion that are relatively difficult to circumvent, as, for instance, the co-depletion effect and the dilution factor. The separation of proteins by their post-translational modifications engenders incomplete captures due to the lack of specific ligands or too low-affinity constants. Group separations by the affinity to a common ligand are also hampered by a large amount of nonspecific binding. In some instances to reduce this problem, Collins et al. (2005) proposed an improved approach to render the separation of phosphopeptides more specific by combining phosphoprotein separation first and then phosphopeptide enrichment with enhanced efficiency analysis than from current single-fractionation methods.

Since none of these methods is ideal, a possible approach could be to rationally associate methodologies, especially those that are complementary. Below are reported three different possibilities, but the reader can organize additional possibilities as a function of the problem to resolve.

3.6.1 Associating Protein Subtraction with Group Separation

It is not always as simple to just remove a few high-abundance proteins, to resolve all the questions related to proteome deciphering. This is due to the extremely high complexity of the protein composition of biological fluids and cell extracts. Removing even 100 proteins from several thousands or even several hundred thousands (if isoforms and other truncation products are considered) does not change a lot in the complexity of the initial sample. This is why group fractionation may be an interesting option for a well-targeted group of proteins.

The most obvious approach is to make a lectin fractionation after a high-abundance protein depletion. The interaction of glycoproteins with lectins is operated in physiological conditions of pH and of ionic strength that are fully compatible with the flow through fraction from immunodepletion. This principle can easily be extended to a large number of lectins used singularly or in a sequence or as a complex mixed-bed packing. In the latter case the elution by category can be accomplished by introducing in the column sequential solutions of different competing sugars. Dayarathna et al. (2008) have described this strategy as M-LAC in view to improve conditions for the analysis of specified diseases. Lectin selection comprises concanavalin A, wheat germ agglutinin, and jacalin. After protein loading, the elution was performed by competitive displacement with appropriate sugars. The authors described this technology as an important improvement over the current methods to detect significant changes in proteome profiles. The initial sample treatment was performed by using an immunodepletion column with a blend of antibodies against the six most abundant proteins from human serum, and the depleted flow-through was injected into the lectin columns. This platform was subsequently used for the analysis of glycoproteomes with high recovery and good reproducibility (Kullolli et al., 2010). The use of spectral counting could produce semiquantitative data crucial for evaluating abundance differences among clinical samples.

This combination is one of the most representative examples for glycoprotein discrimination after the removal of species that form the major abundances. An example is given by Zhao et al. (2006) for the investigation of misregulated species in the serum of pancreatic cancer patients. From the initial serum, the twelve most abundant proteins have been removed first by using a mixed bed of specific immobilized antibodies. Then from this depleted sample, a fractionation followed using three distinct columns of lectins, each one capable of capturing a specific family of glycans: wheat germ agglutinin, *Sambuccus nigra* agglutinin, and *Maackia amurensis* lectin. Using current proteomic methods available for proteomic analysis, the authors claimed they had found two downregulated proteins. While it is hard to say why these lectins were chosen over others, the study clearly showed that a fractionation of proteins of the same group (glycoproteins) after a preliminary treatment for high-abundance protein depletion was quite efficient for the purpose of the study.

In another study, in order to overcome the difficulties involved in identifying potential cancer-specific biomarkers due to the large concentration difference and the extremely large number of gene products and derivatives, Ueda et al. (2007) circumvented the question by first immunodepleting serum from the first six major proteins. Then the obtained fraction was subfractionated by lectin chromatography. More than thirty proteins were thus identified as being misexpressed from lung adenocarcinoma patients. In particular, significant differences were found in the α-1,6-fucosylation level underlying the role of a fractionation of a depleted sample from where no differences could be observed. The depletion allowed increasing the probability of discovering gene products that otherwise would not be detectable because they are hidden by high-abundance species, while chromatography with lectins enhanced the signal of low-abundance glycoproteins. Lectin chromatography used alone would not be effective enough without the process of depletion because of a massive presence of immunoglobulins that are all glycosylated by a variety of glycans. As the authors indicated, this kind of glycoproteomic strategy should accelerate the discovery of pertinent markers of metabolic disease, especially cancer.

In a sort of confirmation, Comunale et al. (2009) found that fucosylated glycans were increased in the serum of patients with hepatocellular carcinoma. This effect was observed after having depleted the initial serum from twelve major proteins, followed by lectin affinity chromatography using *Aleuria aurantia* lectin. Several fucosylated proteins were found useful for a possible marker for diagnostic upon an investigational work covering 300 patients. Among fucosylated proteins, hemopexin was demonstrated as being a good target with a level of both sensitivity and specificity higher than 90%. A number of other published papers could be mentioned combining depletion and glycoprotein fractionation along with their merits (for additional examples see: Dayarathna et al., 2008; Mortezai et al., 2010).

It is also interesting to mention other types of group fractionation after a depletion step. For instance, depletion of major proteins from plasma was followed by a group fractionation using metal ion affinity chromatography by Jmeian & El Rassi. (2009b). After the depletion of eight major proteins by means of immunosorbents, protein A and protein G, the fractionation was performed using metal chelated affinity chromatography with zinc, nickel, and copper ions. Under this configuration, the authors reported the reduction of both the dynamic concentration range of serum constituents and of the complexity, thus facilitating the following analysis performed using two-dimensional electrophoresis and various types of mass spectrometry.

3.6.2 Adopting Protein Group Fractionation after Peptide Library Treatment

The combinatorial peptide ligand library, the focus of this book for the reduction of protein dynamic concentration range (as extensively described in the following chapter), allows in a single operation not only decreasing the concentration of high-abundance species, but interestingly and concomitantly increasing the concentration of low-abundance ones, thus rendering them easily detectable. As a result, a number of new protein signals appear subsequent to the treatment, whatever the analytical method. A simple visual representation is given in Figure 3.7 where a yeast extract was treated with a peptide library and then in parallel to the control free-SH groups from proteins were alkylated by means of iodoacetamide. The resulting sample was then reduced by dithiothreitol, and the released reduced proteins were captured by biotin-HPDP (N-[6-(biotinamido)hexyl]-3'-(2'-pyridyldithio)propionamide). Under these conditions, oxidized proteins are labeled and purified using an avidin-affinity column. Results from a two-dimensional electrophoresis revealed a number of additional protein spots that are undetectable from the crude extract. Clearly, the peptide library treatment allowed concentrating low-abundance oxidized species that are specifically indicated within the boxed areas. When using LC-MS/MS, 146 novel gene products could be identified.

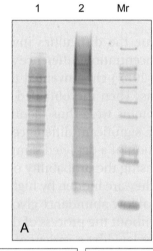

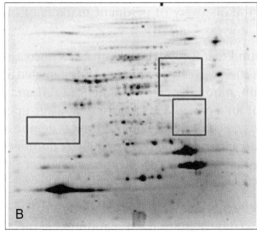 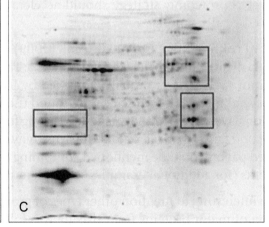

FIGURE 3.7

Analytical electrophoresis results of proteins from *Saccharomyces cerevisiae* before and after treatment with a peptide library.

A: SDS-PAGE of protein samples from initial crude extract (1) and after treatment with the peptide library (2); Mr stands for molecular mass ladder. B & C: Two-dimensional electrophoresis results of oxidized proteins from the same extract. In panel C the initial protein yeast extract was treated with a peptide library and the harvested proteins submitted to alkylation using iodoacetamide at pH 8.8. Proteins were then reduced using dithiothreitol and free cysteine SH reacted with biotin-HPDP. The labeled proteins were then separated by affinity chromatography on avidin-Sepharose and finally submitted to SDS-PAGE. Panel B represents the control (proteins from the same strain submitted to alkylation and reduction but not treated with the peptide library). The boxes represent major areas where novel oxidized proteins appeared upon peptide library treatment. *(Adapted from Le Moan et al., 2007, by permission.)*

A large number of other examples of the enhancement of low- and very low-abundance species are reported (see Chapters 4, 5, and 6). The apparent complexity of peptide library-treated samples is always significantly enhanced because of the compression of the dynamic concentration range. As a consequence, the discrimination of certain species or protein groups could become more challenging due to the presence of a larger number of species with closer characteristics in terms of isoelectric point, molecular mass, capability of getting ionized by mass spectrometry, similar hydrophobic index, comparable post-translational modification, and so on. The rational mode to analyze in depth all these newly revealed species would be to fractionate the peptide library-treated sample with one of the above-described methods based either on their similar interaction properties for an affinity ligand or on their post-translational modifications. Examples are reported in the literature describing not only the principle but also the merit of the approach. In spite of extensive investigations, there are no ideal conditions whereby a given method would apply to every situation; rather, singular interests depend on the context and on the final goal.

In attempting to track lactosylated proteins in heat-treated milk, Arena et al. (2010) captured low-abundance lactosylated proteins after peptide library treatment by means of m-aminophenyl boronic acid grafted on agarose beads. These authors found more than thirty lactosylated proteins in 271 nonredundant sites, highlighting the interest of the combined technologies to discover special proteins. In another study, Stone et al. (2011) reported the results of investigation of salivary phosphoproteome after treatment with peptide library followed by a fractionation by cation exchange and metal ion chelated affinity chromatography. Thus, eighty-five different phosphoproteins were discovered, comprising no less than 217 unique phosphopeptide sites. Stone and colleagues also demonstrated that phosphorylation occurred in a different manner compared to plasma, suggesting different functions that are yet to be elucidated. The utility of combining conventional chromatographic separations such as ion-exchange chromatography and depletion with peptide library sample treatment was reported by Hagiwara et al. (2011); upon treatment a larger number of proteins could be identified with the use of 2D-PAGE and 2D-DIGE.

The use of peptide libraries also provides the possibility of fractionating by itself. In fact, proteins are captured by the solid phase, with a possibility of making progressively stringent fractionated desorptions according to the elution agents used. These possibilities are extensively described in Chapter 4, and recipes are given in Chapter 8. In an example, Léger et al. (2011) described the fractionated elution of peptide library-captured plasma proteins into four groups and demonstrated the low level of redundancy among fractions and the simplification of the sample complexity for an improved way of identifying protein markers for diagnostic purposes (see details of the technology in Section 6.4.5 and Figure 6.7).

As a reminder, a number of relevant references can be found on using fractionated elution to better analyze the treated biological samples. Figure 3.8 shows an example of human serum treated with the peptide library and subfractionated on an anion exchange column. Clearly, additional proteins are evidenced; however, and more importantly, this demonstrates that the single fractions are complementary and can more easily be analyzed by current proteomics means with less overlapping species.

When using the peptide library to enhance low-abundance species, the question that arises concerns the efficiency of a fractionation operated by sequential elution of captured proteins compared to a subsequent fractionation by another tool (e.g., ion- exchange chromatography). Since the mechanisms at the basis of fractionation are not the same, the protein composition of fractions is very different and what one should consider is the least possible protein overlap between collected fractions. To extend the data shown in the previous figure, Figure 3.9 illustrates under the same analytical method fractions obtained by a series of elutions from the peptide library using increasingly stringent desorption mixtures.

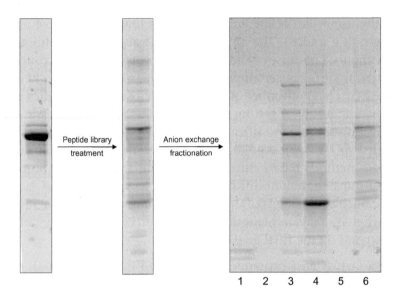

FIGURE 3.8

SDS-polyacrylamide gel electrophoresis of human serum fractions obtained by anion exchange chromatography after treatment with combinatorial peptide ligand library. On the left is the human serum pattern prior to treatment. At the center is the serum after treatment with peptide library (many additional bands are obtained while the albumin band is largely decreased). On the right panel are represented the chromatography fractions.

Fractionation was performed using an initial equilibration of both the column and the serum sample, with 50 mM Tris-HCl, pH 9 containing 1 M urea and 0.22% CHAPS. The flow-through (lane 1) is directly collected while adsorbed proteins are sequentially desorbed using the following buffers: 50 mM HEPES, pH 7 (lane 2), 100 mM sodium acetate pH 5 (lane 3), 100 mM sodium acetate pH 4 (lane 4), 50 mM sodium citrate pH 3 (lane 5), and finally (lane 6) with a hydro-organic mixture constituted of 33.3% isopropanol, 16.6% acetonitrile, 0.1% trifluoroacetic acid in distilled water. Protein staining: Coomassie blue.

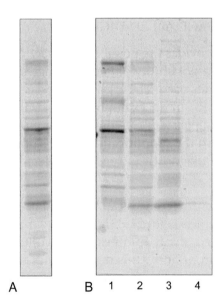

FIGURE 3.9

SDS-polyacrylamide gel electrophoresis of fractions from human serum treated with peptide library. The captured proteins are sequentially eluted and analyzed.

A: Global one-step elution of captured serum proteins.
B: Fractions from sequential elution. First elution was performed using 0.5 M neutral sodium chloride (lane 1); the second fraction was obtained by elution with 2 M thiourea, 7 M urea, 2% CHAPS (lane 2). A third elution followed by using 9 M urea, citric acid to pH 3.3 (lane 3); finally the remaining proteins still on the peptide beads were removed by using a hydro-organic solution composed of 6% v/v acetonitrile, 12% v/v isopropanol, 2% v/v of ammonia and 80% v/v water (lane 4).

3.6.3 The Nonsense of Associating Subtraction and Peptide Library Treatment

With all the described arguments above, for both subtraction methods and reduction of dynamic range using peptide libraries, one would ask a question about the utility of associating these technologies' amplification of low-abundance species using the combinatorial peptide ligand libraries in one sequence or another. As a first approach, this could sound logical because the first removes the problem of high-abundance proteins, while the second gives the advantage of concentrating very low representative polypeptides. In reality, subtraction methods and peptide libraries' enrichment technology are partially redundant, on the one hand, and, more importantly, they are technically incompatible, on the other hand. They are partially redundant because the first removes in principle a few or several well-defined major proteins (with the inconvenience of co-removing nonspecifically adsorbed proteins), and the second largely reduces the concentration of the same proteins (see the mechanism of action in Chapter 4). Therefore this quite common characteristic largely reduces the technical interest to use them in combination.

It should also be emphasized that the use of peptide libraries clearly compresses the dynamic concentration range, with a very significant concentration of rare species that are otherwise undetectable because of their extremely low expression level and hence an extreme dilution. On the contrary, subtraction as, for instance, an immunodepletion does not change the situation of low-abundance proteins that remain extremely dilute. This dilution is significantly larger after treatment, rendering their already difficult detection even more so. From a technical standpoint, these distinct operations are also incompatible for the simple reason that the depletion involves a very small biological sample (e.g., about 20–50 µL of human serum) whereas the peptide library requires a relatively large sample for a significant and consistent enhancement of low-abundance species. In other words, if one wants to build a peptide library treatment after having removed the most abundant species, one should make several operations of depletion to collect a depleted sample large enough to proceed for the reduction of dynamic concentration range and the enhancement of diluted species by means of combinatorial peptide ligand libraries.

In practice, this operation is time consuming because several cycles of depletions are necessary that add to the very high cost of the immunodepletion sorbents. This being said, several authors have compared these two very different approaches side-by-side with a number of interesting results. They have been described as having some level of complementarity if the treated samples are submitted to trypsin digestion and then peptides are subfractionated by isoelectric focusing (Ernoult et al., 2010) (Figure 3.10).

In another study, Beseme et al. (2010) reported that the visibility for undetectable proteins is enlarged with both techniques; however, they found marked differences in the composition of respective treated samples. By two-dimensional electrophoresis, 427 and 557 spots were counted from samples treated, respectively, by immunodepletion or by peptide library, suggesting that the latter method can be more efficiently used to access low-abundance proteins with a high degree of reproducibility required for the quest of biomarker candidates in clinical studies.

Several other studies compared the performance of various sample treatments in view of reducing the suppression effect of high-abundance proteins and the enrichment for very dilute species (see, for example, Millioni et al., 2012; Krief et al., 2012; Cumova et al., 2012). Cumova et al. report the results of a comparative investigation in view of finding the most adapted pretreatment method (immunodepletion or peptide library) to human plasma, with the objective of discovering protein marker signatures upon the use of bortezomid as a drug treatment for multiple myeloma. The first

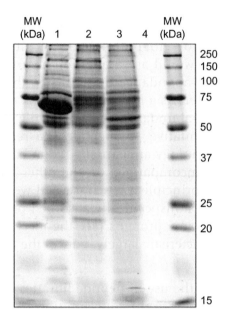

FIGURE 3.10

SDS-polyacrylamide gel electrophoresis of human plasma immunodepleted of the six major proteins or treated with peptide ligand library. Lane 1: untreated human plasma; lane 2: immunodepleted plasma; lane 3: peptide library treated sample. Lane 4 (no detectable proteins) represents a second elution by SDS buffer of proteins from peptide library beads. Protein bands were stained by Deep Purple. *Results from Ernoult et al. (2010) by permission.*

observation is that the peptide library allows detecting a much larger number of protein spots by two-dimensional electrophoresis compared to immunodepletion. The second observation made on the two patient groups under study—chemoresistant *versus* sensitive group to bortezomid—revealed fifteen differently expressed spots corresponding to ten different gene products (seven proteins were downregulated while three others were upregulated). The interest of the peptide library was clearly underlined in this study. Although immunodepletion and peptide library treatments are not fully compatible as described above, it may make sense to remove well targeted proteins before special peptide library treatments. This is the case when some high abundance proteins such as albumin could represent a problem for the treatment of serum in presence of high concentration of lyotropic salts. In this case it is interesting first to remove the problematic proteins and then treat the depleted sample with peptide ligand library for dynamic concentration compression. For specific details see Section 4.5.4.

3.7 References

Adamczyk B, Tharmalingam T, Rudd PM. Glycans as cancer biomarkers. *Biochim Biophys Acta.* 2012;1820:1347–1353.

Afkarian M, Bhasin M, Dillon ST, et al. Optimizing a proteomics platform for urine biomarker discovery. *Mol Cell Proteomics.* 2010;9:2195–2204.

Ahmed N, Barker G, Oliva K, et al. An approach to remove albumin for the proteomic analysis of low abundance biomarkers in human serum. *Proteomics.* 2003;3:1980–1987.

Akparov VK, Stepanov VM. Phenylboronic acid as a ligand for biospecific chromatography of serine proteinases. *J Chromatogr.* 1978;155:329–336.

An HJ, Froehlich JW, Lebrilla CB. Determination of glycosylation sites and site-specific heterogeneity in glycoproteins. *Curr Opin Chem Biol.* 2009;13:421–426.

Andac CA, Andac M, Denizli A. Predicting the binding properties of cibacron blue F3GA in affinity separation systems. *Int J Biol Macromol.* 2007;41:430–438.

Andersson L, Porath J. Isolation of phosphoproteins by immobilized metal (Fe^{3+}) affinity chromatography. *Anal Chem.* 1986;154: 250–254.

Arena S, Renzone G, Novi G, et al. Modern proteomic methodologies for the characterization of lactosylation protein targets in milk. *Proteomics*. 2010;10:3414–3434.

Attwood PV. PN bond protein phosphatases. *Biochim Biophys Acta*. 2013;1834:470–478.

Aye TT, Mohammed S, van den Toorn HW, et al. Selectivity in enrichment of cAMPdependent protein kinase regulatory subunits type I and type II and their interactors using modified cAMP affinity resins. *Mol Cell Proteomics*. 2009;8:1016–1028.

Bain J, Plater L, Elliott M, et al. The selectivity of protein kinase inhibitors: A further update. *Biochem J*. 2007;408:297–315.

Barnett JP, Scanlan DJ, Blindauer CA. Fractionation and identification of metalloproteins from a marine cyanobacterium. *Anal Bioanal Chem*. 2012;402:3371–3377.

Basset C, Dedieu A, Guérin P, Quéméneur E, Meyer D, Vidaud C. Specific capture of uranyl protein targets by metal affinity chromatography. *J Chromatogr A*. 2008;1185:233–240.

Besant PG, Attwood PV. Detection and analysis of protein histidine phosphorylation. *Mol Cell Biochem*. 2009;329:93–106.

Beseme O, Fertin M, Drobecq H, Amouyel P, Pinet F. Combinatorial peptide ligand library plasma treatment: Advantages for accessing low-abundance proteins. *Electrophoresis*. 2010;31:2697–1704.

Binz HK, Amstutz P, Pluckthun A. Engineering novel binding proteins from nonimmunoglobulin domains. *Nat Biotechnol*. 2005;23:1257–1268.

Bjorhall K, Miliotis T, Davidsson P. Comparison of different depletion strategies for improved resolution in proteomic analysis of human serum samples. *Proteomics*. 2005;5:307–317.

Brittain SM, Ficarro SB, Brock A, Peters EC. Enrichment and analysis of peptide subsets using fluorous affinity tags and mass spectrometry. *Nat Biotechnol*. 2005;23:463–468.

Burton DR. Immunoglobulins G: functional sites. *Mol Immunol*. 1985;22:161–206.

Burton N, Cavallini B, Kanno M, Moncollin V, Egly JM. Expression in Escherichia coli: Purification and properties of the yeast general transcription factor TFIID. *Protein Expr Purif*. 1991;2:432–441.

Butterfield DA, Perluigi M, Sultana R. Oxidative stress in Alzheimer's disease brain: New insights from redox proteomics. *Eur J Pharmacol*. 2006;545:39–50.

Butterfield DA, Owen JB. Lectin-affinity chromatography brain glycoproteomics and Alzheimer disease: Insights into protein alterations consistent with the pathology and progression of this dementing disorder. *Proteomics Clin Appl*. 2011;5:50–56.

Cantin GT, Shock TR, Park SK, Madhani HD, Yates JR. Optimizing TiO2-based phosphopeptide enrichment for automated multidimensional liquid chromatography coupled to tandem mass spectrometry. *Anal Chem*. 2007;79:4666–4673.

Cellar NA, Karnoup AS, Albers DR, Langhorst ML, Young SA. Immunodepletion of high abundance proteins coupled on-line with reversed-phase liquid chromatography: A two-dimensional LC sample enrichment and fractionation technique for mammalian proteomics. *J Chromatogr B*. 2009;877:879–885.

Clifton J, Huang F, Gaso-Sokac D, Brilliant K, Hixson D, Josic D. Use of proteomics for validation of the isolation process of clotting factor IX from human plasma. *J Proteomics*. 2010;73:678–688.

Collins MO, Yu L, Husi H, Blackstock WP, Choudhary JS, Grant SG. Robust enrichment of phosphorylated species in complex mixtures by sequential protein and peptide metal-affinity chromatography and analysis by tandem mass spectrometry. *Sci STKE*. 2005;2005:pl6.

Comunale MA, Wang M, Hafner J, et al. Identification and development of fucosylated glycoproteins as biomarkers of primary hepatocellular carcinoma. *J Proteome Res*. 2009;8:595–602.

Crumeyrolle-Arias M, Buneeva O, Zgoda V, et al. Isatin binding proteins in rat brain: In situ imaging, quantitative characterization of specific [3H]isatin dinding, and proteomic profiling. *J Neurosci Res*. 2009;87:2763–2772.

Cumova J, Jedličková L, Potěšil D, et al. Comparative plasma proteomic analysis of patients with multiple myeloma treated with bortezomib-based regimens. *Klin Onkol*. 2012;25:17–25.

Cussenot O. Growth factors and prostatic tumors. *Ann Endocrinol (Paris)*. 1997;58:370–380.

Dalle-Donne I, Giustarini D, Colombo R, Rossi R, Milzani A. Protein carbonylation in human diseases. *Trends Mol Med*. 2003;9:169–176.

Davies SP, Reddy H, Caivano M, Cohen P. Specificity and mechanism of action of some commonly used protein kinase inhibitors. *Biochem J*. 2000;351:95–105.

Dayarathna MK, Hancock WS, Hincapie M. A two step fractionation approach for plasma proteomics using immunodepletion of abundant proteins and multi-lectin affinity chromatography: Application to the analysis of obesity, diabetes, and hypertension diseases. *J Sep Sci*. 2008;31:1156–1166.

Di Girolamo F, Righetti PG. "Proteomineering" serum biomarkers: A study in blue. *Electrophoresis*. 2011;32:3638–3644.

Drake RR, Schwegler EE, Malik G, et al. Lectin capture strategies combined with mass spectrometry for the discovery of serum glycoprotein biomarkers. *Mol Cell Proteomics*. 2006;5:1957–1967.

Dunlap C, Carr P. The effect of mobile phase on protein retention and recovery using carboxymethyl-coated zirconia stationary phases. *J Liq Chromatogr*. 1996;19:2059–2076.

Durham M, Regnier FE. Targeted glycoproteomics: Serial lectin affinity chromatography in the selection of O-glycosylation sites on proteins from the human blood proteome. *J Chromatogr A*. 2006;1132:165–173.

Edelmann LJ. Strong cation exchange chromatography in analysis of post-translational modifications: Innovations and perspectives. *J Biomed Biotechnol*. 2011;2011:936508.

Egly JM, Miyamoto NG, Moncollin V, Chambon P. Is actin a transcription initiation factor for RNA polymerase B? *EMBO J*. 1984;3:2363–2371.

El Rassi Z, Horvath C. Metal chelate-interaction chromatography of proteins with iminodiacetic acid-bonded stationary phases on silica support. *J Chromatogr*. 1986;359:241–253.

El Khoury G, Rowe LA, Lowe CR. Biomimetic affinity ligands for immunoglobulins based on the multicomponent Ugi reaction. *Methods Mol Biol*. 2012;800:57–74.

Eliasson M, Olsson A, Palmcrantz E, et al. Chimeric IgG-binding receptors engineered from staphylococcal protein A and streptococcal protein G. *J Biol Chem*. 1988;263:4323–4327.

Ernoult E, Bourreau A, Gamelin E, Guette C. A proteomic approach for plasma biomarker discovery with iTRAQ labelling and OffGel fractionation. *J Biomed Biotechnol*. 2010;2010:927917.

Fonslow BR, Niessen SM, Singh M, et al. Single-step inline hydroxyapatite enrichment facilitates identification and quantitation of phosphopeptides from mass-limited proteomes with MudPIT. *J Proteome Res*. 2012;11:2697–2709.

Fountoulakis M, Takács B. Design of protein purification pathways: Application to the proteome of Haemophilus influenzae using heparin chromatography. *Protein Expr Purif*. 1998;14:113–119.

Fu Q, Garnham CP, Elliott ST, Bovenkamp DE, Van Eyk JE. A robust, streamlined, and reproducible method for proteomic analysis of serum by delipidation, albumin and IgG depletion, and two-dimensional gel electrophoresis. *Proteomics*. 2005;5:2656–2664.

Gould BJ, Hall PM, Cook JG. A sensitive method for the measurement of glycosylated plasma proteins using affinity chromatography. *Ann Clin Biochem*. 1984;21:16–21.

Govorukhina NI, Reijmers TH, Nyangoma SO, van der Zee AG, Jansen RC, Bischoff R. Analysis of human serum by liquid chromatography-mass spectrometry: Improved sample preparation and data analysis. *J Chromatogr A*. 2006;1120:142–150.

Granger J, Siddiqui J, Copeland S, Remick D. Albumin depletion of human plasma also removes low abundance proteins including the cytokines. *Proteomics*. 2005;5:4713–4718.

Gronborg M, Kristiansen TZ, Stensballe A, Andersen JS. A mass spectrometry-based proteomic approach for identification of serine/threonine-phosphorylated proteins by enrichment with phospho-specific antibodies: Identification of a novel protein, Frigg, as a protein kinase A substrate. *Mol Cell Proteomics*. 2002;1:517–527.

Grönwall C, Sjöberg A, Ramström M, et al. Affibody-mediated transferrin depletion for proteomics applications. *Biotechnol J*. 2007;2:1389–1398.

Gupta G, Lowe CR. An artificial receptor for glycoproteins. *J Mol Recognit*. 2004;17:218–235.

Hagiwara T, Saito Y, Nakamura Y, Tomonaga T, Murakami Y, Kondo T. Combined use of a solid-phase hexapeptide ligand library with liquid chromatography and two-dimensional difference gel electrophoresis for intact plasma proteomics. *Int J Proteomics*. 2011;2011:1–11.

Hale JE, Beidle DE. Purification of humanized murine and murine monoclonal antibodies using immobilized metal-affinity chromatography. *Anal Biochem*. 1994;222:29–33.

Ham BM. *Proteomics of Biological Systems: Protein Phosphorylation Using Mass Spectrometry Techniques*. Hoboken, NJ: Wiley; 2012:1–350.

Hershko A. The ubiquitin system for protein degradation and some of its roles in the control of the cell-division cycle (Nobel lecture). *Angew Chem Int Ed Engl*. 2005;44:5932–5943.

Hinerfeld D, Innamorati D, Pirro J, Tam SW. Serum/Plasma depletion with chicken immunoglobulin Y antibodies for proteomic analysis from multiple Mammalian species. *J Biomol Tech*. 2004;15:184–190.

Hjerpe R, Aillet F, Lopitz-Otsoa F, Lang V, England P, Rodriguez MS. Efficient protection and isolation of ubiquitylated proteins using tandem ubiquitin-binding entities. *EMBO Rep*. 2009;10:1250–1258.

Hofsteenge J, Muller DR, Debeer T. New-type of linkage between a carbohydrate and a protein C-glycosylation of a specific tryptophan residue in human. RNase Us. *Biochemistry*. 1994;33:13524–13530.

Huang L, Harvie G, Feitelson JS, et al. Immunoaffinity separation of plasma proteins by IgY microbeads: Meeting the needs of proteomic sample preparation and analysis. *Proteomics*. 2005;5:3314–3328.

Helenius A, Aebi M. Roles of N-linked glycans in the endoplasmic reticulum. *Annu Rev Biochem*. 2004;73:1019–1049.

Hnasko R, Ben-Jonathan N. Prolactin regulation by heparin-binding growth factors expressed in mouse pituitary cell lines. *Endocrine*. 2003;20:35–44.

Hwang HJ, Quinn T, Zhang J. Identification of glycoproteins in human cerebrospinal fluid. *Methods Mol Biol*. 2009;566:263–276.

Jmeian Y, El Rassi Y. Liquid-phase-based separation systems for depletion, prefractionation and enrichment of proteins in biological fluids for in-depth proteomics analysis. *Electrophoresis*. 2009a;30:249–261.

Jmeian Y, El Rassi Z. Multicolumn separation platform for simultaneous depletion and prefractionation prior to 2-DE for facilitating in-depth serum proteomics profiling. *J Proteome Res*. 2009b;8:4592–4603.

Jonasson T, Ohlin AK, Gottsater A, Hultberg B, Ohlin H. Plasma homocysteine and markers for oxidative stress and inflammation in patients with coronary artery disease—A prospective randomized study of vitamin supplementation. *Clin Chem Lab Med*. 2005;43:628–634.

Juhasz P, Lynch M, Sethuraman M, et al. Semi-targeted plasma proteomics discovery workflow utilizing two-stage protein depletion and off-line LC-MALDI MS/MS. *J Proteome Res*. 2011;10:34–45.

Kagedal L. Metal chelate chromatography. In: *Protein Purification: Principles, High Resolution Methods, and Applications*. New York: VCH Press; 1989:227.

Kirkland PA, Humbard MA, Daniels CJ, Maupin-Furlow JA. Shotgun proteomics of the haloarchaeon Haloferax volcanii. *J Proteome Res.* 2008;7:5033–5039.

Krapfenbauer K, Fountoulakis M. Improved enrichment and proteomic analysis of brain proteins with signaling function by heparin chromatography. *Methods Mol Biol.* 2009;566:165–180.

Krief G, Deutsch O, Zaks B, Wong DT, Aframian DJ, Palmon A. Comparison of diverse affinity based high-abundance protein depletion strategies for improved bio-marker discovery in oral fluids. *J Proteomics.* 2012;75:4165–4175.

Kullolli M, Hancock WS, Hincapie M. Automated platform for fractionation of human plasma glycoproteome in clinical proteomics. *Anal Chem.* 2010;82:115–120.

Kuno A, Kato Y, Matsuda A, et al. Focused differential glycan analysis with the platform antibody-assisted lectin profiling for glycan-related biomarker verification. *Mol Cell Proteomics.* 2009;8:99–108.

Kweon HK, Håkansson K. Selective zirconium dioxide-based enrichment of phosphorylated peptides for mass spectrometric analysis. *Anal Chem.* 2006;78:1743–1749.

Léger T, Lavigne D, Le Caër JP, et al. Solid-phase hexapeptide ligand libraries open up new perspectives in the discovery of biomarkers in human plasma. *Clin Chim Acta.* 2011;412:740–747.

Lei T, He QY, Wang YL, Si LS, Chiu JF. Heparin chromatography to deplete high-abundance proteins for serum proteomics. *Clin Chim Acta.* 2008;388:173–178.

Le Moan N, Guerrier L, Tacnet F, Toledano M. Personal communication. Gordon Research conference on "oxidative stress and disease". Ventura (CA, USA), 2007.

Levin Y, Jaros JA, Schwarz E, Bahn S. Multidimensional protein fractionation of blood proteins coupled to data-independent nanoLC-MS/MS analysis. *J Proteomics.* 2010;73:689–695.

Li S, Dass C. Iron(III)-immobilized metal ion affinity chromatography and mass spectrometry for the purification and characterization of synthetic phosphopeptides. *Anal Biochem.* 1999;270:9–14.

Lin B, White JT, Wu J, et al. Deep depletion of abundant serum proteins reveals low-abundant proteins as potential biomarkers for human ovarian cancer. *Proteomics Clin Appl.* 2009;3:853–861.

Lionneton F, Drobecq H, Soncin F. Expression and purification of recombinant mouse Ets-1 transcription factor. *Protein Expr Purif.* 2001;21:492–499.

Liu XC, Scouten WH. New ligands for boronate affinity chromatography. *J Chromatogr.* 1994;687:61–69.

Ma Z, Ramakrishna S. Electrospun regenerated cellulose nanofiber affinity membrane functionalized with Protein A/G for IgG purification. *J Membr Sci.* 2008;319:23–28.

Mahn A, Lienqueo ME, Salgado JC. Methods of calculating protein hydrophobicity and their application in developing correlations to predict hydrophobic interaction chromatography retention. *J Chromatogr A.* 2009;1216:1838–1844.

Mahn A, Reyes A, Zamorano M, Cifuentes W, Ismail M. Depletion of highly abundant proteins in blood plasma by hydrophobic interaction chromatography for proteomic analysis. *J Chromatogr B.* 2010;878:1038–1044.

Mamone G, Picariello G, Ferranti P, Addeo F. Hydroxyapatite affinity chromatography for the highly selective enrichment of mono- and multi-phosphorylated peptides in phosphoproteome analysis. *Proteomics.* 2010;10:380–393.

Mann M, Ong SE, Gronborg M, Steen H, Jensen ON, Pandey A. Analysis of protein phosphorylation using mass spectrometry: deciphering the phosphoproteome. *Trends Biotechnol.* 2002;20:261–268.

Manning G, Whyte DB, Martinez R, Hunter T, Sudarsanam S. The protein kinase complement of the human genome. *Science.* 2002;298:1912–1934.

Martosella J, Zolotarjova N, Liu H, Nicol G, Boyes BE. Immunoaffinity depletion of 20 high-abundance human plasma proteins. Removal of approximately 97% of total plasma protein improves identification of low abundance proteins. *Origins.* 2005;21:17–23.

Mayor T, Deshaies RJ. Two-step affinity purification of multiubiquitylated proteins from Saccharomyces cerevisiae. *Methods Enzymol.* 2005;399:385–392.

Medvedev A, Kopylov A, Buneeva O, Zgoda V, Archakov A. Affinity-based proteomic profiling: Problems and achievements. *Proteomics.* 2012;12:621–637.

Millioni R, Tolin S, Puricelli L, et al. High abundance proteins depletion vs low abundance proteins enrichment: Comparison of methods to reduce the plasma proteome complexity. *PLoS One.* 2011;6:e19603.

Mortezai N, Harder S, Schnabe C, et al. Tandem affinity depletion: A combination of affinity fractionation and immunoaffinity depletion allows the detection of low-abundance components in the complex proteomes of body fluids. *J Proteome Res.* 2010;9:6126–6134.

Murphy VE, Johnson RF, Wang YC, et al. Proteomic study of plasma proteins in pregnant women with asthma. *Respirology.* 2006;11:41–48.

Nawrocki J, Dunlap C, Lic J, et al. Chromatography using ultra-stable metal oxide-based stationary phases for HPLC. *J Chromatogr A.* 2004;1028:31–62.

Nelson CA, Szczech JR, Dooley CJ, et al. Effective enrichment and mass spectrometry analysis of phosphopeptides using mesoporous metal oxide nanomaterials. *Anal Chem.* 2010;82:7193–7201.

Ni RJ, Shen Z, Yang CP, Wu YD, Bi YD, Wang BC. Identification of low abundance polyA-binding proteins in Arabidopsis chloroplast using polyA-affinity column. *Mol Biol Rep.* 2010;37:637–641.

Nuhse TS, Bottrill AR, Jones AM, Peck SC. Quantitative phosphoproteomic analysis of plasma membrane proteins reveals regulatory mechanisms of plant innate immune responses. *Plant J.* 2007;51:931–940.

O'Brien T, Buithieu J, Nguyen TT, et al. Separation of high-density lipoproteins into apolipoprotein E-poor and apolipoprotein E-rich subfractions by fast protein liquid chromatography using a heparin affinity column. *J Chromatogr.* 1993;613:239–246.

Oda Y, Nagasu T, Chait BT. Enrichment analysis of phosphorylated proteins as a tool for probing the phosphoproteome. *Nat Biotechnol.* 2001;19:379–382.

Olajos M, Szekrenyes A, Hajos P, Gjerde DT, Guttman A. Boronic acid lectin affinity chromatography (BLAC). 3. Temperature dependence of glycoprotein isolation and enrichment. *Anal Bioanal Chem.* 2010;397:2401–2407.

Olsson A, Eliasson M, Guss B, et al. Structure and evolution of the repetitive gene encoding streptococcal protein G. *Eur J Biochem.* 1987;168:319–324.

Peng J. Evaluation of proteomic strategies for analyzing ubiquitinated proteins. *BMB Rep.* 2008;41:177–183.

Peng J, Schwartz D, Elias JE, et al. A proteomics approach to understanding protein ubiquitination. *Nat Biotechnol.* 2003;21:921–926.

Pieper R, Su Q, Gatlin CL, Huang ST, Anderson NL, Steiner S. Multi-component immunoaffinity subtraction chromatography: An innovative step towards a comprehensive survey of the human plasma proteome. *Proteomics.* 2003;3:422–432.

Porath J, Olin B. Immobilized metal ion affinity adsorption and immobilized metal ion affinity chromatography of biomaterials: Serum protein affinities for gel immobilized iron and nickel ions. *Biochemistry.* 1983;22:1621–1630.

Posewitz MC, Tempst P. Immobilized gallium(III) affinity chromatography of phosphopeptides. *Anal Chem.* 1999;71:2883–2892.

Qian WJ, Kaleta DT, Petritis BO, et al. Enhanced detection of low abundance human plasma proteins using a tandem IgY12-SuperMix immunoaffinity separation strategy. *Mol Cell Proteomics.* 2008;7:1963–1973.

Ramstrom M, Zuberovic A, Gronwall C, Hanrieder J, Bergquist J, Hober S. Development of affinity columns for the removal of high-abundance proteins in cerebrospinal fluid. *Biotechnol Appl Biochem.* 2009;52:159–166.

Rho JH, Roehrl MH, Wang JY. Tissue proteomics reveals differential and compartment-specific expression of the homologs transgelin and transgelin-2 in lung adenocarcinoma and its stroma. *J Proteome Res.* 2009;8:5610–5618.

Righetti PG, Boschetti E, Citterio A. Protein Equalizer Technology: The quest for a "democratic proteome". *Proteomics.* 2006;6:3980–3992.

Roche S, Tiers L, Provansal M, et al. Depletion of one, six, twelve or twenty major blood proteins before proteomic analysis: The more the better. *J Proteomics.* 2009;72:945–951.

Roque AC, Gupta G, Lowe CR. Design, synthesis, and screening of biomimetic ligands for affinity chromatography. *Methods Mol Biol.* 2005a;310:43–62.

Roque AC, Lowe CR. Lessons from nature: On the molecular recognition elements of the phosphoprotein binding-domains. *Biotechnol Bioeng.* 2005b;91:546–555.

Rosenfeld R, Bangio H, Gerwig GJ, et al. A lectin array-based methodology for the analysis of protein glycosylation. *J Biochem Biophys Methods.* 2007;70:415–426.

Rush J, Moritz A, Lee KA, Guo A. Immunoaffinity profiling of tyrosine phosphorylation in cancer cells. *Nat Biotechnol.* 2005;23:94–101.

Scholten A, Poh MK, van Veen TA, van Breukelen B, Vos MA, Heck AJ. Analysis of the cGMP/cAMP interactome using a chemical proteomics approach in mammalian heart tissue validates sphingosine kinase type 1-interacting protein as a genuine and highly abundant AKAP. *J Proteome Res.* 2006;5:1435–1447.

Schulze WX. Proteomics approaches to understand protein phosphorylation in pathway modulation. *Curr Opin Plant Biol.* 2010;13:280–287.

Seam N, Gonzales DA, Kern SJ, Hortin GL, Hoehn GT, Suffredini AF. Quality control of serum albumin depletion for proteomic analysis. *Clin Chem.* 2007;53:1915–1920.

Selvaraju S, El Rassi Z. Liquid-phase-based separation systems for depletion, prefractionation and enrichment of proteins in biological fluids and matrices for in-depth proteomics analysis—An update covering the period 2008–2011. *Electrophoresis.* 2012;33:74–88.

Shen Y, Kim J, Strittmatter EF, et al. Characterization of the human blood plasma proteome. *Proteomics.* 2005;5:4034–4045.

Shi Q, Gibson GE. Oxidative stress and transcriptional regulation in Alzheimer disease. *Alzheimer Dis Assoc Disord.* 2007;21:276–291.

Shimada Y, Fukuda T, Aoki K, et al. A protocol for immunoaffinity separation of the accumulated ubiquitin-protein conjugates solubilized with sodium dodecyl sulfate. *Anal Biochem.* 2008;377:77–82.

Shores KS, Knapp DR. Assessment approach for evaluating high abundance protein depletion methods for cerebrospinal fluid (CSF) proteomic analysis. *J Prot Res.* 2007;6:3739–3751.

Singhal RP, Ramamurthy B, Govindraj N, Sarvar Y. New ligands for boronate affinity chromatography: Synthesis and properties. *J Chromatogr.* 1991;543:17–38.

Stadtman ER, Levine RL. Chemical modification of proteins by reactive oxygen species. In: Dalle–Donne I, Scaloni A, Butterfield DA, eds. *Redox Proteomics: From Protein Modifications to Cellular Dysfunction and Diseases.* Hoboken, NJ: John Wiley & Sons; 2006:3–23.

Steel LF, Trotter MG, Nakajima PB, Mattu TS, Gonye G, Block T. Efficient and specific removal of albumin from human serum samples. *Mol Cell Proteomics.* 2003;2:262–270.

Stone MD, Chen X, McGowan T, et al. Large-scale phosphoproteomics analysis of whole saliva reveals a distinct phosphorylation pattern. *J Proteome Res.* 2011;10:1728–1736.

Subramanian S. Dye-ligand affinity chromatography: The interaction of Cibacron Blue F3GA with proteins and enzymes. *CRC Crit Rev Biochem.* 1984;16:169–205.

Sultana R, Perluigi M, Newman SF, et al. Redox proteomic analysis of carbonylated brain proteins in mild cognitive impairment and early Alzheimer's disease. *Antioxid Redox Signal.* 2010;12:325–336.

Sun X, Chiu JF, He QY. Application of immobilized metal affinity chromatography in proteomics. *Expert Rev Proteomics.* 2005;2:649–657.

Sun X, Chiu JF, He QY. Fractionation of proteins by immobilized metal affinity chromatography. *Methods Mol Biol.* 2008;424:205–212.

Thingholm TE, Jensen ON, Larsen MR. Analytical strategies for phosphoproteomics. *Proteomics.* 2009;9:1451–1468.

Tian Y, Zhang H. Glycoproteomics and clinical applications. *Proteomics Clin Appl.* 2010;4:124–132.

Tu C, Rudnick PA, Martinez MY, et al. Depletion of abundant plasma proteins and limitations of plasma proteomics. *J Proteome Res.* 2010;9:4982–4991.

Ueda K, Katagiri T, Shimada T, et al. Comparative profiling of serum glycoproteome by sequential purification of glycoproteins and 2-nitrobenzenesulfenyl (NBS) stable isotope labeling: A new approach for the novel biomarker discovery for cancer. *J Proteome Res.* 2007;6:3475–3483.

Vasudev NS, Ferguson RE, Cairns DA, Stanley AJ, Selby PJ, Banks RE. Serum biomarker discovery in renal cancer using 2-DE and prefractionation by immunodepletion and isoelectric focusing; increasing coverage or more of the same? *Proteomics.* 2008;8:5074–5085.

Vestergaard M, Tamiya E. A rapid sample pretreatment protocol: Improved sensitivity in the detection of a low-abundant serum biomarker for prostate cancer. *Anal Sci.* 2007;23:1443–1446.

Voute N, Bataille D, Girot P, Boschetti E. Characterization of very dense mineral oxide-gel composites for fluidized-bed adsorption of biomolecules. *Bioseparation.* 1998;8:121–129.

Voute N, Fortis F, Guerrier L, Girot P. Performance evaluation of zirconium oxide based adsorbents for the fluidized bed capture of Mab. *Int. J. Biochromatogr.* 2000;5:49–65.

Walsh CT, Garneau-Tsodikova S, Gatto GJ. Protein post-translational modifications: The chemistry of proteome diversifications. *Angew Chem Int Ed Engl.* 2005;44:7342–7372.

Warder SE, Tucker LA, Strelitze TJ, et al. Reducing agent-mediated precipitation of high-abundance plasma proteins. *Anal Biochem.* 2009;387:184–193.

Wells L, Vosseller K, Hart GW. Glycosylation of nucleocytoplasmic proteins: Signal transduction and O-GlcNAc. *Science.* 2001;291:2376–2378.

Whiteaker JR, Zhang H, Eng JK, et al. Head-to-head comparison of serum fractionation techniques. *J Proteome Res.* 2007;6:828–836.

Wisniewski JR. Tools for phospho- and glycoproteomics of plasma membranes. *Amino Acids.* 2010;41:223–233.

Wissing J, Godl K, Brehmer D, et al. Chemical proteomic analysis reveals alternative modes of action for pyrido[2,3-d]pyrimidine kinase inhibitors. *Mol Cell Proteomics.* 2004;3:1181–1193.

Wissing J, Jänsch L, Nimtz M, et al. Proteomics analysis of protein kinases by target class selective prefractionation and tandem mass spectrometry. *Mol Cell Proteomics.* 2007;6:537–547.

Xu Y, Wang BC, Zhu YX. Identification of proteins expressed at extremely low level in Arabidopsis leaves. *Biochem Biophys Res Commun.* 2007;358:808–812.

Yadav AK, Bhardwaj G, Basak T, et al. A systematic analysis of eluted fraction of plasma post immunoaffinity depletion: Implications in biomarker discovery. *PLoS One.* 2011;6:e24442.

Yang F, Yin Y, Wang F, Zhang L, Wang Y, Sun S. An altered pattern of liver apolipoprotein A-I isoforms is implicated in male chronic hepatitis B progression. *J Proteome Res.* 2010;9:134–143.

Yang Z, Harri LE, Palmer-Toy DE, Hancock WS. Multilectin affinity chromatography for characterization of multiple glycoprotein biomarker candidates in serum from breast cancer patients. *Clin Chem.* 2006;52:1897–1905.

Zhang S, Sun Y. Further studies on the contribution of electrostatic and hydrophobic interactions to protein adsorption on dye-ligand adsorbents. *Biotechnol Bioeng.* 2001;75:710–717.

Zhang H, Zha X, Tan Y, Hornbeck PV. Phosphoprotein analysis using antibodies broadly reactive against phosphorylated motifs. *J Biol Chem.* 2002;277:39379–39387.

Zhao J, Simeone DM, Heidt D, Anderson MA, Lubman DM. Comparative serum glycoproteomics using lectin selected sialic acid glycoproteins with mass spectrometric analysis: Application to pancreatic cancer serum. *J Proteome Res.* 2006;5:1792–1802.

Zheng H, Hu P, Quinn DF, Wang YK. Phosphotyrosine proteomic study of interferon alpha signaling pathway using a combination of immunoprecipitation and immobilized metal affinity chromatography. *Mol Cell Proteomics.* 2005;4:721–730.

Zhi W, Purohit S, Carey C, Wang M, She JX. Proteomic technologies for the discovery of type 1 diabetes biomarkers. *J Diabetes Sci Technol.* 2010;4:993–1002.

Zhou M, Lucas DA, Chan KC, et al. An investigation into the human serum "interactome". *Electrophoresis.* 2004;25:1289–1298.

Zhou H, Tian R, Ye M, et al. Highly specific enrichment of phosphopeptides by zirconium dioxide nanoparticles for phosphoproteome analysis. *Electrophoresis.* 2007;28:2201–2215.

Zolotarjova N, Martosella J, Nicol G, Bailey J, Boyes BE, Barrett WC. Differences among techniques for high-abundant protein depletion. *Proteomics.* 2005;5:3304–3313.

partially by the lack of chirality of the resulting molecules. Nonetheless, this approach represented a sort of breakthrough for the technology regarding drug synthesis for two reasons: (i) it was possible to synthesize thousands of diversomers in a limited amount of time for an extensive drug screening and (ii) it represented a sort of revolution in the approach to standardize chemical synthesis. In contrast to finding methods to prepare structures corresponding to a given predetermined model, the objective was to almost blindly create populations of molecules of analogous structures for a full screening against a single disease or multiple targets.

One popular chemical approach is to graft various substituents in different positions of the same molecular scaffold. Although the approach is simple, it requires the help of relatively sophisticated technologies because the grafting chemical reaction needs to be performed at a very high yield; it also needs to be extremely reliable while grafting substituents to the right positioning of the scaffold (see the examples in Figure 4.1). The chemistry as it is known today offers a large number of possibilities. However, since it is out of the question to make billions of molecules that would need to be tested one by one, a preliminary selection of initial scaffold has to be made. This selection is based on known basic properties so as to address a single or a few targets where already known drug structures are used. It is within this context that benzodiazepine libraries could be built (Bunin & Ellman, 1992; Herpin et al., 2000) as well as libraries of hydantoins (Boeijen et al., 1998), tetrahydroisoquinolinone (Lamb et al., 2001), peptoids (Kwon & Kodadek, 2008), and a number of others.

As an example of success, small molecules from combinatorial libraries were used to search structures with cytotoxic effect on cancer cells with strong consensus structure. The determination was performed under ultra-high-throughput and could be exploited for cell or biochemical assays (Townsend et al., 2010).

Combinatorial chemistry approaches therefore contribute to the drug discovery process. From a few dozen compounds of the primary concepts, the number can rise to millions of diversomers, forming very large libraries from which the most effective drug candidates can be retrieved. Focused

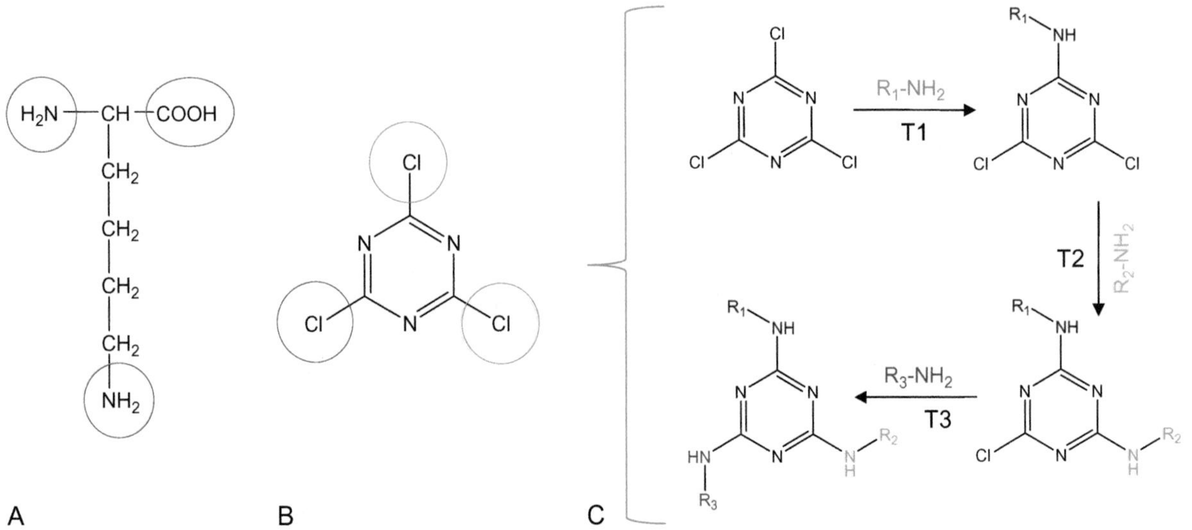

FIGURE 4.1
Examples of scaffolds for chemical combinatorial synthesis. "A" represents a molecule of lysine where there are three points for chemical diversity. The groups are generally protected and deprotected selectively prior to a grafting reaction. This method guarantees the oriented positioning. "B" represents a trichloro-s-triazine molecule where the points of diversity are also three. They do not need to be selectively protected since their reactivity depends on temperature. For instance, the blue circle is the positioning that reacts at low temperature, the green circle indicates a reactivity at intermediate temperatures, and the violet circle is the positioning that reacts only at high temperatures. On the right "C" is illustrated an example of the preparation of a diversomer with three different chemical substitutions on the trichloro-s-triazine molecule. Considering that the number of primary amines available to be grafted on trichloro-triazine molecule is very large, the number of combinations is also very large.

combinatorial libraries are also prepared with the objective of discovering active chemicals for predefined organs and/or disorders such as those of the central nervous system (Zajdel et al., 2009).

It should also be mentioned that combinatorial chemistry was and still is used for preparing ligands for affinity chromatography of proteins. Among the known approaches the one that deserves attention because of the proximity to what is described below is the preparation of so-called combinatorial biomimetic affinity ligands. This is the way to design and make a large number of chemical molecules by mimicking nature. First, the docking site of a protein with its natural partner (e.g., insulin and its natural receptor or protein A with the Fc fragment of immunoglobulins G) is analyzed by X-ray crystallography, and the involved interaction regions are identified (Sproule et al., 2000). A complementary theoretical ligand is thereby conceived and then synthesized by using a starting scaffold. Thus a very large number of related compounds could be made. The chemical approach consists in grafting various substituents in different positions of the same molecular scaffold and evaluating the resulting molecules for their ability to interact with the target protein. Two ways are generally adopted: (i) solid-phase synthesis on a trichloro-s-triazine scaffold (this is a two-component reaction) and (ii) solid-phase multicomponent reaction processes producing a large variety of subtle changes. With the second approach (also called the Ugi reaction as described by El Khoury et al., 2012) that uses an aldehyde sorbent derivative and considering that the available molecules involved in the reaction (primary amines, carboxylic acid, and isocyanadies) are extremely large, billions of ligand combinations can be made.

Combinatorial chemistry started with the seminal work of Merrified on peptide synthesis performed by solid state. Later the combinatorial way to get all possible structures from a given number of amino acids by the so-called split-and-mix approach was proposed in 1991 by Furka et al. and Lam in 1997 rendering possible the preparation of millions of peptides with a limited number of synthesis steps. When this type of chemistry strategy is used on a solid phase, the final result is an assembly of beads where each diversomer or peptidomer is attached on distinct beads, thus forming a sort of mixed bed of distinct affinity resins.

Most chemical libraries are constructed by using solid supports. The popular supports adopted are beaded chromatographic resins that can be prepared or modified for the purpose. In a review, Cargill & Lebl, (1977) reported the common approaches for making libraries of compounds. These authors described the synthesis of a variety of compounds on discrete substrates. Block of arrays, beads, teabags, polymeric pins, and polymeric sheets are among the technologies illustrated. The synthesis could be performed as a combinatorial approach in parallel or simultaneously. Parallel synthesis is also described as being applied to chips of microfabricated components.

4.2 COMBINATORIAL PEPTIDE LIBRARIES

Here the combinatorial peptide library refers to a collection of peptides of a given length (the same number of amino acids) obtained from a predefined selection of amino acids that are used in order to compose peptides in all possible combinations. For example, if one started from ten amino acids, the number of tripeptide combination structures obtained would be 1000 (10^3). When the number of amino acids is larger and the length of the final peptide is extended to five or six or more, the number of combinations becomes really huge.

The history of the preparation of peptides commenced with the seminal work of Merrifield in 1965 and 1986 and then the process was adopted by many authors under different forms and refinements. However, the breakthrough in the preparation of combinatorial peptide libraries started with Furka et al. (1988, 1991), who for the first time described a novel principle of synthesis as a way to simplify their preparation. Instead of elongating peptide chains by a sequential addition of amino acids using all possible combinations within a preselected group of amino acids, the process was based on a simple mathematical model that would reduce dramatically the number of synthesis steps and more importantly cut down the number of reaction vessels. The process is described below and is based on the principle of the so-called split, combine, and pool method developed as reaction cycles. The number of cycles defines the number of amino acids composing the peptide chain, the peptide chain being built by solid-phase synthesis on beaded supports. Soluble peptides are then recovered by hydrolysis of the linker between the solid support and the first amino acid constituting the peptide chain. This process was developed concomitantly with the combinatorial chemistry processes for preparing diversomers for pharma screening (see the previous paragraph) and quite rapidly became a standard method in the preparation of all possible combinations of peptides. It was with Lam et al. (1991) that peptide libraries obtained by combinatorial synthesis were used for the first time as a possible source of affinity ligands for separating proteins from complex mixtures.

Amino acids and peptides have long been used in affinity chromatography as possible specific ligands for the separation of single proteins or families of them. Let us recall that very early studies were performed in 1970 by Porath's group (Porath & Fryklund, 1970a; Porath & Fornstedt, 1970b), who reported Sephadex-coupled β-alanine and agarose coupled to arginine for the purification of venom proteins and human plasma components, respectively. As affinity chromatography largely developed over time, bringing a very strong contribution to protein isolation by using a variety of ligands, the number of examples adopting amino acid ligands multiplied in the last two decades. Most, if not all, single amino acids have been described as ligands for affinity purification of proteins; examples are illustrated by proline (Pass et al., 1973), valine (Koerner et al., 1975), lysine (Traas et al., 1984), tyrosine (Akiyama et al., 1985), phenylalanine (Larcher et al., 1982), arginine (Tsuji et al., 1996), threonine (Janecek et al., 1996), and tryptophan (Tsuboi et al., 1998), just to name a few illustrating their propensity to interact with proteins even on a monomeric form. The efficacy of these ligands attached to chromatographic supports may be explained by their chemical structure that comprises more than one interaction site, better known in chromatography under the term *mix-mode ligands*. Instead of using a single amino acid, dipeptides, tripeptides, or longer peptides would display a larger number of interacting sites, thus increasing the scope of mixed-mode ligands and also contributing to better specificity. A number of papers have been published on the use of these structures (see a review by Tozzi & Giraudi, 2006). In this context, a peptide library is considered an interesting collection of an extremely large number of peptide ligands, which becomes the basis for selecting the most appropriate structure for a given affinity chromatography application. This type of application is described in Section 7.1.

4.2.1 Diversity of Combinatorial Peptides

Based on the number of current natural amino acids, the number of possible diversomers is easily calculated according to the peptide length (see Table 4.1). In reality, due to difficulties of using some amino acids and their propensity to produce errors or induce synthesis issues, the number of amino acids used is lower. For instance, cysteine and methionine are discarded from the synthesis, thus reducing the diversity they could generate by their presence in the structure. Another reason for not using cysteine is related to its particular behavior during the interaction with proteins. It can form covalent disulfur bridges with proteins, thus preventing a proper elution from the beads. Any building block capable of forming covalent interactions is naturally discarded from the synthesis. To this reduction one could also remove a few other amino acids that are not very effective in generating interactions with protein epitopes; this is the case of glycine or alanine. However, they may eventually be used as spacers between amino acid residues for a more optimized distance between interactive sites.

In order to understand the influence of each amino acid composing a peptide for the interaction capability with a given protein, specific studies have been carried out. From these studies amino acids have been classified into two groups: *grand catchers* and *petite catchers*, distinguishing, respectively, those that interact with a large number of proteins with respect to the others interacting quite poorly. A direct consequence is that good libraries could be constructed with a limited number of amino acids, comprising most, if not all, *grand catchers* and a few *petite catchers*. This approach determines the number of diversomers from a library that is obtained in the best economical conditions without significantly compromising the property of the library, at least for the purpose of sample treatment with the objective of reducing the dynamic concentration range. The number of diversomers has another important effect: This is the amount of beads needed to cover the entire library. Since the beads have a predetermined size, the number of diversomers is also correlated to the volume of the beads when one wants to use the entire library. On that point a specific section is devoted to the importance of the bead number for both the preparation and the use of the peptide library (e.g., see Sections 4.5.5 and 8.1.1).

In Table 4.1 the number of peptidomers is calculated as a function of the peptide length and of the number of building blocks; only one category of amino acids (L or D) is considered. If both are used, some additional random stereochemistry is introduced, thereby increasing the number of diversomers. In addition to natural amino acids, it is also possible to replace or use non-natural amino acids, in this way amplifying the diversity of the library (see Figure 4.2 for examples of non-natural amino acids structures).

TABLE 4.1 Number of Combinatorial Peptide Ligands Obtainable by Using a Preselected Number of Diverse Amino Acids and According to the Length of the Targeted Peptides

		Length of Peptide Chain					
		Dipeptide	Tripeptide	Tetrapeptide	Pentapeptide	Hexapeptide	Heptapeptide
Number of amino acids for the construction of the library	1	1	1	1	1	1	1
	2	4	8	16	32	64	129
	3	9	27	81	243	729	2187
	4	16	64	256	1024	4096	16384
	5	25	125	625	3125	15625	78125
	6	36	216	1296	7776	46656	279936
	7	49	343	2401	16807	117649	823543
	8	64	512	4096	32768	262144	2097152
	9	81	729	6561	59049	531441	4782969
	10	100	1000	10000	100000	1000000	10000000
	11	121	1331	14641	161051	1771561	19487171
	12	144	1728	20736	248832	2985984	35831808
	13	169	2197	28561	371293	4826809	62748517
	14	196	2744	38416	537824	7529536	105413504
	15	225	3375	50625	759375	11390625	170859375
	16	256	4096	65536	1048576	16777216	268435456
	17	289	4913	83521	1419857	24137569	410338673
	18	324	5832	104976	1889568	34012224	612220032
	19	361	6859	130321	2476099	47045881	893871739
	20	400	8000	160000	3200000	64000000	1280000000

Considering only a theoretical hexapeptide library made with twenty amino acids, the diversity covers situations spanning from a single amino acid represented in the chain (homogeneous peptide) to a situation where each amino acid is different from the five others. Such a peptide library comprises 64 million diversomers where twenty of them are sequences of a same amino acid. The number of hexapeptides comprising all different amino acids within the same chain is much higher when one considers the fact that the chain comprises only 30% of the overall number of building blocks and the fact that their respective positioning within the chain is combinatorial. In between, there are all other possibilities, with two, three, four, or five repeating units of the same amino acid. This situation indicates that very similar structural and functional situations are present as, for instance, the same chain structure where a glycine is replaced by an alanine or a leucine is replaced by an isoleucine. In both cases, the structural and functional differences are minor. The situations of structural similarities are in large numbers, and their possible redundancy helps temper the lack of diversity when

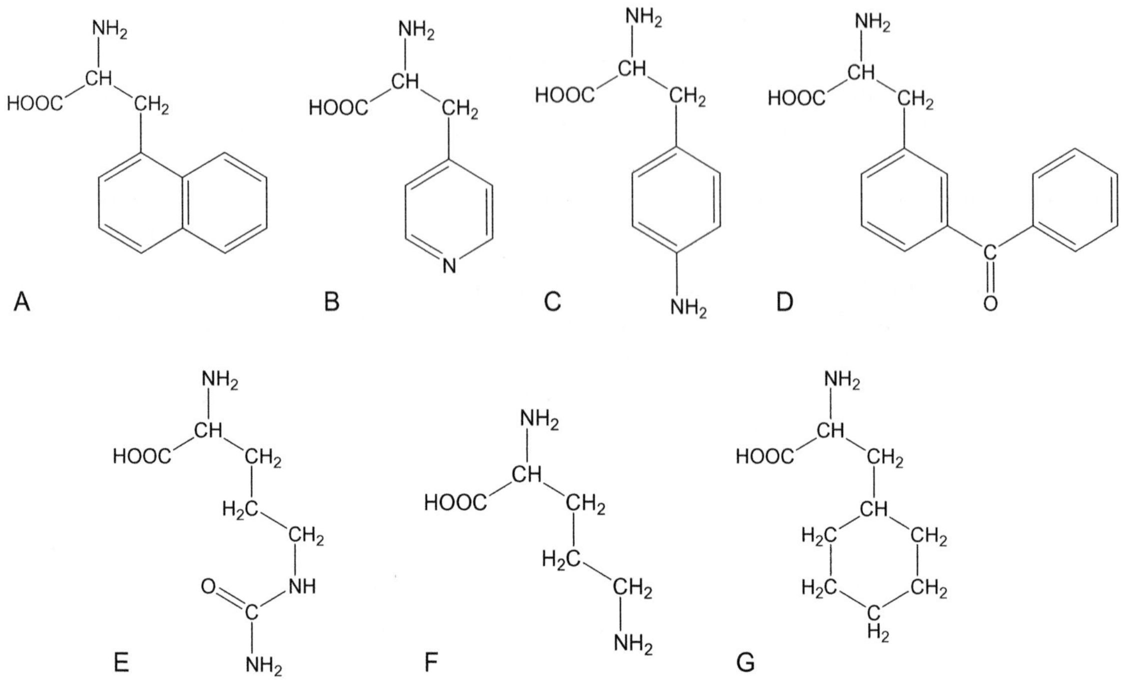

FIGURE 4.2
Examples of non-natural amino acid structures that can be used for the preparation of peptides in addition to natural amino acids. a: naphtyl-alanine; b: 4-pyridyl-alanine; c: 4-aminophenylalanine; d: phenoxyphenylalanine; e: citrulline; f: ornitine; g: cyclohexyl-alanine.

using only small portions of the entire library. Although difficult to formally validate, it has been unambiguously demonstrated that when reducing the volume of a solid-phase library for a given protein sample extract, the final results are very comparable (see Section 4.5.8).

Synthesis errors seem relatively low due to the high yield of peptide synthesis by the introduction of an amino acid at a time. However, errors could be present since even with a 98% grafting yield each step, the overall yield for the split-and-mix synthesis method is around 88% calculated on the basis of the reduced number of synthesis steps to six.

Using current methods of synthesis, each final deprotected peptide of the library has a free primary amine as a distal chemical group. The overall number of primary amines present confers a global cationic character to the library. Nevertheless, this dominant ionic charge can be easily modified by endcapping with various chemical reagents. For instance, acetylation and succinylation modify the dominant ionic charge and as aconsequence the interaction functionality toward the proteins. This aspect will be specifically discussed in Section 4.5.7.

Peptides built on discrete beads through the split-and-recombine procedure can be prepared as a single straight chain, as two parallel chains, as cyclic peptides, or as dendrimers (Maillard et al., 2009). Binary peptidic chains (parallel peptides) can also be prepared starting from cysteine, which is grafted on the solid support using the SH group, thus leaving free the primary amine and the carboxyl groups from which two linear peptide chains can be elongated. On the contrary, dendrimers assembled using amino acids as building blocks are branched peptidic molecules. They are obtained by using di-amino acids every third position so as to form branches. The adopted procedure for library construction while using the same number of amino acids may also engender a different number or shape of peptidomers at the end of the synthesis; therefore different functionality in terms of capability to interact with proteins can be expected. Within the context of protein capture and with the objective of treating the sample in order to reduce the dynamic concentration range, linear peptides are well adapted and relatively easy to make.

4.2.2 Combinatorial Peptoids

Peptoids are oligomers of synthetic molecules having carboxylic and primary amine groups. Peptoids differ from peptides for two different reasons: (i) they have dissimilar side chain structures and (ii) the side chain is attached to the nitrogen atom of the amide group of the backbone. Peptoids can thus be considered chemical entities from N-substituted glycine (Zuckermann et al., 1994; Miller et al., 1995). Long chains can be prepared with a variety of combinatorial properties, depending on the number of building blocks. Figure 4.3 shows the difference between a peptide and a peptoid.

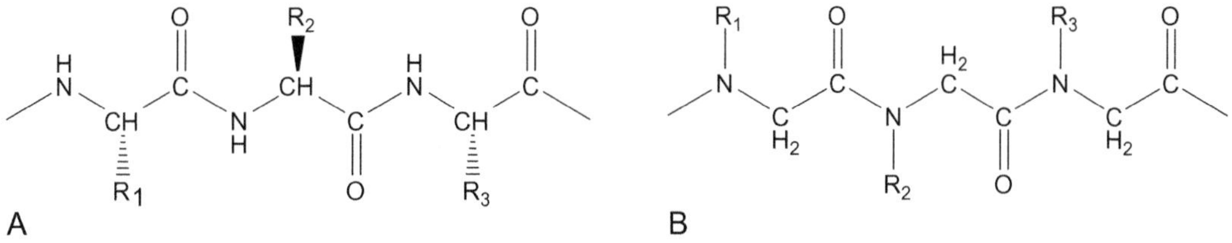

FIGURE 4.3
Schematic representation and structure comparison between a peptide (A) and a peptoid (B). Peptides carry side chains on the carbon of the polymeric backbone; side chains in peptoids are carried by nitrogen atoms since they are composed of N-substituted amines. Another difference is the lack of chirality in peptoids compared to peptides.

A peptoid is generally obtained by first immobilizing a given amino acid and then reacting the free primary amine of the solid phase with bromoacetic acid under nonaqueous conditions and in the presence of a condensation agent. Once this step is completed, the bromine derivative is reacted with a primary amine carrying a different side chain, thereby forming the diversity (see the reaction scheme of Figure 4.4).

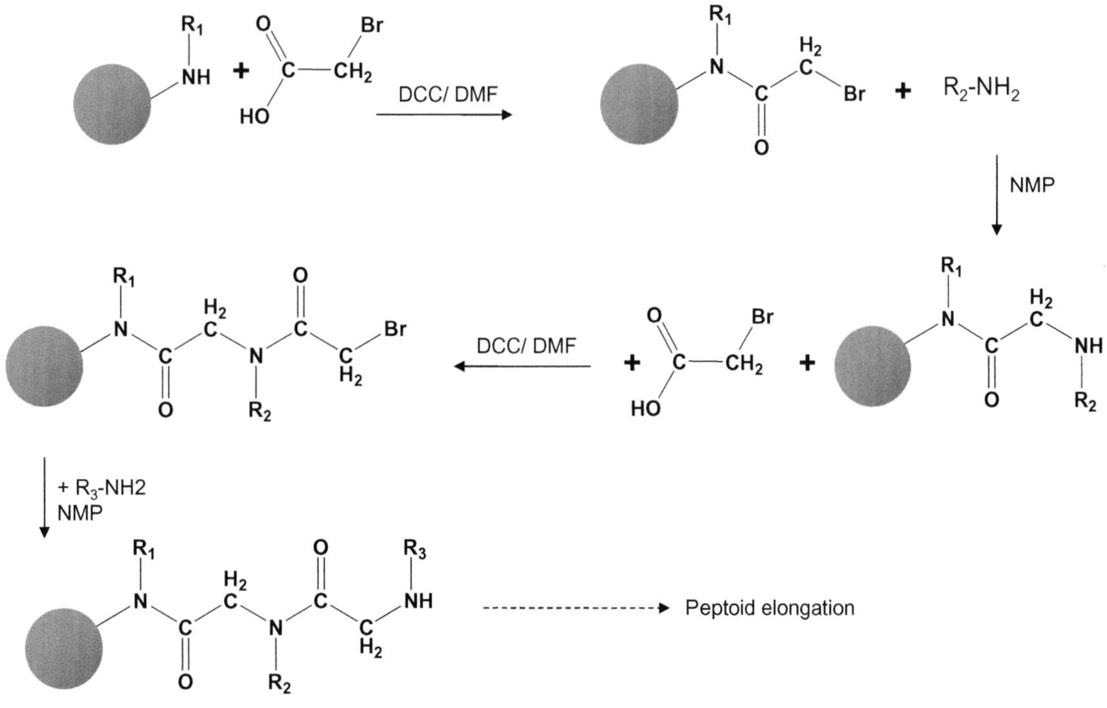

FIGURE 4.4
Schematic representation of reactions for the preparation of a peptoid. The aminated support is reacted first with bromoacetic acid in the presence of dicyclohexyl-carbodiimide in dry dimethylformamide (DMF). Then the brominated derivative is reacted with a primary amine in the presence of N-methylpyridine (NMP). This cycle can be repeated several times to reach the desired length, and the primary amines are selected as a function of the substituent (R_2 and R_3).

Like peptides from *D*-amino acids, peptoids are completely resistant to proteolysis and are therefore interesting for applications in strong proteolytic environments. However, peptoids with their different backbone do not have amide hydrogens involved in the formation of hydrogen bonding within the peptide structure (see Sections 4.4.2 and 4.4.3) and with the partner's polypeptides. Nevertheless, due to the very large number of commercially available amines, the diversity obtainable with peptoids is much larger than that from regular amino acids.

4.3 PREPARATION OF SOLID-PHASE COMBINATORIAL PEPTIDE LIGAND LIBRARIES

Since the proposal of Furka et al. (1988, 1991), several strategies have been designed for the preparation of peptide libraries. This synthesis is obtained by grafting a first amino acid on a solid substrate and then by elongating the peptide by the addition of other amino acids one after the other. Most of the time, the aim of this synthesis is to prepare soluble peptides of a large variety; this implies that once the synthesis is completed, the peptide is released from the solid substrate by the cleavage of the linker. In the technology and related applications described in this book, the peptides are directly used as a solid phase like in affinity chromatography; hence neither special cleavable linkers are used, nor are cleavage methodologies described.

The peptide libraries can be used as they are, as mixed beds, or after having sorted out a selected bead peptide for the separation of one protein or a group of proteins. In this respect, a variety of examples are reported in the literature describing the use of peptide ligands for the purification of proteins (see Samson et al., 1995; Pennington et al., 1996; Lam et al., 1997; Kaufman et al., 2002). This approach was used, for example, to purify human alpha-1-proteinase inhibitor (Bastek et al., 2000), the altered conformation of prion protein (Lathrop et al., 2007), and groups of fusion proteins sharing the same tag directly from a crude extract in the presence of strong denaturing chaotropic agents (Hahn et al., 2010).

Section 7.1 describes the concept and applications of peptide ligands selected from solid-phase libraries for the purification of proteins from crude extracts. The large interest generated by numerous publications witnesses for more than a curiosity approach: the quest for specific ligands for protein isolation is actually a real need in biotechnology. It is in this context that combinatorial peptide ligand libraries on solid chromatography supports have been developed, as described by Lam et al. (1991) and Sebestyen et al. (1995).

Peptide libraries are synthesized directly on chromatographic beaded supports so that the resulting products are libraries of bead ligands in which each bead carries a different structured peptide ligand. From these collections of ligands all mixed together, specific structures can be sorted out for their affinity toward predefined proteins. This approach is very effective for identifying chromatographic ligands as described later in this book (Section 7.1) with a number of peptide ligands used successfully for protein purification. However, even if an ideal peptide is identified, it does not necessarily mean it will interact with the partner protein in solution. Moreover, it may interact with other proteins. Therefore making the peptide first and then grafting it on a solid phase using one of the current immobilization methods may not work well. Actually, the affinity of an immobilized peptide for a given protein is different from the same peptide in solution due to the environmental effects. The oriented exposure of the structure, the ligand density on the solid support, and the nature of the linker are all parameters that modulate the affinity of the peptide for the protein in question.

Overall and in spite of some open questions and possible technical issues, combinations of amino acids under short peptide chains probably represent one of the most powerful sources of affinity ligands for chromatographic separation. This is due to the diversity of amino acid side chains that are composed of different chemical structures, thus displaying various possibilities of interactions. The most common interacting groups are ionic charges (cationic and anionic), hydrophobic aliphatic or aromatic sites, heterocycles, and hydrogen bonding centers. The combination of attractive and repulsive groups modulates the global affinity constant for a protein under predefined physicochemical conditions, with the possibility of a fine-tuned selection as a function of the conditions; interactions could be modulated by mineral cations. The selection of peptide ligands could not only be made for a proper protein capture, but must also be organized with respect to predefined elution conditions. The latter could be achieved, for instance, by an increase of ionic strength, an increase or a decrease of pH, or the intervention of deforming agents. Although the use of the available collection of natural amino acids will generate a large panel of ligand possibilities, the strategy of synthesis and possible use of unnatural amino acids also plays an important role for the diversity and the behavior of resulting peptides. As stated, they could be linear peptides, cyclic peptides, branched peptides, and so on. In this respect, Maillard et al. (2009) have added a number of insights regarding the design of a peptide library, on the one hand, and the strategy of screening, on the other hand. For the purpose of the technology described in this book, two main strategies can be used to prepare peptide libraries: parallel combinatorial synthesis and the split-and-mix approach.

4.3.1 Parallel Peptide synthesis

A number of parallel solid-phase syntheses of peptides have been described in the past several years yielding libraries from a few hundred to hundreds of thousands of compounds. The most popular ways to make parallel synthesis are the so-called reaction block systems. They are composed of arrays of reaction chambers where the synthesis will take place. Reagents are generally introduced at the top of reaction chambers, while the by-products and washing solvents are removed from either the top or the bottom. Operations can be performed manually or automatically.

A number of commercially available devices for parallel synthesis have been introduced by using quite similar approaches with a more or less automation level. Basically, these systems are made up of discrete reaction chambers, like 96-well plates, where the synthesis of diversomers takes place. Parallel peptide synthesis can also be performed by using "tea-bags" (Messeguer et al., 2008) or sets of specifically designed frit-equipped syringes (Krchnak et al., 1996). Alternatives to these approaches are peptide synthesis on polymeric pins or polymeric sheets (Hilpert et al., 2007).

Finally, by using parallel synthesis, microchips could be prepared (Schirwitz et al., 2009). This is performed after having coated the chip with an amino-modified poly(ethylene glycol)methacrylate allowing the covalent attachment of first amino acids. Combinatorial synthesis could thus be implemented to obtain peptide arrays directly on a chip surface. The authors claimed that densities as high as 40,000 peptide spots/cm² could be obtained with minimum labor time to complete the synthesis.

4.3.2 Split-and-Mix Approach

Peptide library synthesis by the split-and-mix method is also known by the term *one-bead-one-compound*. This term means that through this approach it is possible to obtain a mixture of beads where each bead comprises only one type of peptide in billions of chemically grafted copies. In this respect, each bead represents a sort of extremely small-affinity chromatographic bed with a single ligand structure. This method was first described by Furka et al. (1988, 1991) for the preparation of numerous compounds at the time in which the sequencing of amino acids was deciphered by Edman degradation procedure. In 1991, Lam et al. as well as Houghten et al., 1991 similarly described this method of peptide synthesis. The observation was made that, because each bead encounters only one amino acid at a time and the coupling reaction was pushed to completion, each bead carried only one type

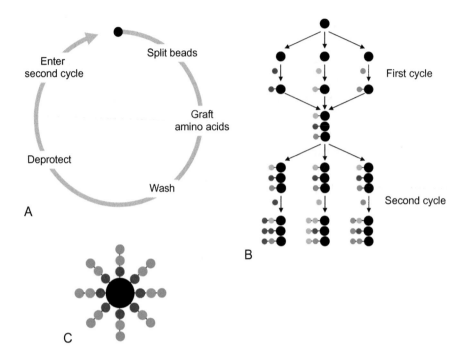

FIGURE 4.5

Schematic representation of the preparation of a combinatorial peptide library. "A" represents one cycle of amino acid grafting on beads (black bowl) followed by a wash and a deprotection before attaching a second amino-protected amino acid in a following cycle. The first cycle (B), being constituted of a series of parallel couplings determined by the number of amino acid used (three in this case: red, green, and blue dots), is followed by a bead blend and split into three equal parts before the attachment of a second amino acid and so on. "C" represents a selected bead from the peptide library of 27 diversomers where there are many tripeptides (eight in this representation) anchored on the solid matrix, but all of them are identical to each other.

of peptide. This fact is better understood while explaining the synthesis principle as exemplified in Figure 4.5. The synthesis is made directly on solid polymeric beads as a solid-phase heterogeneous synthesis. The type of beads is definitely important since its composition must fulfill the requirements of all organic materials used all along the synthesis and washings such as solvents, deprotecting agents, and extreme pHs. In addition, the beads need to comprise primary amines where the first amino acid is attached. From a physical standpoint it is important that the beads do not form aggregates during various synthesis operations and have a similar diameter and hence a similar surface area onto which the synthesis of the polypeptides takes place. Polyacrylate beads have all the attributes to comply with most, if not all, requirements as they are, moreover, swellable in both polar and nonpolar liquids.

Once the number and the type of amino acids involved in the formation of peptides and the length of peptides are defined, a certain amount of beads is taken and aliquots of the same volume are prepared. The number of aliquots must be the same as the number of amino acids selected. The overall volume of the beads must comprise a number of individual beads exceeding by far the number of planned diversomers. For instance, if the goal is to prepare a heptapeptide library using twelve amino acids, the final number of diversomers will be 35.8 millions; thus, the number of initial beads should be at least ten or a hundred times this number. The total volume of beads is calculated on the basis of the bead diameter. This calculation is important because the split-and-mix approach of synthesis is a statistical method of synthesis and its reproducibility is improved as the number of beads is increased. Since each bead should carry a peptide that is different from another bead, 500,000 peptides cannot be synthesized on 100,000 beads.

Each aliquot (in the present case twelve aliquots) is put in a reaction vessel, and the grafting reaction of amino-protected amino acids (one amino acid per vessel) starts. The reaction is pushed to completion, and the excess of reagents is eliminated by extensive washings. Coupled amino acids are then deprotected and all bead aliquots are mixed together thoroughly. At this stage, the library comprises only a single amino acid grafted in each bead. The mixed beads are then split into another twelve aliquots of equal volume (all types of grafted amino acids are represented) put in distinct reaction vessels where a second amino acid is grafted according to the same process. Each aliquot now comprises a mixture of beads with all dipeptides attached. The synthesis cycle is extended to seven coupling steps to achieve the preparation of the heptapeptide library.

Using pure building blocks and pure chemicals, the reactions are generally very close to 100% yield and are very reliable. Nonetheless the obtained peptide library is to be submitted to specific analysis to ensure that its composition corresponds to the initial plan. Since the peptide length is predetermined (seven amino acids in this case), to check that this is consistent with the reality, a mass spectrometry analysis can be performed. By definition. the molar amount of each amino acid involved in the synthesis is the same; this can be experimentally demonstrated by a bulk analysis of amino acids (for instance by HPLC) after total peptide hydrolysis. Another complementary approach could be to use Edman degradation of peptides, and after each step the total diversity of the building blocks is to be found and in comparable amounts. Finally, functional tests can also be performed. Such libraries can be prepared using *D*- or *L*-amino acids; non-natural amino acids can also be used.

The immobilization of the first building block requires a linker to attach amino acids on the solid substrate. Depending on the use of the library, the selection of the linker could be critical. First the linker must withstand the harsh conditions of peptide synthesis and should not bring chemical groups that would induce nonspecific binding of proteins. If the objective is to recover all the library of peptides in solution, the linker must be detachable from the peptide without deteriorating the peptide itself. For stable libraries the use of a six-carbon spacer is common from, for instance, 6-amino hexanoic acid, which is first chemically grafted on a solid substrate carrying free primary amine groups.

The split-and-mix approach for the preparation of peptides has a great advantage over other methods (e.g., parallel synthesis) based on the limited number of distinct reaction chambers. While a parallel synthesis of all combinations of tripeptides from twenty natural amino acids requires 8000 reaction chambers for parallel synthesis, it requires only twenty chambers in the split-and-mix method. The last part of this book, devoted to practical protocols, describes in detail a method for peptide synthesis by the split-and-mix (Section 8.1).

A substantial contribution to this method of preparing a peptide library comes from Lam and coworkers (1991; 1996; 1997; 1998a). This group proposed variations of peptide and peptoid synthesis using all types of amino acids (natural, unnatural). Moreover, they developed a number of applications with a precise focus on the discovery of specific peptides within the libraries capable of interacting with protein partners (Lam et al., 1998b; Liu et al., 2003; Lam et al., 2003). Thus various screening methods have been described (See section 7.1 for more details).

The split-and-mix approach can be modulated in its final functional performance by selecting the amino acids composing the peptides. If, for example, a hydrophobic peptide library is required, amino acids such as isoleucine, valine, leucine, and phenylalanine are selected. In other circumstances, it is possible to make libraries where the diversity of the proximal portion is larger than the diversity of the distal section. To do so, the initial synthesis would involve a larger number of amino acids and hence a larger number of reaction chambers, while for the second portion the number of amino acids would be decreased as the number of reaction chambers is decreased. Through this method, it is possible to endcap the peptide with the same terminal amino acid. To prevent possible degradation with exopeptidases, it is advised to use *D*-amino acids instead of the *L* version, or just to endcap the peptides with a *D* terminal amino acid. Among the natural list of amino acids are those that interact more easily with proteins in solution, and they are already good adsorbents for a number of proteins. They are called "grand catchers" (Bachi et al., 2008). Others are much less active for the interaction with proteins ("petite catchers"). From this concept it is also possible to prepare a library based on the use of only grand catchers.

The simplest peptide library is the one in which only each single natural amino acid is attached to a single bead; this is composed of only twenty diversomers and can be qualified as an amino acid solid-phase library. In spite of its simplicity, it is capable of capturing a large number of proteins (Bachi et al., 2008). If for each individual amino acid the list of proteins captured is determined, specific calculations allow approaching mixtures for the purpose of purifying proteins instead of reducing the dynamic range.

4.4 USE OF PEPTIDE LIGAND LIBRARIES AS A MEANS OF REDUCING PROTEIN DYNAMIC RANGE CONCENTRATION

Proteome deconvolution and analysis of gene expression in biological extracts, especially from blood, represents a real challenge not only because of the large number of polypeptides present, but more importantly because of the extremely large dynamic concentration range of gene products that span over 12 or more orders of magnitude (Pieper et al., 2003a). In addition very often just a few high-abundance proteins represent a large majority of the protein mass (see examples in Table 4.2). From this situation there are three problematic consequences. The first consequence is that, in the presence of massive amounts of a few high-abundance proteins, the detection signal of a number of other species is hidden and so their detection becomes impossible in practice. By comparison, this situation resembles to the desire to observe stars in a clear sky during the day when the light of the most detectable star—the sun—covers the signal of other stars. They exist and are extremely numerous, but they are impossible to be observed. Only when the sun disappears during the night does observation become feasible. The second consequence is the extreme complexity of proteomes that comprises thousands of gene products and, more importantly, probably hundreds of thousands of species if post-translational modifications, fragments, and antibodies are counted. In this respect, the resolution of technical analytical means is critical and most of the time insufficient. The third consequence is the dynamic concentration range that is incompatible with current analytical instrumentation. While the protein concentration difference between the most and the least concentrated ranges over a dozen orders of magnitudes, the instruments reliably cover not much more than 4 orders of magnitude (or perhaps 5 with the most sophisticated MS equipment available today).

TABLE 4.2 Example List of High-Abundance Proteins in Various Biological Extracts

High-Abundance Protein	Origin	Prevalence (%)
Albumin	Serum, CSF	50–60
IgG	Serum, CSF	10–15
Hemoglobin	Blood cell lysate	85–95
Ovalbumin	Egg white	35–45
Lactalbumin	Milk	30–40
β-Lactoglobulin	Milk	15–20
RuBisCO	Plant leaf extract	40–60
Actin	Cell extracts	15–20
Storage proteins	Potato tuber extract	70–80
Toposome and major yolk proteins	Sea urchin coelomic fluid	65–75
Glycolytic enzymes	Skeletal muscle extract	55–65
Storage (11S and 7S globulins)	Soja (soy) beans	60–80
Phloem protein 1 & 2 (PP1 & PP2)	Pumpkin phloem exudates	65–75

When facing this interesting problem, scientists developed their variegated talent to try to find ad hoc solutions. For instance, to resolve the case of the presence of a few high-abundance proteins, selective depletion has been proposed and developed (see, for instance, Pieper et al., 2003b; Greenough et al., 2004; Brewis & Brennan, 2010), as detailed in Section 3.2.2. Concerning the fundamental problem of the very large dynamic range, a compression of the situation has been successfully proposed by a concomitant decrease of the concentration of high-abundance and the concentration increase of the low-abundance species. To overcome the question of resolution due to too numerous, diverse proteins present in an extract, fractionations have been proposed, thus simplifying the composition of each fraction (see, for example, Luque-Garcia & Neubert, 2007; Righetti et al., 2005; Jmeian & El Rassi, 2009; Dayarathna et al., 2008) instead of subtraction or in addition to immunodepletion.

Since protein fractionation and high-abundance protein depletion technologies have already been extensively described in Chapters 2 and 3, respectively, this section will be devoted to the compression of the dynamic concentration range (also called enrichment) even if this expression is not completely appropriate, if not well characterized.

4.4.1 Description of the Principle

The very first report on reduction of the dynamic concentration range by using combinatorial peptide ligand libraries in view of detecting low-abundance proteins was published in a seminal paper by Thulasiraman et al. in 2005. Over time, other papers have been published confirming the first results as well as numerous reviews describing various aspects of the technology (see, for instance, Righetti et al., 2006; 2010; 2011; 2012). This compelling method uses mixed beds of chromatographic beads, all differing from each other by the fact that they carry different peptide ligands. The idea of using such a mixed bed was generated from the affinity chromatographic concept. The concept is generally designed to capture a single protein from a complex mixture. To this end, specific ligands, such as antibodies, are covalently grafted onto chromatographic solid phases and used as biospecific sorbents. If one considers making as many immunoaffinity columns as the diversity of proteins, all proteins could be selectively separated (Figure 4.6A). Theoretically if all these columns had an identical binding capacity and were loaded until column saturation, one would capture all proteins using the same principle, and all proteins would be similarly concentrated.

Naturally, when using all these columns in parallel, the need of loading to reach the saturation of the binding capacity would not be the same. Actually, the immunoaffinity sorbent columns designed for high-abundance proteins, as for instance anti-albumin, would be rapidly saturated and the volume of serum sample used would be very small. Conversely, other immunosorbents against very low-abundance proteins would need a large serum sample to saturate the corresponding specific binder. Nevertheless, at the end of the process all columns would be saturated and could be eluted in parallel. The result of this theoretical (and maybe ideal) approach would be the obtention of the same amount of all proteins but under separate unitary operations. At the end of the process, the protein solutions could be mixed together to get a protein mixture where each species will be present at similar concentration. In other words, one would theoretically reduce the large dynamic range of several orders of magnitudes to just around 1.

Although this process is very compelling for the integral composition of proteomes, it remains at the stage of theory since it will be materially impossible to make thousands, if not tens of thousands, of immunoaffinity columns with similar physicochemical properties and operate them in parallel. If it were achievable to make all possible immunosorbents, instead of using a large number of distinct columns of immunoaffinity sorbents, it would also be feasible to blend all sorbents together at equimolecular proportions and use the mixture of solid phases as a single mixed-bed column (Figure 4.6B). This would be fully possible because the conditions for protein adsorption by each

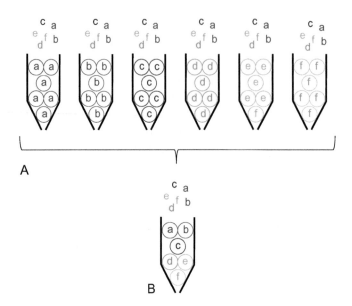

FIGURE 4.6

From affinity chromatography of a single protein to a combinatorial mixed-bed chromatography addressing all proteins at a time.

A: The protein mixture composed of many different proteins (from "a" to "f") is separated specifically and individually by different affinity chromatography columns such as individual immunosorbents. All capture conditions are identical from one column to another.

B: Mixed bed of the same sorbents where each single protein of the mixture is captured by their specific bead partner. Elution is also globally operated.

single immunosorbent bead bait are identical (e.g., physiological buffers). The immunosorbents blend would have to be largely overloaded to try saturating immunosorbents beads designed against very low-abundance species. This overloading condition has no negative effect since, once the corresponding beads are saturated, the excess of the proteins is washed out at the outlet of the column.

While this approach is an ideal but unrealistic situation, the concept already depicts what could be the rational approach to at least reduce the dynamic concentration range of proteins. For the moment it is still a utopian idea to design thousands of single immunoaffinity sorbents. However, while waiting for a possible future development, instead it could be possible to have a very large statistical collection of ligands designed for the capture of proteins. In this respect, the more obvious collection of ligands is a library of peptides. Interestingly, peptide synthesis technology allows making all possible combinations of peptides starting from single amino acids. Moreover, by using the technologies described in Section 4.3.2, it is also possible to obtain individual beads, each carrying a different peptide present in a large number of copies with a total affordable effort. This situation resembles a mixed bed of beads where each individual bead is a distinct affinity chromatography column. Figure 4.6 shows a theoretical route from single affinity to multiple affinity using mixed beds. Only one sample injection is necessary but in large excess so as to try saturating each bead with each corresponding protein partner. Naturally, the binding capacity of each bead will be small, but probably large enough for proteomic analytical determinations. At that stage, the bases of the technology were established.

The first trials that delivered promising results with the use of such mixed beds of peptide libraries under overloading conditions were obtained around 2003 and published for the first time in 2005 by Thulasiraman et al. After 2005, the technology generated an extensive number of published reports to enhance low- and very low-abundance species.

Most applications of these combinatorial peptide ligand libraries in proteomic investigations deal essentially with the detection of species whose concentration is below the sensitivity of current analytical methods. A large number of parameters were to be optimized, however; among them were the ligand density of the peptide ligand within a single bead because this has a direct impact on the specificity and the protein-binding capacity. The diversity of ligands was to be defined by two different parameters: the number of amino acid building blocks and the length of the peptide (see, for instance, this relation in Table 4.1). The size of beads is also extremely important since it determines the number of beads per unit of volume and hence the mini-affinity systems within a given space. The latter determines the minimum volume of the biological sample to be used for a decent reduction of dynamic concentration range. In practice, the impact of sample volume is really large when thinking that the availability of biological extracts is very restricted not only in research but also in diagnostics.

All these parameters will be discussed later; they are mentioned here because the reader should figure out the volume of work that remained to be done from the initial poor level of mechanism of understanding. Whereas the concept of overloading in chromatography would look relatively simple, it becomes very complex with mixed beds (like solid-phase libraries) because of the influence of many physicochemical parameters that will be analyzed one-by-one in the following paragraphs. A very large task laid in front of investigators at the early stages of this technology.

Basically, the practical operations for restricting the dynamic range with the concomitant reduction of high-abundance proteins and enhancement of low-abundance ones were relatively easy to implement because it consisted in placing a peptide library in contact with a biological sample. Under the restrained binding capacity of each individual bead, the saturation is rapidly reached for the most concentrated proteins, and the excess that cannot bind any further is consequently discarded in the flow-through. On the contrary, low-abundance proteins is enriched as long as the sample is available and loaded. After removal of all proteins that are not bound, the composition of proteins retained by the beads is defined by the presence of their specific affinity ligands and the relative concentration of each retained protein species (see all process schemes in Figure 4.7). The resulting modification of the dynamic concentration of protein components of a mixture is represented in Figure 4.8 in which a certain degree of proportionality between species is maintained. Such a behavior is not necessarily the exact representation of the reality because certain low-abundance species are enhanced more than others, but it just introduces the fact that from a sample to another the variation of the concentration of a single species is proportional to its initial concentration under exactly the same treatment conditions. This crucial point will be discussed later in this chapter when the reproducibility and the quantitation aspect are examined.

Considering the library as having at least one representative ligand for each protein of the biological extract, the eluted protein mixture would have the same qualitative composition as the initial sample, but the relative protein concentration range would be largely compressed. This quite compelling, but simplistic, explanation of how a solid-phase combinatorial peptide ligand library works for the

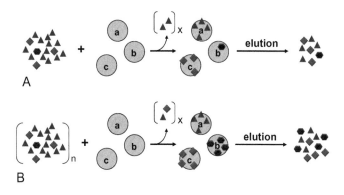

FIGURE 4.7
Schematic representation of two stages of a dynamic concentration range compression involving CPLL. A: the sample load is not large enough to saturate peptide beads "b" and "c"; the process could still be improved. This is the case in "B" where all beads are fully saturated by their corresponding protein partner.

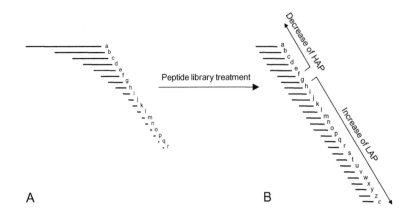

FIGURE 4.8

Schematic representation of the compression of the dynamic concentration range of sample A with concomitant decrease of the concentration of high-abundance proteins (from a to d) and (B) concentration increase of low-abundance proteins (from h to z). The length of each bar represents the concentration of the corresponding protein. Additional proteins (from s to z) not detectable before the process becomes directly detectable.

decrease of the dynamic concentration of protein components is based on the absolute specificity of a peptide bead for a single protein associated with a binding capacity saturation phenomenon. However, the mechanism is probably far more complex. Even if some peptides of the library could be extremely specific for given proteins, many others interact with various proteins.

Since the beginning of its development, the described technology was confronted to existing procedures aimed to improve the low-abundance protein investigations. As described in previous chapters, the technologies used were chromatographic separations, depletion of high-abundance proteins, and narrow gradients for isoelectric fractionation of proteins. Table 4.3 compares the advantages and disadvantages of each technology. Among the important elements of this comparison is the large applicability of the combinatorial peptide ligand library (CPLL) that not only encompasses human plasma and serum, but also extends

TABLE 4.3 Comparison of Main Methods Used to try Reaching Low-Abundance Proteins (LAP)

Competitive Technologies	Main Advantages	Main Disadvantages	Comments
Chromatography fractionation	Easy to implement. Various approaches.	Dilution of proteins. Large number of fractions. Labor intensive.	It is used when no alternative methods allows resolving the problem.
Depletion of HAP	Addresses specifically proteins to remove. Easy to implement.	Very small sample volume. Risk of carryover. Dilution of sample. Species specific.	It is very expensive for a minimum volume of treated sample.
Narrow pH gradients	No sample treatment necessary. Implementable as 2D-PAGE with high resolution.	Massive precipitation of proteins. Only spot by spot analysis.	It is used for special protein identification by spot excision.
Enrichment of protein groups	Easy to implement as affinity separation (e.g., lectins for glycoproteins).	Lack of specificity with non-specific binding.	This technology is available only for very few protein groups.
Reduction of dynamic concentration range	Easy to implement. Applies to all possible species. Allow accessing LAP.	Large sample volume.	Ideal to apply for deep proteome or early-stage biomarker discovery.

throughout other types of extracts and biological fluids as well as microbial and plant extracts. On the contrary and in spite of its large adoption, depletion using antibody ligands is quite limited in performance. In fact, it is confined to the removal of high-abundance proteins with the proven risk of co-depletion and also to the fact that the collected fraction is diluted with consequent additional issues related to the discovery of low-abundance species (see Section 3.2.2).

Liquid chromatography is easy to implement; however, it delivers too many fractions with a large overlapping of species, all of them diluted in larger volumes. Enhancements of protein groups such as phosphopolypeptides and glycoproteins are very focused approaches ignoring all other parts of proteomes. Reduction of the dynamic concentration range is therefore the only one that makes sense when considering the entire proteome; all species are targeted and many undetectable gene products can be enhanced and analyzed (Figures 4.9 and 4.10). In addition, the relative quantitation is preserved, thus maintaining intact the possibility of discovering induced modifications of protein expression, as described later in this chapter.

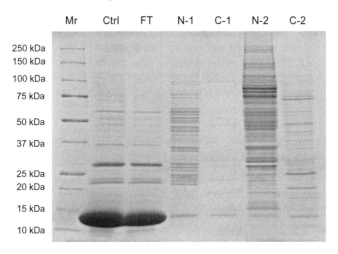

FIGURE 4.9
Example of dynamic range reduction on red blood cell extracted proteins illustrated by SDS-polyacrylamide gel electrophoresis. The initial extract (Ctrl) shows the major band of globin chains at the bottom of migration lane and several medium abundance protein bands. This sample was treated with two peptide libraries in series (terminal primary amine "N" and its carboxylated version "C"), with the obtention of a supernatant of a column flow-through (FT) followed by elution fractions. Note that FT is indistinguishable from Ctrl due to the large overloading. The serial columns with captured proteins were desorbed using two distinct elution solutions: 2 M thiourea, 7 M urea, 2% CHAPS (N-1 and C-1), followed by 9 M urea, 50 mM citric acid pH 3.3 (N-2 and C-2). The proteins collected from the second column are complementary to the first, but in significant lower amounts due to the fact that most of proteins were already captured by the first column. Results obtained during the work performed from red blood protein studies (Roux-Dalvai et al., 2008).

Many examples of biological extracts treated with CPLL have been reported with the benefit of a significant increase of the number of identified proteins, whatever the analytical method used. Concrete applications will progressively be discussed throughout this book. As a summary, Table 4.4 illustrates data on identified gene products after treatment with combinatorial peptide ligand libraries compared to untreated samples.

As it frequently happens, the described technology suffers from some limitations that are intimately related to the concept itself and its reduction to practice compromises. For instance, and quite contrary to what was initially supposed, a single peptide bead frequently captures more than one protein, as has been clearly demonstrated (Boschetti et al., 2007) by SDS-PAGE and mass spectrometry analysis of individual beads of the library. Very interesting observations were made about protein composition differences when comparing the individual eluates from single beads. Among the different captured proteins found, some other proteins were common, suggesting that a single protein might be captured by more than one peptide ligand. With complex mixtures of ligands and proteins, a very large number of situations are present, with affinity constants that might range

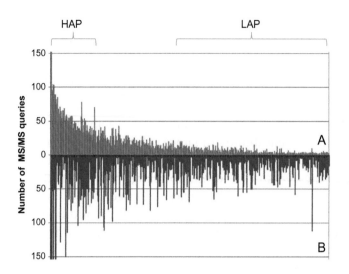

FIGURE 4.10

Experimental demonstration of the dynamic range compression of the detected peptides from a red blood cell lysate before and after treatment with CPLL. The number of MS/MS queries decreases very rapidly when analyzing the untreated sample (part A of the figure). The analysis allowed identifying 540 gene products by decreasing scores. In part B, the number of queries decreased to a much less extent, allowing identification of an additional 558 gene products. A distinction is made between the initial part where high-abundance proteins (HAPs) are present and the low-abundance space (LAP). The latter is particularly rich in number of queries. Results were obtained from the red blood protein studies (Roux-Dalvai et al., 2008).

TABLE 4.4 Examples of Increase of Gene Products Found in Various Biological Extracts Upon Treatment with CPLLs. Identification of Proteins was Performed by LC-MS/MS After Separation by SDS-Polyacrylamide Gel Electrophoresis

Biological Extract	Gene Product Before CPLL	Gene Products After CPLL	Common Gene Products	References
Human saliva	1236	1950	932	Bandhakavi et al., 2009
Human cerebrospinal fluid	476	1149	404	Mouton-Barbosa et al., 2010
Human platelet lysate	197	411	173	Guerrier et al., 2007a
Human bile	141	197	116	Guerrier et al., 2007b
Human urine	96	439	64	Castagna et al., 2005
Human red blood cell lysate	535	1578	470	Roux-Dalvai et al., 2008

between extreme values (low and high). As a result of observations and theoretical considerations, it became clear that the mechanism of reduction of protein concentration range was more complex than anticipated and was dependent on the number of ligands, the affinity constants, and the concentration of each individual protein of the mixture (see Figure 4.11). This observation is not a disadvantage inasmuch as it contributes to the competition of species for the interaction on peptide ligands, thus maintaining the possibility of relative quantification that would be prevented in case of a strict "one-ligand-one-protein" mechanism. Collateral parameters influencing thermodynamic equilibria and affinities between partners are environmental conditions such as temperature, pH, ionic strength, presence of special molecules in solution, and diffusion time, all of which have strong or minor effects on the dynamic phenomena of displacement. Kinetics may also play an important role, particularly when the protein size is large and the affinity for a peptide is weak.

The overall objective of sample treatment with peptide libraries is to reduce the protein dynamic concentration range from an initial situation where the dynamic range is so large that it cannot be explored using current methods and instrumentation. High-abundance proteins are reduced in their

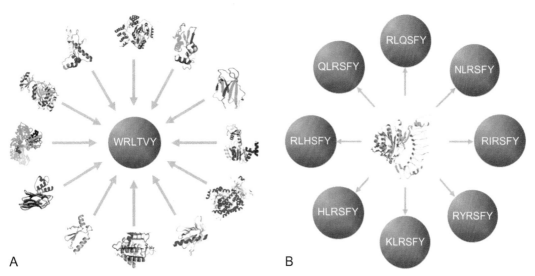

FIGURE 4.11
Schematic representation of the intense protein competition in interacting with peptide beads. Many proteins compete for the same peptide ligand (A) represented by the sequence WRLTVY. Concomitantly, a protein can interact with different peptide beads depending on the affinity constant and the level of competition with other proteins for a given peptide bead (B). In this example, B is represented by a Von Willebrand factor from human plasma. *(Data of panel B are from Buettner et al., 1996)*

concentration, while low-abundance ones are largely concentrated; the extent of the phenomenon depends on the amount of protein extract offered to the mixed-bed ligand library beads. Each important parameter influencing the capture of proteins is extensively presented and discussed in the following paragraphs. However, what could be underlined at this stage is the real capability to decrease the dynamic concentration range of several orders of magnitude independently from the type and origin of the protein extract (see, for example, Figure 4.12).

Various authors have reported 2, 3, and even 4 orders of magnitude reduction. Even if these numbers differ significantly, the effect of CPLL treatment is largely dependent on the amount of proteins offered to a given volume of CPLL beads. Considering the extent of "dynamic concentration compression" and the fact that the most modern instruments and analytical methodologies claim a capability of reading throughout 3 to 5 orders of magnitude, most proteomes could be mapped almost completely if the described technology were associated with the most sensitive analytical approaches. As a result, it is not logical to place one technique against another, as if science were a medieval tournament of gallant chevaliers fighting for the kiss of a grand dame. It is a common opinion that it is much better to be allied and combine the use of the two techniques. Thus we can characterize as fallacious and misleading the idea of Mann and Mann (2011) that there is no need of any CPLL treatment in analysis of the egg white, since in the untreated sample, by using a powerful MS Velos instrument, one can see as many proteins as in a CPLL-treated one. We can bet that if using the Velos on the same egg white (whose proteome is known to be heavily colonized by just six or seven major proteins, which strongly obscure the signals of all other low-abundance species) pretreated with CPLLs, the discovery would have been doubled. The conclusion that "current mass spectrometry technology is sufficiently advanced to permit direct identification of minor components of proteomes dominated by a few major proteins without resorting to indirect techniques, such as chromatographic depletion or peptide library binding" do not make much sense to us!

The price to pay for reducing the dynamic range is the sample volume requirement. It is frequently criticized because biological samples are generally small; however, this attitude is illogical because if one takes a minute amount of sample and treats it, as for instance to remove high-abundance proteins that may occupy 90+% of abundant species, there does not remain much room to

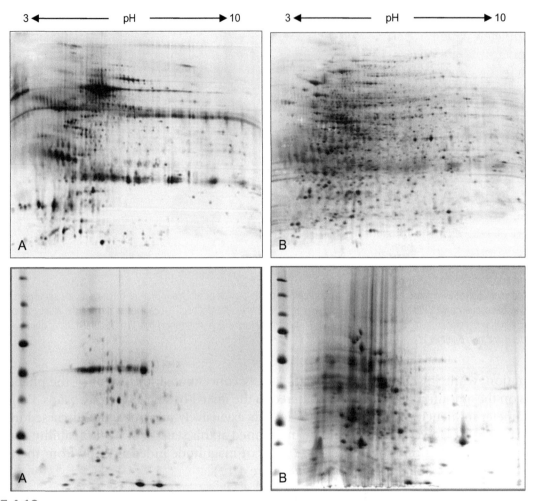

FIGURE 4.12

Two-dimensional gel electrophoresis analysis with the emergence of a very large number of proteins along with their isoforms upon treatment with CPLL at pH 7 ("B" images) compared to the initial nontreated samples ("A" images).

The upper panel represents human urine proteins where albumin and immunoglobulins dominate. The lower panel represents spinach leaf proteins where the dominating protein is Rubisco with all isoforms.

In both cases, the abundant proteins are largely decreased upon treatment, while many undetectable proteins become detectable as a consequence of the enrichment effect. The latter is obtained regardless the molecular mass and isoelectric point of species.

discover gene products that are present at very low trace levels, at least up to the point that another technique of protein "amplification" would be invented (unlikely) on a similar basis as that for nucleic acids. With a small sample, one would never have a chance of finding those low- to very low-abundance proteins that might constitute the reservoir of any possible early-stage biomarker that are present in biological fluids at pico- or sub-pico grams/mL level. On the contrary, the large sample need is a unique and distinct advantage of CPLL, since it is able to handle not just 1, but 10, 50, or 100 mL of sera (or more, as needed), where enough copies of such rare species can be captured and rendered visible.

A very important aspect of this technology is the desorption of captured proteins from CPLL beads. It is not a trivial question because it involves various types of interactions that can be synergistic or antagonistic. In other words, desorbing proteins by weakening certain interaction forces might have a strengthening effect on other types of affinity attractions; therefore the protein recovery would not be complete. This is very annoying since it is from the perfect recovery of captured proteins that all the benefits of the technology can be exploited. A specific section of this chapter is largely devoted to this aspect (see Section 4.6); however, what should be kept in mind here is the verification that all proteins are completely eluted prior to analysis. Protein desorption from the CPLL beads could follow the rules of chromatography with, for instance, a full elution at once or a fractionated elution. The latter is of interest in the case of (i) analysis of groups of proteins or (ii) proteins that are captured by similar interactions. In fact, desorption could be designed in order to collect proteins that are dominantly captured by an ion-exchange effect or by hydrophobic associations or where the dominant interaction is hydrogen bonding. Weakening agents are available for all those distinct operations. There remains, though, a group of proteins that may have no dominant type of interaction but rather a mixed and equilibrated number of various forces. In this case strong deforming agents and/or total denaturation are necessary. Finally, as detailed below, full digestion of the captured proteins could also be performed directly on the beads and the resulting peptides analyzed by mass spectrometry after fractionation by either high-performance chromatography (reverse-phase associated or not with cation-exchange chromatography), capillary electrophoresis, or isoelectric focusing—all technologies described in Chapter 2.

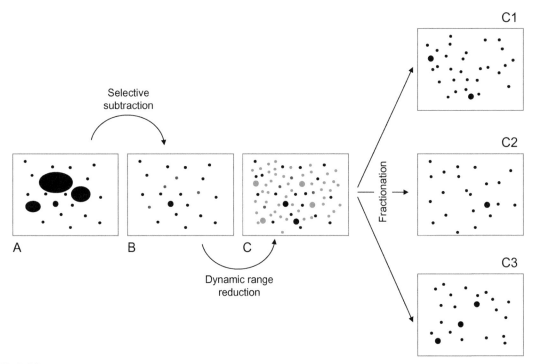

FIGURE 4.13
Schematic idealized approach to decipher the composition of proteomes. The initial sample "A" comprises high-abundance proteins that prevent the detection of a number of other species (by signal suppression for instance); it also comprises many proteins that are very dilute and undetectable by current means. To resolve these issues, a first operation would consist in removing high-abundance species (B): Some hidden proteins are thus detectable; however, dilute species are still below the detection threshold. The second operation is the compression of the dynamic concentration range with consequent enrichment of low-abundance species (C). At this stage the protein composition could become too complex to perform a global protein identification. To increase the resolution of analytical methods, fractionation of complex mixtures into C1, C2, and C3 into complementary groups could be performed.

Overall, this complex situation would suggest operating all of these approaches singularly within an entire process. To stay with the astronomy paradigm, the first treatment would consist in removing the most prominent star (the sun) and observing the sky during a clear night by selectively removing high-abundance species. However, this operation is not powerful enough to reduce the dynamic concentration range within the window of the capabilities of current analytical methods and instrumentation. Therefore the second operation is to be considered: using powerful telescopes possibly outside the atmosphere to dig deeper in the universe. In the case of proteomes, this would ideally consist in a profound compression of the difference in the concentration of proteins composing a sample. The direct consequences of these two operations are that many other species become detectable, but as a counterpart the resulting complexity of the treated sample is magnified. Although this corresponds to the reality of the biological composition, the resolution among species might become critical. Only a fractionation (third operation) would thus render the composition of the treated sample easily analyzable by current instruments. These three operations appear perfectly streamlinable and orthogonal to resolve the entire dilemma. Figure 4.13 depicts a theoretical approach for the profound deconvolution of any complex proteome. Nevertheless, in this ideal vision, the procedures are not perfect, and other issues are identified that render the picture a bit less performant than expected. All these concepts will be progressively detailed in the following sections.

4.4.2 Interaction Mechanisms

The interaction between proteins in solution and grafted peptides develops thanks to several complex elementary interactions, some of which are synergistic and others antagonistic (see Section 4.4.3). Today the phenomenon of polypeptide interaction is largely investigated because it is at the origin of protein communication and signaling (to know more see, for instance, Levy & Pereira-Leal, 2008; Rubinstein & Niv, 2009; et al., 2011; Boja & Rodriguez, 2011; Sanz-Pamplona et al., 2012). The most important molecular interactions comprise ionic interactions such as ion exchange. It is actually known that in protein–protein or protein–peptide associations, ionic charges frequently have a dominant influence over other types of molecular interactions (Gibas et al., 2000). However, other interactions are present such as hydrophobic associations, hydrogen bonding, and a number of other weak interactions as depicted by Righetti et al. (2006).

Figure 4.14 illustrates schematically one linear hexapeptide taken randomly, exactly as developed on the beads; possible interaction centers are shown. While secondary amide bonds among amino acids constituting the peptide contribute to hydrogen bonding, the side chain groups from each amino acid are very active in the variety of interaction developments in terms of diversity and strength. Taken together, all of these individual involvements develop a quite global significant interaction effect. The side chains of amino acids are characterized by the presence of anionic (glutamic acid and aspartic acid) or cationic (arginine, lysine and histidine) charges and non-ionic hydrocarbon chain sites from isoelucine, leucine, and phenylalanine, contributing to the formation of hydrophobic associations.

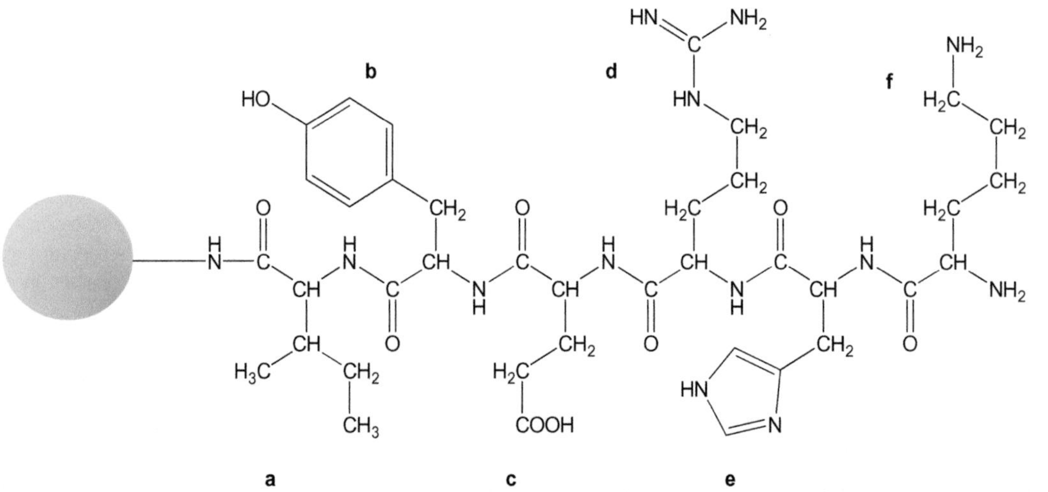

FIGURE 4.14
Example of a selected CPLL bead where a hexapeptide is grafted. It is composed of valine "a," tyrosine "b," glutamic acid "c," arginine "d," histidine "e," and lysine "f." All amino acid side chains can contribute differently to the global interaction with the partner protein. For instance, the "a" residue could generate hydrophobic associations while glutamic acid and arginine could contribute for electrostatic interactions. Tyrosine can form hydrogen bonding with carbonyl groups of the captured protein.

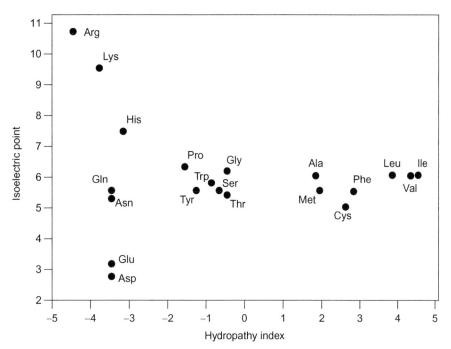

FIGURE 4.15
A two-dimensional map of the distribution of single amino acids on the basis of charge *versus* hydrophobicity. Note that all mono-amino, mono-carboxylic acids (i.e., those without ionizable side chains) have nearly identical pI values, centered around pH 6.0.

Amino acids, by their fundamental ionic and hydrophobic properties, can be positioned on a surface crossing over the isoelectric point (pH at which the net charge is zero) and the hydropathy index as illustrated in Figure 4.15. Since the peptide library is combinatorial, all possible situations are represented: They go from hexa-glycine (glycine is at the center of the graph) with almost no interaction possibilities, to hexa-aspartic acid (very acidic peptide), to hexa-arginine (very alkaline peptide) to hexa-leucine (very hydrophobic peptide). To complete the picture, the terminal amino acid has a primary amine as a free terminal group, conferring to the entire construct a minute cationic character depending on environmental pH. Nevertheless, the fact remains that globally the hexapeptide library is a bit cationic compared to what it would be if the terminal amine were endcapped (e.g., acetylated). Figure 4.16 illustrates two titration

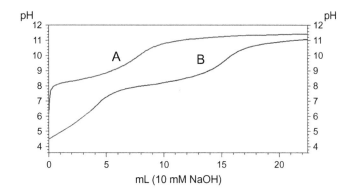

FIGURE 4.16
Titration curves of solid-phase peptide libraries. 250 mg of each library was washed with 100 mM HCl and then washed several times with distilled water to remove the excess of the acid. After having added 2 mL of 1 M KCl, the suspension was titrated with 10 mM sodium hydroxide with continuous pH monitoring.

A: primary amine terminal hexapeptide library.
B: carboxylated hexapeptide library obtained as described in Section 8.2.

curves related to the same library either as such (curve A) or modified by succinylation of free amines (tracing B). It is clearly evidenced that the physicochemical ionizable properties are changed with potentially different properties at the functional level. This particular aspect will be discussed later.

Proteins are constituted of quite long chains of amino acids and hence can be easily folded as a result of internal interaction between portions of the polypeptide chain. Since proteins are massively folded, the interaction with grafted peptides happens mostly with exposed external portions of the polypeptide sequence. They could be largely random or constituted of groups of the same or similar amino acids spatially close to each other conferring, for instance, locally strong charge effects when most of the amino acids of the cluster are acidic or alkaline.

Similarly to affinity chromatography involving peptide ligands, the interaction between hexapeptides of CPLL and corresponding partner proteins in solution is a probabilistic term that can be described by the mass action law. When taking simply two partners (the ligand A peptide and the ligate B protein), their association could be described as follows:

$$A + B \xrightarrow{k_{ass}} AB$$

Equation 1

where k_{ass}, and the corresponding k_{diss}, when the reaction is reversible, are association and dissociation rate constants with the dimension $[M^{-1}s^{-1}]$. They indicate how fast the concentration or dissociation of AB complexes increases or decreases when the concentrations of A and B are $[A]$ and $[B]$.

This equilibrium can also be written as:

$$\frac{d[AB]}{dt} = k_{diss}[AB]$$

Equation 2

where k_{diss} is the dissociation rate constant with the dimension s^{-1}. The dissociation rate constant k_{diss} indicates the fraction of AB complexes that dissociates per second. In affinity chromatography, values for k_{diss} typically range from 10^{-8} to $10^{-2}s^{-1}$, corresponding to half-lives of protein complexes from a few minutes to several hours.

According to the law of mass action, the position of the equilibrium is dependent on the concentration of components and the reaction constants (k_{diss}, k_{ass}) and can be expressed as the association equilibrium constant K_A:

$$K_A = \frac{[AB]}{[A][B]} = \frac{k_{ass}}{k_{diss}}$$

Equation 3

with the dimension $[M^{-1}]$. The dissociation equilibrium constant K_D is the reciprocal unit of the equilibrium association constant with the dimension $[M]$:

$$K_D = \frac{[A][B]}{[AB]} = \frac{k_{diss}}{k_{ass}}$$

Equation 4

Within the situation of ligand libraries in contact with complex proteomes comprising proteins of very different concentrations, a large variety of affinity constants are present. Moreover, the state of affairs is further complicated by the fact that some protein species compete for the same peptide bait.

The specificity here remains a statistical term, defined as the number of proteins that bind a peptide in the presence of other molecules associated with numerous displacement effects, dependent on both the affinity constants and the concentration of proteins present in the mixture. Thus the simplest representation of the capture of a protein (C) by a peptide partner assuming single-component, monovalent associations, is represented by Equations 5 and 6. When the concentrations of the protein in the mobile phase and stationary phase are C_m and C_s and the maximum binding capacity $C_{s,max}$, the rate expression is:

$$\frac{dC_s}{dt} = k_1 C_m \frac{C_{s,max} - C_s}{C_s} - k_{-1} C_s \qquad \text{Equation 5}$$

$$K_D = \frac{k_{-1}}{k_1} = \frac{C_m \left(C_{s,max} - C_s \right)}{C_s} \qquad \text{Equation 6}$$

Both expressions are equivalent to the Langmuir type isotherm (Equation 7):

$$\frac{C_s}{C_{s,max}} = \frac{C_m}{K_D + C_m} \qquad \text{Equation 7}$$

The direct consequence of this situation is that the concentration of components is critical not only for affinity interactions but, more importantly in the present context, for the reduction of the dynamic concentration range. While under normal conditions of affinity chromatography the C_s value decreases over time because the component is captured, under large overloading conditions (the presently described process with CPLL) this term does not change at least all along the capture process until reaching the saturation of the partner peptide ligand bead.

Thermodynamic equilibria are, moreover, dependent on a number of external parameters such as temperature, pH, ionic strength, and the presence of additives. For example, at constant temperature, the pH affects the ionic interaction toward the modification of affinity constants, influencing at the same time the global ionization of both peptide ligands (when they comprise ionic amino acids), on the one hand, and the proteins of the mixture, on the other hand. Although for some proteins the affinity constant may increase, for others it may decrease with consequent changes in competition effects among proteins for given peptide ligands.

In a nutshell, phenomena of affinity recognition and displacements effects associated with ligand diversity suggest that the binding event should occur via the Model 1 depicted in Figure 4.17A and described by Righetti et al. (2010), where the captured protein interact on the grafted peptide involving a single accessible epitope probably located on the external surface of the protein. However, this model does not seem fully consistent with the strong binding of proteins to the peptide beads, which in some instances are desorbed using only very stringent chemical agents (see Section 4.6). As an alternative, Model 2 is proposed (Figure 4.17B). In this case, the adsorbed protein interacts with the help of several peptides (they are the same within a bead) in various exposed polypeptide regions, resulting in a cooperative effect and hence a strong binding. Naturally this possibility is unlikely because a single protein does not necessarily have more than one identical epitope. An explanation of this possibility can be found in the fact that not all six amino acids of the peptide bait are needed for proper binding, but a minimum of three, as demonstrated by Simó et al. (2008), where it is shown that protein capture intensifies with the elongation of the peptide from 1 amino acid to a hexapeptide. However, this behavior is not linear, but it becomes asymptotic rather rapidly; with a length of a tri-peptide, the number of different captured proteins tends already to plateau. Thus some exposed protein sequences could interact with the entire hexapeptide, while others could interact with portions of hexapeptide baits located at the distal (or proximal) moiety or even at the center of the peptide sequence.

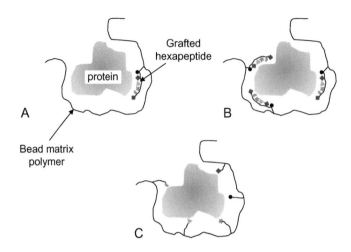

FIGURE 4.17

Models representing the interaction between CPLL beads and one captured protein.

A: Stoichiometric binding of the protein to the grafted hexapeptide sequence.
B: Multiple interactions of several hexapeptide ligands with various epitopes of the same protein.
C: Multiple interactions of portions of grafted peptides (even single amino acids) with different parts of the protein surface.

If one considers that single amino acids are capable of capturing proteins (Bachi et al., 2008), the number of possibilities increases since it could interact with different points of the protein as depicted on Model 3 (Figure 4.17C). All these models are static and are not sufficient to describe the dynamics of the interaction phenomenon that comprises an intense competition among proteins from a biological sample and the peptides of a single bead, especially under large overloading conditions. It is observed that the intensive competition of proteins for given peptide baits is in favor of low-abundance proteins rather than high-abundance ones. As a speculative approach, one could argue that concentrated species may not be shaped to be stickier than average because their large concentration would otherwise favor aggregation phenomena under normal conditions. Moreover, the massive amount of protein load influences the displacement effects overlapping with the bead saturation effect. When looking at a single bead, as soon as the bead is saturated by the most concentrated protein partner another phenomenon overlaps: This is the displacement effect by other proteins that were too dilute to compete effectively. This displacement continues as long as the loading is performed, giving an advantage to low-abundance proteins that are progressively concentrated, consequently becoming easily detectable.

It has long been assumed that each bead, carrying a single type of peptide ligand, would bind to only a single protein or proteins that would share the same epitope complementary to the peptide bait. However, in a combinatorial peptide situation, there are numerous cases where peptide ligands differ from each other by only one amino acid. Frequently, this small difference, especially when it is limited to neutral amino acids such as glycine and when this change is located close to the solid phase, does not significantly change the biorecognition effect. This phenomenon has been demonstrated when searching hexapeptides for the affinity capture of a given protein as described by Huang et al. (1996) and by Miyamoto et al. (2008). The hypothesis of having single peptides interacting highly specifically with single proteins from complex mixtures has been a fascinating idea, but too simplistic to explain how a peptide library works within the context of reducing the dynamic concentration range when the system is used under large overloading conditions.

The reality of the interaction of a library of peptide ligands with an extremely diverse protein mixture appears much more complex. Although situations have been demonstrated where one single peptide ligand interacts specifically with a single or very few proteins (Nozaki & Tanford, 1967; Buettner et al., 1996; Bastek et al., 2000), the most representative situation is when one single peptide ligand interacts with a large number of proteins and when a single protein interacts with a multitude of peptide ligands (Righetti et al., 2006). For instance, during the selection of hexapeptides as affinity baits for the von Willebrand Factor, several hexapeptides were selected, all of them able to capture this protein. They were of similar structure sharing at least the sequences of three amino acids. Different sequences correlated with variations of affinity constants are shown in Table 4.5.

On the given list there are four peptides with an identical sequence of five amino acids, four other peptides having a tetrapeptide identical sequence, and the last four peptides sharing a sequence of three amino acids. Interestingly, a minor change in composition results in a quite large difference in behavior of capture and desorption resistance to sodium chloride and to acetic acid. For instance, peptides No. 6 and 9 differ only by the replacement of a valine by an isoleucine, two amino acids of the same hydrophobic category structurally relatively similar and of very close isoelectric points, as well as pK_a and pK_b. In spite of this minor difference, the interacting property with the same protein is significantly modified (capture capability and sodium chloride susceptibility). The association constant, which is related to the structure of the peptide ligand, also varies as a function of the peptide density grafted on the beaded support. Additional details will be given in Chapter 7, Section 7.1, where peptide ligand libraries are used to identify the best peptides for protein purification.

Similarly, a given hexapeptide can be used as affinity bait for several proteins. It has actually been demonstrated (Boschetti et al., 2007) that the content of one single bead (potentially carrying a single hexapeptide structure) after contact with a crude protein extract, is constituted of a few proteins—some dominant and other minor species—all captured by different interaction intensities from mild to harsh agents to dissociate the complexes to recover the proteins for analysis.

TABLE 4.5 Selection of Specific Hexapeptides for the Purification of von Willebrand Factor. Binding And Elutions were Determined in Column. The Various Peptide Columns were Loaded with Human Plasma-Derived Factor VIII Concentrate Containing Human Albumin (Adapted From Huang et al. 1996).

Peptide No.	Identified Peptide	% of Found von Willebrand Factor		
		Flow-Through	0.1 M NaCl Eluate	2% ac Acid Eluate
1	HLRSFY	0.75	0.28	32.7
2	KLRSFY	1.03	9.83	26.94
3	NLRSFY	19.96	11.88	6.98
4	QLRSFY	29.45	11.69	2.2
5	RVRSFY	0.09	0.08	55.78
6	RFRSFY	0.2	0.07	47.96
7	RYRSFY	1.65	0.14	51.72
8	RIRSFY	1.92	4.09	44.78
9	RLNSFY	0.26	1.5	47.68
10	RLQSFY	0.46	0.54	50.68
11	RLKSFY	0.6	3.59	45.58
12	RLHSFY	1.77	2.89	50.72

Low-Abundance Proteome Discovery

Even though the possibilities of interaction are extremely large, the model depicted above probably does not describe the entire reality. There are proteins that do not find a peptide partner and are therefore lost from the generalized capture. Changes in pH for the protein capture reduce significantly the number of unrecognized proteins, but some others remain refractory to the interaction with the immobilized peptide library. This phenomenon constitutes a limitation of the technology not because proteins that are detectable before library treatment are no longer detectable in the processed sample, but more importantly because this phenomenon could be extended to some very low-abundance proteins that are not normally detectable and represent exactly the target. More extensive capture is possible by enlarging the pH range as well as the lyotropic environment as discussed in Sections 4.5.3, 4.5.4 and 4.8.

4.4.3 Type of Bonds Generating Protein Capture

Bonds that are established between a protein in solution and its peptide partner from the grafted hexameric population are the classical ones that stabilize the conformation of macromolecules, such as weak interactions that are at least one order of magnitude less than that of covalent bonds. These weak interactions describe how atoms or groups of atoms are attracted to or repelled from each other to minimize the energy of conformation (van Holde et al., 1998; Karshikoff, 2006). They can be grouped into: ion–ion, hydrogen bonding, hydrophobic associations, dipole–dipole and dispersion. The permanent dipole-permanent dipole, "permanent dipole-induced dipole," and "induced dipole-induced dipole" interactions are grouped under the name of van der Waals interactions. They are also called attractive London dispersion forces and are among the weakest ones when taken individually. Figure 4.18 depicts molecular interactions and their dependencies.

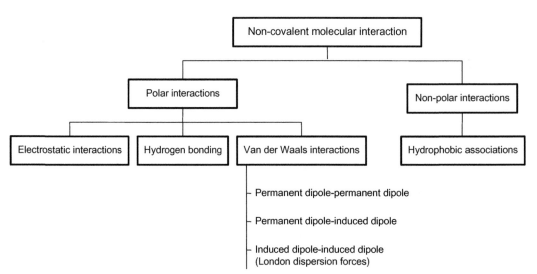

FIGURE 4.18
Classification of noncovalent bonds that are present in proteins contributing to their internal cohesiveness and also involved in protein–protein interaction. Individually, they are weak interaction (at least one order of magnitude weaker than covalent bonds), but all together can easily engender very strong molecular interactions.

All these interactions among atoms or groups of atoms, are, in general, distance-dependent interactions, with the energies being inversely proportional to the distance (or power of distance) separating the two groups. As the power of the inverse distance dependency increases, the interaction approaches zero more rapidly as the distance increases (Righetti et al., 2006). Most of the time protein associations are considered especially with respect to solid-phase interactions. However, directly opposing this attraction is steric repulsion, which does not allow two atoms to occupy the same space

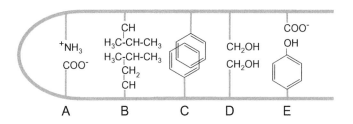

FIGURE 4.19

Simple schematic representation of intramolecular interaction among side chains of amino acids within a single polypeptide.

A: Electrostatic interactions; B: Aliphatic hydrophobic associations; C: Aromatic hydrophobic associations; D: van-der-Waals interactions; E: Hydrogen bonding.

Disulfur bridges between cysteines are not considered because they are covalent bonds.

at the same time. This repulsion occurs at very short distances. Together, the attractive dispersion and repulsive exclusion interactions define an optimum distance separating any two neutral atoms at which the energy is minimized.

The energies associated with relatively long-range interactions are dependent on the environmental medium. The interaction between two charged atoms, for example, becomes shielded in a polar medium and is therefore weakened. The composition of the medium also affects other important molecular interactions, such as hydrogen bonds and hydrophobic interactions. All these interplays are quite well described within the protein structure and can be summarized as shown in Figure 4.19. However, they are also strongly present in protein–protein complexes within the context of biological communication or signalling cascades. These interaction forces are analyzed singularly in the following sections and will also be discussed later in Section 4.6 within the context of aggregate dissociation (e.g., elution of captured proteins).

4.4.3.1 ELECTROSTATIC INTERACTION

Among the amino acids that compose proteins are acids and bases. Acidic amino acids are aspartic and glutamic acids, while in the domain of bases there are lysine, arginine, and histidine. These ionic-building blocks are randomly distributed within a protein and can create clusters of residues where similar ionic charges are grouped together. These charges are relatively weak and are modified by the environmental pH, thus creating a sort of net protein charge resulting from the sum of negative and positive charges. Opposite electrical signs attract each other, thereby creating a so-called electrostatic interaction. Although the presence of electrostatic forces in proteins was hypothesized many decades ago, its contribution and role in protein structure have only relatively recently been elucidated (Nakamura, 1996). It is currently admitted that electrostatic interactions not only are essential components for the 3D protein structure, but also play a role in molecular recognition, one of the major focuses today for understanding biological communications. Each single electrostatic interaction only marginally contributes to the protein structure stability; however, their collective effect acts synergistically to maintain the structure's integrity. The contribution to molecular recognition is generated by electrostatic charges present at the external surface, while the peculiar electric field around the protein active site regulates the catalytic function. Salt bridges are also intimately correlated to the electrostatic interaction within the protein structure.

This type of interaction is probably of great importance for the structure of proteins (chains that fold as a function of the intra- and interchain ionic group's interaction). The residual net charge of a protein is also important since it determines the interplay with other distinct entities that can be other polypeptides, nucleic acids, and small molecules. These interactions are modulated by various parameters. One of them is the modification of the environmental pH (e.g., the pH of the buffer where

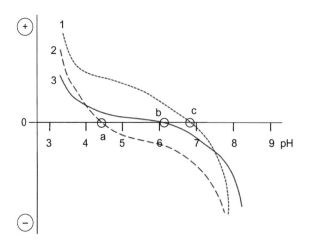

FIGURE 4.20

Typical titration curves of proteins. The variability from protein to protein (from "a" to "c") is due to the amino acid composition. The latter contributes to the isoelectric point of a given protein, which is an intrinsic protein property as is the molecular mass. As a reminder, the isoelectric point is also modulated by post-translational modifications such as sialylation and phosphorylation.

When the environmental pH is below the isoelectric point of a protein, the protein is positively charged. For example, at pH 3 most proteins are positively charged, and this is a way to desorb them based on an electrostatic repulsion mechanism from the beads. When the pH is intermediate (e.g., pH 5 in the present example), one protein is negatively charged (a) and two other proteins (b and c) are positively charged. This situation is very frequently exploited for the separation of mixtures by ion-exchange chromatography (see Chapter 2).

the protein is solubilized): The net charge modification has a direct effect on ionic interactions, with intensification of the attraction or even the repulsion of two species that are normally behaving otherwise. The ionization depends not only on the environmental pH but also on the intrinsic isoelectric point of a protein (see Figure 4.20). For instance at pH 5 proteins 1 and 3 are positively charged (they repulse each other) and protein 2 is negatively charged with electrostatic interaction toward 1 and 3. Ionic liaisons are affected by the presence of small ions such as salts since they are in direct competition against amino acid charged groups. When they are relatively concentrated they can inhibit the formation of ionic linkages, dissociate protein subunits, and denature their biological activity.

The modification of temperature influences the ionic interaction to a small extent, with an intensification of the interaction when temperature decreases. Ionic linkages are the most dominant forces governing the interaction between proteins from complex biological extracts and the hexapeptides of CPLL. Thus within the context of this section where the objective is to explain the electrostatic interaction between proteins and CPLL solid phases, the model adopted is ion-exchange chromatography, a common way to fractionate protein mixtures. The solid phase consists of solid porous beads carrying functional ionizable groups of various origins from very strong acids (e.g., sulfonates) to very strong bases (e.g., quaternary amines). In the present case the solid phase being composed of peptides, the acidic groups are limited to carboxylic acids, and alkaline moieties are essentially guanidine groups from arginine residues, primary amines from lysine, and also histidine side chains.

With regard to the ionic interaction between peptide baits and protein preys, the retention model developed by Kopaciewicz et al. (1983) is applicable. The relative retention k' (or capacity factor) is defined as:

$$k' = \frac{t_R - t_0}{t_0}$$

Equation 8

where t_R is the transit time of a given analyte and t_0 is the column void volume. The dependence of the distribution coefficient of the protein on a solid phase carrying ionizable groups in regard to protein and salt concentration ($K(C_m, I)$) can be empirically described by the following equation, assuming a Langmuir-type adsorption behavior for the protein.

$$K(Cm, I) = \frac{a}{\left[1 + \left(\dfrac{I}{b}\right)\right]^c \cdot \left(1 + d \cdot C_m\right)^2} \qquad \text{Equation 9}$$

where $K(C_m, I)$ is the distribution coefficient depending on salt and protein and C_m and I are the protein and salt concentration, respectively, a, b, c, and d being empirical parameters. This equation shows that the distribution coefficient is inversely proportional to the salt concentration (I) to the power of c. A plot of the distribution coefficient of a model protein in equilibrium with an anion exchanger is shown in Boschetti & Jungbauer (2000).

Although the stoichiometric displacement model does not rigorously enough describe the physical situation, since not all charges are accessible for the protein, the model has been corrected (Whitley et al., 1989) with the introduction of a correction-term compensating for this effect and also for the shielding of charges (Gallant et al., 1995).

Proteins are very diverse in their isoelectric point that can range from as low as pH 2.0 to as high as pH 12, and they interact and possibly adsorb on CPLL beads that have a cationic dominant character as well as on those that have an anionic dominant character, depending on the environmental pH (see Section 4.5.2). However, this rule is valid only at low conductivity conditions because the presence of ions in the environmental buffer may prevent adsorption by the phenomenon of competition with small ions. Additionally, the net charge of protein at a given pH does not always reflect its ability to bind to an ionized solid phase. In fact, as already mentioned, negative and positive charges are randomly distributed only at the surface of a protein. They may be clustered in certain domains, and then they may bind in an oriented manner. This effect may influence the expected electrostatic behavior especially with respect to the CPLL capture. Sufficient protein capture by CPLLs occurs in the presence of physiological buffers; however, better capture is generally reached with buffers of low ionic strength (see Section 4.5.3).

4.4.3.2 HYDROPHOBIC INTERACTION CONTRIBUTION

In water, nonpolar molecules tend to form aggregates. The forces involved in this phenomenon are called hydrophobic interactions or hydrophobic associations, as named within this section. These exothermic processes of association between nonpolar molecules in aqueous environments are from the group of so-called dispersion forces. In water, these hydrophobic associations are concomitant to the water structure around the interaction by means of the formation of hydrogen bondings immobilizing somewhat the water molecules around hydrophobic moieties, thus reducing the entropy.

While molecular mechanisms of hydrophobic associations are extensively discussed for a number of hydrophobic systems, in the context of this book we are interested in the description of such associations for proteins and protein components intramolecularly and intermolecularly. Nonpolar side chains of amino acids tend to avoid the contact with water, hence tending to collapse and contributing to the protein-folding process. This is an important phenomenon that is synergistic to the hydrogen-bonding process discussed below. The most common hydrophobic associations are internal to protein molecules. However, some hydrophobic amino acids are exposed on the external surface and so are available to generate associations with other molecules that could be encountered in solution. Figure 4.21 depicts schematically the situation of hydrophobic association formation. In the unfolded state, the side chains of amino acids are hydrated and exposed, while under the folded configuration they are mostly buried and hidden within the interior of the protein and inaccessible to water.

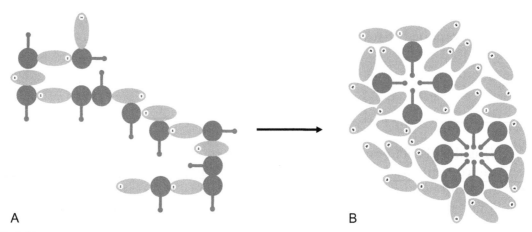

A   B

FIGURE 4.21

Schematic representation of a linear polypeptide (A) constituted of charged (gray ellipses) and non-ionic amino acids (tailed round black bowls). When the protein folds, the hydrophobic amino acids tend to create clusters. This assembly is more pronounced inside the protein structure rather than on its surface.

TABLE 4.6 Main Properties of Natural Amino Acids Composing Proteins

Code	Name	Abbrev	pI	Side chain	pK Side Chain	Hydropathy Index
A	Alanine	Ala	6.01	Apolar, aliphatic	-	1.8
C	Cysteine	Cys	5.05	Polar	8.18	2.5
D	Aspartic acid	Asp	2.85	Acid	3.90	-3.5
E	Glutamic acid	Glu	3.15	Acid	4.07	-3.5
F	Phenylalanine	Phe	5.49	Apolar, aromatic	-	2.8
G	Glycine	Gly	6.06	Apolar	-	-0.4
H	Histidine	His	7.60	Alkaline, heterocyclic	6.04	-3.2
I	Isoleucine	Ile	6.05	Apolar, hydrophobic	-	4.5
K	Lysine	Lys	9.60	Alkaline	10.54	-3.9
L	Leucine	Leu	6.01	Apolar, hydrophobic	-	3.8
M	Methionine	Met	5.74	Apolar	-	1.9
N	Asparagine	Asn	5.41	Polar	-	-3.5
P	Proline	Pro	6.30	Apolar	-	-1.6
Q	Glutamine	Gln	5.65	Polar	-	-3.5
R	Arginine	Arg	10.76	Alkaline	12.48	-4.5
S	Serine	Ser	5.68	Polar	-	-0.8
T	Threonine	Thr	5.60	Polar	-	-0.7
V	Valine	Val	6.00	Apolar, hydrophobic	-	4.2
W	Tryptophan	Trp	5.89	Heterocyclic, aromatic	-	-0.9
Y	Tyrosine	Tyr	5.64	Polar, aromatic	10.46	-1.3

The most hydrophobic amino acids are, in the following order, isoleucine, valine, and leucine. Amino acids with aromatic side groups such as phenylalanine, tryptophan, and tyrosine are also relatively hydrophobic. Paradoxically, however, tyrosine and tryptophan appear to be even more hydrophilic than glycine, threonine, and serine (as measured by hydropathy index using water/octanol partitioning), probably due to the presence of phenolic –OH conferring a polar character with a pK of

10.46 on one hand and a heteroaromatic ring on the other hand (Kyte & Doolittle, 1982) (see the main properties of amino acids in Table 4.6). It is admitted that hydrophobic associations are a little weakened when pH rises and increases when the pH diminishes. These peculiar properties are due to the fact that at high pH a large number of negatively charged groups contribute to increase the global hydrophilicity of the entire protein. On the contrary, at low pH the change of protein conformation generates an intensification of hydrophobic associations. Rearrangements of loop-helix structures are in fact demonstrated as being dependent on pH with, for instance, a direct effect on catalytic residues in the active site of glycinamide ribonucleotide transformylase (Su et al., 1998). The formation of hydrophobic-based rearrangements within a polypeptide chain reduces the possibility of creating hydrophobic associations with other polypeptides. Hence, it can be concluded that pH indirectly affects this type of interaction with combinatorial peptide libraries. The intensity of hydrophobic associations is modified by the temperature: They increase when the temperature is increased up to a certain level (e.g., 45–50 °C) due to an entropic effect and to the attraction forces of van-der-Waals (see the following section), while they decrease at low temperature. In certain cases, however, an increase of temperature has another effect, which is the change in protein conformation with negative effects on hydrophobic attraction. The combination of both effects can induce lower protein solubility.

Hydrocarbon chain associations are weakened by a number of other conditions essentially due to the presence of agents in solution. They are, for instance, alcohols even at low concentrations; detergents and chaotropic agents also contribute to decrease hydrophobic associations as a result of the destructuration of water molecules related to the hydration sphere of proteins.

Intermolecular hydrophobic associations are promoted or enhanced by means of lyotropic salts such as those of the Hofmeister series. For instance, the presence of ammonium sulphate in the initial protein solution triggers the capture of hydrophobic proteins that would otherwise not be adsorbed on the solid-phase peptides. This interaction involves light hydrophobic clusters exposed on the surface of the proteins and hydrophobic counterparts of hexapeptides beads where hydrophobic amino acids are denser (see Section 4.5.4 for protein capture under hydrophobic-promoting effect).

Enhancing and weakening hydrophobic associations is essential when trying to capture proteins and collect them afterward. It is this mechanism that is exploited with CPLLs with, among other interactions, the formation of intermolecular hydrophobic associations between exposed hydrophobic side chains of amino acids and hydrophobic hexapeptides grafted on the beads. A simple wash with a low-ionic-strength buffers removes the lyotropic salt, and a relaxation of the interaction occurs with consequent desorption of proteins that interact with very light hydrophobic associations.

Competitors of hydrophobic groups involved in the complex formation are also deleterious to the protein-peptide interaction; therefore their presence prevents the proper interaction or desorbs proteins that are already adsorbed.

Hydrophobic association phenomena between proteins and a solid-phase sorbent (e.g., CPLL) is based on a similar principle as hydrophobic interaction chromatography. The adsorption mechanism is based on the tendency of apolar molecules to associate in aqueous solutions. The association of apolar groups is characterized by a high-entropy contribution to the free energy of the whole aggregation, while the entalpy contribution is low or even negative.

In the case of proteins, hydrophobic associations with a solid phase act through molecular liaisons involving two hydrophobic moieties: the hydrocarbon side chains of some amino acids and the presence of similar residues grafted on the solid phase. This complex can be produced spontaneously or promoted by the presence of lyotropic salts.

Several mechanistic retention modes in isocratic conditions are described where the effect of salt concentration is similar to the salting-out constant (Melander et al., 1984). Other authors (Staby & Mollerup, 1996) have reported a more rigorous model describing the retention of a protein by a hydrophobic solid phase and the salt concentration in the mobile phase. The effects of modifiers in the mobile phase and of pH are also described. Hydrophobic associations are important for the protein capture by CPLLs, even if their extent in physiological conditions is somewhat lower than ionic interactions.

4.4.3.3 HYDROGEN BONDING INTERACTION CONTRIBUTION

An important interaction in proteins, and hence protein–peptides, is hydrogen bonding that occurs when two electronegative atoms share or compete for a hydrogen atom. The electronegativity of atoms triggers the type of hydrogen bond where the amount of energy involved is different. Electronegativity of atoms naturally increases with the number of electrons that are positioned in the external electron shell. As a result of initial electrostatic attraction, the proton (deshielded hydrogen atom) very closely approaches the acceptor atoms at a smaller distance (typically around 0.3 nm) than the one observed in van-der-Waals interactions (see below). Such an interaction with hydrogen bonding is also observed with aromatic molecules where the electronegativity originates from π-orbitals. Not all molecules have hydrogen atoms able to form hydrogen bonds; however, considering the complex

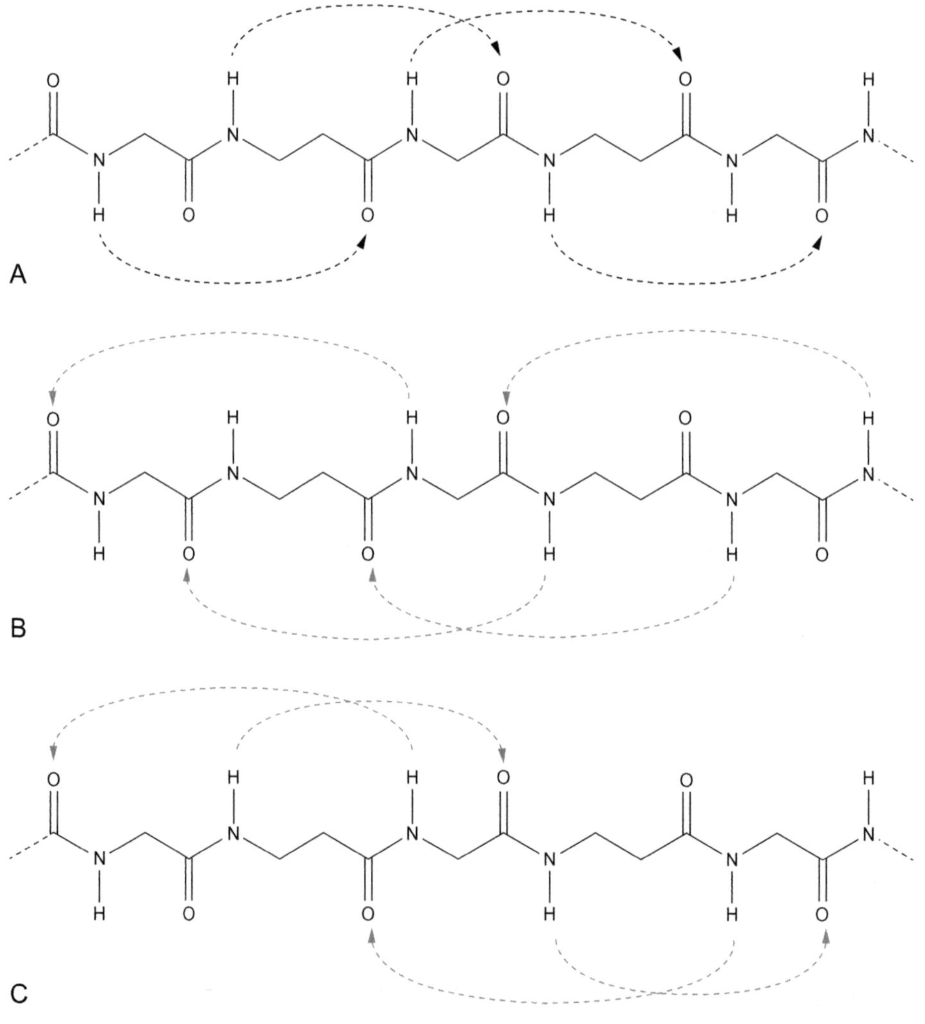

FIGURE 4.22

Examples of forward (A), backward (B), and alternating hydrogen-bonding patterns (C) along the peptide backbone, ensuring the formation or the stability of the helix structure in proteins. *Adapted from Baldauf et al. (2006)*

chemical composition of polypeptides it is interesting to note that a large variety of hydrogen bonds are present. Protein chain positioning is in fact stabilized by hydrogen bonding, and releasing these internal interactions relaxes the entire molecule, with consequent changes in spatial structure. These bonds, which are intramolecular, contribute enormously to stabilize the tertiary structure of proteins.

Within the context of the present discussion, hydrogen bonding is focused between two distinct entities: the immobilized peptide bait from the combinatorial peptide ligand library and the soluble protein prey. Conditions for the formation of hydrogen bonds are satisfied when first the entities involved (the proton donor and acceptor groups) approach quite closely and if the hydrogen atom faces the electron pair. This configuration forms an angle that can vary, largely depending on the spatial situation of the entities involved. If they belong to the same molecule, there could be some restrictions preventing the formation of an ideal hydrogen bond angle; in the case of the interaction of two distinct molecules, the ideal configuration is more easily reached. This is probably the case of most hydrogen bonds between small peptides and a protein. The hydrogen bond angle and the distance between the hydrogen atom and the proton acceptor are correlated to each other: The distance between these entities is the shortest when the ideal hydrogen bond angle approaches the ideality (seems close to 180°). Nonetheless, this does not mean that the energy of interaction corresponds to the minimum. On the contrary, it is the result of a compromise between different forces involved where van-der-Waals interactions are not the least important. The shortest distance between entities forming a hydrogen bond generally produces a strong bond (Figure 4.22). These bonds are stronger between soluble distinct molecules compared to intramolecular functional sites. In this respect the formation of hydrogen bonds between CPLLs and free proteins is quite favorable. There are hydrogen bonds that participate to the stabilization of the secondary structure. They are mostly between the peptide N-H and O=C groups that are coplanar and are at the origin of α-helix and β-sheets (Baldauf et al., 2006). Due to the linear shape of short hexapeptides forming the CPLL, the presence of intramolecular hydrogen bonds has no great probability to be generated. Other hydrogen bonds are present in polypeptides and involve the side chains of the amino acid sequence. Thus interactions occur between=NH and a –OH group; between=NH and the imidazole ring; between=NH and the oxygen of a carboxyl and, finally, between two –OH groups (such as those of Tyr, Thr and Ser).

All these hydrogen-bonding possibilities enormously favor the interaction of the available hexameric ligands grafted on the beads with native proteins in solution (Figure 4.23). In the case of aspartic acid or glutamic acid, the protonated carboxyl groups could act as proton donor, with consequent formation of hydrogen bonds. However, the protonation state of these side chains depends on environmental pH, explaining the dependency of some hydrogen bonding on the pH. This is important because

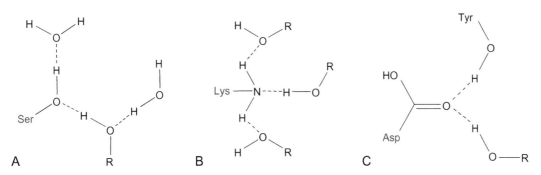

FIGURE 4.23
Examples of hydrogen-bonding formation with chemical functions of amino acids.

A: Formation of hydrogen bonding with –OH residues such as in serine.
B: Involvement of primary amines such as the terminal group of lysine side chain (donor and acceptor).
C: Hydrogen bond to a carbonyl group from aspartic acid and the OH of tyrosine, for example (acceptor only). R=hydrogen.

it allows conceptualizing the mechanisms of association and dissociation of proteins from the partner peptides during practical applications, particularly the elution phase (see Section 4.6). To learn more about hydrogen bonding, the reader can consult the references at the end of this chapter (notably, Jefferey, 1977; Baker and Hubbard, 1984; Janin & Chothia, 1990; White, 2005; Raschke 2006).

4.4.3.4 VAN-DER-WAALS INTERACTION CONTRIBUTION

The van-der-Waals interaction is another important phenomenon that contributes to stabilize of the tridimensional protein structure. Taken individually van-der-Waals interactions are weak attractions between molecules that are in close proximity to each other. They are also known as London dispersion forces. Basically, as two atoms come closer to each other, this attraction increases until they are separated by the van-der-Waals *contact distance*. When two molecules are too close to each other, the potential energy due to repulsion becomes very high; therefore the assembly becomes unstable, and repulsion occurs even when these molecules are neutral. As the molecules move further apart, the potential energy due to repulsion decreases.

This phenomenon is nicely illustrated in the following classical Figure 4.24. Thus as the distance between the molecules increases, the weakly bonded molecules lose their stability and are no longer affected by the van-der-Waals forces due to their large distance apart. Conversely, when the distance between the two molecules decreases, the stabilization is diminished as well due to the electrostatic repulsion between the molecules. This level of repulsion is felt more drastically and more intensely in van-der-Waals interactions than in ionic linkages where the level of repulsion is felt more gradually.

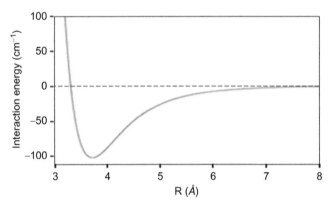

FIGURE 4.24
Relation of interaction energy between two molecules and distance (R) obtained from argon dimer. The interaction energy increases when the distance between two atoms is smaller than 3.8A. Above this value the repulsion decreases and tends to zero.

Energies associated with van-der-Waals interactions are generally small. However, when the surfaces of two large molecules come together, a large number of atoms can be in close contact and the net effect can be substantial. Macromolecules such as proteins contain numerous sites of potential van-der-Waals interactions. In fact, in the coiled-coil protein structure there are the so-called heptad repeats (every seventh amino acid residue) that form, via the side chains, interactions between each alpha helix. When these repeating residues are hydrophobic, van-der-Waals interactions are formed, stabilizing this protein structure. Due to the cumulative effect of these repeated small binding forces the final interaction involved can be very large.

Van-der-Waals forces also play a role in the interaction between proteins and hexapeptides of CPLLs. How and when this attraction–repulsion takes place is difficult to predict, especially because first a number of other forces are involved that may act synergistically or antagonistically and second because of the presence of an extremely large number of situations generated by the combinatorial character of the peptide ligands where all possible configurations are present.

4.5 PEPTIDE LIGAND LIBRARY PROPERTIES IN PROTEIN CAPTURING

Before going into detail about CPLL bead properties, it is clearly understood, from what is described above, that environmental conditions have a strong effect on the protein interaction with their peptide partner. Even a small modification of capture conditions can be detrimental to the repeatability of the experiments. Thus it is highly recommended that one uses samples that are equilibrated with the same conditions of the beads or *vice versa*. In other words, if the capture step is intended to be performed in physiological conditions, it is preferable to use the same buffer to equilibrate the CPLL beads and the sample as underlined in Chapter 8 (protocols of use).

4.5.1 Influence of Peptide Length

The current peptide length for the reduction of the dynamic protein concentration range comprises six amino acids. This length was a good compromise between the capture efficiency of proteins, the cost involved in the synthesis, and the risk of refolding in case of long peptides. This choice was also based on the fact that hexapeptides have long been used as a source of ligands for the purification of proteins (see Section 7.1). Nonetheless, the interaction with proteins is not exclusive of hexapeptides since shorter peptides can capture proteins as well as single amino acids.

The behavior of peptide libraries of different lengths has been examined with a crude cytoplasmic extract of human red blood cells that contains a massive amount of hemoglobin (close to 98% of the overall protein amount) and also comprises a very large number of very low-abundance proteins. The study has been performed using libraries from single amino acids to hexapeptides (see Figure 4.25). Columns filled with different libraries received a massive amount of red blood cell extract that amply oversaturated the columns; after washing to eliminate the excess of unbound proteins, the captured species were separately desorbed and analyzed (Simó et al., 2008). Protein loading was performed under physiological conditions of pH and ionic strength and protein desorption operated using the UCA mixture (8 M urea, 2% CHAPS, and 50 mM citric acid pH 3.3). The analytical determinations performed were SDS-polyacrylamide gel electrophoresis, two-dimensional gel electrophoresis, and mass spectrometry. The latter was carried out after protein separation by SDS-PAGE, gel lane slicing, trypsination, and LC-MS/MS of the obtained peptides.

SDS-PAGE analysis showed that even libraries made with a single amino acid were able to capture a substantial number of different proteins, particularly in the high M_r region (50 to 200 kDa). As the peptide chain was progressively elongated, it was found that a larger pool of proteins was captured, including low M_r proteins that were not captured by the single amino acids column. These results were compared to two controls: (i) the initial extract and (ii) the eluate from a column that did not have any peptides attached to check the level on nonspecific binding. The former showed only very few bands, essentially heavy bands of α- and β-globin chains. The eluate from beads without peptides showed only a very small amount of proteins nonspecifically captured whose 2D gel electrophoresis patterns appeared similar to the initial sample.

The curve obtained by crossing over the number of proteins captured as a function of peptide length shows an asymptotic profile, with the transition toward a plateau already manifested from the 4-mer protein eluate level. The capture of additional species with the extension of the peptide length became smaller and smaller, while the number of diversomers increased exponentially. Considering that amino acids alone can capture a large number of proteins, one can conclude that single monomer structures interact with more than one protein unless protein–protein interaction phenomena are formed on the protein captured by the peptide bait. This does not seem to be a dominant event since by elongating the peptide chain the number of captured proteins does not increase dramatically, while the number of diversomers increases really fast. Amino acids and dipeptide library eluates appear to be rich in species, with relatively large M_r, typically above 40 kDa. It seems that these simple libraries need rather large proteins for efficient binding, since the binding groups are sparse on the polymeric network of the beads

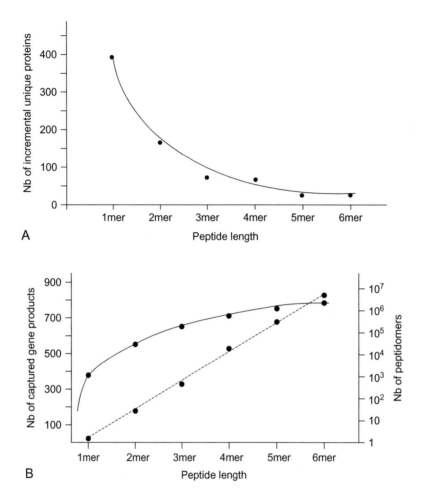

FIGURE 4.25

Schematic representations of the incremental contribution of each peptide library from single amino acids to hexapeptides in the number of captured and identified unique proteins by LC-MS/MS. The upper panel (A) illustrates the incremental contribution of the peptide progressive elongation to the discovery of additional gene products. The lower panel (B) shows that an exponential increase of diverse peptides by elongation of the structure allows only for a modest incremental addition of new gene products.

(typically around 50 μmol/mL), and thus only large-size proteins would have a chance to interact with more than one binding site, rendering the interaction cooperative and thus more stable. As the peptide chain is lengthened above a tri-peptide, the population of proteins captured is enlarged, including low M_r species that are well visible from SDS-PAGE analysis and two-dimensional gel electrophoresis. A relatively large number of species become detectable in the 10 to 20 kDa region and within the pH 7-10 interval.

When comparing the lists of protein captured individually by each library, it is interesting to observe that a core set of proteins is still present while some proteins disappear from column to column and others appear as the peptide elongates. When these libraries with peptides of different length are used as cascade from the amino acid to the hexapeptides, other specific observations can be made. For instance, the common set of captured proteins suggests that these species need only the terminal amino acid of the peptide to be captured. The remaining part of the peptide, especially for the largest peptides, would serve either as a simple spacer or increase the affinity constant for the proteins in question. Species that disappear from one column to the next one with longer peptides may be explained by the fact that they represent the very low-abundance species that are sequestered and thus fully removed (depletion) by the immobilized peptide column. Another possible interpretation is related to possible repulsion phenomena generated by the peptide region that is not involved in the recognition. Beyond the demonstration of the effect of peptide length on the protein capture, the experimental setup could be an interesting tool to separate species according to affinity properties possibly related to dissociation constants as a result of synergistic or antagonistic effects from amino acids composing the peptide ligands.

Study of the importance of peptide length suggests that (i) hexapeptide ligands are required for the capture of proteins from complex extracts because of their ability to increase the probability of partnerships with proteins and (ii) a cascade of peptide libraries of different length may allow discovering additional proteins from a given extract, making this study one of the largest soluble red blood cell proteome descriptions obtained so far. Figure 4.25 summarizes the results obtained by using peptide libraries of different length.

4.5.2 Influence of Environmental pH

Being constituted of charged amino acids, polypeptides are polyelectrolytes that are not insensitive to the environmental pH. Each peptide as well as each single protein is thus an ionizable macromolecule with a net charge, depending on the pH of the buffer where these molecules are dissolved (see the schematic view in Figure 4.20). The pH at which the net charge is zero is called the isoelectric point (pI); at a pH below the pI, the protein is globally positively charged and above the pI it bears a net negative charge. While the pH of the buffer where polypeptides are dissolved regulates the ionization of individual proteins, it also promotes specific behavior: Those having a net charge complementary to others are attracted, while those with the same net charge are repelled. The ionic interaction extent therefore depends on the relative strength of the net charge.

The capture of proteins by a combinatorial peptide library is thus dependent on environmental pH because environmental pH changes the global ionic state and, consequently, it profoundly affects the extent of the interactions. By modulating the ionic net charge of both the grafted peptides and the proteins of the soluble sample, a modification of the affinity constant takes place as described in Section 4.4.2. In the case where the global ionic charge of both the protein and the partner peptide becomes of the same sign and when the other interaction forces are quite weak, the docking does not occur and the protein is not captured by the bead. This phenomenon is very important for several reasons: (i) the selection of the environmental pH drives the capture phenomenon, and (ii) the precision of pH from an experiment to another governs the reproducibility of the process. In addition, at relatively extreme pHs, the change would promote the capture of mostly acidic or mostly alkaline proteins.

The modulation of capturing pH has been used a number of times for enlarging the capability of the peptide library to capture the highest number of species. With the hypothesis that at different pHs the affinity constants would be changed as a consequence of differences in both peptide ligands and proteins ionization, experimental demonstration was made first with models and then with natural protein mixtures. The first benefit of such an approach would render the capture of proteins more effective if operated at least at three different pH values. In practice, serum protein adsorption can be performed at pH 4, 7, and 9, and each captured group fully recovered under classical elution conditions as described by Fasoli et al. (2010). Generally protein capture is operated at pH around 7; the addition of two other pHs allows a significant increase in the number of detectable proteins after treatment with CPLL. In a number of published data, while SDS-PAGE and 2D gel electrophoresis show different complementary patterns, LC-MS/MS clearly indicates that the identified proteins are different. The number of gene products from human serum found at pH 4.0, 7.2, and 9.4 was, respectively, 322, 383, and 309, with a relatively large number of exclusive species found in each condition. In fact, at pH 4.0 the number of proteins found that were absent from captured species at pH 7.2 and 9.4 was 127. At pH 7.2 the exclusive number of unique proteins was 103, and it was 96 at pH 9.4.

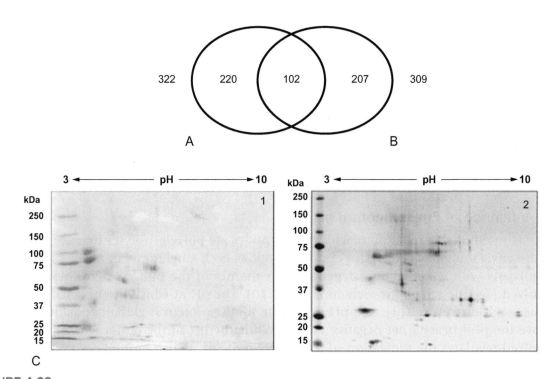

FIGURE 4.26

Protein analysis data from serum treated with CPLL at two different pHs. Elution was performed in both cases by means of 1 M sodium chloride to limit the desorption process to electrostatic interactions.

A and B are the sum of gene products identified by nano-LC-MS/MS respectively captured at pH 4 and pH 9.3.

C represents two 2D electrophoreses of serum protein after treatment with CPLL at pH 4.0 (left plate) and 9.3 (right plate). The protein positioning is very different from a capturing pH to another and appears quite complementary, revealing the different ionization state of proteins during the capturing stage.

Figure 4.26 summarizes some of the experimental findings. The Venn diagram (A) indicates that the number of proteins captured at pH 4 and 9.3 is not very different with a relatively large overlapping of species. However, the examination of protein characteristics allowed interesting observations. For instance, among newly detectable species at acidic pH a large number of proteins were acidic, while among newly captured compounds under alkaline pH were basic proteins. Such a behavior is found for other proteomes as represented in Figure 4.27. Although this phenomenon might be difficult to explain due to the complexity of the underlying equilibrium, speculative explanations have been proposed. For example, pK_a of acidic amino acids, particularly aspartate, is lower in folded proteins as opposed to short peptides (Table 4.7), and this is likely at least partially responsible for the above observations (Thurlkill et al., 2006; Boschetti et al., 2009a). Furthermore, the matrix polymer of the beads supporting the peptides might increase the pK_{as} of the peptides' carboxylic groups, adding to the observed asymmetry of protein–peptide interactions under acidic conditions.

The changes in affinity constant of both combinatorial peptides and proteins in solution deeply modify the phenomenon of displacement effect of species for the same bait peptide and help capturing overall many more proteins than one would expect. In an investigated case, Fasoli et al. (2010) reported that the number of gene products was increased by 65% (from 383 at single neutral pH to 632 upon scanning from pH 4 to pH 9.3). The common proteins found at all pH values represent only 12% of the total; in other words, only a minor portion of proteins were captured equally whatever the pH. The use of combined pHs is also interesting to lessen the losses already

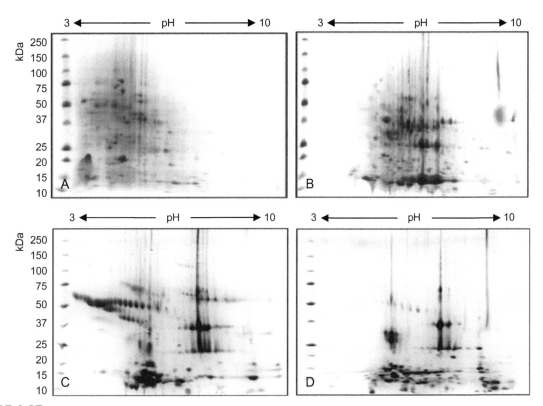

FIGURE 4.27

Two-dimensional electrophoresis of protein extracts after treatment with CPLLs at pH 4.0 (A and C) and 9.3 (B and D).

Upper panels: Spinach leaf protein extract. 150 µg of proteins loaded in 150 µL.

Lower panels: *Bitis arietans arietans* venom. 200 µg of proteins loaded in 150 µL.

Staining with colloidal Coomassie blue. *Adapted from Fasoli et al. (2010)*

reported of proteins that do not adsorb on peptide libraries at a given pH but that are normally detected without any treatment (Fasoli et al., 2010). pH changes are not limited to the influence of electrostatic interactions; they also act under a limited extent on hydrogen bonding and van-der-Waals interactions. For example, interactions between biological receptors and ligands *in vivo*, which are highly specific, are frequently the result of a subtle balance between multiple forces. These phenomena are only marginally reported in the literature. In similar domains, affinity titration curves are used to determine the dissociation constants of lectin-sugar complexes and of their pH-dependence by isoelectric focusing electrophoresis (Ek et al., 1980). However, this approach does not provide information about the complexity of the molecular interaction. In general, models are not extensively described, especially when the ligand structure produces more than one molecular association event. Even considering very simple models such as the affinity of a synthetic hapten for its specific antibody (e.g., N-(P-cyanophenyl)-N'-(diphenylmethyl)-guanidinium acetic acid), the interaction appears to be dependent on the dielectric environment, ionic strength, and pH (Livesay et al., 1999).

Investigations on the electrostatic contribution to the association-free energy neutralizing acidic residues inside the antibody binding site demonstrated the importance of the complementary electrical charge but did not yield any information about other molecular attractions or repulsions that altogether are not easily amenable to modeling in practice. Ion-exchange interactions themselves, in spite of their apparent simplicity, cannot be properly modeled with the steric mass action model. More particularly, the proposed models cannot be extrapolated throughout the entire pH range (Iyer et al., 1999; Shi et al., 2005). In fact the charge regulation effect of coexisting molecular species would need to be incorporated into the model. The situation where there is a large overloading effect on the interaction between a very complex mixture of proteins and a combinatorial library of millions of hexapeptide is particularly unmanageable. Shen & Frey, (2005) used a more complex model where the pH titration behavior of a polypeptide and the associated adsorption equilibrium of various charged forms have been incorporated into a model accounting for the steric hindrance of counterions. This approach clearly illustrates the importance of pH for the interaction of charged species in water. Unfortunately, however, it applies only to a single protein and certainly not to a very complex protein mixture where intense competition effects take place.

TABLE 4.7 pK Values of Ionizable Groups of Amino Acids in Peptides and Proteins (From Grimsley et al., 2009)

Group	pK_a Value in Model Pentapeptides	Average pK_a Value in Folded Proteins
Asp	3.9	3.5 ± 1.2
Glu	4.3	4.2 ± 0.9
His	6.5	6.6 ± 1.0
Cys	8.6	6.8 ± 2.7
Tyr	9.8	10.3 ± 1.2
Lys	10.4	10.5 ± 1.1
C-term	3.7	3.3 ± 0.8
N-term	8.0	7.7 ± 0.5

High pK_a of arginine's side chain (ca.12) ensures that it stays positively charged throughout accessible pH range (Nozaki & Tanford, 1967).

In a more simplified model, study of the ion-exchange equilibrium of albumin in a large pH range revealed that the chromatographic behavior was dominantly controlled by the functional groups in the chromatographic contact regions (Bosma et al., 1998). Similar conclusions were reached when investigating the ion- exchange interaction between two partner proteins (Gibas et al., 2000) underlining the complexity of the situation due to the protonation state of protein partners that do not have the same isoelectric point as in the interaction between free proteins and CPLL solid phases. This is attributed to the similarity of protonation at the macroscopic level when in equilibrium with a solution of a same pH, which contrasts with the microscopic situation where numerous local possibilities at each pH value contributes to the tridimensional protein configuration and hence to the affinity interaction. Small changes of pH may not affect much the overall protonation state of the protein in question; however, even small pH changes produce local alterations of electrostatic potential with consequent modification of the interaction capability with other polypeptide partners and hence with peptide beads when the protein electrostatic cluster is located at the external accessible part of the protein.

This general picture explains the difficulty of modeling the interaction that occurs between proteins of a biological mixture with a peptide library, the latter also displaying a number of other interaction sites depending on their composition. Consequently, this situation does not follow the Langmuir model as recently described for several enzymes (Zhang & Sun, 2002).

Nonetheless, it has been found that electrostatic interactions are the most common forces for the capture of proteins. The protein adsorption at different pHs clearly creates a different capture pattern as seen throughout 2D gel electrophoresis map analyses. This also happens each time an increase of ionic strength with current salts such as sodium chloride takes place, with consequent competition against a large number of captured proteins. Hydrophobic concomitant interactions are not submitted to this rule because this kind of molecular association is almost independent of pH (see explanations of individual molecular interactions in Section 4.4.3).

Although a given protein could be desorbed by either an increase of ionic strength or the help of a hydrophobic competitor, in some situations the protein in question cannot be completely desorbed by either an increase of ionic strength or by separate use of a hydrophobic competitor. The first explains the fact that such a protein may partner with two different peptides, individually insufficient to maintain the overall interaction so that desorption occurs when one of them is weakened. The second case illustrates a different situation in which the peptide acts as a mixed mode hydrophobic and ionic ligand: the interaction is strong enough to keep the protein adsorbed in spite of the use of competitors. Moreover, in this case an increase of ionic strength may even reinforce the hydrophobic association. Figure 4.28 illustrates the pattern of serum proteins eluted by 1 M sodium chloride

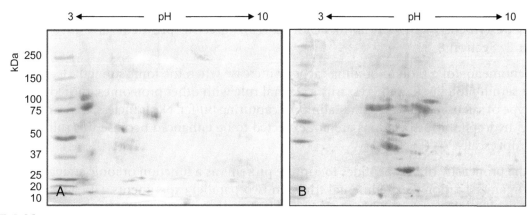

FIGURE 4.28

Two-dimensional electrophoresis of human serum eluates treated with peptide libraries used at pH 4.0. Captured proteins are eluted by two steps: 1 M sodium chloride (A) and then 60% ethylene glycol (B).

The sodium chloride eluate is composed of dominantly acidic proteins as a result of electrostatic dissociation. Conversely, the ethylene glycol eluate is composed of proteins that are randomly disseminated throughout the isoelectric space. No discrimination is detectable at the molecular mass level. Staining with colloidal Coomassie blue. *Adapted from Fasoli et al. (2010)*

(acting on ionic interaction only), followed by an elution with 60% ethylene glycol of proteins captured at pH 4. Clearly it shows that a large number of proteins desorbed with the salt is dominantly of acidic isoelectric point, while hydrophobic proteins are spread throughout the pH range (see Figure 8.14 in Chapter 8). Interestingly, when comparing this behavior to proteins captured at pH 9.3, those desorbed with ethylene glycol show practically the same pattern, while protein desorbed with the saline solution are dominantly alkaline analogously to those found when capturing the serum under acidic conditions (which have predominantly acidic pI values).

Although this study does not give a complete explanation of the interaction process due to the extreme complexity of the situation (multiple ligands, multiple proteins, displacement effects, differences in protein components concentration, multiple types of molecular interactions on a same peptide ligand, overloading conditions), it shows that, in order to capture the greatest number of proteins from a complex sample, it is useful to operate at different pH values. It also demonstrates that when attempting to make comparative pattern studies (e.g., biomarker discovery), it is necessary to keep all thermodynamic environmental parameters such as pH under strict control. Protocol details are given in Section 8.9.

4.5.3 Influence of Ionic Strength

The initial experiments on protein capture by CPLLs were all performed using physiological conditions of pH and ionic strength. Phosphate-buffered saline (PBS) was generally used for this purpose. It was thought that the best way to make a dynamic range compression should be performed under "normal" biological conditions. Quite rapidly, however, it became clear that the ionic strength could have an important effect on the binding capacity. First, trials were performed using human serum. As described by Guerrier et al. (2006), the absolute binding capacity for serum proteins in PBS after incubation with a large excess of proteins was determined to be 12.5 mg protein/mL beads. This number substantially increased (to 19 mg/mL of beads) when sodium chloride was removed from PBS. This phenomenon, attributed to the presence of more active ionic charges at low ionic strength, could be classified as a nonspecific electrostatic interaction. This correlation between ionic strength and binding capacity was confirmed for three different pHs: 5.0, 7.4, and 9.0. However, at this stage no analytical determinations were performed to assess the differences in protein patterns. Low ionic strength CPLL capture was extensively used to find low-abundance proteins as described by Di Girolamo et al. (2011). Specific protocols are also described in Chapter 8, Section 8.8.

This phenomenon of a protein-binding capacity increase when the ionic strength decreases and *vice versa* for serum proteins is not necessarily a general rule with other proteomes; moreover, it depends on the type of salt used to increase the salinity of capturing buffer. Nevertheless, in the case of sodium chloride, hydrophobic associations are not expected to be enhanced because this salt is very low in the lyotropic scale.

To test the propensity of hexapeptides to capture proteins as a function of ionic strength Rivers et al. (2011) organized rational experimental trials. In two parallel experiments, CPLL beads were incubated with skeletal muscle soluble proteins of chicken breast in 20 mM phosphate buffer pH 7.5, in one instance not containing any salt and in a second trial in the presence of 150 mM sodium chloride. The analysis of protein extracts revealed that 107 gene products were exclusively bound at high ionic strength, 41 were bound at low ionic strength, and 74 were bound at both ionic strength values (see Figure 4.29). Clearly, the change in conditions has a profound influence on the composition of captured-eluted proteins. No rules were formally found correlating the ionic strength with the hydrophobicity character of proteins.

Some proteins, notably the glycolytic enzymes such as glyceraldehyde 3-phosphate-dehydrogenase or glycogen phosphorylase, are bound to a greater extent in high ionic strength buffers. Others such as the heat shock proteins and titin bind much more extensively at low ionic strength values. The example of proteins captured independently from the ionic strength was alpha-actinin. Generally, it appears that environmental conditions change the intramolecular interactions of protein structures,

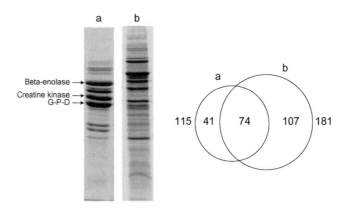

FIGURE 4.29

Skeletal muscle proteins before (a) and after (b) treatment with hexapeptide ligand library. By SDS-PAGE (left panel) it appears that the initial extract comprises several high-abundance proteins such as, for instance, beta-enolase, creatine kinase, and glyceraldehyde 3-phosphate-dehydrogenase, all of them reduced in their concentration during the treatment. The Venn diagram reconstructed from the original reported data (right panel) represents the identified proteins by mass spectrometry. *Adapted from Rivers et al. (2011)*

exposing different or modified epitopes that are the location of the interaction with peptide baits of the solid phase library. Starting from these experimental data, one may find it interesting to depict maps of proteomes after CPLL treatment as a function of both ionic strength and pH throughout two-dimensional functional maps (Righetti et al., 2012).

4.5.4 Influence of Lyotropic Salts

As said in the previous paragraph physiological conditions of pH and ionic strength are currently used for the protein capture by CPLL. In this paragraph it is described a way to enhance the hydrophobic associations while maintaining low the electrostatic interactions. The goal is to try capturing hydrophobic species that could escape the adsorption of peptide library because of their too weak hydrophobicity. To this end the environmental lyophilicity is increased by the presence of high salt levels (typically around 1 M), selected from lyotropic salts of the Hofmeister's series such as ammonium sulfate. The presence of such a salt enhances the hydrophobic associations as described in Section 4.4.3.2 while concomitantly decreasing the electrostatic effect by the high ionic strength.

Ammonium sulfate is also known for its capability to induce protein precipitation by salting out (see Section 8.4), an effect that results from the same phenomenon of removal of water molecules around hydrophobic chemical structures and producing thus aggregates. With this in mind it is important to calibrate the concentration of this salt that should be as concentrated as possible without inducing protein precipitation of the sample. Typically this concentration is around 1M as adopted by Santucci et al. (2013). These authors treated a sample of human serum that was previously treated to reduce the concentration of albumin and IgG with an interesting observation compared to the same samples without ammonium sulfate. First the presence of hydrophobic-promoting conditions allowed capturing proteins that were not found in the non-treated sample. Conversely the absence of ammonium sulfate allowed capturing a number of proteins escaping the hydrophobic association as a corollary of the first phenomenon. Proteins identified within the group where the hydrophobic interaction was induced were of different properties in terms of mass and isoelectric point.

From this analysis one could try to correlate hydrophobic properties of captured proteins with the concentration of lyotropic salts, but this is more difficult since all these proteins are freely circulating in the blood stream and not membranaceous. Moreover even in the presence of hydrophobic-inducing conditions, protein interactions with a peptide library are also conducted by other types of interaction forces (see Section 4.4.3). Nevertheless when analyzing the experimental data one could distinguish at least five proteins Apo A2, Apo C2, Apo F, Apo L1, Apo M whose main function is to bind hydrophobic ligands. These species were only found in the eluate involving ammonium sulfate. Other lipoproteins such as Apo A1, Apo A4, Apo B, Apo C3, Apo D, Apo E, Apo A and Apo H were found predominantly in the same eluate from the hydrophobic capture and, in trace amounts in the eluate performed in the absence of lyptropic salts.

4.5.5 Influence of Sample Loading

The reduction of the protein dynamic range by combinatorial peptide libraries is the result of protein capturing on a solid phase. Thus we will now consider the amount of sample to load on a given volume of peptide library beads. If the latter is predefined and constant, several rules can be followed. The first rule is related to the binding capacity saturation of each bead by the corresponding protein partner(s). The binding capacity of the beads is about 10 mg/mL, or roughly 3 ng of protein per bead of approximately 65 µm diameter. When the amount of protein is increased, the beads become progressively saturated. Below bead bed saturation, the compression of the dynamic range is incomplete or even irrelevant. Figure 4.30 shows this phenomenon where the load was made up of various volumes of a cell culture supernatant incubated with the same volume of peptide beads. Clearly, with very low loads compared to the binding capacity of the beads, no difference was observed between

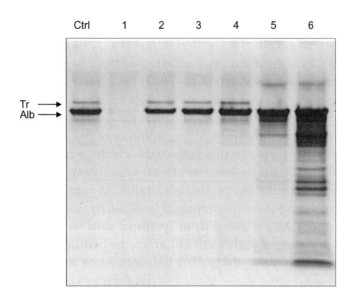

FIGURE 4.30

Cell culture supernatant analyzed by SDS-PAGE, before (Ctrl) and after treatment with hexapeptide ligand library. Three aliquots of CPLL of 1 mL were incubated, respectively, with 1, 10 and 100 mL of lymphocyte culture supernatant.

Lanes 1 to 3: noncaptured proteins as a result of bead oversaturation for these species (from 1 mL to 100 mL sample loading).

Lanes 4 to 6: captured and eluted proteins by means of 6 M guanidine-HCl (from 1 mL to 100 mL sample loading). The desorbed fractions are progressively enriched in low-abundance proteins that were undetectable in the control. Tr: transferrin; Alb: serum albumin. *Adapted from Thulasiraman et al., (2005)*

the initial sample and the eluted species. However, when the volume of the load was 100 times larger, the composition of the eluate was extremely different: Many species undetectable before the treatment became clearly visible in SDS-PAGE. When the loading exceeds the saturation, the competition of species for a given bead becomes more intensive to the profit of species that displace others due to their higher affinity for the peptide bait.

This game is also played well beyond the saturation as observed experimentally when analyzing proteins from red blood cell lysate under different overloading conditions. By using a very large sample (e.g., 5700 mg of proteins) for 1 mL of peptide library beads—in other words with an overload of 570 times the binding capacity—more than 1288 proteins were identified (Roux-Dalvai et al., 2008). In contrast, when using about 1400 mg for the same volume of beads 535 gene products were found. Data analysis from both lists of proteins showed that more low-abundance proteins were found in the larger overloading compared to the second case. For example, very rare embryonic hemoglobin species (ε and ζ chains), which were supposed to be repressed upon cell maturation and thus absent from adult red blood cells, were unambiguously identified when using the largest amount of proteins (5700 mg), whereas they did not show up when using about 1400 mg proteins. Similarly, other rare gene products were found with the larger loading as, for instance, purine-5'-nucleotidase, cyclin-dependent kinase inhibitor, guanine nucleotide-binding protein, and hepatoma-derived growth factor, which were all absent from the second experiment.

In another experiment, where the objective was to enrich for low-abundance glycoproteins, Huhn et al. (2011) found that an increase of the loading sample resulted in a quick saturation with most abundant proteins such as albumin. Many different protein bands were detected from all eluted fractions, with an intensity that was roughly the same for several different loading amounts. However, the authors found that "at very high loading conditions, some additional protein bands are visible." Plasma samples of 250, 500, 750, and 1000 μL diluted 1:1 with loading buffer were mixed with 100 μL CPLL beads and left for three hours while gently shaking. After

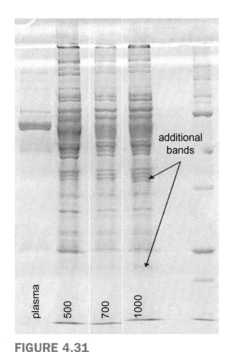

additional
bands

plasma

500

700

1000

FIGURE 4.31

Comparison of TUC (2 mol/L thiourea, 7 mol/L urea, 2% CHAPS) elutions from 100 μL ProteoMiner beads loaded with different amounts of plasma (500, 750, and 1000 μL). The protein markers (right) had masses of 250, 150, 100, 75, 50, 37, 25, and 20 kDa (From the top to the bottom). Additional protein bands are the result of increased overloading conditions. *From Huhn et al. (2012), reproduced upon authorization.*

washing to eliminate protein excess, a three-step elution was applied. The collected protein fractions were analyzed by SDS-PAGE, and the profiles were compared as usual. As expected, the pattern of untreated plasma was dominated by albumin and IgG, and only a few additional bands were detectable visually. On the contrary, a large number of protein bands were clearly visible for the CPLL-treated samples: the enrichment of low-abundance species was reached quite easily, even with a low bead- loading volume. Very importantly at very high loading conditions (e.g., 1000 μL of plasma proteins), some additional protein bands became detectable. With these very low-abundance proteins, the bead-binding capacity was not reached under lower loading (from 250 to 750 μL). With these data another demonstration was made showing that protein profiles can change progressively as the load continues to increase by the fact that other challenging proteins enter into an intense competition with other species already present (see Figure 4.31). This phenomenon is valid not only for large proteins (say, above 20 kDa), but interestingly it was also observed for smaller polypeptides (below 6 kDa), as demonstrated on Figure 4.32.

Although the sample load has a strong impact on the discovery of novel species, when it is increased greatly, some limitations also develop. For instance, progressively the appearance of novel high-abundance proteins can be detected, and therefore novel low-abundance species may be hidden again progressively. This was in fact reported by Rivers et al. (2011) with the demonstration that when the loading was progressively increased, novel proteins became visible as expected, but other novel high-abundance species appeared. The interpretation given by the authors is that at high loads some species such as α-actinin, calmodulins, calpains, and heat shock proteins 90 and 70 might accumulate as superimposed layers on the beads, creating large assemblies of protein complexes. This proposal definitely deserves more attention and stimulates additional investigations; it opens the way for further developments of the already proposed concept of the possible interaction of a sticking to many peptide ligands, as seems to be the case for apolipoprotein A1 (Boschetti & Righetti, 2008). It also opens the way for the possible formation of aggregates of the same captured protein.

The experiment was organized with proteins extracted from chicken pectoral muscles. Six parallel experiments were performed by using increased amounts of protein (20, 50, 100, 250, 500 and 1000 mg) loaded on 100 μL of CPLL beads. After five washings with the incubation buffer, the captured proteins were desorbed and analyzed by SDS-PAGE. It was found that the skeletal muscle exhibits an asymmetry in the dynamic range of protein expression where few proteins, essenvtially glycolytic enzymes involved in muscle contraction, dominate the proteome pattern. Naturally, numerous other low-abundance species are also present but at a level that is impossible to detect prior to sample treatment. After application of CPLL technology while progressively increasing the protein load, the authors found a great compression of the dynamic range, with the distribution of protein abundances becoming shallower, where a good correlation can be appreciated between the appearance of low-abundance species at the increasing sample loads. As described in Section 6.7 (and illustrated in Figure 6.11), the group representing the most concentrated proteins

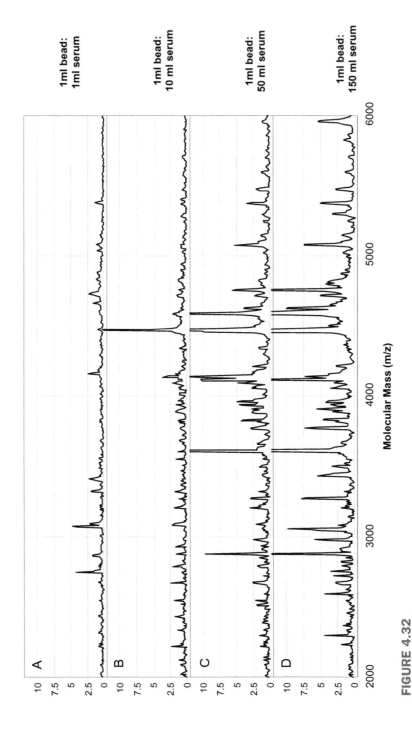

FIGURE 4.32

Effect of protein loading on the enhancement of low-abundance species of low mass. The initial sample was human serum. 1 mL of peptide beads was put in contact with, respectively, 1 mL (spectrum "A"), 10 mL (spectrum "B"), 50 mL (spectrum "C"), and 150 mL (spectrum "D") of human serum and captured species desorbed using 50 mM citric acid pH 3.3. SELDI MS associated with CM ProteinChip array was used for the detection of species of low mass (up to 6 kDa). Many signals increased progressively with the increasing of the load.

before the treatment largely populates the region while leaving void the low-abundance compounds region. In contrast, the bead eluates occupy a progressively larger place. Other interesting phenomena take also place, such as the appearance of novel high-abundance species when the overloading becomes really large.

All these findings verify that one can dig deeper and deeper as long as the protein loading increases. Probing rare proteins is one possible application to improve proteome composition and understanding; another even more interesting possibility, at least from a diagnostic standpoint, is to probe differential protein expression of very rare species applicable to biomarker discovery at an early stage of a disease.

4.5.6 Influence of Library Bead Volume

As a corollary of the previous section, where the volume of the sample was considered, here the volume of beads is discussed in light of its representativeness of the entire library. We will recall that the number of diverse peptides constituting a library depends on the number of selected amino acids and the length of the peptide. The number of diversomers composing the library is easily calculated, as illustrated in the Table 4.1. Figure 4.33 also illustrates the evolution of the diversomers as a function of the number of amino acids and the length of the peptide chain. If one considers, for example, that the number of diversomers is 16.8 million (obtainable by using 15 amino acids for preparing all possible combinations of hexapeptides) and that the bead diameter is 60–70 μm, then the minimum volume occupied by the entire library is 4.5–5.0 mL. Although in theory this is the minimum volume of beads to have the whole library exposed to the biological sample, in reality when sampling 4.5–5.0 mL of these beads out of a bulk, there is no certainty that one will get a representative of each ligand, due to an unavoidable large number of redundancies. Probabilistically, then, to have a better certainty of getting all the diversity, it is at least necessary to sample a larger volume of beads. This situation is clearly explained by Maillard et al. (2009) where it is understood that to get about 90% of the entire library it is necessary to sample 2.43 times the theoretical volume (Figure 4.34). In other words, in the present case a volume of 2.43 × 4.5 = 10.9 mL of beads should be taken. Considering the binding capacity of the beads and the necessity of working under large overloading, such a volume of beads would necessitate a very large protein sample. For example, considering that for a 100 μL of CPLL the amount of serum should at least be of 1 mL, here the required volume of serum would be really large (at least 100 mL).

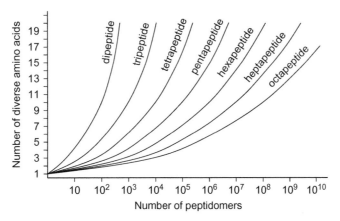

FIGURE 4.33
Number of combinatorial peptides obtainable as a function of the number of amino acids involved and the length of the peptides (from dipeptides to octapeptides).

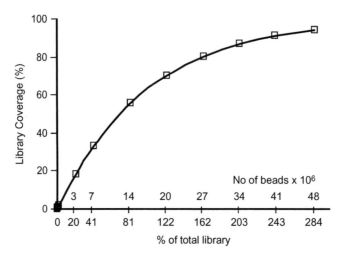

FIGURE 4.34

Statistical representation of library coverage as a function of the volume of beads taken. This applies to a hexapeptide library composed of 16 different amino acids where the total number of diverse peptides is between 16 and 17 million. *Calculated according to Maillard et al. (2009)*

For practical reasons, most reported experiments with a large variety of biological extracts are performed using between 100 and 1000 μL of bead peptide library, with a number of beads significantly lower than the entire library. This represents, respectively, 2.1% or 21% of the entire collection of peptide diversomers. Nevertheless, in spite of this relatively small number of different hexapeptides, the results have always been observed to be satisfactory and very reproducible (see Section 4.5.9). How could this be possible? This dilemma needs to be put in perspective: (i) among amino acids there are similar structures (e.g., glycine and alanine, leucin and isoleucine, aspartic acid and glutamic acid); (ii) different amino acids have similar properties in terms of isoelectric point and hydrophobicity index (see Table 4.5) (iii) the most distal amino acids are those that contribute the most to protein interaction; and (iv) a single hexapeptide captures several proteins, even if with different affinity constants. It has been demonstrated experimentally several times that a given protein can be captured by various peptide structures (Huang et al., 1996; Kaufman et al., 2002) and also that a single peptide structure (single bead) can easily capture various species (Boschetti et al., 2007). Moreover, the serum protein pattern difference found when capturing proteins with a tetrapeptide library and a tripeptide library was very similar (Simó et al., 2008). Therefore, when integrating these different effects, it is reasonable to say that a small proportion of the library does not change the overall behavior of the CPLL for the capture of proteins and the amplification of low-abundance species.

It is within this context that experimental comparative studies focusing exclusively on the reproducibility of sample treatment with the described peptide ligand library have been devised (see Section 4.5.9). Among various experiments, data comparison was made when decreasing the volume of beads and hence their number. These experiments were planned to try understanding if and to which extent a decrease of the volume of bead library would produce different results in terms of protein pattern. Figure 4.35 represents two-dimensional gel electrophoresis of eluates obtained using 20, 50, and 100 μL of ProteoMiner beads treated with, respectively, 0.2 mL, 0.5 mL, and 1 mL of human plasma (the same proportionality between the CPLL volume and the sample volume). Three gels were run for each of the three columns per library volume. The overall spot patterns and protein count from three gel plates did not show significant differences (Li et al., 2009). The data was analyzed either on the entire plate or on a restricted area with similar observations. For example, in a small surface area from pH 6 and 7.5 and masses of about 10 kDa and 50 kDa, it was found that outside of 155 spots detected, there were three spots (2%) with intensity change of over twofold between 20 μL versus 50 μL beads and 20 μL versus 100 μL beads. Principal component

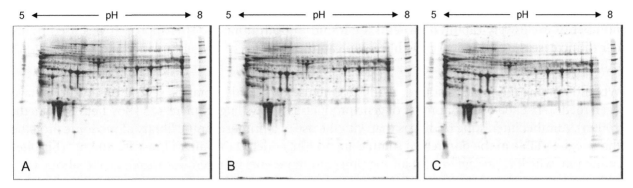

FIGURE 4.35
Two-dimensional gel electrophoresis of eluted plasma proteins from peptide libraries when using various volumes of beads. The plasma volume/beads volume ratio was always the same (1 mL per 100 µL of beads). The volume of beads taken was of 20, 50, and 100 µL for, respectively, "A", "B," and "C." The protein patterns did not significantly differ from each other. All trials were performed in duplicate (not shown).

analyses using SameSpot software showed no significant difference between the three volumes of combinatorial beads.

In another experiment performed on 20 µL and 100 µL libraries using a human serum sample, 382 spots were detected, and it was observed that only three spots (< 1%) had an intensity change of over twofold. Overall from these experiments, it was found that there was no significant difference when decreasing the volume of the CPLL beads to as low as 20 µL.

Since two-dimensional gel electrophoresis gives essentially a qualitative image of the reality, even if spot intensity could be interpreted as a possible variation in the relative concentration of a given protein, additional experiments were performed. This was done by using semiquantitative experiments with SELDI-based mass spectrometry and fully quantitative multiplexed immunoassay assays on a panel of cytokines using Bio-Plex system. Results from these additional experiments confirmed the comparability of patterns obtained when using 20, 50, and 100 µL of CPLL beads (Li et al., 2009).

4.5.7 Modifying the Library and Resulting Effects

The current hexapeptide library used for the reduction of protein dynamic concentration range is a terminal primary amine library. As such, it can easily be modified chemically by grafting different chemical functions. Acetylation is one of the easiest ways to endcap the free primary amines. However, when performed on the unprotected library, the acetylation process takes place not only at the terminal amino group but also on the amine located at the end of the side chain of lysine. The library resulting from such a treatment has different physicochemical properties. For instance, its dominant cationic character is canceled, as is demonstrated on the titration curves illustrated in Figure 4.16. The relatively modest change in the structure of peptides (not on the sequence) also modifies the protein capture as a consequence of affinity constant changes between proteins and peptide partners.

A more in-depth change can be obtained by a succinylation reaction. When exposing the library to succinic anhydride according to current methods used for solid-phase materials (Guerrier et al., 2007a), the dominant cationic character of the library changes to the profit of a dominant anionic character. This change is not only exemplified in the titration curves, but is also proven as quite fundamental with respect to the significant behavior change in the protein sample treatment. After succinylation, it can be seen that about 75–80% of the library remains intact since the modification of the terminal amine changes about 17% of the library composition and 6% change takes place in the lysine side chain. This change is still sufficient to modify the pattern of captured proteins. Several papers have been published comparing the behavior of a primary amine terminal library to a succinylated one, with interesting results. Although a large number of common proteins are captured, a significant number of exclusively captured proteins are found in both libraries. This approach was used in order to increase the capability of the library to capture proteins of low abundance and those that are otherwise lost.

When observing the proteins found, thanks to the use of peptide libraries, several observations have been made. First, the overall number of gene products is always largely increased (Boschetti & Righetti, 2009b). Another interesting finding is that with the use of a primary amine library alone some proteins that are detectable in the nontreated sample are no longer detectable after library treatment. This phenomenon, which is present almost all the time and represents between less than 1% and about 15% of high-medium-abundance proteins (see examples in Table 4.8), could be explained by the fact that several proteins are unable to find their partner peptide or that they are totally displaced by other more affine proteins—or even because their affinity constant for a given peptide is too low. All these hypotheses are of course valid in the selected experimental conditions, including the type of library. However, when comparing the captured proteins from a primary amine terminal library with the corresponding succinylated version, it is possible to reduce the number of "lost" proteins after the treatment.

TABLE 4.8 Protein Identified By LC-MS/MS From Various Protein Extracts Before and After Treatment with Two Peptide Ligand Libraries (Amino and Carboxy Terminal Libraries). Missing Proteins are Species that were Found in the Nontreated Sample and were not Found After Library Treatments

Biological sample	Nb ID Proteins Before CPLL	Nb ID Proteins After CPLL	Nb Missing Proteins	Missing Proteins (%)
Red cell lysate	536	1543	57	3.5
Cerebrospinal fluid	433	1098	61	5.2
Human platelets	197	411	24	5.5
Human urine	134	383	30	7.2
Bile fluid	141	197	25	11.2
Egg white	41	147	1	0.7

Figure 4.36 illustrates three concrete examples. In each case, the number of exclusively found gene products thanks to the use of a second library (e.g., succinylated amino terminal amine) was substantially increased. In two cases it was increased by about 10%, and in the other two cases the increase was larger, with about 30–40% additional proteins. Concomitantly, the reduction of losses was between 20 and 50%, depending on the proteome type. When analyzing the exclusive proteins found by these two variants of the same library, it is soon evident that the proteins belong to two different populations. For instance, while observing the human platelet proteome after peptide library treatment, the proteins exclusively found by using, respectively, a primary amine and a carboxylated version of the same library, and plotted in a space crossing over the isoelectric point and the hydrophobic Gravy index (HGI), gave strikingly different results (see Figure 8.14). Proteins from a primary amine terminal library were mostly located within the acidic space (pI 4–6 range), whereas those from the carboxylated library were mostly located in the opposite area. Few exceptions were observed; however, the fundamental trend followed the isoelectric point of the proteins. Interestingly, the positioning according to the hydrophobic properties was very random, in agreement with the observation that most of the platelet proteins retained and analyzed in this figure are rather hydrophilic; only a few have positive HGI values.

A number of other experiments in this domain have been performed, and published data demonstrate the differences and the complementary performance of carboxylated libraries versus amino-terminal libraries for a number of biological extracts such as bile fluid (Guerrier et al., 2007b), egg white (D'Ambrosio et al., 2008), red blood cell lysate (Roux-Dalvai et al., 2008), and cerebrospinal fluid (Mouton-Barbosa et al., 2010).

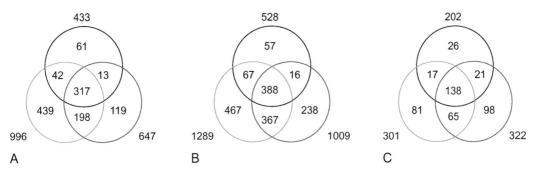

FIGURE 4.36

Proteins identified by LC-MS/MS from different biological extracts after treatment with two types of CPLLs: primary amine terminal peptides (blue circles) and carboxylated version (red circles). All results are plotted as Venn diagrams to visualize the contribution of each type of CPLL and their exclusive performance. Within the same protein extract the comparison is made against the nontreated samples (black circles). In all cases it is interesting to notice that with a single library several high- or medium-abundance proteins detectable in the starting sample are lost. They are largely, but not completely, recovered when using a complementary library.

A: Human cerebrospinal fluid. B: Red blood cell lysate. C: Human platelet extract.

The numbers outside the circles are the counts of unique gene products identified; numbers inside the graphics represent exclusive proteins found in each analyzed group or common proteins (common areas between circles).

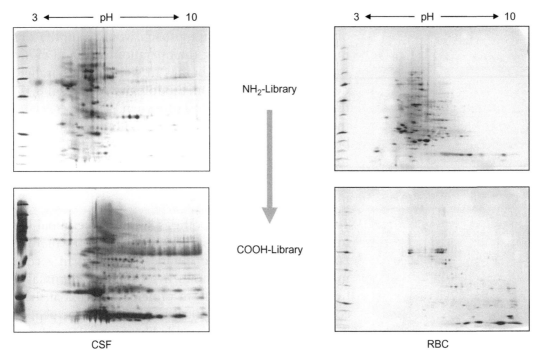

FIGURE 4.37

Two-dimensional electrophoresis analysis of two different proteomes treated with two different types of CPLL.

CSF: human cerebrospinal fluid.

RBC: red blood cell lysate.

The isoelectric focusing gradient was between pH 3 and 10; the second dimension masses were from 250 to 10 kDa. Protein elution from beads was implemented by 9 M urea, 50 mM citric acid up to pH 3.3. Staining with colloidal Coomassie blue.

Figure 4.37 presents examples of differences in two-dimensional electrophoresis patterns when using different libraries. Main differences focus on the isoelectric point of captured proteins: the amino-terminal library capturing dominantly acidic proteins, while carboxy-terminal libraries favoring basic gene products. One can argue that the different behavior is due to an ion-exchange effect of these complementary libraries. This is in fact not a dominant effect: first, because the terminal carboxylic acids and primary amines contribute only marginally to the whole net charge of the library at neutral pH and second because the capture step is performed under physiological conditions (presence of 150 mM of sodium chloride in the buffer), which largely reduces all ion-exchange effects. It should also be added that at neutral pH the ion-exchange effects, even in low ionic strength, are small. In conclusion, all changes are due to the affinity constants of protein peptide interactions modulated by the presence of terminal charged groups: Carboxy-terminal baits would favor a stronger affinity for basic proteins and capture them more efficiently. This mechanism is reversed when operating with a primary amino peptide terminal library. As stated above, the concomitant use of two libraries contributed to increase by about 20% the total number of detectable proteins that otherwise would have been missing. Other chemical modifications are not reported, but they are possible and would confer to the library properties that would probably extend the coverage in terms of protein capture and also would make the libraries more focused for specific applications. Acylation, alkylation, glycosylation, elongation to peptoid structures, and other treatments could easily be possible. Some suggestions of chemical modifications are described in Section 8.2 with possible protocols of use.

4.5.8 Mixing Libraries and Resulting effects

Since chemically modified libraries are used under comparable conditions and give different, complementary protein patterns, it may be interesting to explore how two different libraries would behave once blended. Such an approach would simplify the capturing process by reducing the operation time and elution steps. Instead of using two libraries in series or in parallel (see the description of separate methodologies in Section 8.11), would they globally behave similarly when used after having mixed the library beads? The blend of two libraries doubles the number of diversomers. With the diversomers acting independently from each other with an intensive competition only within the same bead, the final result would be the additive effect of each individual library.

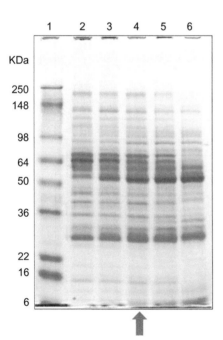

FIGURE 4.38
SDS polyacrylamide gel electrophoresis of human serum protein eluates from two different peptide libraries (amino terminal peptides and carboxylated version) mixed in different proportions. Lane 1: molecular standards. Lane 2: 100% of primary amine library. Lane 3: 75% of amine library and 25% of carboxylated version. Lane 4: 50% of amine library and 50% of carboxylated version. Lane 5: 25% of amine library and 75% of carboxylated version. Lane 6: 100% of carboxylated version. The arrow indicates the optimal blend to obtain the largest number of detectable proteins.

To demonstrate this hypothesis, a carboxy-terminal library was mixed with a primary amine-terminal library in various proportions and the mixed libraries were used in a comparative manner. Figure 4.38 shows five protein patterns as obtained by SDS-PAGE analysis from proteins collected at the outlet of columns constituted of 100% of primary amine peptide library, 75% of primary amine peptide library and 25% of carboxy-terminal peptide library, 50%–50% mixture, 25%-75%, and finally 100% of carboxy-terminal peptide library. The two libraries independently show different patterns and individual protein bands that appear all together when the libraries are mixed in equal proportions. This experiment shows that it is practically possible to save time and sample manipulations when the plan is to use more than a single library. The use of libraries separately could, however, be justified when the sample is very complex. As a result, better analytical results could be obtained with less labourious elution samples as is the case for sequential elutions.

At this stage, an interesting question remains open: Would it be similar or different to use a peptide library (possibly after acetylation of the primary amine terminal group) at acidic pH (e.g., pH 4) and then at alkaline pH (e.g., pH 9) compared to the use of carboxyl- and then amine-terminal libraries at neutral pH? From the ionic standpoint, these two approaches are equivalent since instead of ionizing the library it is the protein mixture that is differently ionized. Nonetheless, as discussed in Section 4.5.2, the change in pH may also modify the protein folding, with little modification of hydrophobicity; therefore some differences are expected. A second aspect to consider involves the use of blended libraries (primary amine terminal and carboxylated version) at three different pHs. A lot of redundancies would be expected, but very probably the number of gene product enhanced would be larger and the number of "lost" species would be further reduced. These alternatives might also induce some differences in the overall protein patterns when fractionation elution is operated. Ultimately, if the objective is to keep the process as simple as possible while enlarging the capability of a library to capture more acidic and more alkaline proteins, it is suggested to keep the pH neutral while mixing two or more complementary libraries.

4.5.9 Reproducibility Considerations

The preceding discussion of the number of beads, the volume of the sample treated per a given number of beads, and thermodynamic considerations prompt a natural question about the reproducibility of the technology both within the same batch of beads and with respect to different batches. Since the technology first developed, reproducibility was described in various reports such as during the discovery of novel proteins from human serum (Sennels et al., 2007) and then for proteins identified from red blood cell lysate (Roux-Dalvai et al., 2008). Consistency demonstrations were deduced from analytical electrophoresis and mass spectrometry data during numerous studies. Interesting evidence of efficiency, recovery, and reproducibility was presented by Sihlbom et al. (2008); these authors concluded that the use of CPLL for low-abundance species enhancement was obtained "in a reproducible fashion."

An even stronger reproducibility assessment was made by Mouton-Barbosa et al. (2010) during the treatment of cerebrospinal fluid. Dedicated scientific reports have also been published in which various approaches have been taken to investigate the question of reproducibility of the sample treatment (Li et al., 2009; Dwivedi et al., 2010; Marco-Ramell and Bassols, 2010; Fröbel et al., 2010; De Bock et al., 2010). The need for reproducibility demonstration came mostly from bead library sampling consistency, especially when small volumes are involved (see the discussion in the preceding section): The representativity and proportions of peptide ligands are not statistically the same. The situation becomes a greater concern when very small volumes of beads and hence of diversomers are involved. In spite of theoretical concerns, all experimental data demonstrated very good reproducibility from where logical explanations were found.

As an example of reproducibility, Figure 4.39 reports protein patterns of eight serum mouse samples obtained by SDS-PAGE. Protein capture by CPLL was performed under physiological conditions, while protein elution was conducted in three distinct phases using increasingly stringent desorbing agents. All eluates from the same elution are remarkably similar, while eluates from different steps are clearly complementary. To complement these data, another experiment was performed by using the same sample—human serum—treated with three different CPLL columns, where remarkably comparable results were shown (Li et al., 2009). These authors found that treatments of serum or plasma with different lots of CPLL beads and different volumes showed similar results when analytical methods such as SDS-PAGE, two-dimensional electrophoresis, mass spectrometry (single and tandem), and immunoassay were used. Interestingly, the signal from a large number of mass spectrometry protein peaks detected by SELDI-TOF over six different experiments performed under the same conditions showed very similar responses (see Figure 4.40). The dispersion of results was rather narrow, with pooled variability coefficients not larger than 20% of the signal intensity, well within the experimental error in daily laboratory practice.

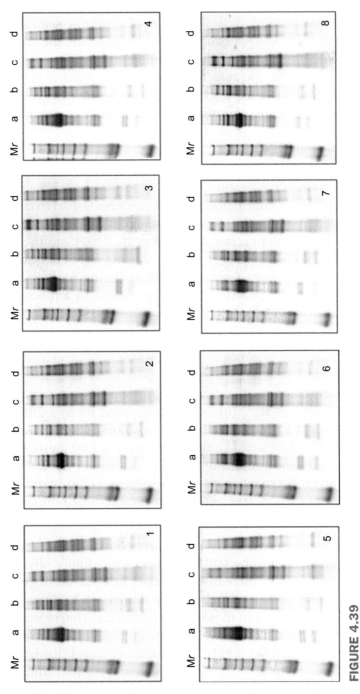

FIGURE 4.39
SDS-PAGE of eight samples of mouse serum before (lanes "a") and after treatment with CPLL. Three successive elutions were performed to desorb the captured proteins: The first was operated by using 1 M sodium chloride (lanes "b"); the second with 3 M guanidine-HCl pH 6; and the third elution using 9 M urea containing 50 mM citric acid pH 3.3. Mr are lanes with molecular mass standards.

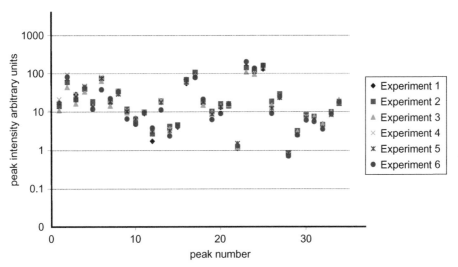

FIGURE 4.40
Variation of mass spectrometry signal intensity of 30 proteins from six different serum treatment replicates with peptide ligand beads. Experiments performed using SELDI-TOF-MS with CM-10 ProteinChip array. The sample treatments were performed by using each time 1000 μL human serum loaded onto 100 μL hexapeptide beads.

Reproducibility was also investigated by using LC-MS/MS to determine whether the number of gene products identified is similar when repeating the same experiment. Figure 4.41 shows data from two different experiments (two different aliquots of CPLL beads) performed in triplicate with very satisfactory results. The little dispersion in this case may not only be due to the CPLL treatment but also to the reproducibility of all other involved steps such as trypsination, liquid chromatography of peptides, and mass spectrometry analysis.

The reproducibility results obtained in very different conditions and even though the CPLL may not comprise all diverse peptides can be explained by using the same arguments discussed in the Section 4.5.6. In another set of experiments, three times 1 mL of human serum was treated with 100 μL of beads. The proteins collected by means of 8 M urea containing 2% CHAPS and 5% acetic acid were analyzed by standard two-dimensional gel electrophoresis against a control, followed by spot count and spot intensity analysis. The control was very different from samples treated with CPLLs; however, the patterns from treated samples were very similar in terms of spot positioning throughout the isoelectric point and the mass ranges. Visual and scanner examination of the maps after the library treatment did not result in significant differences in spot count, positioning, and intensity, thus confirming other data from orthogonal analytical methods (Sennels et al., 2007).

Similarly, Dwivedi et al. (2010) examined the reproducibility of the technology by using LC-MS/MS for peptides obtained from protein eluates from library columns and labeled with different species of iTRAQ reagents. The results thus obtained suggested that the peptide library treatment of human serum would provide the required reproducibility for biomarker discovery based on similar patterns of isolated protein such as albumin, transferrin, fetuin, and peptidyl-glycine α-amidating monooxygenase. Although the proteins were captured with somewhat different efficiencies, the reported data showed that the overall captured repertoire was very similar. In addition, when extending the analysis to individual iTRAQ-labeled peptides from the two runs, evidence was provided that the reporter ions were present in each spectrum. The overall experimental results demonstrated that there were in fact few, if any, verifiable qualitative differences in the peptides detected with confident reporter ion intensities.

From this first paper where iTRAQ technology was used for quantitation purposes, other authors adopted a similar approach (see the next section on quantitation aspects of CPLL). Fertin et al. (2010),

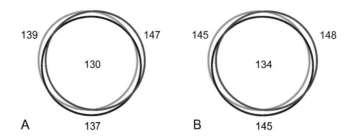

FIGURE 4.41

Venn diagram representation of data from protein identification by nanoLC-MS/MS of serum treated with CPLL. Two different experiments were performed (A and B), each of them in triplicate (red, blue, and green circles). The very narrow dispersion of data with differences that are attributed not only to CPLL but also to LC-MS/MS treatment and other manipulations should be noted as indicated in the text. *Adapted from Li et al. (2009).*

for example, performed an analysis of expression difference of low-abundance proteins in blood proteins from patients with left ventricular remodeling after myocardial infarction in view of identifying biomarker candidates. They found that CPLLs not only allowed increasing the resolution of proteins, but also demonstrated good reproducibility for the use of the technology in protein marker discovery, with specific signals found only in samples treated with peptide libraries. They concluded that the technology associated with their analytical determinations gives access to low-abundance differently expressed proteins to select proper signals as a basis for biomarker discovery in left ventricular remodelling.

Reproducibility studies have been extended beyond human serum or plasma. For instance, Marco-Ramell and Bassols (2010) described the applicability and reproducibility of CPLL technology to bovine and porcine sera where classical immunodepletion cannot be used due to the specificity of species. Using human urine, Santucci et al. (2012) described an entire process of pretreatment and CPLL use, with very reproducible results that form the basis for difference expression studies. Inter- and intra-assay reproducibility data were generated by using normal human urines from different individuals, and analyses were repeated several times, with very good reproducibility data assessed by two-dimensional electrophoresis.

In a more challenging situation, Mouton-Barbosa et al. (2010) evaluated the reproducibility of CPLL treatment of cerebrospinal fluid (CSF). This biological fluid is available in very small samples, and the proteins it contains are very dilute, with a similar dominance of high-abundance proteins as in serum. After having performed the protein analysis from a large pool of CSF, where they discovered 1213 unique gene products, the authors extended the analysis to sample sizes compatible with hospital practices. When using only 2μL and 5 μL of CPLL beads loaded with 1 mL of CSF, they could reproducibly identify 530 proteins (about one-half the total catch compared to the large-scale experiment), more than twice as many as in untreated CSF. This suggests that even miniaturized protocols are efficient and reproducible, thus opening the way for the discovery of biomarkers of brain diseases with a minimized risk of losing protein signals. The number of gene product found was lower than the main experiment (530 versus 1213) as expected due to the much smaller sample used that did not allow for the enrichment of all very low-abundance species. Overall reproducibility results are largely demonstrated with the use of peptide libraries for the treatment of biological samples. Nevertheless, it is essential that all physicochemical parameters remain absolutely identical from experiment to experiment. Attention should be paid to verification that at least the pH and the ionic strength are the same for CPLL equilibration and the biological samples (although, of course, capture times and temperature are just as critical).

4.5.10 Quantitation Aspects

Modification of the dynamic concentration range of a biological extract by using CPLLs was considered an obstacle to proper quantitation of expressed species. In reality, when dealing with the quantitation aspect of protein expression there are two main points to consider here: (i) the relative and (ii) the absolute quantitation of expressed proteins. Both questions, moreover, have to be put in perspective according to the goal of the quantitative determination: a mean either to determine the abundance of selected proteins in a given biological extract or to determine the expression difference between two samples (e.g., a normal case *versus* a diseased one).

The compression of the dynamic concentration range with CPLL is the result of intense competition among proteins for partnering with a given group of peptides. This competition is similar any time the initial samples are comparable (same source, same species, same physicochemical conditions). This phenomenon of competition is highly dependent on the concentration of single protein species; therefore, if two samples, comprising one low-abundance protein of different concentration, are treated with CPLL, the increase of the concentration of such a protein subsequent to dynamic range compression is lower with the sample containing initially lower amounts of the same protein. If then the patterns are compared, the signal related to the protein in question is different. In other words if one protein is overexpressed, it remains more concentrated compared to a control even after the treatment with the peptide library. Naturally this phenomenon is interesting only for species that are not detectable (or of very low-abundance) since for all other proteins the treatment is either irrelevant or even unnecessary. Such behavior leaves the detection of expression difference fully possible. The absolute difference may not be the same when compared to the untreated sample, but the possibility of discovering protein markers by expression difference remains untouched (see many examples of protein marker discovery in Chapter 6). The treated sample is, however, usable only for determining expression difference and not for absolute quantification of the misexpressed protein. Once identified, the latter could always be quantified by classical immunochemical sensitive methods.

The quantification aspect can therefore be performed by using the CPLL treated sample. Figure 4.42 schematically illustrates the alternative procedures. A biological sample treated with a peptide library gives two fractions: The first fraction is the nonadsorbed, and the second comprises the captured group of proteins desorbed from the beads by means of appropriate eluting agents. The amount of proteins captured by the beads compared to the initial sample is small: Based on the use of 1 mL of serum and 100 µL of CPLL beads, the captured proteins represent about 1.0–1.5% of all proteins contained in the initial sample. If the misexpressed protein is detectable only after CPLL, the quantitation is possible on the basis of the volume of eluted sample. Conversely, if the misexpressed protein is detectable in the nontreated material, the determination and calculations can easily be performed using the noncaptured sample. In all cases the general rule is first to discover expression difference

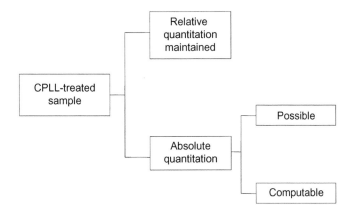

FIGURE 4.42
Possibility of quantitation of misexpressed proteins found after treatment with CPLL. When the misexpressed protein is found only in the treated sample, a direct quantitation can be calculated based on both elution volume and initial sample volume. Alternatively, the formally identified protein can be quantified using, for instance, ELISA-based methods on the initial nontreated sample.

of low-abundance gene products after CPLL treatment and, after identification, accurately determine quantitatively the expression difference directly from nontreated samples.

A number of published reports describe the quantitation aspect of proteins after treatment with combinatorial peptide ligand libraries. For instance, already in 2008, Roux-Dalvai et al. demonstrated the maintenance of proportionality between protein concentrations. To a human protein extract the authors added an exogenous alcohol dehydrogenase and analyzed the signal of peptides after treatment with CPLL. The signal was linear in the interval 100 to 1000 pmol, suggesting for the first time that differential quantitative proteomics could be performed for low-abundance proteins after reduction of the dynamic concentration range.

Other evidence of quantitation came from the work of Hartwig et al. (2009). Here the authors spiked human serum with an entire *E. coli* lysate previously labeled with Cy5, in the 3 to 300 μg concentration range. Samples were then analyzed by two-dimensional electrophoresis, and a central region of the pattern comprising about 450 spots was investigated. It was first found that no measurable amounts of spiked proteins were detected in the flow-through (except for the 300 μg load), indicating an essentially complete binding of the *E. coli* proteins. Moreover, the signal was linear for about two orders of magnitude. It was demonstrated that combinatorial peptide ligand libraries are able to sweep with high efficiency not just a few spiked proteins, but an entire proteome as represented by the *E. coli* lysate. The reported linear recovery over about 2 orders of magnitude suggests the possibility to proceed for differential quantitative analysis of low-abundance proteins detectable after CPLL treatment. Translated in moles, the range of linearity for the signal is assumed to be in the pico- to nano-mole range.

This aspect has been further clarified by Mouton-Barbosa et al. (2010). In this study, four heterologous proteins were spiked (beta-lactoglobulin, beta- and kappa-caseins, and phosphorylase b) into human serum in the nano- up to micro-molar range followed by a treatment with combinatorial peptide library (the experiments were performed in triplicate). After trypsination, the mass spectrometry signals were analyzed, and a growing linear signal as a function of protein concentration over a concentration spanning at least three orders of magnitude was found. Thus it was demonstrated that relative quantitation of proteins is possible after peptide library treatment, as long as the saturation of the beads is not reached. A good reproducibility was reported, with possible application for label-free relative quantitation of sample processed in parallel.

Another significant demonstration was given by Li et al. (2009) when the proteins were captured in fixed and variable volumes of CPLL beads and when several cytokines were spiked into human sera and their recovery was studied as a function of different volumes of capturing beads (20 and 100 μL, respectively). Cytokines were recovered from the beads, with similar efficiency in both bead volumes. The efficiency of concentrating them onto the beads was in general very high (90% for four of them, 60% for two of them, and 20–25% for just one, interleukin 4).

In other cases, the quantitative aspect of low-abundance protein discovery was approached by combining CPLL and isobaric mass tags (iTRAQ). This was the case for the identification of candidate protein markers for diabetic nephrology in type 1 diabetes (Overgaard et al., 2010), performed on a cohort of more than 120 individual plasma samples. Differential expression of several types of lipoproteins was validated by multiplexed immunoassays, confirming the patterns observed by iTRAQ quantitation assays.

Similarly altered protein expressions were determined by using the iTRAQ process in view of qualifying possible protein markers of clinical interest for plasma samples from individuals infected by HIV-1 (Shetty et al., 2011). Infected and healthy samples were treated in parallel with CPLLs first and then subject to iTRAQ labeling followed by LC-MS/MS analysis. A number of proteins up- and downregulated have been found, but the most relevant were apolipoproteins and complement proteins. They were involved in hepatic lipid metabolism and acute phase response signaling pathways.

Wang et al. (2011) investigated the protein expression difference from endometrial carcinoma, a quite common gynecological malignancy in women. First, serum samples of different degrees of malignancies were treated with CPLLs, and then the collected fractions were labeled with isobaric tags for further analysis by tandem mass spectrometry. More than a dozen proteins were found to be modified in their expression and were quantified. When the list was compared to an international database, seven proteins were found for the first time: α1-antichymotrypsin, apolipoprotein A-IV, haptoglobin, histidine-rich glycoprotein, inter-α-trypsin inhibitor chain 4H, orosomucoid, and serpin C1.

Another quantitative proteomic analysis after CPLL treatment followed by iTRAQ was reported to detect pregnancies at risk for preeclampsia (Kolla et al., 2012). Comparison of maternal plasma samples (12-week gestation) between uncomplicated deliveries and those who developed preeclampsia showed several upregulated proteins, including clusterin, fibrinogen, fibronectin, and angiotensinogen and an immune-modulatory molecule galectin-3-binding-protein, most of them known to be associated with preeclampsia. In this study, the quantitative proteomic iTRAQ analysis also suggested that CPLL could be used in conducting differential quantitation investigations.

4.5.11 Combinatorial Libraries versus Random Peptide Mixtures

Although the mechanism of action of combinatorial peptide ligand libraries has been extensively described in this chapter, especially in Section 4.4.1, the following material seeks to demonstrate that the coexistence of diverse bead ligands within the mixed bed of CPLL is essential to properly compress the protein dynamic range of biological extracts. Although this seems obvious, some statements in the literature compare ProteoMiner with current reverse solid phases and with crude solid-phase extraction (SPE), with affirmations that this peptide library acts according to a regular hydrophobic adsorption mechanism whereby the diversity of peptides has only a marginal effect (Keidel et al., 2010). In reality, the molecular mechanisms contrast considerably with this affirmation. First, the molecular mechanisms of interactions are diverse, and it appears that the main interaction is ionic. This is demonstrated by the large protein desorption effect when using buffers enriched with salts. Salted buffers are recommended in fact for a first desorption of proteins (see the protocol sections in Chapter 8). Clearly, not all these proteins are captured by hydrophobic interactions. Supporting this reality is the fact that by reducing the salt content during the adsorption phase the binding capacity of the CPLL increases considerably, underlining the electrostatic effect of interaction as described in Section 4.5.3. Needless to say, hydrophobic interaction does not occur in low-ionic strengths.

A solid-phase library is a mixed bed made up of a collection of beads, each of which carries a different grafted chemical structure. In the CPLL case, each bead carries a different peptide under a quite high-grafted density and each single bead is in principle capable of adsorbing a different protein. However, due to existing peptide affinity similarities for a given protein or various protein epitopes, the protein docking is modulated by the conditions of interactions, the respective concentration of proteins within the mixture, and the level of overloading conditions calling for an intensive displacement effect of analytes. As explained in a recent review (Righetti et al., 2010), the larger the sample loaded for a given volume of beads, the larger the number of low-abundance proteins detected. These conditions, which are necessary, are satisfied when using hydrophobic interaction chromatography or even when using ion-exchange media where all beads are exactly the same.

Another very compelling demonstration that CPLLs are differentiated in their ability to capture diverse proteins is seen in the fact that the same libraries have been extensively used for identifying hexapeptide structures used for the purification of one protein of the mixture (Pennington et al.,

1996; Bastek et al., 2000; Miyamoto et al., 2008; Yan et al. 2009). In other words, once the peptide ligand is identified, it is used as an affinity column where almost only the target protein is captured while all other proteins are eliminated in the flow-through (for details see the opening Section of Chapter 7). This operation is clearly impossible with the use of either hydrophobic solid phases or ion exchangers. Thus, although Keidel et al. (2010) might think that they have given a superb performance of a Rossini opera (Figaro's celebrated cavatina, "Largo al factotum" in the *Barbiere di Siviglia*?), we suspect that their voice has rather the splintered quality of a note sounded on a length of cracked bamboo.

Given the diversity of peptide ligands and their large number of interaction points, one could ask whether it is necessary to have each of them individually grafted on singular beads. It is within this context that specific experiments have been performed. Two different libraries have been made in parallel: The first was a combinatorial library of hexapeptides as described in Section 4.3.2, and the second was composed of the same peptides but all of them were grafted together within the same bead. With these two solid phases, the capture of serum proteins under conditions of overloading was performed, and the collected proteins were analyzed by two-dimensional electrophoresis. Figure 4.43 shows different patterns as expected. Both decreased the level of abundance of albumin (see arrow). However, while the "genuine" CPLL solid phase (one-bead-one-ligand) captures a multitude of low-abundance proteins, the "randomized" library (one-bead-all-ligands) seizes considerably fewer species. Moreover, the one-bead-all-ligands approach generated mostly one high-abundance species identified as apolipoprotein A1 (double arrow).

This phenomenon has several possible explanations, but the most likely one is that when intensive competition takes place for embarking on the same bead, strong "opportunistic" binders present an advantage against many other proteins. This is where the overloading finds a strong advantage with regular CPLL. Actually, the displacement phenomenon of protein binding on all together grafted peptides (all-peptides-one-bead) could be differently interpreted compared to the CPLL beads. It is assumed that within a single bead more proteins participate to local displacement effects, while interacting with more than one ligand or a portion of it within the same bead. Therefore, the saturation phenomenon due to overloading conditions may not be the same as it is in the case of one-peptide-one-bead configuration. As a result, the final composition of eluted proteins from such an immobilized peptide solid phase is largely different from the composition that results from the one-peptide-one-bead library (Boschetti & Righetti, 2008). Interestingly, the all-peptides-one-bead library does not seem to give any advantage to the capture of low-abundance species; on the contrary, it reveals fewer spots in two-dimensional electrophoresis and creates new high-abundance proteins.

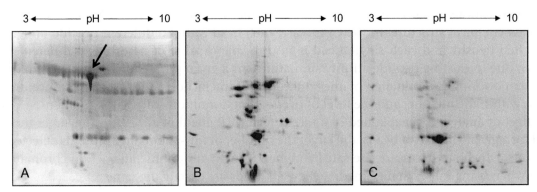

FIGURE 4.43
Two-dimensional maps of human serum proteins (A) treated with either a combinatorial peptide ligand library (each bead with a different peptide ligand) (B) and with a randomized peptide grafted in bulk (all peptides in each bead) (C). First dimension: immobilized pH 3-10 gradient; second dimension: SDS-PAGE in 8 to 18% polyacrylamide slab, in a discontinuous Laemmli buffer. Staining with colloidal Coomassie Blue.

4.6 ELUTION MECHANISM DIVERSITY AND COMPLEXITY

The elution or the recovery of proteins captured by CPLLs is a fundamental step in dynamic range compression technology. Protein desorption must be complete without leaving protein traces on the beads. This operation requires stringent chemical agents used as a single step or as sequential elution steps producing two or more fractions. Such an approach is based on the use of desorbing agents, each of them targeting the dominant physicochemical interactions between bait peptides and prey proteins. The interaction mechanisms involved in the protein–peptide docking are the result of various forces and can generate very strong noncovalent bonds depending on the number of synergistic attractions and the atomic distances. Molecular interactions are largely dependent on environmental conditions; however, conditions that weaken some interactions may reinforce others. Therefore finding a common "denominator" is not always easy.

Figure 4.44 depicts possible peculiar situations where the desorption step may be challenging. Full denaturation of proteins generally produces a complete desorption; unfortunately, it also destroy the functional property of the protein precluding biological determinations such as enzymatic assays. Immunochemistry could also be a problem when the recognition of the protein in question by its corresponding antibody is generated by the spatial situation of the recognition epitope. Table 4.9 presents a list of protein desorption methods with relative comments and limitations. All these methods can be used either singularly or sequentially depending on the context. However, when a single elution is desired, the desorption solution must be very effective against most, if not all, proteins. Conversely, when a sequential desorption is chosen, the process can start with relatively mild conditions preferably targeting one kind of physicochemical interactions and then using solutions of progressively increased stringency.

An alternative to protein elution is the direct on-bead protein digestion with the collection of peptides that are then analyzed after fractionation. This procedure is very attractive when the goal is just

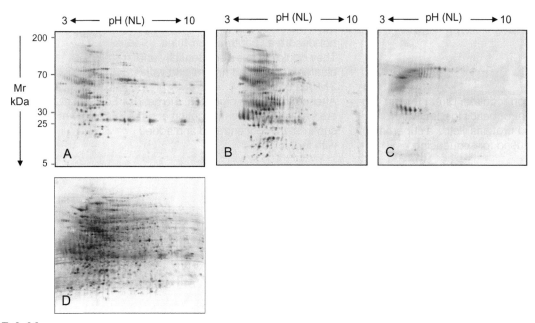

FIGURE 4.44
Two-dimensional electrophoresis of human urine proteins captured by CPLL and sequentially eluted with increasingly stringent agents. First eluent ("A") is 2 M thiourea, 7 M urea, 2% CHAPS; second eluent ("B") is 9 M urea, 50 mM citric acid pH 3.5; third eluting agent is 6 M guanidine-HCl pH 6. The proteins are progressively desorbed, and the last one is considered strong enough to strip out all the captured proteins. The protein spots are complementary, each indicating that eluting agents could be relatively selective. This also demonstrates that individual eluting agents even if very stringent are unable to desorb all proteins at the same time. Only SDS elution ("D") desorbs all proteins at once. 2D-PAGE analysis conditions: The pH gradient was between 3 and 10 nonlinear; protein detection with silver staining.

TABLE 4.9 Elution Methods Commonly used for the Recovery of Proteins Captured by CPLL Beads

Elution Method	Comments	References
Single elution with 10% sodium dodecyl sulphate containing 50 mM DTT	Followed or not by LC-MS/MS (gel slices followed by trypsination) or western blot.	Candiano et al., 2009
2 M thiourea-7 M urea-4% CHAPS-50 mM cysteic acid. 9 M urea-4%CHAPS-100 mM acetic acid, pH 3. 9 M urea-4%CHAPS-100 mM ammonia, pH 11. 6 M guanidine-HCl, pH 6.0. Various sequences (see references)	Thiourea-urea-CHAPS-cysteic acid fully compatible with 2D-PAGE analytical method. Eluates from urea-CHAPS and guanidine-HCl require a clean-up step to reduce the presence of salts.	Guerrier et al., 2008
9 M urea containing 2% CHAPS at pH 3.0-3.5 with 50 mM citric acid (or other strong acids). 6 M guanidine-HCl pH 6. 2 M thiourea-7 M urea-4% CHAPS-50 mM cysteic acid 9 M urea-4%CHAPS-100 mM ammonia, pH 11.	Described methods are fully compatible with this application.	Thulasiraman et al., 2005 Castagna et al., 2005
See 2D-PAGE above.	Eluates from all methods require a clean-up step to reduce the presence of salts.	D'Amato et al., 2009 Guerrier et al,. 2008
20 mM Tris containing 7 M urea, 2 M thiourea and 4% CHAPS pH 8.5. (sodium carbonate could also be used instead of Tris).	The elution buffer is compatible with cyanine dye labelling protocols.	Hartwig et al., 2009
0.2 M glycine-HCl, 2% NP-40, pH 2.4; 0.1 M acetic acid, 2% NP-40; 1 M NaCl, 2% NP-40; or 0.1 M acetic acid containing 40% ethylene glycol.	Harsh elution conditions may denature captured proteins rendering them unable to react with antibodies. Each indicated elution method may not desorb all captured proteins. They could be used sequentially and eluates pooled and neutralized with 3 M Tris base.	Unpublished data
No elution required. Direct trypsin treatment of beads loaded with captured proteins yields a supernatant directly used for sequencing.	After direct trypsin digestion, the beads are separated by centrifugation and the supernatant concentrated and added with formic acid.	Fortis et al., 2006 Fonslow et al., 2011

to make a global identification of low-abundance proteins present in a proteome. This section gives information on the elution variations that will later be detailed in Chapter 8 devoted to protocols. Whatever the elution protocol, the user is invited to check if all proteins have been fully desorbed by simply boiling the stripped beads in SDS-PAGE buffer for a few minutes and migrating the supernatant on a polyacrylamide plate. Revelation of residual protein bands could possibly be performed with sensitive stains.

4.6.1 Single Protein Elution or Stripping

To be effective, a single-step elution must comprise eluting agents capable of addressing all interactions between partners (peptides from CPLL and proteins from the sample). Thus an increase in ionic strength is necessary, along with agents capable of destroying the water structure and a deforming agent of protein structure. The simplest solution is 6 M guanidine-HCl that assembles all these properties. Alternatively, mixtures of molecules are possible such as 8 M urea containing 2–4% CHAPS and an acid such as 0.5% acetic acid or 50 mM citric acid. Urea dissociates hydrogen bonding, CHAPS is a detergent against hydrophobic associations, and the acidic pH acts on the protein structure along with the ionization of proteins, thus annihilating electrostatic interactions. Citric acid is preferred to acetic acid in case some bivalent cations stabilize the capture interactions. These elution procedures are not fully compatible with isoelectric focusing analysis of proteins or with the first dimension of 2D gel electrophoresis because the ionic strength of 6 M guandine-HCl is to high and because the acids present (acetic or citric acid) condense around pH 4 in the focusing step, disturbing the proper formation of the pH gradient. Alternatively these acids can be replaced by 25 mM of cysteic acid (Farinazzo et al., 2009). Due to its very low pI value (1.80), cysteic acid does not interfere with the first isoelectric focusing step performed in immobilized pH gradients, since it migrates spontaneously toward the anodic compartment, thereby vacating the gel and allowing a normal isoelectric focusing of proteins. Similar behavior is expected from formic acid; however, it has the potential to modify proteins by formylation at the levels of threonine and serine residues as well as at the ε-amine of the lysine side chain.

A completely different, powerful dissociation agent is lithium or sodium dodecyl sulfate (SDS) at a concentration of 4–10% as commonly used to solubilize proteins prior to polyacrylamide gel electrophoresis. This detergent associates very tightly to proteins by hydrophobic associations, exposing strongly charged sulfate groups and conferring to proteins a negative global charge. It is this phenomenon that participates in the dissociation of all proteins from the beads. Unfortunately, with this elution procedure operated at boiling temperature, it is essential to get rid of SDS prior to analysis of the proteome, especially two-dimensional electrophoresis. Actually the first dimension of 2D gel electrophoresis can be successful only if all traces of SDS are eliminated. Recipes for this purpose are described in Section 8.15. The only analytical method compatible with the presence of this detergent is SDS-PAGE. In practice, the SDS eluate from CPLLs can be directly processed and applied to the determination of protein identification of a whole extract after having sliced the gel lanes, followed by trypsination and finally LC-MS/MS analysis (see section 8.25). This procedure is well known and largely adopted as described in detail by Roux-Dalvai et al. (2008) and by Mouton-Barbosa et al. (2010).

4.6.2 Fractionated Protein Elution

Desorbing proteins under a sequential treatment serves different goals: (i) it simplifies the composition of each fraction compared to a global elution composition, allowing an easier analysis of the proteins contained in each fraction; (ii) it allows desorbing all proteins otherwise impossible at once because of contradictory interaction forces; and (iii) it is the way to desorb proteins under nondenaturing conditions for biological tests. In the third case, the collected fractions can be reassembled if needed.

Successive elution steps have been reported since the beginning of the technology development. Some of them started with a quite concentrated solution of sodium chloride (Thulasiraman et al., 2005) to desorb proteins dominantly captured by electrostatic interaction, arguing that it was a good nondenaturing procedure for an in-depth analysis of proteomes by all means, including the detection of biological functions. Proteins predominantly captured by hydrophobic association were desorbed by an aqueous solution of 60% ethylene glycol in a similar manner as in hydrophobic chromatography of proteins. Tightly adsorbed proteins because of various concurrent physicochemical forces were removed by means of a deforming solution composed of 0.2–0.4 M glycine-HCl buffer, pH 2.5, similar to the one largely employed in immunoaffinity chromatography. Finally, all other strongly captured proteins were desorbed by a 6 M guanidine-HCl solution, pH 6.0.

In another example, the captured proteins were desorbed by a sequence involving in the following order: (i) 7 M urea, 2 M thiourea, and 2% CHAPS; (ii) 9 M urea pH 3.5 by acetic acid; and (iii) a mixture composed of 6% acetonitrile-12% isopropanol-10% of 17 M ammonia-72% distilled water (Guerrier et al., 2007a; 2007b). In other examples the sequence was composed of three successive elutions: (i) 1 M sodium chloride; (ii) 3 M guanidine-HCl, pH 6.0; and (iii) 9 M urea, 50 mM citric acid pH 3–3.5.

With regard to how plant proteins are to be desorbed, D'Amato et al. (2010) proposed first 1 M sodium chloride, followed by 7 M urea, 2 M thiourea, 2% CHAPS, and then completed with 9 M urea, 50 mM citric acid pH 3.3. The use of other desorbing agents did not result in additional recovery of proteins; this suggests that all proteins were completely separated from the beads. With all these different elution methods used in sequence, one would ask the question about the comparative efficiency of desorbing agents and how they should in practice be used. It is in this context that Candiano et al. (2009) completed a compelling work by using urinary proteins, which served as a quite good model due to the presence of a large variety of proteins, a lot of them common with blood serum. It was found that a sequence of 7 M urea, 2 M thiourea, 2% CHAPS followed by 9 M urea, 50 mM citric acid pH 3.3 did not desorb all proteins. Actually, a strong stripping agent such as 6 M guanidine-HCl pH 6 contributed to collect additional proteins as illustrated in Figure 4.44. Only the addition of these three elutions could match with a two-dimensional electrophoresis map obtained using 10% SDS elution at boiling (panel D of the same figure).

We may therefore conclude that there is no ideal situation for desorbing absolutely all captured proteins because there are no highly specific agents designed to break singular molecular interactions. pH does not have an absolute effect on protein ionization; it also modulates hydrophobic associations as a collateral result of protein deformation where molecular distances may no longer be ideal for formation of the association of moieties. Urea affects not only hydrogen bonding, but also somewhat hydrophobic associations as a result of the modification of water structure around hydrophobic bonds, an important parameter for the stabilization of this kind of associations. Even if unexpected, lyotropic salts from the Hofmeister series also have an effect on hydrogen bonding, as described by Nucci & Vanderkooi (2008).

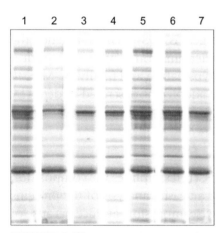

FIGURE 4.45

SDS-PAGE of serum protein eluates upon CPLL treatment in view of determining the elution efficiency of different nondenaturing chemical agent mixtures. The elution control (lane 1) was in this case one of the best eluting agents 9 M urea, 50 mM citric acid pH 3-3.5. Lane 2: elution performed with 1 M sodium chloride containing 2% of Nonidet NP40. Lane 3: elution performed with 0.1 M acetic acid containing 2% of Nonidet NP40. Lane 4: elution performed with 0.2 M glycine-HCl pH 2.5. Lane 5: elution performed with 0.2 M glycine-HCl pH 2.5 containing 60% ethylene glycol as hydrophobic displacer. Lane 6: elution performed with 0.2 M glycine-HCl pH 2.5 containing 2% of Nonidet NP40. Lane 7: elution performed with 0.2 M glycine-HCl pH 2.5 containing 2% Triton X-100.

4.6.3 Mild Elution Proposals Preserving Biological Activity

Sequences of mild but complementary elution steps that preserve the protein integrity for further biological determinations have also been experienced. This was the case with the desorption of cytokines captured by the beads and then quantified by their corresponding antibodies. After a large exploration of potential desorbing mixtures on human serum model and comparing to 9 M urea, 50 mM citric acid pH 3 (UCA), it appeared that sodium chloride alone and also added with NP40 was not able to desorb all proteins. Acetic acid alone or added with the detergent NP40 also was unable to desorb all captured proteins compared to UCA. Glycine buffer at pH 2.5 appeared to be more effective than acetic acid and even better when in the presence of non-ionic surfactants such as NP40 or Triton X-100 (see Figure 4.45).

Displacement of captured proteins with consequent elution could possibly be performed by specific competitive agents while preserving very mild nondenaturing conditions. They are naturally amino acids. However, since the interaction is the result of several synergistic single interactions by more than one amino acid composing the peptide bait, a single amino acid is not effective enough to allow the elution step. Mixtures of amino acids might be the solution to the problem. This proposal has been positively experienced with urinary proteome analysis by Candiano et al. (2012). It consists in challenging the CPLL beads with a mixture of four amino acids in different proportions (lysine, arginine, aspartic acid, and glutamic acid), two of which are qualified as "grand catchers" within the context of determining the contribution of each natural amino acid for the capture of proteins from a red blood cell lysate (Bachi et al., 2008). Such a mixture allowed recovering proteins very efficiently in a native form while avoiding the use of denaturing agents. Upon this elution approach at least 3300 spots could be visualized in a two-dimensional electrophoresis map compared to a control (in which only 600 spots were visible). Interestingly, it was possible to distinguish isoforms of several abundant proteins that were otherwise undetectable.

4.6.4 Direct On-bead Protein Digestion

The proteins captured by CPLL beads could be treated directly with proteases to produce characteristic peptide maps with the objective of identifying gene products present in the bulk of captured proteins (Thulasiraman et al., 2005). Generally, to this purpose, trypsin is used, and peptide sequence identification is then assigned to the original protein. Direct protein digestion is not very popular because it prevents the visualization of a protein pattern before and after sample treatment. Therefore classical separation/analysis methods such as SDS-PAGE, IEF, and 2D gel electrophoresis are not applicable to peptides. Only high-performance mass spectrometry analysis could be made with the objective of identifying en masse all possible proteins present. In practice, this approach is too simplistic because the peptide mixture is extremely complex; this mixture needs to be fractionated by available methods for small or medium-size molecules like high-performance liquid chromatography methods or any other peptide fractionation mean described in Chapter 2.

Since the liquid-phase composition could be a problem for the following mass spectrometry analysis, reverse-phase chromatography (RPC) involving volatile solvents is the preferred method. In the presence of very high peptide complexity and to serve the goal of better reliability for mass spectrometry data, two chromatographic methods could be sequenced as, for instance, a cation exchange (CEX) and a RPC. The first step produces several fractions that are individually subfractionated by RPC. CEX could be replaced by isoelectric focusing, a more orthogonal separation method to RPC by means of capillary electrophoresis (Wang et al., 2007) or even by the separation of peptides under discrete pI ranges by Off-Gel fractionators (Ernoult et al., 2010).

What is important for the present discussion is the direct digestion of protein captured by the beads. This is a relatively easy technology, especially when the volume of beads involved is very small and when the amount of collected proteins is lower than the amount required for current whole protein analyses. It is necessary to equilibrate the beads with the appropriate buffer and then use an amount of trypsin that is generally larger than that in solution digestion because of a possible partial capture of trypsin by the CPLL beads themselves. Parallel routes of protein identification after elution or directly on the beads are illustrated in Figure 4.46. Two detailed protocols are given in Chapter 8. Reference papers reporting the use of this approach are from Fonslow et al. (2011), who describe improving metrics for low-abundance protein detection, from Meng et al. (2011), who deal with HeLa cell extracts and human plasma samples and more recently by Yates, 2013.

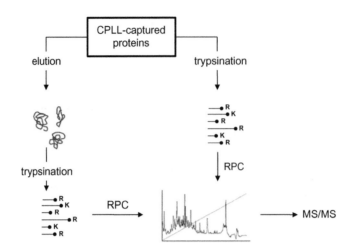

FIGURE 4.46

Identification of gene products following two different ways. CPLL-captured proteins are first desorbed from the beads (left route), and then they are treated with trypsin to produce peptides. The latter are then analyzed by mass spectrometry. The alternative route (right) proposes to directly trypsinize the captured proteins on the beads. When the proteome is very complex, the first route comprises a protein fractionation before trypsin treatment by, for instance, SDS-PAGE, while the second route comprises two chromatographic separations of peptides using orthogonal approaches (e.g., cation exchange and reverse phase). Other peptide separation methods can alternatively be applied (see text).

4.7 SELECTION OF ELUTION PROCESSES AS A FUNCTION OF ANALYTICAL METHODS

4.7.1 General Considerations

Reduction of the dynamic protein concentration range makes sense only if an extensive (global or individual) analysis of the resulting proteome is made. A number of approaches are available for proteome analysis, and they must be compatible with the composition of treated protein extract. The latter is in fact at least different from the initial samples because of its protein composition in number and concentration. It may contain chemicals used for CPLLs elution that could not be compatible with the following analytical method. It could comprise proteins that are partially or totally denatured, with multiple effects on the methods involving antibodies as a way to qualify or quantify polypeptides. All these issues are important enough to report several of these practical issues, with suggestions to increase the success rate of analytical determinations.

In most cases, proteomic analyses currently used are not fully compatible with the protein solution directly desorbed from CPLL beads. Two main practical possibilities are available: (i) the collected samples are equilibrated and/or concentrated and adjusted with a buffer compatible with the analytical determination, or (ii) protein desorption is customized to comply with the following analysis. In the first case, the protein eluate(s) is(are) desalted by extensive dialysis and then lyophilized. They could also be desalted by gel filtration or through dedicated membranes under centrifugation and concomitantly concentrated if needed. Precipitations using different agents could also apply (all these protocols are detailed in Chapter 8).

In the second instance, there are various possibilities to elute proteins captured by CPLL beads as described in the previous section and thus render the solution directly compatible with the analysis to follow. This is the case, for example, with protein desorption using SDS-Laemmli buffer under boiling conditions where the sample can be used directly for SDS-PAGE analysis. Details and suggestions on individual tandem technologies are given below.

4.7.2 CPLL Associated with Two-dimensional Electrophoresis

Two-dimensional electrophoresis is probably the most popular way to globally analyze proteomes. At the same time it allows separation of protein spots within a two-dimensional space, crossing over two fundamental protein properties (isoelectric point and molecular mass), thus also evidencing the presence of isoforms distinguishable only by their post-translational modification characteristics. Comparative imaging analysis also allows a quite precise idea of the expression difference of a single spot or a group of spots between a normal sample and a diseased sample. Last but not least, it makes a distinction between an intact protein and a fragment owing to the fact that these entities issued from the same gene product are at least different in their mass. A detailed protocol is described in Section 8.26.

In a way, two-dimensional gel electrophoresis represents a convenient method to multiplex the protein expression difference with possible applications in diagnostics. Among all described protein elution methods from CPLL, only one is fully compatible with two-dimensional electrophoresis and uses a mixture comprising 7 M urea, 2 M thiourea, and 2-4% CHAPS. Unfortunately, this solution is unable to desorb all captured proteins from the beads and will give an incomplete image of the proteome after enrichment of low-abundance species. To this end, a little modification could resolve the question, which is the addition of 25 mM of cysteic acid to the mixture as described by Farinazzo et al. (2009). Using all other elution solutions, it is mandatory to make an intermediate sample treatment to remove inappropriate components. They are,, for instance, salts, ionizable desorbing agents, ionic detergents, strong acids, and strong bases that are all incompatible with the first isoelectric focusing dimension migration. To this end, several treatments are suggested, such as precipitation of proteins as described in Sections 8.4 and 8.5. Dialysis followed by lyophilization or concentration dialysis with appropriate devices is also usable. These methods allow adjusting the protein concentration for an optimized protein load for the 2D-PAGE analysis. Many examples streamlining CPLL with 2D gel electrophoresis have been reported not only for purposes of pure analytical composition determination of proteomes with a full view on low-abundance species but also for the discovery differences in protein expression as a way to qualify a disease.

In the first domain it should be mentioned the extensive work performed for deciphering the red blood cell proteome, where 2D gel plates demonstrated the extensive increase of protein spots throughout the two-dimensional space for all fractions obtained from the sequential elution of CPLL-captured proteins. An image of the situation is given in Figures 4.9, 4.37, and 4.44, where the complementarity of the desorption methods is also clearly demonstrated. In another example, CPLL-collected proteins were confronted by two-dimensional electrophoresis in order to access low-abundance proteins in human plasma, with an analysis of proteins in a restricted acidic pH range (Beseme et al., 2010). The authors compared the CPLL technology results to untreated plasma and to immunodepletion with a substantial increase in the number of spots: 157 from native plasma, 427 from immunodepletion, and 557 from combinatorial library. With regard to a similar objective of enhancing and then detecting low-abundance proteins, Fahiminiya et al. (2011a) used two-dimensional electrophoresis following CPLL sample treatment to human, porcine, and equine follicular fluid. The protein composition of this fluid is quite complex, with the presence of very-high-abundance proteins as in the blood plasma. The authors concluded by emphasizing that the majority of proteins in follicular fluid are of high-abundance and that only a treatment of the sample with CPLL (and to a lower extent with immunodepletion) permitted reaching low-abundance species with the possibility of also revealing differentially expressed ones. This enrichment method combined with two-dimensional gel electrophoresis has been advantageously used for further investigations in follicular fluid (Fahiminiya et al., 2011b).

In the case of human plasma treated with two peptide libraries in a sequence with the elutions also performed sequentially, Léger et al. (2011) collected fractions that were dialyzed or directly diluted with the buffer or the solution that is currently used for 2D-PAGE first-dimension rehydration buffer (namely, 7 M urea, 2 M thiourea, 5% glycerol, 2% CHAPS, 0.35% sulfobetaine-10, 10% isopropanol, 12.5% isobutanol saturated in water, 0.35% Triton X100, 0.2% Biolytes 3–10, 0.002% bromophenol blue, and 100 mM dithiothreitol). Throughout a large number of plasma samples and hence many fractions, Leger demonstrated the reliability of 2D-PAGE following CPLL treatment with the possibility of detecting expression differences for diagnostic utility.

A specific aspect of two-dimensional gel electrophoresis analysis is the so-called 2D-DIGE (Differential gel electrophoresis) where two samples at a time are analyzed on the same plate. To this end, the samples are separately labeled with cyanine fluorochrome dyes having the same mass but capable of excitation by different wavelengths. The labeled samples are then mixed prior to two-dimensional separation under similar conditions as regular 2D-PAGE. The difference between samples is operated directly when observing the final plate under two different wavelengths. How would the elution from CPLL impact this technology? Since the protein samples are to be stained with a cyanine dye, the labeling conditions not only need to be very precise but also must be performed under appropriate conditions. Elution agents from CPLL are not compatible with the labeling method; thus for this kind of operation and with the objective of saving labor, protein elution can be performed using 20 mM Tris-hydroxymethylaminomethane containing 7 M urea, 2 M thiourea and 4% CHAPS, pH 8.5, or using 20 mM sodium carbonate containing 7 M urea, 2 M thiourea and 4% CHAPS, pH 8.5. For details see Sections 8.14, 8.25, and 8.28. Neither method, however, guarantees that the protein desorption is absolutely complete.

Table 4.10 gives a list of possible elution methods that are directly or indirectly compatible with 2D gel electrophoresis with relative comments. Detailed elution recipes are also given in Chapter 8. All the above examples prove the significant developments expected from the association of sample treatment with CPLL and two-dimensional gel electrophoresis analysis for both low-abundance

TABLE 4.10 Compatibility of CPLL Protein Elution Methods with 2D gel Analysis

Elution Method	Eluate Treatment Prior 2D gel	Comments
Single elution with 10% sodium dodecyl sulphate containing 50 mM DTT.	Sodium dodecyl sulphate is not compatible with first dimension of 2D-PAGE.	Upon elimination of sodium dodecyl sulphate some proteins could precipitate depending on the nature of the initial extract.
2 M thiourea-7 M urea-4% CHAPS-50 mM cysteic acid. 9 M urea-4% CHAPS-100 mM acetic acid, pH 3. 9 M urea containing 2% CHAPS at pH 3.0-3.5 with 50 mM citric acid	The eluate is directly compatible with 2D-PAGE analysis. What needs to be eliminated is acetic acid and citric acid for pH gradients between 3 and 10.	Cysteic acid does not affect the isoelectric migration phase. Both could be directly used when the pH gradient is narrow and starts above pH 4-5.
6 M guanidine-HCl, pH 6.0.	Guanidine-HCl must be eliminated as described above.	After elimination of guanidine HCl proteins are precipitated and dissolved in IEF compatible solution.
20 mM Tris containing 7 M urea, 2 M thiourea and 4% CHAPS pH 8.5. 20 mM sodium carbonate containing 7 M urea, 2 M thiourea and 4% CHAPS pH 8.5.	Only for 2D-DIGE. These eluates are used directly after the protein labelling with Cy-dyes.	These eluates are not compatible with regular 2D-PAGE.
0.2 M glycine-HCl, 2% NP-40, pH 2.4. 1 M acetic acid, 2% NP-40. 1 M NaCl, 2% NP-40. 0.1 M acetic acid containing 40% ethylene glycol.	Proteins from these eluates are generally cleaned by current precipitation processes or other clean-up procedures.	They are partial eluents for sequential protein desorption. Individually, they do not allow eluting quantitatively captured proteins.

protein detection and the identification of specific groups of proteins that are not directly distinguishable by LC-MS/MS, such as protein fragments, isoforms (e.g., proteins with glycan variations), and allergens.

4.7.3 CPLL Associated with MALDI or SELDI

Top-down mass spectrometry by means of MALDI and/or SELDI is not fully compatible with chemical agents such as sodium dodecyl sulfate or urea, two molecules that are largely used for the elution of CPLL-captured proteins. Fortunately the sensitivity of these analytical methods is high, and protein solutions eluted from CPLL frequently requires an extensive dilution. The latter reduces the amount of noncompatible compounds with consequent direct use of protein samples. When this is not the case, a simple dialysis or precipitation according to one of the methods described in Section 8.5 suffices.

The association of MALDI with CPLL has been discussed within the context of serum profiling, with the objective of early-stage disease discovery (Callesen et al., 2009). The sample treatment with different types of solid phases directly followed by MALDI analysis is considered advantageous because it is rapid and susceptible to be easily adapted to high-throughput configuration. The adsorption of proteins on dedicated solid phase prior to MALDI analysis also contributes to the elimination of nonproteinaceous material, thus removing background and interfering problems. The fractionation by sequential protein desorption comprising subproteomes is also interesting for a more focused research of expression differences.

SELDI analyses after CPLL protein extract treatments have also been repeatedly reported. In one instance, Sihlbom et al. (2008) reported the direct compatibility of CPLL and this type of mass spectrometry analysis that associates an additional feature for fractionation into subproteomes when using actively adsorptive chip surfaces. The authors reported the possibility of easily detecting low-abundance proteins with the potential for improving diagnostic approaches. The direct compatibility of these two technologies, with consequent improvement of the detectability of novel proteins, suggests a better delivery of data exploitable in diagnostic applications. This approach was also supported by Devarajan et al. (2008), who underlined the flexibility of multiple-chip chemistry for retentate chromatography applied to CPLL-treated samples and the high sensitivity of the method, along with the possibility of extending the investigation to characterize antibody-antigen association and devising protein–protein interaction studies. More recently, Frobel et al. (2010) recognized this approach as very appropriate for biomarker discovery, introducing a simplified workflow and demonstrating its suitability for large-scale and reproducible serum proteome profiling studies. A number of other published reports describe the merits and the application of this tandem technology for the detection of potential biomarkers applied to a variety of diseases.

Fakelman et al. (2010) and Felix et al. (2011) found a way to probe CPLL-treated samples from pancreatitis and pancreas cancer by using a high-throughput protocol. CPLLs are here used in a 96-well filtration plate allowing the parallel treatment of many samples at a time followed by SELDI technology, which is also based on a high-throughput configuration. This is clearly also to be attributed to the fact that no intermediate treatment of eluates is necessary prior to loading the MS array. The authors conclude that the technology can be used for extensive investigations of serum low-abundance protein comparative expression. For non-small cell lung cancer protein marker detection, Monari et al. (2011) found that a satisfactory screening method is reached by the conjunction of CPLL and SELDI MS. Also in this case the use of IMAC-30 and of H50 ProteinChip arrays was possible without preliminary treatment of CPLL eluates due to the tolerance of the detection system and to the dilution of the eluted protein samples.

4.7.4 CPLL Associated with SDS-PAGE and Then LC-MS/MS

Mapping globally a proteome from complex samples by using LC-MS/MS is a very popular method. Proteins are first globally digested with trypsin, and all resulting peptides analyzed by mass spectrometry. Peptide sequences are generally specific for their original protein, and in that way it is possible to retrieve the name of the proteins from dedicated data banks. However, considering that the number of peptides generated by a single protein is large and that the number of proteins from a given proteome is also very large, overall the total number of peptides is so huge that a fractionation before mass spectrometry analysis and identification is highly recommended. This separation is operated by fractionating the protein mixture before trypsin breakdown or at the peptide level after global trypsination or at both stages. The latter case is discussed in Section 4.7.6. As far as the separation is operated on the protein sample, eluates from CPLL need to be compatible with the selected fractionation method. Although a large number of fractionation methods is available (see Chapter 2), the one that is extensively used for the purpose of this paragraph is polyacrylamide gel electrophoresis in the presence of SDS. When the elution from CPLL is operated by using a SDS solution, the sample can be utilized directly without any preliminary treatment. In case the eluate results from other methods (see Section 4.6), it is recommended that one make a preliminary dialysis against the SDS solution adopted for the electrophoretic migration and then analyze the sample as usual.

The proteins eluted with noncompatible chemical agents could easily be precipitated by one of the methods proposed in Chapter 8 and then the pellets resolubilized directly with Laemmli buffer and boiled for few minutes prior to electrophoretic migration.

4.7.5 CPLL Associated with Immunochemical-based Methods

Analytical methods based on the use of antibodies work best when the proteins to be analyzed are in their native form. In the presence of denaturing agents, the molecular recognition may be prevented by (i) the fact that one of the two parties is inactivated or (ii) the protein epitopes that are supposed to be recognized by the antibody are changed in their spatial conformation. These situations are encountered when the presence of harsh desorbing agents is required to elute proteins from CPLLs. The ideal way to render peptide library treatments and immunochemical detection fully compatible is to use one or more eluents that are sufficiently mild to maintain the protein integrity, but strong enough for the total protein desorption. Section 4.6.3 suggests using desorbing agents in one step or sequences composed of salts in the presence of mild non-ionic detergents and acidic buffers associated with detergents or competitors of hydrophobic associations. Section 8.17 gives technical details and suggests recipes.

Generally, ELISA methods are used once a protein expression difference is detected as a quantitative confirmation (Chen et al., 2011; Zhi et al., 2011; Yang et al., 2010). However, in some cases this antibody-based method is useful, as described in Li et al. (2009) where the detectability of treated cytokines clearly depends on the mode of elution. Immunoblot analyses performed, for instance, on a 2D gel electrophoresis plate are another aspect of immunochemistry detection after having treated the initial sample for low-abundance protein enhancement (Shahali et al., 2012). In this case it appears that no dedicated or specifically designed elution methods are necessary to make the detection that is only possible after elecrophoteric migration.

4.7.6 CPLL Associated with MudPIT Analysis

This is a typical example in which all proteins captured by CPLL are directly trypsinized on beads and then the obtained peptides are fractionated and analyzed by tandem mass spectrometry. Under this configuration, it is naturally not necessary to elute the captured proteins by any means described in a preceding section, but more interestingly to treat the beads after protein capture directly with a solution of trypsin (Figure 4.46). With the elimination of the elution, the risk involved in seeing very minor protein not eluted is removed. Trypsin acts on adsorbed proteins like any solid-state chemical of biochemical reaction since it diffuses within the pores of the beads, thus degrading all possible proteins. However, this situation involves taking some precautions since trypsin itself could be captured by CPLL beads and thus be subtracted from the solution, therefore hampering its action as a proteolytic enzyme. The precaution here is thus to use a little more trypsin compared to trypsination in solution. For precaution it is also recommended that the enzymatic reaction be carried out overnight to be sure that trypsin diffuses completely within the bead pores so that all captured proteins are hydrolyzed. The completion of protein hydrolysis is otherwise checked by analyzing a desorbed sample of proteins by SDS-PAGE to see if intact proteins are still present.

In spite of the great efficiency of this approach, examples of the use of a direct trypsination on CPLL beads followed by MudPIT (multidimensional protein identification technology) analysis are not numerous. Although on-bead direct trypsination is always possible, there are times when it is highly recommended—for example, when dealing with very small biological samples treated with small volumes of beads. The amount of proteins collected could be so minute that to prevent possible losses during elution and dialysis, a direct trypsination has a lot of advantages. The limitation of this approach is that the collected peptides can only be used for mass spectrometry analysis after one or two separation steps to decomplexify the composition of the samples. The methods currently used are reverse-phase chromatography in miniaturized columns, the resolution of peptides being developed by hydro-organic gradients. More than with a single column, the extensive peptide complexity can be better managed by the use of a strong cation exchanger, followed by reverse-phase chromatography. This is in fact the general philosophy of MudPIT, as described by Wolters et al. (2001) and also discussed in Chapter 2. An alternative to chromatography is the fractionation by the isoelectric focusing by either using capillary electrophoresis systems according to Wang et al. (2007) or by means of multi-well devices, such as the Off-Gel fractionator as proposed by Waller et al. (2008). Compared to cation-exchange chromatography, the isoelectric fractionation requires longer time (12–72 hours) for the separation of species by groups of narrow isoelectric point ranges. Its advantage is the potential to discover more species, probably because this method is orthogonal to RPC with less overlapping of species present in each fraction.

The reproducibility of MudPIT technology has been especially criticized with respect to low- and very low-abundance proteins because many identifications are dependent on a single peptide, making the task of data interpretation doubtful. However, with the association of CPLL, this issue is largely diminished. It is within this context that Fonslow et al. (2011) described the improvement of this approach by two different reasons: first because of the increased abundance of rare species, with more peptides available to confirm the identification; and second because the identification of a non-treated sample is hampered by the presence of very-high abundance proteins. Better quality peptides were reported, along with the quality of MS/MS spectra acquired due to increases in the precursor intensity of peptides. The authors suggested that the association of CPLL and MudPIT opens a new avenue to improve proteome knowledge.

The detection of early-stage protein markers has been reported by Meng et al. (2011), who used the full digestion of CPLL-captured proteins followed by multidimensional peptide separation and mass spectrometry. The authors believed that this task could be accomplished only if the question of high-abundance protein presence was resolved and with a large number of samples. They addressed the issue with the use of CPLL associated with isobaric label-based 2D LC-MS/MS in a high-throughput protocol. With about 6% of differentially expressed plasma proteins identified, they demonstrated the effectiveness of their protocol when applied to large cohorts of patients with complex diseases. At the same time, the approach allows reducing the dynamic concentration range (access to low-abundance species), identifying the differently expressed signals (isobaric label of peptides), and substantially facilitating protein identifications (peptide sequence).

In an investigation of very low-abundance protein signatures of type 1 diabetes, this technology was applied to serum samples with the objective of discovering type 1 diabetes biomarkers (Zhi et al., 2010). After full proteolysis, the peptides were separated and analyzed by LC-MS. The authors described this approach as "a preferred method for the discovery of biomarkers."

In conclusion, it should be stressed that each situation needs to be considered with care. Table 4.9 lists selected elution conditions as a function of the type of analysis that follows the elution. It is also important to emphasize that what should drive the study is the final goal because, for instance, the development of knowledge of proteome composition may require different approaches than those used to determine the expression difference of a singular protein or a group of proteins between samples.

4.8 TOWARDS A TWO-DIMENSIONAL PROTEIN CAPTURE?

In sections 4.5.3 and 4.5.4 the use of CPLL throughout a large pH range and also within a lyotropic field suggests that a sort of two-dimensional protein capture can be approached. The use of acidic pHs and alkaline pHs proved that not only alkaline species can be segregated from acidic proteins respectively, but also allowed increasing the overall number of dilute species captured and thus identified. Most of experiments performed by scanning a large pH range were performed under physiological ionic strength. Interestingly the use of a wide range of pH for capture could be superimposed to the presence of more or less massive amounts of ammonium sulfate as a mean to change the environment and to increase the propensity of capturing hydrophobic species. This approach first suggested by Rivers et al., (2011) and experimentally demonstrated for pH (Fasoli et al., 2011) and for hydrophobic index (Santucci et al., 2013), opens towards various applications. The most obvious is the use of lyotropic salts not only under neutral conditions but also at different pHs promoting the capture of hydrophobic proteins from the lowest to the highest possible isoelectric points. It is without doubts that such a processing mode would enlarge the capability of protein capture resolving thus the questions related to proteins escaping the capture as discussed by Boschetti & Righetti, (2009b). Moreover this approach would capture additional low-abundance proteins as this was the case when just scanning a relatively large portion of pH. Potential variations can also be introduced as for instance the use of various concentrations of lyotropic salts to exacerbate the potentialities of hydrophobic interactions. Under these conditions some proteins of the sample would probably precipitate, but this is not necessarily a big problem because once separated by centrifugation they could be resolubilized and captured under other conditions. In both cases the number of captured proteins would be potentially enlarged with a better knowledge of the overall protein composition. The applicability of these experimental approaches is clearly addressable to species that are reluctant to be captured by CPLL but of interest for clinical analysis or for the detection of protein trace in biological samples as described in Chapters 5 and 6.

4.9 References

Akiyama T, Kadooka T, Ogawara H. Purification of the epidermal growth factor receptor by tyrosine-Sepharose affinity chromatography. *Biochem Biophys Res Commun.* 1985;131:442–448.

Bachi A, Simó C, Restuccia U, et al. Performance of combinatorial peptide libraries in capturing the low-abundance proteome of red blood cells. 2. Behavior of resins containing individual amino acids. *Anal Chem.* 2008;80:3557–3565.

Baker EN, Hubbard RE. Hydrogen bonding in globular proteins. *Prog Biophys Mol Biol.* 1984;44:97–179.

Baldauf C, Gunther R, Hofmann HJ. Theoretical prediction of the basic helix types in α, β-hybrid peptides. *Biopolymers.* 2006;84:408–413.

Bandhakavi S, Stone MD, Onsongo G, Van Riper SK, Griffin TJ. A dynamic range compression and three-dimensional peptide fractionation analysis platform expands proteome coverage and the diagnostic potential of whole saliva. *J Proteome Res.* 2009;8:5590–5600.

Bastek PD, Land JM, Baumbach GA, Hammond DH, Carbonell RG. Discovery of alpha-1-proteinase inhibitor binding peptide from the screening of a solid phase combinatorial library. *Sep Sci Technol.* 2000;35:1681–1706.

Beseme O, Fertin M, Drobecq H, Amouyel P, Pinet F. Combinatorial peptide ligand library plasma treatment: Advantages for accessing low-abundance proteins. *Electrophoresis.* 2010;31:2697–2704.

Boeijen A, Kruijtzer JA, Liskamp RM. Combinatorial chemistry of hydantoins. *Bioorg Med Chem Lett*. 1998;8:2375–2380.

Boja ES, Rodriguez H. The path to clinical proteomics research: Integration of proteomics, genomics, clinical laboratory and regulatory science. *Korean J Lab Med*. 2011;31:61–71.

Boschetti E, Jungbauer A. Separation of antibodies by liquid chromatography. In: Ahuja S, ed. Separation Science and Technology; vol. 2. Handbook of Bioseparations. San Diego Academic Press; 2000:535–632.

Boschetti E, Lomas L, Righetti PG. Romancing the "hidden proteome", Anno Domini two zero zero six. *J Chromatogr A*. 2007;1153:277–290.

Boschetti E, Righetti PG. The ProteoMiner in the proteomic arena: A non-depleting tool for discovering low-abundance species. *J Proteomics*. 2008;71:255–264.

Boschetti E, Bindschedler L, Tang C, Fasoli E, Righetti PG. Combinatorial peptide ligand libraries and plant proteomics: A winning strategy at a price. *J Chromatogr*. 2009a;1216:1215–1222.

Boschetti E, Righetti. The art of observing rare protein species in proteomes with peptide ligand libraries. *Proteomics*. 2009b;9:1492–1510.

Bosma JC, Wesselingh JA. pH dependence of ion exchange equilibrium of proteins. *AIChE J*. 1998;44:2399–2409.

Brewis IA, Brennan P. Proteomics technologies for the global identification and quantification of proteins. *Adv Protein Chem Struct Biol*. 2010;80:1–44.

Buettner JA, Dadd CA, Baumbach GA, Masecar BL, Hammond DJ. Chemically derived peptide libraries: A new resin and methodology for lead identification. *Int J Pept Protein Res*. 1996;47:70–83.

Bunin BA, Ellman JA. A general and expedient method for the solid-phase synthesis of 1,4-benzodiazepine derivatives. *J Am Chem Soc*. 1992;114:10997–10998.

Callesen AK, Madsen JS, Vach W, et al. Serum protein profiling by solid phase extraction and mass spectrometry: A future diagnostics tool? *Proteomics*. 2009;9:1428–1441.

Candiano G, Dimuccio V, Bruschi M, et al. Combinatorial peptide ligand libraries for urine proteome analysis: Investigation of different elution systems. *Electrophoresis*. 2009;30:2405–2411.

Candiano G, Santucci L, Bruschi M, et al. "Cheek-to-cheek" urinary proteome profiling via combinatorial peptide ligand libraries: A novel, unexpected elution system. *J Proteomics*. 2012;75:796–805.

Cargill GF, Lebl M. New methods in combinatorial chemistry and parallel synthesis. *Curr Opin Chem Biol*. 1977;1:67–71.

Castagna A, Cecconi D, Sennels L, et al. Exploring the hidden human urinary proteome via ligand library beads. *J Proteome Res*. 2005;4:1917–1930.

Chen G, Zhang Y, Jin X, et al. Urinary proteomics analysis for renal injury in hypertensive disorders of pregnancy with iTRAQ labeling and LC-MS/MS. *Proteomics Clin Applic*. 2011;5:300–310.

D'Amato A, Bachi A, Fasoli E, et al. In-depth exploration of cow's whey proteome via combinatorial peptide ligand libraries. *J Proteome Res*. 2009;8:3925–3936.

D'Amato A, Bachi A, Fasoli E, et al. In-depth exploration of *Hevea brasiliensis* latex proteome and "hidden allergens" via combinatorial peptide ligand libraries. *J Proteomics*. 2010;73:1368–1380.

D'Ambrosio C, Arena S, Scaloni A, et al. Exploring the chicken egg white proteome with combinatorial peptide ligand libraries. *J Proteome Res*. 2008;7:3461–3474.

Dayarathna MK, Hancock WS, Hincapie M. A two step fractionation approach for plasma proteomics using immunodepletion of abundant proteins and multi-lectin affinity chromatography: Application to the analysis of obesity, diabetes, and hypertension diseases. *J Sep Sci*. 2008;31:1156–1166.

De Bock M, de Seny D, Meuwis M-A, et al. Comparison of three methods for fractionation and enrichment of low molecular weight proteins for SELDI-TOF-MS differential analysis. *Talanta*. 2010;82:245–254.

Devarajan P, Ross GF. SELDI technology for identification of protein biomarkers. *Meth Pharmacol Toxicol* 2008;251–271.

Di Girolamo F, Boschetti E, Chung MC, Guadagni F, Righetti PG. "Proteomineering" or not? The debate on biomarker discovery in sera continues. *J Proteomics*. 2011;74:589–594.

Dwivedi RC, Krokhin OV, Cortens JP, Wilkins JA. An assessment of the reproducibility of random hexapeptide peptide library based protein normalization. *J Proteome Res*. 2010;9:1144–1149.

Ek K, Gianazza E, Righetti PG. Affinity titration curves: Determination of dissociation constants of lectin-sugar complexes and of their pH-dependence by isoelectric focusing electrophoresis. *Biochim Biophys Acta*. 1980;626:356–365.

El Khoury G, Rowe LA, Lowe CR. Biomimetic affinity ligands for immunoglobulins based on the multicomponent Ugi reaction. *Methods Mol Biol*. 2012;800:57–74.

Ernoult E, Bourreau A, Gamelin E, Guette C. A proteomic approach for plasma biomarker discovery with iTRAQ labelling and OFFGEL fractionation. *J Biomed Biotechnol*. 2010;2010:927917.

Fahiminiya S, Labas V, Dacheux J, Gérard N. Improvement of 2D-PAGE resolution of human, porcine and equine follicular fluid by means of hexapeptide ligand library. *Reprod Domest Anim*. 2011a;46:561–563.

Fahiminiya S, Labas V, Roche S, Dacheux J-L, Gérard N. Proteomic analysis of mare follicular fluid during late follicle development. *Proteome Science*. 2011b;9:54–56.

Fakelman F, Felix K, Büchler MW, Werner J. New pre-analytical approach for the deep proteome analysis of sera from pancreatitis and pancreas cancer patients. *Arch Physiol Biochem*. 2010;116:208–217.

Farinazzo A, Fasoli E, Kravchuk AV, et al. En bloc elution of proteomes from combinatorial peptide ligand libraries. *J Proteomics*. 2009;72:725–730.

Fasoli E, Farinazzo A, Sun CJ, et al. Interaction among proteins and peptide libraries in proteome analysis: pH involvement for a larger capture of species. *J Proteomics*. 2010;73:733–742.

Felix K, Fakelman F, Hartmann D, et al. Identification of serum proteins involved in pancreatic cancer cachexia. *Life Sci*. 2011;88:218–225.

Fertin M, Beseme O, Duban S, et al. Deep plasma proteomic analysis of patients with left ventricular remodeling after a first myocardial infarction. *Proteomics Clin Appl*. 2010;4:654–673.

Fonslow BR, Carvalho PC, Academia K, et al. Improvements in Proteomic Metrics of Low Abundance Proteins through Proteome Equalization Using ProteoMiner Prior to MudPIT. *J Proteome Res*. 2011;10:3690–3700.

Fortis F, Guerrier L, Areces LB, et al. A new approach for the detection and identification of protein impurities using combinatorial solid phase ligand libraries. *J Proteome Res*. 2006;5:2577–2585.

Fröbel J, Hartwig S, Paßlack W, et al. ProteoMiner and SELDI-TOF-MS: A robust and highly reproducible combination for biomarker discovery from whole blood serum. *Arch Physiol Biochem*. 2010;116:174–180.

Furka A, Sebestien F, Asgedom M, Dibo G. More peptides by less labour. In: *Proc. 10th Int. Symp. Med. Chem*. Budapest, Hungary, 1988 August 15–19:288.

Furka A, Sebesryen F, Asgedom M, Dibo G. General method for rapid synthesis of multicomponent peptide mixtures. *Int J Pept Protein Res*. 1991;37:487–493.

Gallant SR, Kundu A, Cramer SM. Modeling non-linear elution of proteins in ion-exchange chromatography. *J Chromatogr A*. 1995;702:125–142.

Gibas CJ, Jambeck P, Subramaniam S. Continuum electrostatic methods applied to pH-dependent properties of antibody–antigen association. *Methods*. 2000;20:292–309.

Greenough C, Jenkins RE, Kitteringham NR, et al. A method for the rapid depletion of albumin and immunoglobulin from human plasma. *Proteomics*. 2004;4:3107–3111.

Grimsley GR, Scholtz GM, Pace CN. A summary of the measured pK values of the ionizable groups in folded proteins. *Protein Sci*. 2009;18:247–251.

Guerrier L, Thulasiraman V, Castagna A, et al. Reducing protein concentration range of biological samples using solid-phase ligand libraries. *J Chromatogr B*. 2006;833:33–40.

Guerrier L, Claverol S, Fortis F, et al. Exploring the platelet proteome via combinatorial hexapeptide ligand library. *J Proteome Res*. 2007a;6:4290–4303.

Guerrier L, Claverol S, Finzi L, et al. Contribution of solid-phase hexapeptide ligand libraries to the repertoire of human bile proteins. *J Chromatogr*. 2007b;1176:192–205.

Guerrier L, Righetti PG, Boschetti E. Reduction of dynamic protein concentration range of biological extracts for the discovery of low-abundance proteins by means of hexapeptide ligand library. *Nature Protocols*. 2008;3:883–890.

Hahn R, Seifert M, Greinstetter S, et al. Peptide affinity chromatography media that bind N(pro) fusion proteins under chaotropic conditions. *J Chromatogr*. 2010;1217:6203–6213.

Hartwig S, Czibere A, Kotzka J, et al. Combinatorial hexapeptide ligand libraries: An innovative fractionation tool for differential quantitative clinical proteomics. *Arch Physiol Biochem*. 2009;115:1–6.

Herpin TF, Van Kirk KG, Salvino JM, Yu ST, Labaudinière RF. Synthesis of a 10,000 member 1,5-benzodiazepine-2-one library by the directed sorting method. *J Comb Chem*. 2000;2:513–521.

Hilpert K, Winkler DF, Hancock RE. Peptide arrays on cellulose support: SPOT synthesis, a time and cost efficient method for synthesis of large numbers of peptides in a parallel and addressable fashion. *Nat Protoc*. 2007;2:1333–1349.

Houghten RA, Pinilla C, Blondelle SE, et al. Generation and use of synthetic peptide combinatorial libraries for basic research and drug discovery. *Nature*. 1991;354:84–86.

Huang PY, Baumbach GA, Dadd CA, et al. Affinity purification of von Willebrand factor using ligands derived from peptide libraries. *Bioorg Med Chem*. 1996;4:699–708.

Huhn C, Ruhaak LR, Wuhrer M, Deelder AM. Hexapeptide library as a universal tool for sample preparation in protein glycosylation analysis. *J Proteomics*. 2011;75:1515–1528.

Iyer H, Tapper S, Lester P, Wolk B, Reis RV. Use of the steric mass action model in ion-exchange chromatographic process development. *J Chromatogr A*. 1999;832:1–9.

Janecek J, Dobrova Z, Moravec V, Naprstek J. Improved method for rapid purification of protein kinase from streptomycetes. *J Biochem Biophys Methods*. 1996;31:9–15.

Janin J, Chothia C. The structure of protein–protein recognition sites. *J Biol Chem*. 1990;265:16027–16030.

Jefferey GA. An introduction to hydrogen bonding. New York: Oxford University Press; 1977.

Jmeian Y, El Rassi Z. Liquid-phase-based separation systems for depletion, prefractionation and enrichment of proteins in biological fluids for in-depth proteomics analysis. *Electrophoresis*. 2009;30:249–261.

Karshikoff A. Non covalent interactions in proteins. Imperial College Press; 2006.

Kaufman DB, Hentsch ME, Baumbach GA, et al. Affinity purification of fibrinogen using a ligand from a peptide library. *Biotechnol Bioeng*. 2002;77:278–289.

Keidel EM, Ribitsch D, Lottspeich F. Equalizer technology. Equal rights for disparate beads. *Proteomics*. 2010;10:1–10.

Koerner K, Rahimi-Laridjani I, Grimminger H. Purification of biosynthetic threonine deaminase from *Escherichia coli*. *Biochim Biophys Acta*. 1975;397:220–230.

Kolla V, Jen P, Moes S, et al. Quantitative Proteomic (iTRAQ) Analysis of 1st Trimester Maternal Plasma Samples in Pregnancies at Risk for Preeclampsia. *J Biomed Biotechnol*. 2012;2012:1–8.

Kopaciewicz W, Rounds MA, Fausnaugh J, Regnier FE. Retention model for high-performance ion-exchange chromatography. *J Chromatogr*. 1983;266:3–21.

Krchnak V, Cabel D, Lebl M. MARS: Multiple automated robotic synthesizer for continuous flow of peptides. *Pept Res*. 1996;9:45–49.

Kwon YU, Kodadek T. Encoded combinatorial libraries for the construction of cyclic peptoid microarrays. *Chem Commun (Camb)*. 2008;44:5704–5706.

Kyte J, Doolittle RF. A simple method for displaying the hydropathic character of a protein. *J Mol Biol*. 1982;157:105–132.

Lam KS, Salmon SE, Hersh EM, et al. A new type of synthetic peptide library for identifying ligand-binding activity. *Nature*. 1991;354:82–84.

Lam KS, Lake D, Salmon SE, et al. A one-bead one-peptide combinatorial library method for B-cell epitope mapping. *Methods*. 1996;9:482–493.

Lam KS. Application of combinatorial library methods in cancer research and drug discovery. *Anticancer Drug Des*. 1997;12:145–167.

Lam KS, Lebl M, Krchnàk V. The "One-Bead-One-Compound" combinatorial Library. *Methods Chem Rev*. 1997;97:411–448.

Lam KS, Lebl M. Synthesis of a one-bead one-compound combinatorial peptide library. *Methods Mol Biol*. 1998a;87:1–6.

Lam KS. Determination of peptide substrate motifs for protein kinases using a "one-bead one-compound" combinatorial library approach. *Methods Mol Biol*. 1998b;87:83–86.

Lam KS, Lehman AL, Song A, et al. Synthesis and screening of "one-bead one-compound" combinatorial peptide libraries. *Methods Enzymol*. 2003;369:298–322.

Lamb ML, Burdick KW, Toba S, et al. Design, docking, and evaluation of multiple libraries against multiple targets. *Proteins*. 2001;42:296–318.

Larcher G, Bouchara JP, Annaix V, Symoens F, Chabasse D. Purification and characterization of a fibrinogenolytic serine proteinase from *Aspergillus fumigatus* culture filtrate. *FEBS Lett*. 1982;308:65–69.

Lathrop JT, Fijalkowska I, Hammond D. The bead blot: A method for identifying ligand–protein and protein–protein interactions using combinatorial libraries of peptide ligands. *Anal Biochem*. 2007;361:65–76.

Léger T, Lavigne D, Le Caër JP, et al. Solid-phase hexapeptide ligand libraries open up new perspectives in the discovery of biomarkers in human plasma. *Clin Chim Acta*. 2011;412:740–747.

Levy ED, Pereira-Leal JB. Evolution and dynamics of protein interactions and networks. *Curr Opin Struct Biol*. 2008;18:349–357.

Li L, Sun CJ, Freeby S, et al. Protein sample treatment with peptide ligand libraries: Coverage and consistency. *J Proteomics Bioinformatics*. 2009;2:485–494.

Liu R, Marik J, Lam KS. Design, synthesis, screening, and decoding of encoded one-bead one-compound peptidomimetic and small molecule combinatorial libraries. *Methods Enzymol*. 2003;369:271–287.

Livesay D, Linthicum S, Subramaniam S. pH dependence of antibody: Hapten association. *Mol Immunol*. 1999;36:397–410.

Luque-Garcia JL, Neubert TA. Sample preparation for serum/plasma profiling and biomarker identification by mass spectrometry. *J Chromatogr A*. 2007;1153:259–276.

Maillard N, Clouet A, Darbre T, Reymond J-L. Combinatorial libraries of peptide dendrimers: design, synthesis, on-bead high-throughput screening, bead decoding and characterization. *Nature Protocols*. 2009;4:132–142.

Mann K, Mann M. In-depth analysis of the chicken egg white proteome using an LTQ Orbitrap Velos. *Proteome Science*. 2011;9:7.

Marco-Ramell A, Bassols A. Enrichment of low-abundance proteins from bovine and porcine serum samples for proteomic studies. *Res Vet Sci*. 2010;89:340–343.

Melander WR, Corradini D, Horváth C. Salt-mediated retention of proteins in hydrophobic-interaction chromatography. Application of solvophobic theory. *J Chromatogr*. 1984;317:67–85.

Meng R, Gormley M, Bhat VB, Rosenberg A, Quong AA. Low abundance protein enrichment for discovery of candidate plasma protein biomarkers for early detection of breast cancer. *J Proteomics*. 2011;75:366–374.

Merrifield RB. Automated synthesis of peptides. *Science*. 1965;150:178–185.

Merrifield B. Solid phase synthesis. *Science*. 1986;232:341–347.

Messeguer J, Cortés N, García-Sanz N, et al. Synthesis of a positional scanning library of pentamers of N-alkylglycines assisted by microwave activation and validation via the identification of trypsin inhibitors. *J Comb Chem*. 2008;10:974–980.

Miller SM, Simon RJ, Ng S, et al. Comparison of the proteolytic susceptibilities of homologous L-amino acid, D-amino acid, and N-substituted glycine peptide and peptoid oligomers. *Drug Dev Res*. 1995;35:20–32.

Miyamoto S, Liu R, Hung S, Wang X, Lam KS. Screening of a one bead-one compound combinatorial library for beta-actin identifies molecules active toward Ramos B-lymphoma cells. *Anal Biochem*. 2008;374:112–120.

Monari E, Casali C, Cuoghi A, et al. Enriched sera protein profiling for detection of non-small cell lung cancer biomarkers. *Proteome Science*. 2011;9:55–65.

Mouton-Barbosa E, Roux-Dalvai F, Bouyssié D, et al. In-depth exploration of cerebrospinal fluid combining peptide ligand library treatment and label-free protein quantification. *Mol Cell Proteomics.* 2010;9:1006–1021.

Nakamura H. Roles of electrostatic interaction in proteins. *Q Rev Biophys.* 1996;29:1–90.

Nozaki Y, Tanford C. Examination of titration behavior. *Methods Enzymol.* 1967;11:715–734.

Nucci NV, Vanderkooi JM. Effects of salts of the Hofmeister series on the hydrogen bond network of water. *J Mol Liq.* 2008;143:160–170.

Overgaard AJ, Thingholm TE, Larsen MR, et al. Quantitative iTRAQ-based proteomic identification of candidate biomarkers for diabetic nephropathy in plasma of type 1 diabetic patients. *Clin Proteomics.* 2010;6:105–114.

Pass L, Zimmer TL, Laland SG. The use of affinity chromatography in determining the sites of protein-protein interaction relative to the binding sites of substrates in gramicidin S synthetase. *Eur J Biochem.* 1973;40:43–48.

Pennington ME, Lam KS, Cress AE. The use of a combinatorial library method to isolate human tumor cell adhesion peptides. *Mol Divers.* 1996;2:19–28.

Pieper R, Gatlin CL, Makusky AJ, et al. The human serum proteome: Display of nearly 3700 chromatographically separated protein spots on two-dimensional electrophoresis gels and identification of 325 distinct proteins. *Proteomics.* 2003a;3:1345–1364.

Pieper R, Su Q, Gatlin CL, et al. Multi-component immunoaffinity subtraction chromatography: An innovative step towards a comprehensive survey of the human plasma proteome. *Proteomics.* 2003b;3:422–432.

Porath J, Fryklund L. Chromatography of proteins on dipolar ion adsorbants. *Nature.* 1970a;226:1169–1170.

Porath J, Fornstedt N. Group fractionation of plasma proteins on dipolar ion exchangers. *J Chromatogr.* 1970b;51:479–489.

Raschke TM. Water structure and interactions with protein surfaces. *Curr Opin Struct Biol.* 2006;16:152–159.

Righetti PG, Castagna A, Antonioli P, Boschetti E. Pre-fractionation techniques in proteome analysis: The mining tools of the third millennium. *Electrophoresis.* 2005;26:297–319.

Righetti PG, Boschetti E, Lomas L, Citterio A. Protein Equalizer technology: The quest for a "democratic proteome". *Proteomics.* 2006;6:3980–3992.

Righetti PG, Boschetti E, Kravchuk A, Fasoli E. The proteome buccaneers: How to unearth your treasure chest via combinatorial peptide ligand libraries. *Expert Rev Proteomics.* 2010;7:373–385.

Righetti PG, Boschetti E, Candiano G. Mark Twain: How to fathom the depth of your pet proteome. *J Proteomics.* 2012;75:4783–4791.

Rivers J, Hughes C, McKenna T, et al. Asymmetric proteome equalization of the skeletal muscle proteome using a combinatorial hexapeptide library. *PLoS ONE.* 2011;6:e28902.

Roux-Dalvai F, Gonzalez de Peredo A, Simó C, et al. Extensive analysis of the cytoplasmic proteome of human erythrocytes using the peptide ligand library technology and advanced mass spectrometry. *Mol Cell Proteomics.* 2008;7:2254–2269.

Rubinstein M, Niv MY. Peptidic modulators of protein-protein interactions: Progress and challenges in computational design. *Biopolymers.* 2009;91:505–513.

Samson I, Kerremans L, Rozenski J, et al. Identification of a peptide inhibitor against glycosomal phosphoglycerate kinase of *Trypanosoma brucei* by a synthetic peptide library approach. *Bioorg Med Chem.* 1995;3:257–265.

Santucci L, Candiano G, Bruschi M, et al. Combinatorial peptide ligand libraries for the analysis of low-expression proteins. Validation for normal urine and definition of a first protein map. *Proteomics.* 2012;12:509–515.

Santucci L, Candiano G, Petretto A, et al. Combinatorial ligand libraries as a two-dimensional method for proteome analysis. *J Chromatogr A.* 2013; in press.

Sanz-Pamplona R, Berenguer A, Sole X, et al. Tools for protein–protein interaction network analysis in cancer research. *Clin Transl Oncol.* 2012;14:3–14.

Schirwitz C, Block I, König K, et al. Combinatorial peptide synthesis on a microchip. *Curr Protoc Protein Sci.* 2009;18:1–13.

Sebestyén F, Szalatnyai T, Durgo JA, Furka A. Binary synthesis of multicomponent peptide mixtures by the portioning-mixing technique. *J Pept Sci.* 1995;1:26–30.

Sennels L, Salek M, Lomas L, et al. Proteomic Analysis of Human Blood Serum Using Peptide Library Beads. *J Proteome Res.* 2007;6:4055–4062.

Shahali Y, Sutra JP, Fasoli E, et al. Allergomic study of cypress pollen via combinatorial peptide ligand libraries. *J Proteomics.* 2012;77:101–110.

Shen H, Frey DD. Effect of charge regulation on steric mass-action equilibrium for the ion-exchange adsorption of proteins. *J Chromatogr A.* 2005;1079:92–104.

Shetty V, Jain P, Nickens Z, et al. Investigation of plasma biomarkers in HIV-1/HCV mono- and coinfected individuals by multiplex iTRAQ quantitative proteomics. *OMICS.* 2011;15:705–717.

Shi Q, Zhou Y, Sun Y. Influence of pH and ionic strength on the steric mass-action model parameters around the isoelectric point of protein. *Biotechnol Prog.* 2005;21:516–523.

Sihlbom C, Kanmert I, von Bahr H, Davidsson P. Evaluation of the combination of bead technology with SELDI-TOF-MS and 2-D DIGE for detection of plasma proteins. *J Proteome Res.* 2008;7:4191–4198.

Simó C, Bachi A, Cattaneo A, et al. Performance of combinatorial peptide libraries in capturing the low-abundance proteome of red blood cells. 1. Behaviour of mono- to hexapeptides. *Anal Chem.* 2008;80:3547–3556.

Sproule K, Morrill P, Pearson JC, Burton SJ, Hejnaes KR, Valore H, Ludvigsen S, Lowe CR,. New strategy for the design of ligands for the purification of pharmaceutical proteins by affinity chromatography. *J Chromatogr B.* 2000;740:17–33.

Staby A, Mollerup J. Solute retention of lysozyme in hydrophobic interaction perfusion chromatography. *J Chromatogr A*. 1996;734:205–212.

Su Y, Yamashita MM, Greasley SE, et al. A pH-dependent stabilization of an active site loop observed from low and high pH crystal structures of mutant monomeric glycinamide ribonucleotide transformylase at 1.8 to 1.9 A. *J Biol Mol*. 1998;281:485–499.

Thulasiraman V, Lin S, Gheorghiu L, et al. Reduction of concentration difference of proteins from biological liquids using combinatorial ligands. *Electrophoresis*. 2005;26:3561–3571.

Thurlkill RL, Grimsley GR, Scholtz JM, Pace CN. pK value of the ionizable groups of proteins. *Protein Sci*. 2006;15:1214–1218.

Townsend JB, Shaheen F, Liu R, Lam KS. Jeffamine derivatized TentaGel Beads and PDMS Microbead Cassettes for ultra-high throughput in situ releasable solution-phase cell-based screening of OBOC combinatorial small molecule libraries. *J Comb Chem*. 2010;12:700–712.

Tozzi C, Giraudi G. Antibody-like peptides as a novel purification tool for drugs design. *Curr Pharm Des*. 2006;12:191–203.

Traas DW, Hoegee-de Nobel B, Nieuwenhuizen W. Factors influencing the separation of glu-plasminogen affinity forms I and II by affinity chromatography. *Thromb Haemost*. 1984;52:347–349.

Tsuboi Y, Takahashi M, Ishikawa Y, Okada H, Yamada T. Elevated bradykinin and decreased carboxypeptidase R as a cause of hypotension during tryptophan column immunoabsorption therapy. *Ther Apher*. 1998;2:297–299.

Tsuji A, Edazawa K, Sakiyama K, et al. Purification and characterization of a novel serine proteinase from the microsomal fraction of bovine pancreas. *J Biochem (Tokyo)*. 1996;119:100–105.

van Holde KE, Johnson WC, Ho PS. Principles of Physical Biochemistry. Upper Saddle River, NJ: Prentice Hall; 1998:9–11.

Waller LN, Shores K, Knapp DR. Shotgun proteomic analysis of cerebrospinal fluid using off-gel electrophoresis as the first-dimension separation. *J Proteome Res*. 2008;7:4577–4584.

Wang W, Guo T, Rudnick PA, et al. Membrane proteome analysis of microdissected ovarian tumor tissues using capillary isoelectric focusing/reversed-phase liquid chromatography-tandem MS. *Anal Chem*. 2007;79:1002–1009.

Wang Y-S, Cao R, Jin H, et al. Altered protein expression in serum from endometrial hyperplasia and carcinoma patients. *J Hematol Oncol*. 2011;4:15–23.

White SH. How hydrogen bonds shape membrane protein structure. *Adv Protein Chem*. 2005;72:157–172.

Whitley RD, Wachter R, Liu F, Wang NH. Ion-exchange equilibria of lysozyme, myoglobin and bovine serum albumin. Effective valence and exchanger capacity. *J Chromatogr*. 1989;465:137–156.

Wolters DA, Washburn MP, Yates 3rd JR. An automated Multidimensional protein identification technology for Shotgun proteomics. *Anal Chem*. 2001;73:5683–5690.

Yan H, Gurgel PV, Carbonell RG. Characterization of the Hexamer Peptide Affinity Ligands that Bind the Fc Region of Human Immunoglobulin G. *J Chromatogr A*. 2009;1216:910–918.

Yang Y, Cheng G, Zhao H, Jiang X, Chen S. Differential proteomics analysis of plasma protein from *Escherichia coli* infected and clinical healthy dairy cows. *Xumu Shouyi Xuebao*. 2010;41:1191–1197.

Yates JR. The revolution and evolution of shotgun proteomics for lerge scale proteome analysis. *J. Am. Chem. Soc*. 2013, in press.

Zajdel P, Pawłowski M, Martinez J, Subra G. Combinatorial chemistry on solid support in the search for central nervous system agents. *Comb Chem High Throughput Screen*. 2009;12:723–739.

Zhang S, Sun Y. Steric mass-action model for dye–ligand affinity adsorption of protein. *J Chromatogr A*. 2002;957:89–97.

Zhi W, Purohit S, Carey C, et al. Proteomic technologies for the discovery of type 1 diabetes biomarkers. *J Diabetes Sci Technol*. 2010;4:993–1002.

Zhi W, Wang M, She J-X. Selected reaction monitoring (SRM) mass spectrometry without isotope labeling can be used for rapid protein quantification. *Rapid Commun Mass Spectrom*. 2011;25:1583–1588.

Zuckermann RN, Martin EJ, Spellmeyer DC, et al. Discovery of nanomolar ligands for 7-transmembrane G-protein-coupled receptors from a diverse N-(substituted)glycine peptoid library. *J Med Chem*. 1994;37:2678–2685.

CHAPTER 5

Plant Proteomics and Food and Beverage Analysis via CPLL Capture

5.1 INTRODUCING GLOBAL PROTEIN ANALYSIS IN FOOD

Proteomics investigations with continuously improved technologies started with the elucidation of proteomes and then expanded to identification of differences between healthy and pathological situations. The purpose was to discover novel biomarkers for medical diagnosis that would have much higher sensitivities and specificities than those in use since the 1960s. With the advent of new technologies and the improvement of genomic data banks, proteomics investigations extended toward a variety of domains, one of them being plant and food science. This science covers nutrition studies conducted with modern technologies, including all those currently used in proteomics. The term *foodomics* (Herrero et al., 2012) has been coined to describe food science that uses modern biology approaches and technologies such as genomics, transcriptomics, proteomics, and/or metabolomics. But we much prefer the longer established term *food science.* For all these domains mass spectrometry and electrophoresis methods serve as powerful tools for investigating exogenous contaminants (e.g., pesticides), food allergens, and toxins. Food science encompasses food genomics (Rist et al., 2006), food allergies (Aiello et al., 2011), nutraceutical science (Chen et al., 2006; Bernal et al., 2011), and genetically modified food (Goodmann and Tetteh, 2011; D'Alessandro and Zolla, 2012).

Deep genomic investigations started years ago, and hence extensive data are now available. However, genomics has far from yielded complete information. For example, even if, by comparison, molecular components among different plants appear orthologous, their function could be very different (Rakwal and Agrawal, 2003). Therefore functional genomics aimed at assigning a role to each gene remains an uncompleted task. Transcriptomics, at the next level, has made great advances, even if it is not informative enough about protein expression and even less helpful in explaining post-translational modifications. It is for this reason that proteomics appears irreplaceable in adding to biological/biochemical knowledge, especially with respect to food science where numerous parameters interact. Actually, it is probably the sole option available for identifying and understanding post-translational modifications, the interaction of proteins with metabolites (e.g., transport function), and protein–protein interactions. Proteomics also allows determining whether food protein components have beneficial or detrimental effects on the organism that receives them.

One goal in current food science is to enhance the level of understanding of food components through focusing on the functions of single species. In 2004 protein profiles derived from proteomic technologies were established for assessing food composition, its origins and possible adulteration (Carbonaro, 2004). Other goals of interest include how to improve health by using genetically modified food with targeted pharmacological properties rather than prescribing drugs (Dureja et al., 2003). Today the situation is more complex because genetically modified organisms may be added to food, inducing unexpected, undesirable effects. A transgenic organism is considered safe by health authorities for human consumption if, by its documented history, it is found to be equivalent to its conventional counterpart. In addition, encoded recombinant information should not generate toxic or allergenic reactions (El Sanhoty et al., 2004). As reported by Simó et al. (2010), investigations of protein expression of transgenic food are of particular interest because of their potential ability to induce allergic effects or provide unwanted enzymatic or inhibitory activity as well as possible toxic effects. In this respect, such transgenic organisms require targeted studies in protein expression alterations that are necessary to decipher when they are present even in a very limited amount. The example presented by Simó et al. (2010) was soya beans where no expression differences could be detected. However, it has been reported that variations in storage proteins ratio (which represent more than 80% of the proteome with a large preponderance of 11S- and 7S- globulins) could impact functional and nutritional properties with possible implications in safety (Natarajan et al., 2006). Within this context, proteomics investigations are increasingly used to screen and differentiate transgenic plants in order to either evaluate food safety or identify transgenic components in food (Luo et al., 2009). Expression changes are not exclusively the result of transgenicity; they can also be species-specific, with possible implications for human allergies as described in Emami et al. (2010).

Why, then, is it so important to continue investigations of new transgenic plants? There are two reasons. The first is related to the economy as a whole. Certain plants are expensive and not sufficiently productive to cover the world request; they need too much care and water and have poor resistance to adventitious agents and thus require massive use of pesticides. To respond to these issues, some genetically oriented modifications are necessary for some current plants (García-Canas et al., 2011).

The second reason justifying the modification of plant genomes is to enhance the expression of specifically targeted molecules that are useful to alleviate given diseases or to improve the quality of life. Whatever the justification, transgenic versus nontransgenic plants intended for use as food components must be analyzed to demonstrate expression differences and possible effects on the receiving host (García-Lopez et al., 2009). The proteome equivalence between a regular organism and its genetically modified counterpart thus becomes a real investigational work in proteomics. For example, Simó et al. (2010). recently examined the soybean in a comparative study. In this study, no detectable changes were observed by using capillary electrophoresis followed by mass spectrometry. However, the proteome analysis did not consider specific sample-enrichment treatments allowing further investigations of low- and very low-abundance gene product detection.

Globally, food science assembles modern disciplines and technologies to keep track of all the above food components, origin, and effects. Thus genomics, transcriptomics, proteomics, and metabolomics are part of this field, with the aim of discovering the composition of foods, noting differences among samples, exploring nutrition properties, and managing the questions surrounding the genuineness, quality, expected effects, and evidence of agents harmful to human health. The complexity of this endeavor is therefore evident not only at the level of study disciplines such as genomics versus proteomics or metabolomics, but also in terms of differential expression between species and of selection of investigation technologies. One possible approach to resolve this complex task is a global analytical attack involving proteins from a few or many organisms depending on the food composition. It is a very ambitious approach since it must federate technologies of protein extraction that are not necessarily the same when dealing with plant extracts compared to microorganism and/or animal extracts. Another immense technical challenge involves how to deal with the protein dynamic concentration range when the food combines components from animals, plants, and other organisms all of which have their own extremely large differences in the concentration of high-abundance versus low-abundance proteins.

Since vegetables are a very important part of human food, special focus also needs to be placed on the protein content and analysis of vegetables. In fact, plant proteomics constitutes almost a world of its own when we consider the difficulty involved in accessing and analyzing these proteomes (Jorrìn et al., 2006; Jorrìn-Novo et al., 2009).

To reduce the complexity that is so prevalent in the food science world and hence to simplify scientific thinking, in this chapter we have limited the discussion to three main areas, explored, of course, with the combinatorial peptide ligand library (CPLL) technique: (a) plant and recalcitrant tissue proteomics; (b) proteomics of large-consumption foodstuffs; and (c) the proteomics of alcoholic and nonalcoholic beverages. The major focus is on low-abundance species and on exploration of each tissue in order to find hidden allergens (if any) and new proteins endowed with important nutritional value.

5.1.1 Sample Treatment Prior to Detection of Proteins

The protein content of foodstuffs is extremely variable. Moreover, when food is the result of preparations involving blends from vegetable and/or animal origin, the complexity reaches levels that can be quite difficult to manage. In all cases the overall protein analysis requires a pretreatment of the initial material at least to remove non-proteinaceous components and then, if necessary, adopt treatments for unmasking very dilute species. Finally the specific removal of some proteins or the enrichment of trace components might be desirable.

5.1.2 Extraction and Pretreatment

Posing the largest challenge is the extraction of plant proteins (vegetable-based food) since they are generally accompanied by vegetal pigments, polysaccharides, lipids, and polyphenols. In addition, numerous plant proteins are heavily glycosylated, thus adding difficulties for analyses such as difficult ionization in mass spectrometry and abnormal migrations in two-dimensional electrophoresis. In this respect, plant proteomics teaches how to make a proper cleanup of these samples, with minimal protein loss. Since soluble protein concentration is generally very low in differentiated plant tissues (Rose et al., 2004), and since these tissues contain a number of undesirable metabolites that interfere with protein separation and analysis (Gengenheimer, 1990), before initiating a proteome investigation several points need to be considered. The most common approaches are extraction and precipitation (see Carpentier et al., 2008). For the extraction phase the following general rules apply: (a) with highly viscous material, such as eggs (white and yolk) a dilution with energetic agitation is recommended; (b) protein extraction should be performed in relatively low ionic strength to prevent the solubilization of nucleic acids and in the presence of small amounts of chelating agents such as EDTA; (c) for the extraction of proteins within the cell walls, some amounts of non-ionic detergent (less than 0.5–1%) and urea (less than 3 M) should be added to the extraction buffer. Finally, since plant extracts contain a quite powerful protease charge, the addition of inhibitors is highly recommended.

Undesirable material is removed as part of a second operation immediately following the initial crude extraction. Precipitation and/or selective extraction can also be considered, depending on the context. For instance, the removal of lipids requires special attention since their presence can render the subsequent proteomics analysis ineffective. The presence of detergents in the extraction solution should disaggregate lipids while solubilizing hydrophobic proteins. However, if large amounts of lipids are present, as for instance in seeds, chemical delipidation prior to protein extraction might be necessary (Pihlström et al., 2002). In this case, lipid removal is achieved by extraction with hydrocarbon solvents (Van Renswoude and Kemps, 1984) or of mixtures thereof (Radin, 1981), or even containing chlorinated solvents (Wessel and Flugge, 1984). Partial, but useful, delipidation with ethanol or acetone can also be adopted (Menke and Koenig, 1980). Precipitation of proteins using trichloroacetic acid associated with acetone and containing reducing agents allows the collection of most proteins as pellets with a supernatant comprising a lot of undesired materials that is eliminated by centrifugation or filtration (Méchin et al., 2007). Most of these methods are detailed in Chapter 8 with specific recipes.

Interfering substances can be discarded by using a Tris-HCl solution saturated with phenol, followed by protein precipitation with ammonium acetate or ammonium sulphate. This option is especially recommended when the analysis of proteins is based on two-dimensional electrophoresis. As a variant, the combination of phenol extraction followed by a precipitation with ammonium acetate in methanol has been described by Faurobert et al. (2007). A less popular method is protein precipitation using a chloroform–methanol mixture with water in 1–4–3 proportions (Wessel and Flugge, 1984). The following step, which is the resolubilization of precipitated proteins, is not always easy depending on the nature of proteins. Quite frequently, the presence of zwitterionic detergents and chaotropes is necessary (Isaacson et al., 2006).

Animal protein extraction is probably easier than plant protein extraction. In food they are generally present as tissues and are involved in membranes. The cell membrane is the result of a complex structure involving proteins and lipids. Lipids could be structurally different and contain phosphoric esters. Proteins generally have a central hydrophobic segment that is fully involved in the architecture of the membrane and two extremes that are either exposed to cytoplasm or to the extracellular domain. These proteins are so tightly involved in the membrane structure that it is difficult to displace them except when the construct is destroyed by means of buffers containing detergents and possibly chaotropic agents such as urea (Cilia et al., 2009). Reducing agents, selected cations, sugars, and other chemical agents may also be beneficial to protein solubilization.

All these preliminary operations for plant and animal protein extraction not only remove undesired materials, but also contribute to concentrate proteins that are present in very low amounts and in quite a large number. They are also beneficial when the sample is to be treated with CPLLs. In fact, the presence of polyphenols and other polyacidic substances (mucins, nucleic acids, etc.) prevents the proper capture of proteins by peptide libraries. However, the presence of solubilizing agents such as highly concentrated chaotropes and sodium dodecyl sulfate (SDS) must also be eliminated prior to CPLL treatment by, for instance, protein precipitation or dialysis.

When devising protein extraction methods, the compatibility of an initial sample treatment with further processing, such as, for instance, protein fractionation and proteomic analysis, should be kept in mind. Protein extracts that are to be submitted to two-dimensional electrophoresis analysis should preferably be neutral and in low ionic strength solutions without traces of ionic detergents such as sodium dodecyl sulphate. If the proteome analysis is performed by top-down mass spectrometry, the protein sample also needs to be cleaned up before being loaded on the MS probe (Rey et al., 2010). When using protein arrays (e.g., antibodies), the protein extract should not comprise denaturing agents that prevent the correct formation of the antibody–antigen complex. In bottom-up mass spectrometry where peptides are directly analyzed, there are other rules to follow as, for instance, the problem of nonspecific binding of molecules on the inner capillary wall resulting in poor separation efficiency as a consequence of band broadening (Castro-Puyana et al., 2012). Conversely, when using reverse-phase chromatography, alkaline peptides may be extremely tightly adsorbed by the solid phase and frequently escape the MS analysis.

5.1.3 High-Abundance Protein Depletion versus Dynamic Range Normalization

Evidencing low-abundance proteins from whatever extract is always challenging. There are two main reasons for this difficulty: (i) the low amount of these proteins falls under the detection limits of the current analytical tools and methods or (ii) few high-abundance proteins are so largely present that the detectability of dilute species is prevented due to a signal subtraction effect. In food science there are many examples of products that comprise proteins of high abundance: ovalbumin in eggs, RuBisCO in leaves, actin in animal tissues, and casein in milk, to name just a few. In many tissue extracts and biological fluids, few proteins represent the large mass of proteins. This is illustrated in Table 4.2. To facilitate analytical determinations, a relatively common process is to remove all or most of the high-abundance proteins (depletion).

Although this argument has been treated at length in Chapter 3, Section 3.2, we will briefly recall some basic concepts here. Depletion technologies started with the removal of high-abundance proteins from human serum (Zolotarjova et al., 2005; Echan et al., 2005; Shi et al., 2012). First, current affinity sorbents were used to remove albumin and immunoglobulins G to end up to the generalized use of immunosorbents to remove more specifically several proteins at the same time (Polaskova et al., 2010). Such a process is limited to the upper part of the protein expression scale and as a consequence minor components that represent the largest number of gene products remain as dilute as they were before the treatment and, worse, are frequently more dilute. Therefore, these proteins are still undetectable, unless they are submitted to a concentration process. Even so, due to the very limited volumes of biological samples submitted to immunosubtraction, the amount of these low-abundance species is frequently below the sensitivity of analytical methods. Moreover, with immunosorbents for depletion there is a constant risk of protein co-depletion (Stempfer et al., 2008) that has nothing to do with the antigen–antibody interaction. This phenomenon can be devastating (Shen et al., 2005), with a loss of hundreds of co-depleted proteins (Yadav et al., 2011). The immunodepletion process is not only protein- but also species-specific and cannot be used as a general tool in food science. In fact, contrary to what is largely applied to human serum, specific antibodies against numerous proteins of different origin are not necessarily available.

Examples of immunosubtraction in plant proteome analysis are rather scarce in the literature. As an example, Hashimoto and Komatsu (2007) applied this method in proteomic analysis of rice seedlings during cold stress. When analyzing leaf blades and sheaths via differential 2D electrophoresis mapping, among the 250–400 protein spots detected, 39 proteins were found to change in abundance after cold stress, with 19 proteins increasing and 20 proteins decreasing. Since ribulose bisphosphate carboxylase/oxygenase (RuBisCO) accounted for about 50% of the total leaf proteins, to minimize this problem Hashimoto and Komatsu prepared an antibody–affinity column to trap RuBisCO. After this immunodepletion, however, only four proteins were newly detected after cold stress, thus reinforcing the notion that immunosubtraction does not dramatically improve protein discovery. Thus, the way to reveal low-abundance species is consequently restricted to the use of tools that are capable of enriching for low-abundance proteins while decreasing the protein dynamic range or that are designed to capture pre-designated species (see next two sections).

This tool seems to be largely restricted to the use of the combinatorial peptide ligand library (CPLL) technique, whose application to plant and foodstuff analysis is extensively covered in this chapter. Since CPLLs have been amply described in this book, we will go straight to the applications and skip their description, while referring the reader to some basic articles that amply cover the subject (Thulasiraman et al., 2005; Righetti et al., 2006; Roux-Dalvai et al., 2008; Boschetti and Righetti, 2009; Righetti et al., 2010). Additionally, there are other methods capable of enriching low-abundance species. However, they do not compress the dynamic concentration range, but rather focus on specific groups of proteins such as phosphoproteins (Gronborg et al., 2002; Larsen et al., 2005) and glycoproteins (Dayarathna et al., 2008). Aptamers have also been used on microarrays for the enrichment and identification of low-abundance targets (Ahn et al., 2010).

In the following sections, we will deal with three main topics: (a) plant proteomics and related vegetable food sources; (b) foods of animal origin; and (c) analysis of alcoholic and nonalcoholic beverages. Such topics, per se, would of course require several books to be properly covered. Here we will only cover examples in which the CPLL technique has been applied, simply to show how this method can substantially improve the discovery of the respective proteomes, with special emphasis on low-abundance species. As such, references will be limited to the examples provided in this chapter.

5.2 PLANT AND RECALCITRANT TISSUES PROTEOMICS

Why plant proteomics? Is it just an exciting topic among plant biologists, or is there a genuine approach to tackling the issues that plague plant biology in the real world—understanding the phenotype; digging deep into the mechanisms behind plant and crop growth and productivity against the ravages of abiotic and biotic stresses; finding the means to reduce food shortages and prevent hunger, starvation, and death; and preserving the biodiversity and our environment? There is a definite need to work to solve each of these issues not singly but by integrating proteomics data with other omics technologies, such as transcriptomics and metabolomics, and by working closely with plant physiologists, breeders, and bioinformaticians. Moreover, plants, as sessile organisms, have developed very specific adaptive mechanisms that cannot be deduced from studies with other organisms. Studying plants is rendered even more difficult when we consider that there exist about 300,000 flowering plant species on the planet, and that lessons learned from one species might not necessarily help solve a question about other species (Rossignol et al., 2006).

Without doubt, feeding the world population is becoming an increasingly severe problem. It was even more difficult in the past, as this anecdote involving the art world illustrates. The French impressionists produced a big revolution in painting when these artists moved out of their studios and began to represent the surrounding nature "en plain air." One of their followers, however, the expressionist Chaïm Soutine, resisted this trend. A refugee from Smilavichi, a tiny village in poverty-stricken Byelorussia, Soutine moved to Paris in 1913 in an attempt to find a better life. Obsessed by the lack of food he had experienced in Byelorussia, he started painting carcasses of animals that had been taken from slaughterhouses. His canvasses depict entire hindquarters of cows, or slaughtered poultry, rendered with thick, bloody-red strokes of a spatula, so vivid that one could almost imagine the artist taking a bite and quelling his own hunger. Unfortunately, France soon entered World War I, and with the war came famine throughout Europe, and so our artist remained as hungry as he had been in his hometown. Eating a succulent steak, even in the modern capitalistic era, might put a dent in your wallet and will surely lead to a a bigger damage to nature, since the subtraction of land to agriculture due to cattle grazing in the open range takes away so much land that would instead be devoted to agriculture, thereby resulting in additional burden and famine in the Third World.

The approach of poverty-stricken countries in the Third World is to achieve subsistence out of plant-derived foodstuffs. For the Mayans, for example, corn was the main staple and was practically their only foodstuff. Yet corn proteins do not sustain life, as nutritionists found out in modern times: The main family of proteins (the storage proteins zeins) are very poor in lysine, an essential amino acid. The Navajo Indian unknowingly used to complement their poor diet with beans, a rich source of good proteins and the second sacred plant for this tribe. Much later, in the early 1960s, nutritionists at the Department of Nutrition and Food Science of MIT devised a simple remedy to the poor content of lysine: the Incaparina, a cornflower supplemented with robust doses of free lysine. MIT's nutritionists taught poor Guatemalan campesinos not to waste a single drop of the cooking water for making tortillas or other corn food, lest free lysine would go down the drain.

These two examples highlight some of the severe problems plaguing today's hungry. On the one hand, there is the problem of consuming meats, with their high cost in taking land away from agriculture; on the other hand, there is the problem that plant products might not have an optimal amino acid composition for feeding humans. The problems call for proper knowledge of their proteomes and for intervention by geneticists to improve the nutritional value of these crops. (See Figure 5.1 for a summary of this discussion.)

There is a dichotomy in present-day civilization: *the hungry world*, in which about one billion people have to struggle daily to find a minimum amount of food for survival, and the *hunger-creating world*, in which the increasing population and the shortage of utilizable and fertile land are among the

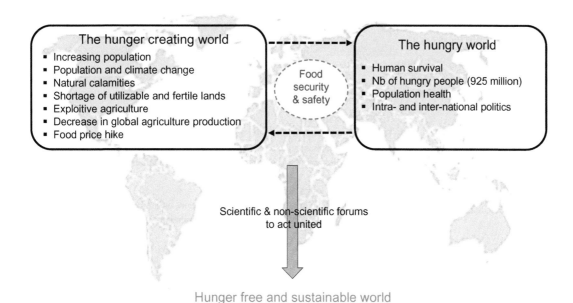

FIGURE 5.1
The vicious cycle of the global food crisis. Food security is a multifaceted problem affected by multiple factors worldwide. Here are shown the major issues that are the cause of food crisis, as well as those created by the food crisis. These problems are both social and scientific in nature, and therefore, there is a strong need to present a united front in creating a hunger-free and sustainable world (by courtesy of Drs. G.K. Agrawal and R. Rakwal).

biggest dilemmas. In this gloomy scenario (exacerbated by the utilization of land and crops not for human consumption but for producing fuel for cars; e.g., through the production of ethanol from maize and other crops), plant science in all of its facets, including proteomics, genomics, and breeding for better crops, may be the only solution to global food shortages. Therefore, plant genomics and proteomics will surely merit much greater efforts and attention from the scientific community. Fortunately, beginning in 2000, this field has experienced steady growth, as illustrated in Figure 5.2, which shows its various growth phases: pre-, initial, and progressive. The pre-stage can be considered the beginning of proteomics where one-dimensional electrophoresis and two-dimensional electrophoresis techniques were applied to separate proteins and to their identification using N-terminal Edman sequencing. The initial stage started with the genome revolution from the year 2000 onward. Since the publication of the draft genome sequences of two plants, *Arabidopsis thaliana* (weed and dicot model) (The Arabidopsis Genome Initiative, 2000) and rice (*Oryza sativa* L., cereal crop and monocot model (Goff et al., 2002; Yu et al., 2002) in 2000 and 2002, respectively, plant proteomics research has experienced rapid growth (Cànovas et al., 2004; Agrawal et al., 2006, 2008). In this initial phase, the *Arabidopsis* scientific community began working toward the proteome of this model plant by establishing thea Multinational Arabidopsis Steering Committee Proteomics subcommittee (MASCP, www.masc-proteomics.org).

Plant proteomics has now moved into the progression stage, with plant researchers involved in enriching the scientific community publishing reviews on rice, plants, and protein phosphorylation,

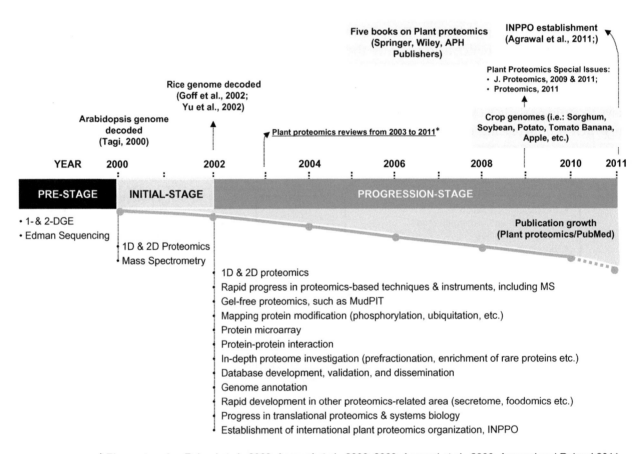

* Rice proteomics: Rakwal et al., 2003; Agrawal et al., 2006, 2009; Agrawal et al., 2008, Agrawal and Rakwal 2011
* Plant proteomics: Canovas et al., 2004; Jorrin-Novo et al., 2007, 2009; Rossignol et al., 2006
* Plant phosphorylation: Kersten et al., 2006; Kersten et al., 2009.

FIGURE 5.2

Timeline of plant proteomics development. Details are in the main text (by courtesy of Drs. G.K. Agrawal and R. Rakwal).

as well as five books on plant proteomics. The later part of the 2000 decade also saw the development of a global initiative on plant proteomics that led to the establishment of the International Plant Proteomics Organization (INPPO, www.inppo.com). Now that more plant genomes have been sequenced, from model to non-model (Feuillet et al., 2010; Agrawal et al., 2011), there is no turning back to the utilization of proteomics approaches in various aspects of plant biology research. Just like HUPO (Human Proteome Organization) and its initiatives (e.g., the PPP—Plasma Proteome Project) have helped further the field of human proteomics, it is hoped that INPPO will have the same beneficial effects on advancing plant science as a whole.

Is there any need to further improve the ability to dig deep in plant proteomes? Definitely yes, as the examples below illustrate. It is not uncommon, in mammalian proteomics, to be able to survey almost the entire gene expression in a single sweep. Perhaps the most impressive example has been that given by Geiger et al. (2012) who surveyed the proteome of 11 human cell lines and described the expression of 11,731 proteins, and on average 10,361±120 proteins in each cell line—a substantial achievement indeed. Things are not that easy in plant proteomics, though. Even in the case of organisms whose genome has been fully decoded, the situation is not as brilliant as in the mammalian field. For instance, the genome of *Arabidopsis* has been found to comprise roughly 26,000 genes. Yet, even the most extensive investigation summarizing the data obtained by exploring eight different tissues (primary leaf, leaf, stem, silique, seedling, seed, root, or inflorescence) yielded the global identification of only 2943 proteins, products of barely 663 different genes (Bourguignon and Jaquinod, 2008).

Individual organelles have not fared much better. For instance, in the case of chloroplasts, whose proteome has been estimated to comprise roughly 4500 proteins, only 690 proteins could be correctly detected and identified. In the case of the mitochondrion, whose proteome has been predicted at 3200 proteins, only 547 nonredundant species could be assessed. The summary of all proteins described so far in *Arabidopsis* (< 3500 species) remains dramatically low as compared to the 26,000 genes in its genome (Bourguignon and Jaquinod, 2008). No doubt, one of the major problems, especially in plant tissues, is the fact tht the most abundant proteins present in the sample are repeatedly analyzed and thus prevent identification of the low-abundance proteins. As an example, the most abundant protein in plants, RuBisCO, is the one most frequently identified in 2D polyacrylamide gel mapping: Giavalisco et al. (2005) reported that this protein, in its 2D maps, represented a total of 366 spots, amounting to 12.5% of all assigned spots (numbering 2943 over 6000 visible spots). Clearly, ways and means of reducing the interference of HAPs (high-abundance proteins) in favor of LAPs (low-abundance proteins) are sorely needed. The few examples given below on the use of CPLLs will drive this point home.

5.2.1 Spinach Proteomics

Spinach (*Spinacia oleracea*) is an edible flowering plant in the Amaranthaceae family. It is native to central and southwestern Asia. It is an annual plant that grows as tall as 30 cm. In temperate regions spinach may overwinter. The leaves are alternate, simple, ovate to triangular-based, variable in size from about 2–30 cm long and 1–15 cm broad, with larger leaves at the base of the plant and small leaves higher on the flowering stem. Spinach has a high nutritional value and is extremely rich in antioxidants, especially when fresh, steamed, or quickly boiled. It is a rich source of vitamin A (and especially high in lutein), vitamin C, vitamin E, vitamin K, magnesium, manganese, folate, betaine, iron, vitamin B_2, calcium, potassium, vitamin B_6, folic acid, copper, protein, phosphorus, zinc, niacin, selenium, and omega-3 fatty acids. Recently, opioid peptides called rubiscolins have also been found in spinach. Given its importance in human nutrition, it is surprising that almost no studies have been devoted to exploration its proteome.

One such study, however, has recently been performed by Fasoli et al. (2011a). They have mapped the cytoplasmic proteome of spinach leaves with the help of commercially available combinatorial

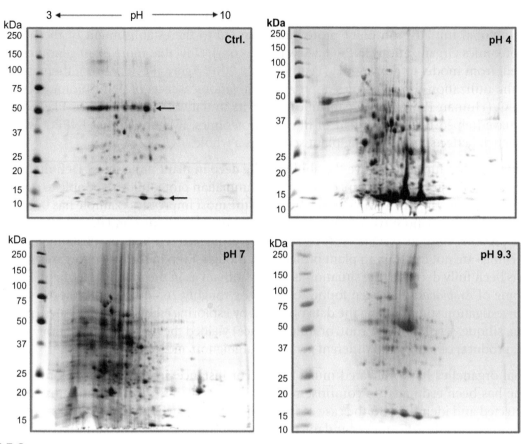

FIGURE 5.3

Two-dimensional maps of the untreated control (Ctrl) and the three CPLL eluates from different pH captures, pH 4.0, 7.0, and 9.3. In all cases, the pH gradient was a nonlinear pH 3-10 IPG strip, and the second dimension was an 8–18% polyacrylamide gradient. In each panel, the bands on the left track represent the Mr markers, in the 10 to 250 kDa interval. The sample load was 150 μL of eluates and control (corresponding to 200 to 250 μg total protein) re-swollen in the dry IPG strips. All maps stained by micellar Coomassie Brilliant Blue. The arrows in the control map indicate the large and small RuBisCO subunits. *(From Fasoli et al., 2011a, by permission).*

peptide ligand libraries as well as with home-made ligand beads prepared in house. The protein capture was performed at three pH values (4.0, 7.0, and 9.3), and elution was carried out in 4% boiling SDS, 20 mM DTT (dithiothreitol). The number of gene products found, exclusively thanks to the use of CPLLs and classified as low abundance, was particularly high, with 208 proteins versus 114 found in the initial extract with an overall gene product count of 322 (Figure 5.3). Two main phenomena allowed the researchers to detect so many more proteins: the massive reduction of the concentration of RuBisCO and the enrichment of many rare proteins initially present in trace amounts made possible by the use of a relatively large initial volume of spinach extract. The use of CPLLs at three different pHs enlarged the harvesting of species that are particularly acidic and alkaline (see Section 4.5.2 and a general recipe in Section 8.9). Among the enriched cellular components were ribosomal proteins and their complexes. Many of those exclusively found after CPLL treatment were 30S, 40S, 50S, and 60S.

Other low-abundance proteins found in leaf spinach were, as expected, those proteins related to the chloroplast category and those operating in photosynthetic organelles (e.g., inorganic pyrophosphatase-1 and phosphoglycerate kinase). From the same category, but involved in the CO_2 fixation and carbohydrate metabolism, were glutamate decarboxylase 2, glucose-1-phosphate adenylyltransferase, phosphoglucomutase, fructose bisphosphate aldolase 2, ribulose-phosphate 3-epimerase, and sucrose-phosphate synthase. In spite of numerous novel gene products found upon treatment with the peptide library, many others could not be identified because genomic data was not available.

From Figure 5.3, one can appreciate how much richer in spots are the 2D electrophoresis maps after CPLL treatment, especially those of the captures performed at pH 7.0 and 4.0. One can also note the dramatic decrement of the RuBisCo spots, marked by arrows in the control panel.

5.2.2 Olive (*Olea europaea*) Seed and Pulp Proteomics

The olive tree (*Olea europaea*) is a long-life tree from which humans have obtained wood and oil for more than 5000 years (Rallo et al., 2005). It is the only tree of the *Oleaceae* family with edible fruit. Olive fruits have been used for thousands of years to produce olive oil and table olives. Generally, in the manufacture of olive oil the whole olive fruit is subjected to a mechanical press, thus passing both pulp and seed components to the olive oil. The consumption of olive oil has been associated with decreased incidence of major illnesses such as cardiovascular diseases, cancer, and Alzheimer's disease in the Mediterranean area (De la Lastra Romero, 2011).

Olive fruit has been investigated in depth to determinate its composition, its nutritional and sensory values, and its great benefits for human health. Some compounds such as fatty acids, polyphenols, or sterols have been investigated in detail (Savarese et al., 2007; Haddada et al., 2007). Despite their highly informative value, their nutritional value (Rodríguez et al., 2008), and their suggested role in food stability (Koidis and Boskou, 2006; Georgalaki et al., 1998a,b) and allergenicity (Esteve et al., 2012d), proteins in the olive have scarcely been investigated in comparison with its other components (Alché et al., 2006; Esteve et al., 2011). Although information on the amount of proteins that can pass from the fruit to the oil during oil extraction is very scarce, in 2001 proteins were for the first time established as minor components in olive oils (Hidalgo et al., 2001), demonstrating the ability of transferring a minor part of proteins from the fruit to the oil.

Olive fruit is an especially complex matrix due to its lipid nature and its large amount of interfering compounds, which have hampered the extraction of proteins from this fruit. Many efforts have been focused on developing an extraction procedure of proteins from both olive seed and pulp (Montealegre et al., 2013). But especially in the case of olive pulp, the results have not been completely successful, yielding just one band when Coomassie blue staining was used. Esteve et al. (2012a) have recently tackled the investigation of olive pulp and seed proteome using the CPLL approach. As a result, a large number of compounds have been identified: 61 in the seed (vs. only four reported in current literature) and 231 in the pulp (vs. 56 described thus far). In the seed, it highlights the presence of seed storage proteins, oleosins, and histones, of which as many as 14 different unique species have been listed. In the pulp, the allergenic thaumatin-like protein (Ole e 13) was confirmed, among the other 231, as the most abundant protein in the olive fruit.

5.2.3 Avocado Proteomics

Avocado, the fruit of the tropical tree *Persea americana*, is native to Mexico but is now grown and consumed in many other parts of the world. The oil obtained from pressing the avocado fruit was used in Mexican folk medicine in the sixteenth century (Argueta-Villamar et al., 1994) and today is employed for uses in food, cosmetics, and health care products (Swisher, 1988). In addition, avocado oil has been proposed as a domestic source of cooking oil to help improve the nutritional status of populations in some developing countries. The consumption of both the avocado fruit and oil has been associated with several health benefits, notably with decreases of total serum cholesterol, LDL-cholesterol, and triglycerides (Lopez-Ledesma et al., 1996), control of blood pressure (Salazar et al., 2005), and inhibition of certain types of cancers (Lu et al., 2005).

The composition of the avocado fruit has been extensively studied, leading to a good characterization of small-size compounds such as fatty acids and sterols (Plaza et al., 2009). To date, however, no work has been done on identifying the avocado pulp proteome; the only reports that have appeared are on the avocado seed (Sanchez-Romero et al., 2002) and root (Acosta-Muñiz et al., 2012) proteins. On the other hand, avocado is a known source of allergens that can elicit diverse IgE-mediated reactions, including anaphylaxis in sensitized individuals. Those adverse reactions are well described in the literature, mainly in the context of latex sensitization (Barre et al., 2009), otherwise termed *latex-fruit syndrome* (Lavud et al., 1995; Ahlroth et al., 1995; Moller et al., 1998). In addition, there have been some cases of cross-reactivity with banana (Ito et al., 2006) and melon (Rodriguez et al., 2000). Nonetheless, despite different reactive bands having been detected by immunoblot, only the main allergen, *Pers a 1*, has been studied in depth (Sowka et al., 1998; Diaz-Perales et al., 2003), and just one report has suggested the presence of a second allergen, *Pers a 4* (van Ree et al., 1992). The remaining allergens remain uncharacterized.

The most critical step in any proteomics study is sample extraction and preparation, which is typically more problematic for plant tissues than for other organisms because of the relatively low concentrations of soluble proteins, the presence of proteases, and the high concentration of compounds that severely interfere with protein separation and analysis, including cell wall and storage polysaccharides, lipids, phenolic compounds, and a broad array of secondary metabolites. The main difficulty in avocado protein extraction is the presence of a high content of oil, up to 32%, a similar percentage as that observed in the olive pulp, whose analysis has been reported in Section 5.2.2. However, in the case of olive seed and pulp, although the use of CPLLs has substantially improved the identification of proteins present therein, the overall catch remains rather poor: In no case was a total count of more than 250 protein species achieved. It was felt that extra effort was needed to study these recalcitrant tissues and in general any plant proteome.

This extra effort has materialized in more than one extraction protocol, that is, under native and denaturing conditions, in both cases followed by capture with CPLLs of the extracted proteins, as shown in Figure 5.4. The extraction under denaturing conditions is performed in 3% SDS, which is

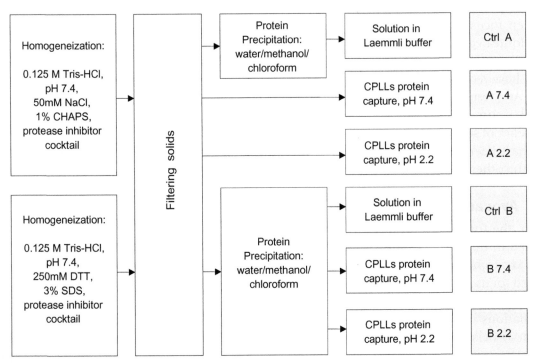

FIGURE 5.4

General scheme for protein extraction from recalcitrant plant tissues. Solubilization is performed under native, physiological conditions (extract A) as well as under denaturing conditions in 3% SDS (extract B). In both cases, CPLL capture is carried out at two pH values, 7.4 and 2.2 (the latter under conditions mimicking reversed-phase capture). In the case of a SDS-denatured sample, the surfactant is eliminated through precipitation in cold methanol/chloroform. Alternatively, SDS could be diluted to 0.1% in the presence of 1% CHAPS, conditions compatible with subsequent CPLL treatment. *(From Esteve et al., 2012b, by permission).*

anathema in CPLL treatments, since it would completely inhibit the capture. Yet, there are two ways to go about it: One is to remove the SDS via the classical acetone/methanol precipitation, and the other is to dilute the SDS from 3 to 0.1% in the presence of another compatible surfactant such as 1% CHAPS, conditions that would be compatible with the CPLL technology. As shown in Figure 5.4, a substantial amount of extra work is involved, yet the results have been outstanding: About 1300 unique gene products have been detected, an increment of more than one order of magnitude as compared to the findings in olive pulp proteome.

Figure 5.5 details the various discoveries obtained with the treatments: The unique species identified in the control were 236 versus 796 in the global CPLL treatment, 250 being in common in the two

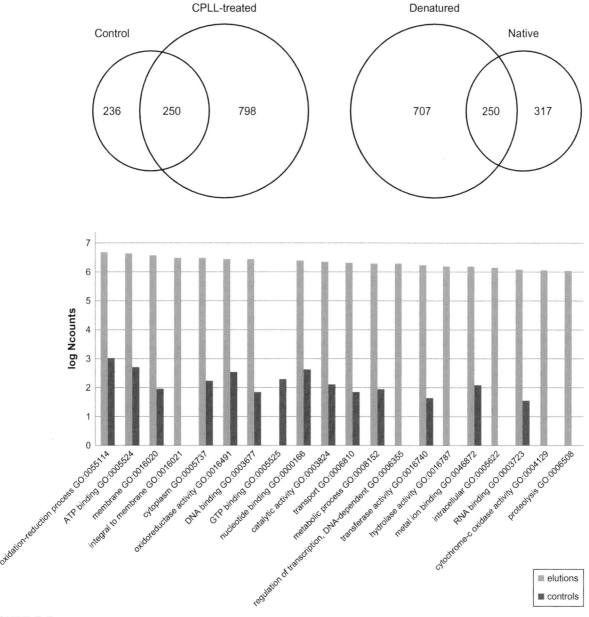

FIGURE 5.5

Upper panel: Venn diagrams showing the total identifications of avocado proteins in either the control or the CPLL-treated samples (left diagrams). The Venn diagrams to the right show the relative contributions to the total IDs of the denatured versus the native sample extractions. Lower panel: GO analysis of the proteins detected in various metabolic processes in the control versus CPLL-treated samples. In the latter case, five additional metabolic processes can be described, which are not visible in the control sample and most likely represent low- to very-low abundance proteins, whose visibility has been substantially enhanced via the CPLL technology. *(From Esteve et al., 2012b, by permission).*

sets. The Venn diagram to the right in the figure shows how much more the denaturing conditions extraction contributed to the total discovery as compared to the native protocol: The unique catch has been more than double. The bar graph of gene ontology (GO) analysis (the lower panel in Figure 5.5) also shows some outstanding data. Not only all the various GO categories are considerably more populated in the CPLL-captured samples, but also five new metabolic processes, not discernible in the control, could be described (Esteve et al., 2012b).

5.2.4 Banana Proteomics

Musa spp., comprising banana and plantain, is grown extensively in many developing countries and is considered to be one of the most important sources of energy in the diet of people living in tropical humid regions. Because of its antioxidant and cell antiproliferative activities, consumption of bananas has been associated with reduced risk of chronic diseases such as cardiovascular diseases and cancer (Sun et al., 2002).

To date, no in-depth work has been focused on identifying the banana fruit proteome; some reports have appeared just on the banana meristem (Samyn et al., 2007; Carpentier et al., 2007) and leaf (Vertommen et al., 2011) proteomes. However, the articles published on banana proteins reflect the scientific community's interest in this subject (Anderson et al., 1970). Some years ago the Global *Musa* Genomics Consortium (GMGC), an international network of scientists who are applying genomics tools to the banana to ultimately improve its breeding and management (http://www.musagenomics.org), was created. In comparative genomics, *Musa* is seen as an interesting model for understanding genomic evolution in relation to biotic and abiotic stresses.

Other genomic groups have started sequencing banana mRNAs and have submitted their results to the National Center for Biotechnology Information (NCBI, http://www.ncbi.nlm.nih.gov) from which it is possible to download a still incomplete EST database of different species of *musa*. Moreover, one can find some reports on protein changes associated with banana fruit ripening, showing the differential protein accumulation during maturation and the presence of specific proteins of ripe fruit (Dominguez-Puigjaner et al., 1992; Peumans et al., 2002; Choudhury et al., 2008) and how storage and transport at low temperatures affect the activity of some enzymes (Der Agopian et al., 2011). Other works have been focused on the purification, identification, and/or characterization of single proteins such as a polygalacturonase (Dominguez-Puigjaner et al., 1992), different chitinases (Peumans et al., 2002; Clendennen et al., 1997; Ho and Ng, 2007), a thaumatin-like protein (Barre et al., 2000; Leone et al., 2006; Ho et al., 2007), and a β-1,3-glucanase (Peumans et al., 2000).

Some of these proteins have been studied because of their interesting activities. For example, class I chitinase and thaumatin-like protein have demonstrated to be involved in the hypersensitive response of plants for conferring a significant protection toward a broad range of phytopathogens, including fungi, bacteria, and viruses. The expression of thaumatin-like protein genes in transgenic banana plants has demonstrated enhanced resistance to phytopathogens (Mahdavi et al., 2012). Among banana proteins, up to five allergens have been identified (*Mus a 1-5*) that correspond to the proteins profilin (Reindl et al., 2002) [18], class I chitinase (Sanchez-Monge et al., 1999), lipid transfer protein (LTP) (Palacin et al., 2011), thaumatin-like protein, and β-1,3-glucanase (Aleksic et al., 2012), respectively. Among them, both class I chitinase (Mikkola et al., 1998) and β-1, 3-glucanase (Barre et al., 2000) have been proposed as being responsible for the so-called latex-fruit syndrome, whereas profilin is involved in the cross-sensitization between pollen and plant-derived foods (Vieths et al., 2002).

Since fresh banana pulp contains approximately 20% of carbohydrates and only 1% of proteins, this fruit has been traditionally considered as a difficult matrix for protein extraction, being a target in studies of optimization of protein extraction methodologies. Fruits, as is true of every biological source, contain highly-abundant proteins, which are often of limited interest to proteome analysis, whereas other proteins may be orders-of-magnitude less abundant, although still of high importance. The protein extraction protocols from banana fruit are based on conventional methods, which are quite tedious and not completely satisfactory since some contaminants are often co-extracted and low-abundance proteins are usually lost.

Here, too, the combinatorial peptide ligand libraries after different types of protein extractions for searching the very low-abundance proteins in banana, have been applied (Esteve et al., 2013). The use of advanced mass spectrometry techniques and the *musa* mRNAs database in combination with the Uniprot_viridiplantae database allowed identification of 1131 proteins. Among this huge amount of proteins are found several known allergens, such as *musa a 1*, pectinesterase, and superoxide dismutase, and new allergens may have also been detected. Additionally, several enzymes involved in degradation of starch granules and strictly correlated to ripening stage were identified. These results constitute the largest description so far of the banana proteome.

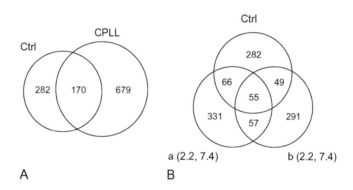

FIGURE 5.6
A: Venn diagram of the total number of identified protein banana species in the controls (Ctrl) versus those detected after CPLL treatment. B: Venn diagram of the species identified in control (Ctrl) and in eluates from native ("a") and denaturing extraction protocol ("b"). *(From Esteve et al., 2013, by permission).*

Figure 5.6 shows the Venn diagrams summarizing the contributions of the different fractions to the total identifications. Out of a total number of 1131 identified proteins, the various captures with the CPLL beads permitted identification of 849 proteins, while the controls allowed identifying 452 proteins, 170 species being common. As would be expected, the various extractions employing both native and denaturing conditions, followed by CPLLL treatment, permitted a higher amount of identifications, totaling 849 unique gene products, whereas the extractions in the absence of such treatment (controls) permitted identification of just 452 species. This result highlights the fact that just 170 identified proteins were common between controls and CPLL-treated samples. This confirms that the use of two parallel protein extraction methods via different chemical, physiological, and denaturing conditions (just as adopted for the avocado proteome) could be a useful procedure to increase the amount of identified species.

It has also been confirmed that a proper CPLL treatment allows detection, in general, of at least twice the species than those in the control (although on many occasions up to 500% more species could be detected in CPLL-captured samples). In addition, gene ontology analysis was performed by filtering the found accession numbers against the QuickGO protein database. The resulting statistics, consisting of the percentage of accession numbers related to each GO category, are plotted in the pie chart shown in Figure 5.7. A high number of identified proteins is implicated in enzymatic processes.

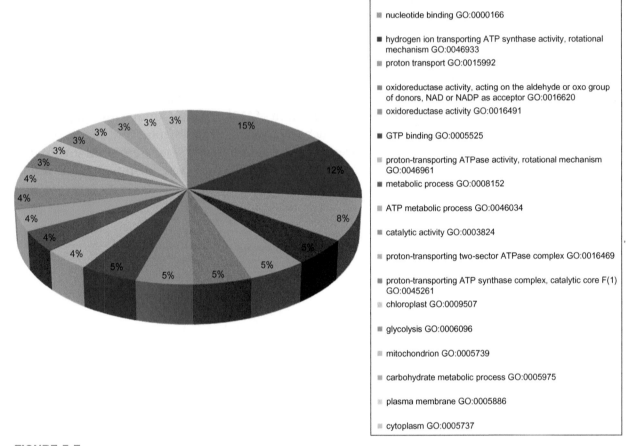

- oxidation-reduction process GO:0055114
- ATP binding GO:0005524
- nucleotide binding GO:0000166
- hydrogen ion transporting ATP synthase activity, rotational mechanism GO:0046933
- proton transport GO:0015992
- oxidoreductase activity, acting on the aldehyde or oxo group of donors, NAD or NADP as acceptor GO:0016620
- oxidoreductase activity GO:0016491
- GTP binding GO:0005525
- proton-transporting ATPase activity, rotational mechanism GO:0046961
- metabolic process GO:0008152
- ATP metabolic process GO:0046034
- catalytic activity GO:0003824
- proton-transporting two-sector ATPase complex GO:0016469
- proton-transporting ATP synthase complex, catalytic core F(1) GO:0045261
- chloroplast GO:0009507
- glycolysis GO:0006096
- mitochondrion GO:0005739
- carbohydrate metabolic process GO:0005975
- plasma membrane GO:0005886
- cytoplasm GO:0005737

FIGURE 5.7
Gene ontology (GO) analysis of total banana proteins identified. The main GO categories enriched are related to oxido-reduction process, ATP binding, nucleotide binding, proton transport, and metabolic processes. *(From Esteve et al., 2013, by permission).*

In fact, the main GO categories enriched are related to the oxido-reduction process, ATP binding, nucleotide binding, proton transport, and metabolic processes.

Given the importance of bananas in human nutrition (for instance, banana and apple homogenates are among the first foods fed to infants during weaning), these findings might be important in ferreting out valuable data bearing on nutritional aspects. Surely in such a long list of 1131 species detected (vs. just a handful of known proteins up to the present time,) there must be quite a few proteins with highly positive nutritional values and health benefits, which will be the job of nutritionists to evaluate and enucleate from the present data. Some of these proteins (and peptides therefrom) might be of importance and perhaps marketed in nutraceutics formulations, a field that is now greatly expanding. Just to give an example of the importance of this fruit in human nutrition, here it is worth recalling some statistics: Banana exports in 2010 were as high as 17 million tons, much higher than apples (8 million tons) and oranges (6 million tons).

These few examples show the power of the CPLL technique in determining the low-abundance proteome of plants. It is expected that the technique will gain more adherents, in view of the remarkable progress in the field.

There is also an interesting example of studies on the latex of *Hevea brasiliensis*, which has provided extensive coverage not only of its proteome but also of yet-unreported allergens. Even in this case,

CPLLs proved their worth: A total of 300 unique gene products were identified in this fluid, whose proteome had been largely unknown until this investigation (D'Amato et al., 2010a). In the search for unknown allergens (see Section 6.8), control latex and eluates from the ligand libraries have been fractionated by two-dimensional mapping, blotted and confronted with sera of 18 patients.

In addition to the already known and named Hevea major allergens, several others were unambiguously detected, as for instance, heat shock protein (81 kDa), proteasome subunit (30 kDa), protease inhibitor (8 kDa), hevamine A (43 kDa), and glyceraldehyde-3-phosphate dehydrogenase (37 kDa). Yet this plant is definitely not important in human nutrition, although its latex kept turning the wheels of the Allied Army in Word War II, as Righetti (2009) illustrated in an article celebrating the bicentennial of electrophoresis. The examples given here also show that by exploiting more than one extraction protocol (typically only under native conditions, since denaturing extractions were considered to be incompatible with CPLLs), it is possible to expand the proteome coverage of any plant or fruit product, achieving the description of at least 1000 unique gene products versus barely 200 to 300 species in all previous investigations (including the olive pulp and stone, in which, unfortunately, these improved protocols had not been applied).

5.3 FOOD AND BEVERAGES PROTEOMICS: EXAMPLES AND POTENTIAL APPLICATIONS

Although food is necessary to the survival of living organisms, it may also represent a number of risks inherent to its composition and origin. With all the advances in industrialization, food composition has become more and more complex, involving a large number of ingredients of vegetal and animal sources. To this reality it is also important to add that the primary material for food preparation can also come from genetically modified organisms for reasons that, as mentioned above, are related to improved resistance to given attacks or because the expression of specific components is intended to enhance selected pharmaceutical properties (nutraceuticals).

Within this complex situation, global analysis can contribute to an understanding and identification of the protein composition of genuine nutriments. Alternatively, comparative proteomic analysis, such as side-by-side pattern observations, allows improving knowledge of composition and variability. Finally, determining the presence of allergens could help to verify the presence of food components that are very difficult to identify when their presence in the food is very low. This last point is extremely important considering that food allergens are quite common and they act at the trace level.

5.3.1 Targeting Specific Aspects: Human Health, Allergens, Additives, Food Traceability, Transgenes

Proteins in food can also have functions other than just being a source of energy. Food can contain proteins beneficial to human health. There are, for example, proteins with antibacterial properties such as lysozyme, a well-known small protein found in egg white (You et al., 2010) and milk (Wiesner and Vilcinskas, 2010; Ibrahim et al., 2011). The action of lysozyme in milk can also be enhanced by the presence of lactoferrin and lactoperoxidase. These two proteins are extracted from bovine milk and then re-injected in specific preparations to improve the antimicrobial properties (Zimecki and Kruzel, 2007).

Proteins with anticoagulant properties are present in echiuroid worms (Jo et al., 2008). Antihypertensive properties are found in tuna proteins (Lee et al., 2010). To complete the list, proteins with antioxidant properties, and hence anti-aging effects, are present in many living organisms, especially in vegetables (Ningappa and Srinivas, 2008). In food preparation, all the above-mentioned pharmaceutical properties can be conferred by the simple addition of selected animal or plant tissues or of (semi) purified proteins. Here the proper analysis of resulting food in terms of consistency and effects is performed by proteomics analysis of the whole food preparation; the preferable method is use of sample preparation tools that allow simple access to species that are present in trace amounts. It is known, however, that not all proteins are good. Food can be a source of adventitious proteins detrimental to individual or collective health such as polypeptidic toxins. It can contain powerful allergens of protein origin, with very dramatic effects. Their stability against proteolysis that normally operates in the digestive tract induces direct sensitization once in the intestinal tract, generating sometimes dangerous systemic allergic reactions upon ingestion (Astwood et al., 1996).

Exact information on the prevalence of food allergy is not yet available, though it is generally acknowledged that the rate rises over time. Allergenic proteins are found in many types of food, but most common are allergens in nuts, fruits, fish, shellfish, seeds, various vegetables, eggs, and milk. Reactions can be extremely rapid and intense, even if the protein allergens are present in just trace amounts. Thus, before ingesting food preparations, it is essential to have exact knowledge of the composition, which can only be accessed through a global proteomic analytical approach. Since protein allergens can come from bacterial, animal, and plant sources, a generalized protein analysis is mandatory. In a recent publication Waserman and Watson (2011) pointed out the dangers of some food components and provided data on pathophysiology, diagnosis, and food allergy management. Nevertheless, prevention is essential and can be approached by appropriate proteomic analysis.

A specific source of allergy that is not yet well documented is related to transgenic organisms that could be used as food additives. Even if the idea that protein expression of genetically modified food might have allergic effects is still just speculation, an in depth analysis including very low-abundance proteins generally undetectable appears more than useful, at least as long as knowledge in this domain is not well developed. CPLLs might positively contribute to this enterprise (D'Amato et al., 2009 and 2010a, Shahali et al., 2012).

Over the years, there has been a marked shift of emphasis to topics involving food analysis. In the 1970s, a number of federal laws were passed in the United States that dealt with minimizing exposure to toxic chemicals. The Clean Air Act (1970), for example, designated a listing of hazardous air pollutants, and the Safe Drinking Water Act (1974) addressed the issue of minimizing or eliminating toxic contaminants from drinking water. The mindset accompanying these federal laws was that of a "chemical of the month" for public and regulatory scrutiny: One chemical after another was detected and found to possess toxicity at some level that would suggest they be branded, monitored, and removed from use in ways that added to their environmental concentrations. DDT and other "legacy" pesticides were in that group, as were volatile organic compounds (VOCs) and semivolatile organic compounds (SVOCs), widespread contaminants in water, hazardous air pollutants (HAPs), artificial sweeteners (e.g., cyclamates), naturally occurring chemicals such as aflatoxins, nitosamines, mercury, and other heavy metals, and many others. Pesticides in particular were scrutinized, and many were removed from registration or greatly curtailed in use: chlordane, toxaphene, Alar, and ethyl parathion, to name a few. Accordingly, the vast majority of papers published in those years dealt with analysis of such pollutants in foodstuffs, including the presence of microbial pathogens. Thus, the emphasis was on the toxicological aspects of food.

Starting in the mid-1990s, however, an increased emphasis on the identification and optimization of health-beneficial constituents in foods has become evident in the published literature. This trend is reflected, for instance, in one of the major journals devoted to foodstuff analysis (the *Journal of Agricultural Food Chemistry*). The number of papers on antioxidants in foods rose dramatically, as did reports on healthy foods such as red wine and olive oil. Papers on lycopene and health-beneficial bioactives in foods (such as phytonutrients and nutraceuticals exceed those reporting on chemicals with detrimental effects. The interest in diet and health has focused on populations that seem to be thriving, living longer, with lower incidence of cancer and cardiovascular disease, and on how their diets may be a major contributor to better health and longevity. The Mediterranean diet has stimulated scientific interest in the constituents of olive oil, red wine, fish, whole grains, and other "healthy" foods that are mainstays in this diet (Seiber and Kleinschmidt, 2012; Tomás-Barberán et al., 2012). A recent review describes how different foodstuffs can mitigate inflammation in human tissues, thus attenuating and lessening processes leading to chronic diseases (Wu and Schauss, 2012). Yet, the vast majority of papers most heavily emphasize metabolites and small molecules. Thus, the major aim of proteomics is how to focus more on a challenging but important domain—the protein and peptide content of food—and to evaluate their beneficial effects as well. Following are some examples of these analyses, with specific reference to low-abundance gene products.

5.3.2 Selected Proteome Analyses of Food Products

Rice is a very important natural food resource for humans and, as such, it deserves special attention with regard to food safety and function. A nice review of the matter has been recently published (Agrawal and Rakwal, 2011). Although the study of rice proteins is largely pursued for their importance in a variety of domains, the authors placed special focus on the extraction of proteins, which was designated as critical to proteomic studies. In this respect, a special sample treatment is described for the largest protein harvesting before proteomic investigations. Elimination of nonproteinaceous material with extraction of difficult proteins is reviewed, with attention to how proteins are precipitated and then reconstituted.

Once the proteins have been extracted, the selection of low-abundance species is another crucial step in understanding protein families that are expressed in low copies. An example is that of RNA-binding proteins that play a central role in regulating seed storage proteins (Crofts et al., 2010). Several strategies are employed to extract this group of proteins; a well-established one is affinity chromatography using PolyU Sepharose (Xu et al., 2007). Among the RNA-binding proteins in rice seeds, 14-3-3 proteins were identified, as in the case of *Arabidopsis thaliana* (Hajduch et al., 2010). In this plant the number of RNA-binding proteins is predicted to be a group of more than 200 species that is by itself indicative of the amount of work that remains to be done in low-abundance protein detection and identification in rice.

Among many other examples of proteomics analysis of important food products are potato tubers (Khalf et al., 2010). They actually constitute a large part of the diet in northern countries under different forms. As such, this plant has stimulated specific genomic modifications in view of better resistance against insects (Khalf et al., 2010). To this end, protease inhibitors have been encoded in plants to prevent insect aggressions. The impact of these manipulations to endogenous proteases has been recently investigated for tubers from potato lines expressing a broad spectrum of protease inhibitors in view of detecting undesired effects. Protein extraction of lyophilized tuber powder was performed in Tris-HCl buffer containing sucrose, $MgCl_2$, and PMSF in the presence of 1% polyvinylpyrrolidone. From the initial extract, proteins were precipitated in methanol/chloroform and separated by centrifugation. They were finally redissolved in 0.5% IPG buffer containing 8 M urea, 2% CHAPS, and 60 mM DTT. Mono-dimensional as well as two-dimensional electrophoresis separations were performed, followed by immunodetection and mass spectrometry. The obtained patterns did not show significant quantitative variation among the lines. However, for few proteins that were independent on transgene expression, some differences appeared. They seemed to be related to the patatin storage protein complex. On the contrary, protease inhibitor profiles seemed substantially unchanged, suggesting no intrinsic allergenic potential with substantial equivalence to nontransgenic standard tubers. Here again the analysis was restricted to the protein extract, with no focus on the low-abundance protein enrichment where the expression could have been significantly modified.

5.3.3 Enhancing the Visibility of the Low-Abundance Proteome in Food Products with CPLLs

One essential source of nutrients for animals and humans is milk, which represents a key raw material for food preparations. It comprises proteins of nutritional and health interest possessing, among other effects, bone development/protection, cellular growth, and immunosystem stimulation. Bovine milk proteins become part of regular daily food under different forms and can be modified during processing transformations. As described by Arena et al. (2010), thermal treatments modify proteins by the formation of lactose adducts. These modifications need to be monitored because important protein functions might be altered upon physicochemical treatments. The authors have thus analyzed the proteome with the help of CPLLs in order to evidence specific transformations that are of low-abundance. After enrichment and protease digestion, the resulting peptides were submitted to a separation on a phenyl-boronic acid sorbent to harvest lactosylated forms (see Sections 3.3.2 and 3.6.2). Large modifications were observed not only in major milk proteins, but also in those classified as very low-abundance species that could have important functions in infant nutrition, defense against external agents, and development. Protein lactosylation revealed after CPLL treatment has recently been confirmed and analyzed by mass spectrometry (Siciliano et al., 2012). The authors concluded that the described modifications, in particular to proteins that are resistant to digestion and act at the intestinal level, could enhance the allergic response since lactose molecules grafted to proteins have already been described as hapten-like antigens (Karamonova et al., 2003).

As is probably true of all biological materials, milk comprises a large number of low-abundance species that are undetectable with current methodologies and instruments with no preliminary treatment. In a general study the enrichment process involving CPLLs was applied to bovine milk after having eliminated fats and casein, and the results of investigation were quite impressive (D'Amato et al., 2009). From 84 gene products found in the untreated milk, 65 additional proteins of presumably low abundance were identified. Among them were numerous enzymes involved in sugar metabolism such as UTP—glucose-1-phosphate uridylyltransferase, glucosidases, aldose 1-epimerase, fructose-1,6-bisphosphatase and tissue alpha-L-fucosidase. Figure 5.8 shows three 2D PAGE maps of untreated whey (Ctrl) compared to the eluates from two different peptide libraries (ProteoMiner and succinylated peptide library). It is of interest to note how the unmodified library captures many more acidic species, whereas the succinylated one displays also a large number of additional neutral to quite alkaline proteins, all of them not visible in the control. It was also very interesting to use the CPLL extract for the detection of potential allergens. In this respect isoelectric separations were blotted with the serum of two dozens of patients allergic to bovine milk. A polymorphic alkaline protein was observed with a strong positive signal identified as an immunoglobulin fragment, a minor protein that had been largely amplified by the treatment with peptide libraries. Few but other potent allergic signals were also found at very acidic protein levels, but they had not been formally identified. The individual allergic protein pattern was always largely heterogeneous underlining the fact that allergy is probably not restricted to a single protein and that a given protein could react differently from patient to patient.

To complete the picture of bovine milk, a recently reported review included a list of more than 500 proteins with the aim of understanding their function (D'Alessandro et al., 2011). Several expected functional families were found such as transport, immunosystem response, and lipid metabolism. Interestingly, other groups were found such as proteins involved in anatomical development and cellular proliferation. Clearly, the findings from this important dietary product are far from being all discovered and are presumably related to species that are of very low-abundance and probably biologically very active. More recently, an in-depth investigation was performed on donkey's milk, which today is categorized as among the best mother's milk substitutes for allergic newborns, due to its much reduced or absent allergenicity, coupled with excellent palatability and nutritional

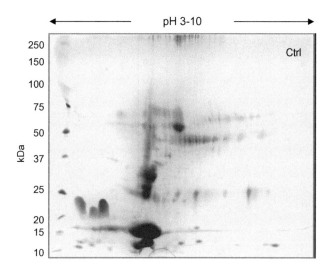

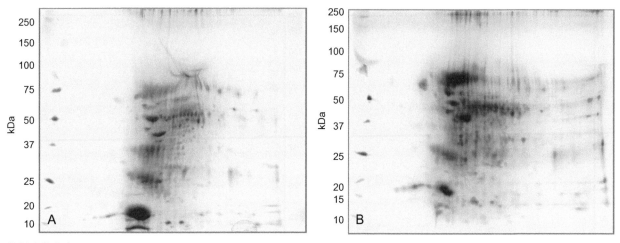

FIGURE 5.8

Two-dimensional maps of: Ctrl=milk whey starting sample (upper panel); A=eluate (7 M urea, 2 M thiourea, 2%CHAPS and 25 mM cysteic acid, pH=3.18) after CPLL treatment. B=eluate (7 M urea, 2 M thiourea, 2% CHAPS and 25 mM cysteic acid, pH=3.18) after succinylated CPLL treatment. All samples adsorbed onto the IPG pH 3-10 strips in 150 μL volume (200 μg protein concentration). Staining with colloidal Coomassie blue. *(From D'Amato et al., 2009, by permission).*

value. By exploiting the CPLL technology, and treating large volumes (up to 300 mL) of defatted, de-caseinized (whey) milk, Cunsolo et al. (2011a; 2011b) identified 106 unique gene products, constituting by far the largest description so far of this precious nutrient.

Figure 5.9 shows two-dimensional electrophoresis maps of the untreated sample (Ctrl) versus three peptide library eluates after capture at pH 4.0 (4E), pH 7.0 (7E), and pH 9.3 (9E). Here, too, one can appreciate how many more spots are displayed in the three eluates, especially those at pH 7.0 and 9.3, with additional acidic spots found only in the pH 4.0 eluate. Due to poor knowledge of the donkey's genetic asset, only 10% of the proteins could be ascertained via the *Equus asinus* database; the largest proportion (70%) could be designated by homology with the proteins of *Equus caballus*. By grouping all of the identified proteins in donkey's milk on the basis of their putative molecular function, it was possible to reveal that the most abundant classes are represented by proteins belonging to the categories of protein binding, ion binding, enzyme inhibitors, and hydrolase and therefore involved in the nutrient transport and the immune system response. The well-known low allergenic properties of donkey's with respect to cow's milk seem to be mainly related to the remarkable differences in

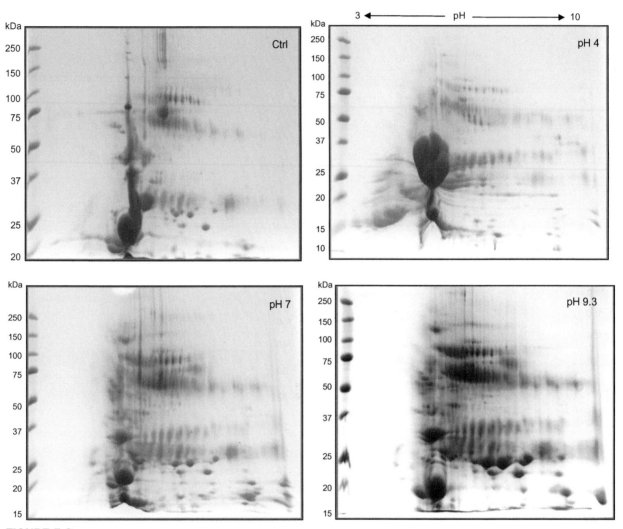

FIGURE 5.9

Two-dimensional map analyses of three eluates after CPLL capture at pH 4.0 (4E), 7.0 (7E), and 9.3 (9E) as compared to control (Ctrl) donkey's milk. In all cases, a total of 150 μg protein was applied to each IPG strip, in the pH interval 3–10. Staining with Coomassie brilliant blue. *(Modified from Cunsolo et al., 2011a).*

the primary structure of their proteins, which determine deep divergences between the amino acid sequences of IgE-binding linear epitopes of cow's milk allergens and the corresponding domains present in donkey's milk proteins. The importance of the above findings on cow's and donkey's proteomes should not be underestimated. Considering that just eight major proteins constitute the vast majority of bovine (but also human) milk (α_{s1}-, α_{s2}-, β-, κ-, γ-caseins, together with β-lactoglobulin, α-lactalbumin, and serum albumin that infiltrates from blood to the mammary gland), this leaves little room for searching for the low-abundance proteome, which in fact could be explored in depth only with the solid-phase combinatorial peptide ligand library technology.

With the objective of the largest protein identification, this technology was also applied to other products of food interest as, for example, egg white (D'Ambrosio et al., 2008) and egg yolk (Farinazzo et al., 2009). Eggs are largely used as food or food components throughout the world. They bring essential elements to one's nutrition, among them proteins with various important functions. They are extensively used in the food industry as gelling, emulsifying, and foaming agents. In addition to their nutrition importance, egg components are used in the cosmetic and pharma industries. The large majority of investigations performed on egg proteins (white and yolk separately) is limited to detectable species without preliminary sample treatment. However, CPLLs were used to investigate the composition of the deep proteome with a significant degree of success. The egg white proteome is composed of a large number of proteins where 85 to 90% is represented by ovalbumin, ovotransferrin, ovomucoid, lysozyme, and ovomucin (Guérin-Dubiard et al., 2006). The remaining proteome includes many minor undetectable proteins that are poorly known but essential to biological processes. From both yolk and egg white, the number of identified gene products upon the use of peptide libraries was more than doubled (to 148 for egg white and to 255 for yolk) thus largely increasing the repertoire of known proteins from eggs and hence contributing to a better understanding of nutritional effects.

As an example, Figure 5.10 gives the SDS-PAGE profiling of untreated egg white (Ctrl) as well as that of the flow-through (FT) after peptide library treatment (the two profiles being identical, since only the most abundant proteins appear in these two samples). Conversely, the three eluates from the two types of beads (A, regular hexapeptide library, B succinylated peptide library) exhibit many more bands, especially in the 10 to 30 kDa region, as is customary after CPLL treatment. Just as an example on the depth of the CPLL treatment, we recall here that in 2006 Guérin-Dubiard et al. described barely 16 species. This number was substantially increased to 76 by Mann in 2007, by exploiting very advanced and sophisticated instrumentation (LC-ESI Fourier-transform, ion cyclotron resonance MS) and by performing MS/MS and MS^3 experiments. Yet, with simple instrumentation and a facile experimental approach, the CPLL technology allowed easily doubling the discovery.

By the same token, Figure 5.11 displays the SDS-PAGE profiling of untreated egg yolk proteins (Ctrl) as well as of the flow-through (FT) after peptide library treatment, followed by the eluates from three different libraries (A: regular hexapeptide library; B: carboxylated version of the same; C: tertiary amino terminus version). Here, too, the number of extra bands, covering the entire Mr range 10-100 kDa, can be perceived at a glance. Unfortunately, eggs could produce allergic adverse effects with individual reactions and other health risks, including asthma. These reactions are attributed to the presence of

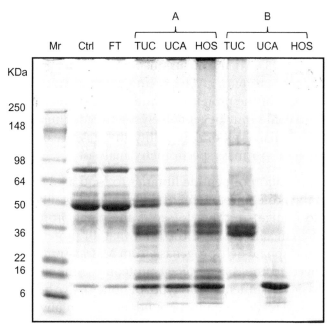

FIGURE 5.10
SDS-PAGE analysis of egg white after treatment with two different libraries (native A and succinylated B hexapeptides) and eluted in three fractions. Sample load for all tracks: 10 μL. Samples: Mr: markers (the scale is in kDa); Ctrl: starting egg white; FT: flow through; TUC: eluates in 2 M thiourea, 7 M urea, 3% CHAPS; UCA: eluates in 8 M urea, 50 mM citric acid, pH 3.3; HOS: eluates in hydro-organic solvent. Staining with colloidal Coomassie Blue. *(Modified from D'Ambrosio et al., 2008).*

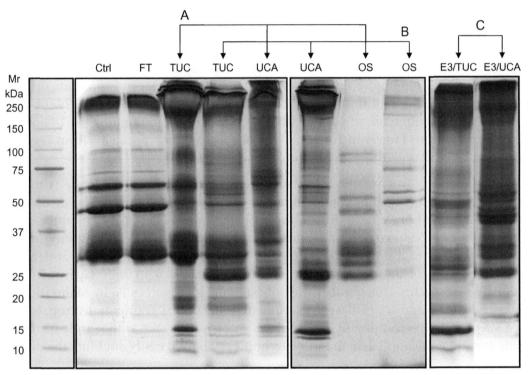

FIGURE 5.11

SDS-PAGE analysis of egg yolk treated with two peptide libraries: native "A" hexapeptides, succinylated "B" hexapeptides and quaternarized hexapeptides "C". Samples: Ctrl=egg yolk plasma control before library treatment; FT=flow-through after peptide library capture; TUC: eluates by 2 M thiourea, 7 M urea, 3% CHAPS; UCA: eluate by 9 M Urea, 50 mM citric acid and 2%CHAPS; OS: eluate by 6%v/v acetonitrile, 12%v/v isopropanol, 10%v/v ammonia 20% and 72% water. All samples loaded at 50 μg/lane; staining with colloidal Coomassie Blue. *(From Farinazzo et al., 2009, by permission).*

ovalbumin, but other proteins present in much smaller amounts could have even more sensitization effects. Curiously, Mann and Mann (2011) re-investigated the egg white proteome with the use of the LTQ Orbitrap Velos mass spectrometer and identified just about the same number of proteins as reported in D'Ambrosio et al. (2008). Their conclusions were that one does not need CPLLs to amplify the signal of the low-abundance proteins, a Velos seemingly doing the job in the untreated material. However, it turns out that most of the proteins reported by them are not on the list of D'Ambrosio et al. (2008) and 44 of them could only be tentatively identified. Thus, one wonders if German eggs are genetically different form Italian eggs, or if, in reality, the presence of the five very high-abundance proteins does indeed hamper proper mass spectrometry analysis.

As a last example of the application of peptide libraries to food analysis, Di Girolamo et al. (2012) have recently investigated the proteome content of different unifloral honeys (from chestnut, acacia, sunflower, eucalyptus, and orange) in order to see if any plant proteins present would allow the proteo-typing of these different varieties. It turned out that all proteins identified (except one, the enzyme glyceraldehyde-3-phosphate dehydrogenase from *Mesembryanthemum crystallinum*) were not of plant origin but belonged to the *Apis mellifera* proteome. Among the total proteins identified (eight, but only seven as basic constituents of all types of honey), five belonged to the family of major royal jelly proteins 1–5 and were also the most abundant ones in any type of honey, together with α-glucosidase and defensin-1. It thus appears that honey has a proteome resembling the royal jelly proteome (but with considerably fewer species), except that its protein concentration is lower by 3 to 4 orders of magnitude as compared to royal jelly. Attempts to identify additional plant (pollen, nectar) proteins via peptidome analysis were unsuccessful. The only other report available on the honey proteome could identify barely one protein (Won et al., 2008).

5.3.4 Proteome Analyses of Nonalcoholic Beverages

As stated in the introduction, we will limit this survey to beverages that have been explored by the peptide library methodology, in comparison with untreated samples, as reported in the literature. We will only deal with proteomes, although in such beverages a peptidome screening could also be performed with CPLLs. For peptide analyses, one is referred to a series of surveys published almost biannually by Kašička (1999, 2001, 2003, 2006, 2008, 2010, 2012), which can be used as Ariadne's thread to help the readers out of the labyrinth. These investigations demonstrated the unique performance of CPLLs, such as: (a) the ability to handle large sample volumes (one liter and more) in an easy and user-friendly protocol; (b) sample enhancement factors up to 4 orders of magnitude; and (c) extremely high detection sensitivities, reaching as little as 1 µg protein/L of sample, a sensitivity rarely obtained by present-day methodologies. Needless to say, these results were far superior to those obtained with conventional techniques, due to their inability to properly concentrate and "normalize" the relative concentration of the proteomes under analysis.

There are at least two reasons for exploring the trace proteome of nonalcoholic beverages: (a) in order to certify the genuineness of such products and find out if they contain proteins of the vegetable extracts they have been prepared from; and (b) in order to screen for the presence of any potential allergens in such beverages. Indeed, such analyses have been made on almond milk and orgeat syrup (Fasoli et al., 2011b), coconut milk (D'Amato et al., 2012), as well as a cola drink (D'Amato et al., 2011a), Ginger Ale (Fasoli et al., 2012a) and even white-wine vinegar (Di Girolamo et al., 2011). In the case of almond milk, 137 unique protein species were identified, the deepest investigation so far of the almond proteome. In the case of orgeat syrup, a handful of proteins (just 13) were detected, belonging to a bitter almond extract. In both cases, the genuineness of such products was verified. In contrast, cheap orgeat syrups produced by local supermarkets and sold as their own brands were found not to contain any residual proteins, suggesting that they were likely produced only with synthetic aromas and no natural plant extracts.

Figure 5.12 displays the SDS-PAGE profiles of orgeat syrup (two tracks to the left) and of almond' milk. In the case of orgeat, no single protein band is visible in the control, untreated sample; as for almond's milk, whereas the control lane shows two major bands at 23 and 35 kDa, plus a few other zones, the two eluates after capture at pH 7.0 and 9.3 exhibit a very extensive band profile from 10 to 100 kDa.

In the case of coconut milk, a grand total of 307 unique gene products could be listed, 200 discovered via CPLL capture, 137 detected in the control, untreated material, and 30 species in common between the two sets of data. This is by far the most extensive mapping of this nutritious beverage, in which, to date, only a dozen proteins are known—those belonging to the high- to very-high abundance class. This unique set of data could be the starting point for nutritionists and researchers involved in nutraceutics for enucleating proteins responsible for some of the unique beneficial health effects attributed to coconut milk.

An interesting detective story can be told here about the "invisible" proteome of a cola drink (D'Amato et al., 2011a), stated to be produced with a kola nut extract. Indeed, a few proteins in the Mr 15 to 20 kDa range could be identified by treating large beverage volumes (one liter) and performing the capture with CPLLs at very acidic pH values (pH 2.2) under conditions mimicking reverse-phase adsorption (Figure 5.13). It turned out that only three proteins were identified: one belonging to the kola nut and the other two identified as agave species. This beverage label stated that the cola drink had been generously sprinkled with 6% agave syrup, honor saved all around. Conversely, things did not go so well (for the producer) in the case of a ginger beverage (Fasoli et al., 2012a). Although found only in traces, the presence of five grape proteins and one apple protein could be confirmed, but there was not even the faintest trace of any ginger root proteins. The first two findings are correct,

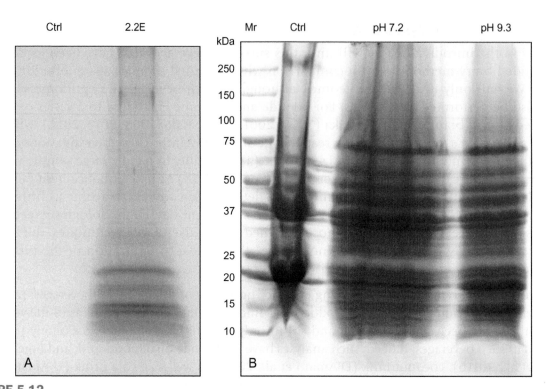

FIGURE 5.12

SDS-PAGE profiles of orgeat syrup (panel A) and of almond's milk (panel B). A:Ctrl orgeat extract=100 μL of control orgeat solution before CPLL treatment. 2.2E:50 μL 2% SDS eluate of orgeat solution treated with a home-made-CPLL library at pH=2.2.

B: Ctrl: almond's milk control before CPLL treatment. pH 7.2: eluate 2% SDS eluate after capture of the proteins at pH 7.2 on a 1:1 mixed library of regular hexapeptides and a home-made-CPLLs. pH 9.3: 2% SDS eluate after protein capture at pH 9.3 on the same mixed library. *(Modified from Fasoli et al., 2010).*

as the producer stated that this beverage had been reinforced with 12% grape juice and 6% apple juice, but the absence of even traces of ginger proteins did not permit classification of this beverage as a ginger extract on a proteomics scale. It was thus concluded that either the *Zingiber officinalis* was present in traces or only its flavors had been added and not any root extract.

In another investigation on nonalcoholic beverages, the trace proteome of white-wine vinegar has been identified via capture with home-made combinatorial peptide ligand libraries under conditions mimicking reverse-phase capture—that is, at pH 2.2 in the presence of 0.1% trifluoroacetic acid. A total of 27 unique gene products could be tabulated, of which 10 were specific to the database *Vitis vinifera*, 13 were found in the general database Uniprot_viridiplantae, and 4 in Swiss Prot_all entries. The most abundant species detected, on the basis of spectral counts, appears to be the whole-genome shotgun sequence of line PN40024, scaffold_22 (a protein of the glycosyl hydrolase family, indicated by an arrow in Figure 5.14) (Di Girolamo et al., 2011). Two hypotheses have been set forward regarding these 27 surviving proteins (to the strongly acidic, pH 2.2, environment of vinegar): the fact that this set of proteins still persists in solution would suggest that they are resistant to denaturation, a phenomenon that typically occurs with proteins which have a tightly packed hydrophobic core and are thus insensitive to pH denaturation. The other interesting aspect of this set of 27 survivors is the fact that they seem to be also resistant to proteolytic attack, since their apparent Mr values (see Figure 5.14) are well distributed in the 12 to 70 kDa range. Conversely, when the ultra-trace proteome of a cola drink was explored, the minute traces of the only tree surviving proteins were found just above the leading/terminating boundary of the Laemmli discontinuous buffer in the SDS-PAGE gel, where typically only protein fragments are confined (see Figure 5.13). This finding suggests that they were all degraded species.

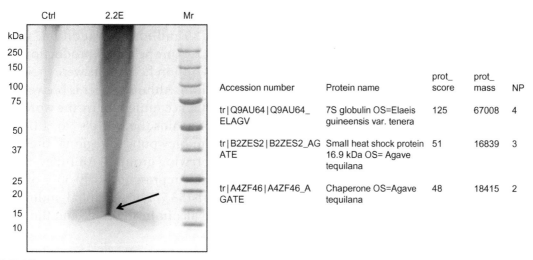

Accession number	Protein name	prot_score	prot_mass	NP
tr|Q9AU64|Q9AU64_ELAGV	7S globulin OS=Elaeis guineensis var. tenera	125	67008	4
tr|B2ZES2|B2ZES2_AGATE	Small heat shock protein 16.9 kDa OS= Agave tequilana	51	16839	3
tr|A4ZF46|A4ZF46_AGATE	Chaperone OS=Agave tequilana	48	18415	2

FIGURE 5.13

SDS-PAGE of control (Ctrl) and library eluate (2.2E) at pH 2.2 of the Cola drink.

Ctrl lane: 20 μL of cola solution before library treatment; 2.2E lane: 20 μL of the concentrated 4% SDS eluate of cola solution treated with a home-made library. Mr: molecular mass standards. Micellar Coomassie blue staining. The arrow indicates the diffuse zone of proteinaceous material centred at around 15 kDa. The table to the right gives the IDs of the three proteins identified in the 15-20 kDa region of the gel. Abbreviations: prot: protein; NP: number of peptides. *(Modified from D'Amato et al., 2012).*

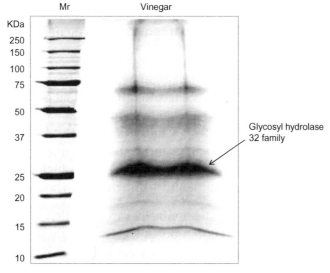

FIGURE 5.14

SDS-PAGE of white-wine vinegar. Mr: molecular mass ladder. Laemmli discontinuous buffer. Gel: 8-18% polyacrylamide gradient. Run: three hours at 300 V. Sample load: 25 μL containing ca. 50 μg protein. Staining with micellar Coomassie Brilliant Blue. The major protein (the whole-genome shotgun sequence of line PN40024, scaffold_22, a protein belonging to the glycosyl hydrolase 32 family) is indicated with an arrow. *(Modified from Di Girolamo et al., 2011).*

In summary, the cumulated data suggest that this could be the starting point for investigating the myriad beverages that in the last decades have invaded the shelves of supermarkets the world over. Their genuineness and natural origin have never been properly assessed. The CPLL technology might become an easy and reproducible official test, if it is adopted or recommended by regulatory agencies.

5.3.5 Proteome Analyses of Alcoholic Beverages

Alcoholic beverages represent a very large proportion of the global food market and is in continually expanding (Giribaldi and Giuffrida, 2010). For example, a 2009 survey of worldwide wine production reported a grand total of 270 million hectoliters, with Italy producing 50 million hectoliters, followed by France (46 million) and Spain (35 million). In addition, these three countries produced nearly 50% of the world total and 80% of Europe's total production. The data for beer are even more astonishing: In 2008, more than 1.8 billion hectoliters were brewed worldwide, producing total global revenues in excess of $295 billion. Although beer is brewed in a total of 169 countries worldwide, more than 92% of total output is accounted for by the world's 40 biggest beer-producing countries, among which China is and remains the world's No. 1 beer nation. Although many thousands of small producers, ranging from brewpubs to regional breweries, can be listed throughout the world, in reality the five largest brewing groups—AB InBev, SAB Miller, Heineken, Carlsberg, and China Resource Brewery Ltd.—now represent nearly 50% of the world beer market (http://www.researchandmarkets.com/reports/53577/beer_global_industry_guide. htm). Beer is the world's oldest and most widely consumed alcoholic beverage and the third most popular drink after water and tea.

Given these statistics, it would be highly desirable for producers and customers alike to have an easy method for controlling the origin of these beverages and to assess whether their origin as stated in the label is accurate or whether they have been counterfeited and are thus a fraud. In the absence of proper controls, the loss for both the producers and customers could be huge. This acutely felt problem was recently addressed by Hamburg in an editorial in *Science* (2011): "Ensuring the safety and quality of food products has never been more complicated. Societies around the world face increasingly complex challenges that require harnessing the best available science and technology on behalf of consumers.... We must also develop new science to protect the safety of our food supply." The problem of food traceability (and of certifying its genuineness) is becoming an increasingly serious challenge.

To complicate the matters further, modern wines might be quite different from those drunk by our ancestors. One reason for the difference lies in the fact that the residual grape proteins that survived the fermentation process slowly aggregate, leading to amorphous sediments or flocculates, causing turbidity (Kwon, 2004; Moreno-Arribas et al., 2002). A haze or deposit in bottled wine indicates that the product is unstable, has a low commercial value, and is therefore unacceptable for sale. Therefore, it has become customary, especially in white wine, to remove the residual proteins remaining in the finished product, so as to prevent haze formation and sediment in the bottled wines available for sales.

Among the fining agents, one of the most popular is casein, which is derived from bovine milk. However, caseins are also known as major food allergens and therefore, according to Directive 2007/68/EC of the European Community (EC), "any substance used in production of a foodstuff and still present in the finished product" must be declared in the label, especially if it originates from allergenic material. Because caseins are nearly insoluble at the pH of white wines and they form insoluble complexes with phenolic compounds, they are considered to be almost completely coagulated and thus are eliminated by precipitation after treatment. Therefore, no winemaker has reported the presence of caseins in their fined product (although this mandatory labeling was first postponed to the end of December 2010 and then extended until the end of June 2012, likely due to protest from wine producers). Yet, classical chemistry laws suggest that traces of caseins should remain even after their massive co-precipitation with residual grape proteins. Unfortunately, the official ELISA test of the EC has a too-low-sensitivity limit of 200 µg casein per liter (Weber et al., 2009)—in other words not enough to detect traces of it. When an entire bottle of white wine (750 mL) was treated with ProteoMiner at pH 3.3, the results were quite exceptional in that as little as 1 µg casein could be assessed as a residue after the treatment (Cereda et al., 2010). The major result of this investigation was its proof that the CPLL technique had a sensitivity 200 times higher than that of the current ELISA test, a non-negligible accomplishment.

Curiously, however, in a paper that appeared simultaneously, Monaci et al. (2010), when tackling the same topic, stated "when fined wine samples were considered, the lowest added concentration for which the peptide marker could be detected was 50 µg/mL" (the peptide marker referring to casein digests, as identified by mass spectrometry). This seems to be a terribly low detection sensitivity, 50,000 times lower than that reported by Cereda et al. (2010). These authors continued in their search of traces of caseins in white wines and reported yet another method for tracing residual milk allergens, this time based on the use of a single-stage Orbitrap MS instrument (Monaci et al., 2011). Yet, the improvement in detectability was not spectacular; in their own words, the "minimum detectable added caseinate concentrations, i.e. those corresponding to response with S/N=3, were estimated between 39 and 51 µg/mL."

Recently, Palmisano et al. (2010) published an extensive investigation on grape proteins present in white chardonnay. They adopted a multiplexed glycopeptide enrichment strategy in combination with tandem mass spectrometry in order to analyze the glycoproteome of this brand of white wine, thus identifying a total of 28 glycoproteins and 44 glycosylation sites. The identified glycoproteins were from grape and yeast origin. In particular, several glycoproteins derived from grape, like invertase and pathogenesis-related (PR) proteins, and from the yeast, were found after the vinification process. Bioinformatic analysis revealed sequence similarity between the identified grape glycoproteins and known plant allergens. These data, however, could be obtained only by analyzing a wine that had not been treated with any fining agent.

Aware of this limitation, D'Amato et al. (2011b) analyzed some bottles of Recioto wine (a dessert wine made from partly dehydrated grapes left for longer periods on the plants) and of Garganega wine, a white table wine produced from the same grapes picked at ripening, with none of them treated with any fining agent. A CPPL capture at four different pH values (pH 2.2, 3.8, 7.2, and 9.3) was performed. The combined data on the discoveries in the four CPLL eluates, as well as in the collected bottle sediment, allowed identification of 106 unique gene products belonging to *Vitis vinifera*, as well as an additional 11 proteins released by the *S. cerevisiae* used in the fermentation process. Among the residual grape proteins detected in the Recioto wine, about 30% were categorized as medium- to high-abundance species versus 70% low-abundance ones. This finding once again proved the power of the CPLL technique in enhancing the signal of trace components (interestingly, in the untreated sample, only seven proteins could be identified). The results are shown in Figure 5.15: The capture at pH=2.2 under reversed-phase conditions harvested a much higher quantity of proteins (at least three times as much as assessed by densitometry of the Coomassie-stained bands) and enriched particularly two bands at around 25–30 kDa, seen only as faint zones in all other SDS-PAGE tracks.

The next logical step was to extend these investigations to red wines (Wigand et al., 2009), either to detect traces of fining agents or the entire grape-proteomic asset in nontreated samples. In the case of red wines, one would expect to find ovalbumin or entire egg white, since these are the customary fining agents adopted. However, in an extensive investigation of Italian red wines, D'Amato et al. (2010b) found that they did not contain traces of egg albumins but again of bovine caseins, especially α- and κ-caseins, with essentially no residual grape proteins, except for traces of thaumatin fragments. With another surprising finding: the wines analyzed (in northern Italy, around the Lake Garda, 2009 harvest) contained (albeit in traces) abnormal amounts of proteins originating from fungal infection, such as from *Botryotinia fuckeliana*, *Sclerotinia sclerotiorum*, and *Aspergillus aculeatus*. This finding suggests that there were adverse meteorological conditions at harvest time in that year. Traces of egg-white proteins in red wines were indeed detected by Tolin et al. (2012) but in wines that they had treated with different amounts of ovalbumin simply to determine the detection limits of the MS analyses.

Following these extensive investigations of white and red wines, a study on the beer proteome was also due. This came from Fasoli et al. (2010) who assessed this proteome again via CPLL capture at

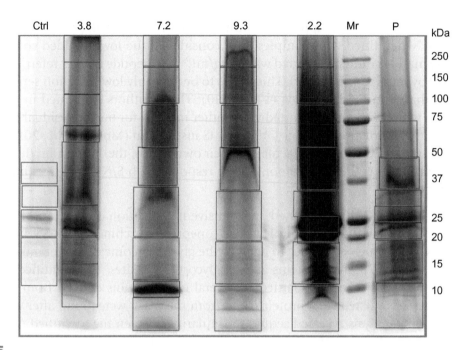

FIGURE 5.15

SDS-PAGE of Recioto wine samples. Ctrl: 20 μL of control wine before library treatment. Lanes 3.8, 7.2, 9.3, and 2.2 are, respectively, 4% SDS protein eluates after library treatments at pH 3.8, 7.2, 9.3, and 2.2. Mr: molecular mass markers. P: precipitate. Micellar Coomassie Brilliant Blue staining. The gel segments in red boxes are those cut out for protein digestion and identification via MS analyses. *(From D'Amato et al., 2011, by permission).*

three different pH (4.0, 7.0, and 9.3) values. Through mass spectrometry analysis of the recovered fractions, after elution of the captured populations in 4% boiling SDS, they could categorize such species in 20 different barley protein families and 2 maize proteins, the only ones that had survived the brewing process (the most abundant ones being Z-serpins and lipid transfer proteins). In addition to those, they could identify no less than 40 unique gene products from *Saccharomyces cerevisiae*, as routinely used in the malting process. These latter species must represent trace components, as in previous proteome investigations barely two such yeast proteins could be detected.

In conclusion, one could ask what could be the possible mechanism for protein survival in the industrial product. Perhaps a general survival mechanism begins to emerge, namely, the fact that most likely the proteins that remain in solution after the industrial processing are either small-size species or large-Mr components that have been degraded to smaller fragments possibly by proteases acting during the fermentation/industrial manipulations. Confirmation of the first mechanism of survival comes from the Recioto wine proteome: Out of 106 unique gene products found, the vast majority were small Mr proteins, ranging from 10 to 35 kDa. A verification of the second survival mechanism is given in Figure 5.16: In the case of the champagne proteome, Cilindre et al. (2008), using 2D electrophoresis mapping, found no less than 16 spots of vacuolar invertase-1 ranging in size from 20 to 75 kDa (the latter being the native Mr), with the native, undegraded form representing less than 5%.

5.3.6 Proteome Analyses of Aperitifs

The natural extension of the proteome investigation of nonalcoholic and alcoholic beverages would be to explore the trace proteome of aperitifs, that is, alcoholic beverages that are halfway in between wines (with 11–13% alcohol content) and hard liquors (e.g., cognac, whiskey, grappa, and the like, typically with more than 40% alcohol). Aperitifs generally have alcohol content of around 20–22% (although their alcohol content can be as low as about 15%) and are in general made with herbal infusions. It is widely held that the first aperitif was produced in 1786 in Turin, Italy, when Antonio Benedetto Carpano invented vermouth (which is probably a corruption of wormwood, the common

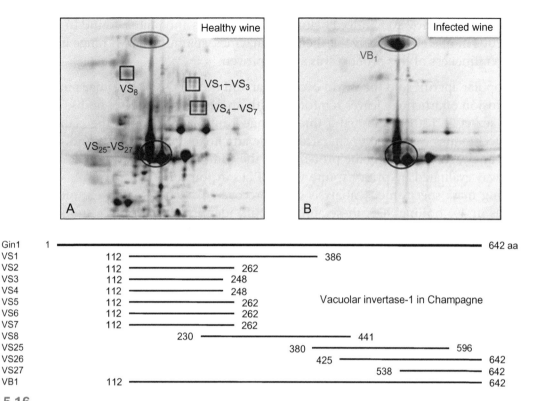

FIGURE 5.16

Two-dimensional maps of control (A) and treated (with *Botryotinia fuckeliana*, panel B) Champagne. The squares and circles identify the major spots of invertase-1 (GIN1), while the lines underneath (delimited by numbers) give the amino acid sequences of the fragments of invertase detected (blue delimited areas). It can be seen that only a small percentage of undegraded enzyme is found at the native Mr value (magenta circles), whereas the vast majority is seen as degraded fragments. *(Courtesy of Dr. C. Cilindre, University of Reims, France).*

name of *Artemisia absinthium*) in this city. In Milan and the surrounding region (Lombardia), toward the end of the nineteenth and the beginning of the twentieth century, following upon the Industrial Revolution, as the new rich bourgeoisie supplanted the old aristocracy and new social classes emerged, two novel aperitifs won national and then international acclaim: Fernet Branca and bitter ("amaro" in Italian) Campari. Their widespread popularity stemmed from their large consumption in pubs and social gatherings and from the medical observation that they favored food intake and stimulated digestion. (These bitter liquors were classified as *cholagogues*, i.e., medicinal agents promoting the discharge of bile from the system.) Interestingly, the idea that bitter aperitifs aid digestion has been fully confirmed in recent literature, since bitter-taste receptors do exist in the gut, as well as the mouth. Janssen et al. (2011) indeed found that, when treating mice with compounds that activate these receptors, a rise in a hunger hormone called α-ghrelin, promoting food intake, could be detected.

Since in general all producers declare that their aperitifs are made with herbal infusions, it was of interest to investigate their trace proteome, in order to verify the truth of their claims. With this in mind, Fasoli et al. (2102b) decided to analyze a popular aperitif in northern Italy, Amaro Braulio, in search of residual proteins, if any. This aperitif is made with an infusion of 13 mountain herbs and berries, among which four are officially indicated in the label: *Achillea moschata*, juniper (*Juniperus communis* subsp. *alpina*) berries, absinthe (*Artemisia absinthium*), and gentian (*Gentiana alpina*) roots. Through capture with solid-phase combinatorial peptide ligand library at pH 7.0 and 2.2, these authors identified 72 unique gene products, including the PR5 (parasite resistance) allergen *Jun r 3.2*, a 25 kDa species from *Juniperus rigida*. Due to the paucity of data on these alpine herbs, it was difficult to attribute these proteins to the specific plant extracts presumably present in this beverage. Most of the species identified, however, indeed belong to alpine herbs and plants, found in a habitat between 1000 and 2000 m of elevation. Most of them are enzymes, spanning an Mr range from 10 to 65 kDa. Thus, the genuineness of this product was amply proven.

Another popular aperitif sold the world over, Cynar, did not fare as well. Cynar purports to be made with an infusion of artichoke leaves reinforced with an extract of 13 secret herbs. To err on the side of safety, Saez et al. (2013) adopted a three-pronged approach to verify the claim. First, different extraction techniques were used to characterize the artichoke's proteome; second, a home-made infusion was analyzed; and finally the proteome of the commercial drink was checked. The artichoke proteome was evaluated by prior capture with CPLLs at four different pH (2.2, 4.0, 7.2, and 9.3) values. Using mass spectrometry analysis of the recovered fractions, after elution of the captured populations in 4% boiling SDS, these authors could identify a total of 876 unique gene products in the artichoke extracts, 18 in the home-made infusion, and no proteins at all in the Italian Cynar liqueur. This finding therefore cast severe doubts on the procedure stated by the manufacturer (i.e., that the aperitif was produced by an infusion of artichoke leaves plus 13 different herbs). In the event extracted proteins from the artichoke are present at only a trace level, it would thus appear that CPLLs could be a formidable tool for investigating this commercial drink in order to protect consumers from adulterated products.

5.4 CONCLUSIONS

We can report two conclusions. The first one regards the survival mechanism of proteins in industrially treated food as, for instance, beverages, and the second one regards what one can do by exploiting the CPLL technique.

For the first conclusion, all large-size-proteins (e.g., the ubiquitous 7S globulins, 61 to 67 kDa) remain in solution only as fragments, typically around 15–20 kDa. They are degraded for various reasons the major one being the action of proteases during industrial treatment. Proteins that survive in very acidic solutions (e.g., vinegar, pH ca. 2.3) are most probably hydrophobic ones that resist unfolding and thus avoid the aggregation and flocculation that occur during the unfolding process. Another point to consider is the presence of protein allergens: They are most likely present even in beverages, but they have much reduced allergenicity due to the fact that they are also partially cleaved and/ or post-translationally modified, through glycation, for example, which could mask the allergenic sequences. This assessment could explain why the most pestiferous and ubiquitous allergen, the lipid transfer protein, which is so heavily present in beers (it has excellent foaming properties), has never given an anaphylactic shock in the German or Czech population at large, which is known to gulp down 160 liters per capita per annum! Finally it should be said that most of the undegraded proteins that survive in solution are small-size species, typically 10 to 20 kDa.

The second reason is related to what one could learn from this methodology as it applies to food science. Clearly, CPLLs can amplify the signal of traces of proteins by 3 to 4 orders of magnitude even when used in small amounts (50 to 100 μL of peptide beads) for very large sample volumes (1 liter and more). In addition, CPLLs allow detection of as little as 1 μg/L of proteins, which means they can easily operate in the low- to very low-abundance regions where proteins can hardly be detected even with the most sophisticated MS instruments. Moreover, CPLL beads have a very high sweeping efficiency that even for traces of proteins in solution is typically of the order of 85 to 90%. This capture is best performed through reverse-phase adsorption at pH 2.2 in the presence of TFA. In this last case, special libraries can easily be designed to serve this objective, as already reported.

In summary, then, the CPLL technique is easy, user-friendly, rapid, very efficient, and not so expensive that it should continue to be ignored.

5.5 References

Acosta-Muñiz CH, Escobar-Tovar L, Valdes-Rodriguez S, et al. Identification of avocado (*Persea americana*) root proteins induced by infection with the oomycete. Phytophtora cinnamomi using a proteomic approach. *Physiol Plant.* 2012;144:59–72.

Agrawal GK, Jwa NS, Iwahashi Y, Yonekura M, Iwahashi H, Rakwal R. Rejuvenating rice proteomics: Facts, challenges, and visions. *Proteomics.* 2006;6:5549–5576.

Agrawal GK, Rakwal R. Plant Proteomics: Technologies, Strategies, and Applications. In: Agrawal GK, Rakwal R, eds. Hoboken, NJ: John Wiley & Sons; 2008.

Agrawal GK, Jwa NS, Rakwal R. Rice proteomics: End of phase I and beginning of phase II. *Proteomics.* 2009;9:935–963.

Agrawal GK, Rakwal R. Rice proteomics: A move toward expanded proteome coverage to comparative and functional proteomics uncovers the mysteries of rice and plant biology. *Proteomics.* 2011;11:1630–1649.

Ahlroth M, Alenius H, Turjanmaa K, Makinen-Kiljunen S, Reunala T, Palosuo T. Cross-reacting allergens in natural rubber latex and avocado. *J Allergy Clin Immunol.* 1995;96:167–173.

Ahn JY, Lee SW, Kang HS, et al. Aptamer microarray mediated capture and mass spectrometry identification of biomarker in serum samples. *J Proteome Res.* 2010;9:5568–5573.

Aiello D, De Luca D, Gionfriddo E, et al. Review: Multistage mass spectrometry in quality, safety and origin of foods. *Eur J Mass Spectrom (Chichester, Eng).* 2011;17:1–31.

Alché JD, Jiménez-López JC, Wang W, Castro-López AJ, Rodríguez-García MI. Bichemical characterization and cellular localization of 11S type storage proteins in olive (Olea europaea L.) seeds. *J Agric Food Chem* 2006;54:5562–5570.

Aleksic I, Popovic M, Dimitrijevic R, et al. Molecular and immunological characterization of Mus a 5 allergen from banana fruit. *Mol Nutr Food Res.* 2012;56:446–453.

Anderson LB, Dreyfuss EM, Logan J, Johnstone DE, Glaser J. Melon and banana sensitivity coincident with ragweed pollinosis. *J Allergy.* 1970;45:310–319.

Arena S, Renzone G, Novi G, et al. Modern proteomic methodologies for the characterization of lactosylation protein targets in milk. *Proteomics.* 2010;10:3414–3434.

Argueta-Villamar A, Cano L, Rodarte M. *Atlas de las Plantas de la Medicina Tradicional Mexicana.* México: Ed. Instituto Nacional Indigenista; 1994.

Astwood JD, Leach JN, Fuchs RL. Stability of food allergens to digestion in vitro. *Nat Biotechnol.* 1996;14:1269–1273.

Barre A, Culerrier R, Granier C, et al. Mapping of IgE-binding epitopes on the major latex allergen Hev b 2 and the cross-reacting 1,3β-glucanase fruit allergens as a molecular basis for the latex-fruit syndrome. *Mol Immunol.* 2009;46:1595–1604.

Barre A, Peumans WJ, Menu-Bouaouiche L, et al. Purification and structural analysis of an abundant thaumatin-like protein from ripe banana fruit. *Planta.* 2000;211:791–799.

Bernal J, Mendiola JA, Ibáñez E, Cifuentes A. Advanced analysis of nutraceuticals. *J Pharm Biomed Anal.* 2011;55:758–774.

Boschetti E, Righetti PG. The art of observing rare protein species in proteomes with peptide ligand libraries. *Proteomics.* 2009;9:1492–1510.

Bourguignon J, Jaquinod M. An overview of the Arabidopsis proteome. In: Agrawal GK, Rakwal R, eds. *Plant Proteomics.* Hoboken, NJ: John Wiley & Sons; 2008:143–164.

Cánovas FM, Dumas-Gaudot E, Recorbet G, Jorrin J, Mock HP, Rossignol M. Plant proteome analysis. *Proteomics.* 2004;4:285–298.

Carbonaro M. Proteomics: present and future in food quality evaluation. *Trends Food Sci Technol.* 2004;15:209–216.

Carpentier SC, Panis B, Vertommen A, et al. Proteome analysis of non-model plants: A challenging but powerful approach. *Mass Spectrom Rev.* 2008;27:354–377.

Carpentier SC, Witters E, Laukens K, Van Onckelen H, Swennen R, Panis B. Banana (*Musa ssp.*) as a model to study the meristem proteome: acclimation to osmotic stress. *Proteomics.* 2007;7:92–105.

Castro-Puyana M, García-Cañas V, Simó C, Cifuentes A. Recent advances in the application of capillary electromigration methods for food analysis and Foodomics. *Electrophoresis.* 2012;33:147–167.

Cereda A, Kravchuk AV, D'Amato A, Bachi A, Righetti PG. Proteomics of wine additives: mining for the invisible via combinatorial peptide ligand libraries. *J Proteomics.* 2010;73:1732–1739.

Chen L, Remondetto GE, Subirade M. Food protein-based materials as nutraceutical delivery systems. *Trends Food Sci Technol.* 2006;17:272–283.

Choudhury SR, Roy S, Saha PP, Singh SK, Sengupta DN. Characterization of differential ripening pattern in association with ethylene biosynthesis in the fruits of five naturally occurring banana cultivars and detection of a GCC-box-specific DNA-binding protein. *Plant Cell Rep.* 2008;27:1235–1249.

Cilia M, Fish T, Yang X, McLaughlin M, Thannhauser TW, Gray S. A comparison of protein extraction methods suitable for gel-based proteomic studies of aphid proteins. *J Biomol Tech.* 2009;20:201–215.

Cilindre C, Jégou S, Hovasse A, et al. Proteomic approach to identify champagne wine proteins as modified by Botrytis cinerea infection. *J Proteome Res.* 2008;7:1199–1208.

Clendennen SK, Lopez-Gomez R, Gomez-Lim M, Arntzen CJ, May GD. The abundant 31-kilodalton banana pulp proteins is homologous to class-III acidic chtiniases. *Phytochemistry.* 1997;47:613–619.

Crofts AJ, Crofts N, Whitelegge JP, Okita TW. Isolation and identification of cytoskeleton-associated prolamine mRNA binding proteins from developing rice seeds. *Planta.* 2010;231:1261–1276.

Cunsolo V, Muccilli V, Fasoli E, Saletti R, Righetti PG, Foti S. Poppea's bath liquor: The secret proteome of she-donkey's milk. *J Proteomics.* 2011a;74:2083–2099.

Cunsolo V, Muccilli V, Saletti R, Foti S. Review: Applications of mass spectrometry techniques in the investigation of milk proteome. *Eur J Mass Spectrom (Chichester, Eng).* 2011b;17:305–320.

D'Alessandro A, Zolla L, Scaloni A. The bovine milk proteome: Cherishing, nourishing and fostering molecular complexity. An interactomics and functional overview. *Mol Biosyst.* 2011;7:579–597.

D'Alessandro A, Zolla L. We are what we eat: Food safety and proteomics. *J Proteome Res.* 2012;11:26–36.

D'Amato A, Bachi A, Fasoli E, et al. In-depth exploration of cow's whey proteome via combinatorial peptide ligand libraries. *J Proteome Res.* 2009;8:3925–3936.

D'Amato A, Kravchuk AV, Bachi A, Righetti PG. Noah's nectar: The proteome content of a glass of red wine. *J Proteomics.* 2010b;73:2370–2377.

D'Amato A, Bachi A, Fasoli E, et al. In-depth exploration of *Hevea brasiliensis* latex proteome and "hidden allergens" via combinatorial peptide ligand libraries. *J Proteomics.* 2010a;73:1368–1380.

D'Amato A, Fasoli E, Kravchuk AV, Righetti PG. Going nuts for nuts? The trace proteome of a cola drink, as detected via combinatorial peptide ligand libraries. *J Proteome Res.* 2011a;10:2684–2686.

D'Amato A, Fasoli E, Kravchuk AV, Righetti PG. Mehercules, adhuc Bacchus! The Debate on Wine Proteomics Continues. *J Proteome Res.* 2011b;10:3789–3801.

D'Amato A, Fasoli E, Righetti PG. Harry Belafonte and the secret proteome of coconut milk. *J Proteomics.* 2012;75:914–920.

D'Ambrosio C, Arena S, Scaloni A, et al. Exploring the chicken egg white proteome with combinatorial peptide ligand libraries. *J Proteome Res.* 2008;7:3461–3474.

Dayarathna MK, Hancock WS, Hincapie M. A two step fractionation approach for plasma proteomics using immunodepletion of abundant proteins and multi-lectin affinity chromatography: Application to the analysis of obesity, diabetes, and hypertension diseases. *J Sep Sci.* 2008;31:1156–1166.

De la Lastra Romero CA. An up-date of olive oil bioactive constituents in health: molecular mechanisms and clinical implications. *Curr Pharm Des.* 2011;17:752–753.

Der Agopian RG, Goncalvez Peroni-Okita FH, Solares CA, et al. Low temperature induced changes in activity and protein levels of the enzymes associated to conversion of starch to sucrose in banana fruit. *Postharvest Biol Technol.* 2011;62:133–140.

Di Girolamo F, D'Amato A, Righetti PG. Horam nonam exclamavit: sitio. The trace proteome of your daily vinegar. *J Proteomics.* 2011;75:718–724.

Di Girolamo F, D'Amato A, Righetti PG. Assessment of the floral origin of honey via proteomic tools. *J Proteomics.* 2012;75:3688–3693.

Diaz-Perales A, Blanco C, Sanchez-Monge R, Varela J, Carrillo T, Salcedo G. Analysis of avocado allergen (Prs a 1) IgE-binding peptides generated by simulated gastric fluid digestion. *J Allergy Clin Immunol.* 2003;112:1002–1007.

Dominguez-Puigjaner E, Vendrell M, Ludevid MD. Differential protein accumulation in banana fruit during ripening. *Plant Physiol.* 1992;98:157–162.

Dureja H, Kaushik D, Kumar V. Developments in nutraceuticals. *Indian J Pharm.* 2003;35:363–372.

Echan LA, Tang HY, Ali-Khan N, Lee K, Speicher DW. Depletion of multiple high-abundance proteins improves protein profiling capacities of human serum and plasma. *Proteomics.* 2005;5:3292–3303.

El Sanhoty R, El-Rahman AA, Bögl KW. Quality and safety evaluation of genetically modified potatoes spunta with Cry V gene: compositional analysis, determination of some toxins, antinutrients compounds and feeding study in rats. *Nahrung.* 2004;48:13–18.

Emami K, Morris NJ, Cockell SJ, Golebiowska G, Shu QY, Gatehouse AM. Changes in protein expression profiles between a low phytic acid rice (*Oryza sativa* L. Ssp. *japonica*) line and its parental line: A proteomic and bioinformatic approach. *J Agric Food Chem.* 2010;58:6912–6922.

Esteve C, Cañas B, Moreno-Gordaliza E, Del Rio C, García MC, Marina ML. Identification of olive (*Olea europaea*) pulp proteins by matrix-assisted laser desorption/ionization time-of-flight mass spectrometry and nano-liquid chromatography tandem mass spectrometry. *J Agric Food Chem.* 2011;59:12093–12101.

Esteve C, D'Amato A, Marina ML, García MC, Citterio A, Righetti PG. Identification of olive (*Olea europaea*) seed and pulp proteins by nLC-MS/MS via combinatorial peptide ligand libraries. *J Proteomics.* 2012a;75:2396–2403.

Esteve C, D'Amato A, Marina ML, Concepción García M, Righetti PG. In-depth proteomic analysis of banana (*Musa spp.*) fruit with combinatorial peptide ligand libraries. *Electrophoresis.* 2013;34:207–214.

Esteve C, D'Amato A, Marina ML, Concepción García M, Righetti PG. Identification of avocado (*Persea americana*) pulp proteins by nanoLC-MS/MS via combinational peptide ligand libraries. *Electrophoresis.* 2012;33:2799–2805.

Esteve C, Montealegre C, Marina ML, Garcia MC. Analysis of olive allergens. *Talanta.* 2012;92:1–14.

Farinazzo A, Restuccia U, Bachi A, et al. Chicken egg yolk cytoplasmic proteome, mined via combinatorial peptide ligand libraries. *J Chromatogr A.* 2009;1216:1241–1252.

Fasoli E, Aldini G, Regazzoni L, Kravchuk AV, Citterio A, Righetti PG. Les Maîtres de l'Orge: the proteome content of your beer mug. *J Proteome Res.* 2010;9:5262–5269.

Fasoli E, D'Amato A, Kravchuk AV, Boschetti E, Bachi A, Righetti PG. Popeye strikes again: The deep proteome of spinach leaves. *J Proteomics.* 2011a;74:127–136.

Fasoli E, D'Amato A, Kravchuk AV, Citterio A, Righetti PG. In-depth proteomic analysis of non-alcoholic beverages with peptide ligand libraries. I: Almond milk and orgeat syrup. *J Proteomics.* 2011b;74:1080–1090.

Fasoli E, D'Amato A, Citterio A, Righetti PG. Anyone for an aperitif? Yes, but only a Braulio DOC with its certified proteome. *J Proteomics.* 2012b;75:3374–3379.

Fasoli E, D'Amato A, Citterio A, Righetti PG. Ginger Rogers? No, ginger ale and its invisible proteome. *J Proteomics.* 2012a;75:1960–1965.

Faurobert M, Pelpoir E, Chaïb J. Phenol extraction of proteins for proteomic studies of recalcitrant plant tissues. *Methods Mol Biol.* 2007;355:9–14.

Feuillet C, Leach JE, Rogers J, Schnable PS, Eversole K. Crop genome sequencing: Lesions and rationales. *Trends Plant Sci.* 2010;16:77–88.

García-Canas V, Simó C, León C, Ibáñez E, Cifuentes A. MS-based analytical methodologies to characterize genetically modified crops. *Mass Spectrom Rev.* 2011;30:396–416.

García-Lopez MC, García-Canas V, Marina ML. Reversed-phase high-performance liquid chromatography-electrospray mass spectrometry profiling of transgenic and non-transgenic maize for cultivar characterization. *J Chromatogr A.* 2009;1216:7222–7228.

Geiger T, Wehner A, Schaab C, Cox J, Mann M. Comparative proteomic analysis of eleven common cell lines reveals ubiquitous but varying expression of most proteins. *Mol Cell Proteomics.* 2012;11:M111.014050.

Gengenheimer P. Preparation of extracts from plants. *Methods Enzymol.* 1990;182:174–193.

Georgalaki MD, Bachmann A, Sotiroudis TG, Xenakis A, Porzel A, Feussner I. Characterization of a 13-lipoxygenase from virgin olive oil and oil bodies of olive endosperms. *Fett-Lipid.* 1998a;100:554–560.

Georgalaki MD, Sotiroudis TG, Xenakis A. The presence of oxidizing enzyme activities in virgin olive oil. *J Am Oil Chem Soc.* 1998b;75:155–159.

Giavalisco P, Nordhoff E, Kreitler T, et al. Proteome analysis of Arabidopsis thaliana by two-dimensional gel electrophoresis and matrix-assisted laser desorption/ionisation-time of flight mass spectrometry. *Proteomics.* 2005;5:1902–1913.

Giribaldi M, Giuffrida MG. Heard it through the grapevine: proteomic perspective on grape and wine. *J Proteomics.* 2010;73:1647–1655.

Goff SA, Ricke D, Lan TH, et al. A draft sequence of the rice genome (*Oryza sativa L. ssp. Japonica*). *Science.* 2002;296:92–100.

Goodmann RE, Tetteh AO. Suggested improvements for the allergenicity assessment of genetically modified plants used in foods. *Curr Allergy Asthma Rep.* 2011;11:317–324.

Guérin-Dubiard C, Pasco M, Mollé D, Désert C, Croguennec T, Nau F. Proteomic analysis of hen egg white. *J Agric Food Chem.* 2006;54:3901–3910.

Haddada FM, Manai H, Oueslati I, et al. Fatty acid, triacylglycerol, and phytosterol composition in six Tunisian olive varieties. *J Agric Food Chem.* 2007;55:10941–10946.

Hajduch M, Hearne LB, Miernyk JA, et al. Systems analysis of seed filling in Arabidopsis: using general linear modeling to assess concordance of transcript and protein expression. *Plant Physiol.* 2010;152:2078–2087.

Hamburg MA. Advancing regulatory science. *Science.* 2011;331:987–988.

Hashimoto M, Komatsu S. Proteomic analysis of rice seedlings during cold stress. *Proteomics.* 2007;7:1293–1302.

Herrero M, Simó C, García-Cañas V, Ibáñez H, Cifuentes A. Foodomics: MS-based strategies in modern food science and nutrition. *Mass Spectrom Rev.* 2012;31:49–69.

Hidalgo FJ, Alaiz M, Zamora R. Determination of peptides and proteins in fats and oils. *Anal Chem.* 2001;73:698–702.

Ho VS, Ng TB. Chitinase-like proteins with antifungal activity from emperor banana fruits. *Protein Pept Lett.* 2007;14:828–831.

Ho VSM, Wong JH, Ng TB. A thaumatin-like protein antifungal protein from the emperor banana. *Peptides.* 2007;28:760–766.

Ibrahim HR, Imazato K, Ono H. Human lysozyme possesses novel antimicrobial peptides within its N-terminal domain that target bacterial respiration. *J Agric Food Chem.* 2011;59:10336–10345.

Isaacson T, Damasceno CM, Saravanan RS, et al. Sample extraction techniques for enhanced proteomic analysis of plant tissues. *Nature Protoc.* 2006;1:769–774.

Ito A, Ito K, Morishita M, Sakamoto T. A banana-allergic infant with IgE reactivity to avocado, but not to latex. *Pediatr Int.* 2006;48:321–323.

Janssen S, Laermans J, Verhulst PJ, Thijs T, Tack J, Depoortere I. Bitter taste receptors and α-gustducin regulate the secretion of ghrelin with functional effects on food intake and gastric emptying. *Proc Natl Acad Sci U S A.* 2011;108:2094–2099.

Jorrín JV, Maldonado AM, Castillejo MA. Plant proteome analysis: A 2006 update. *Proteomics.* 2007;7:2947–2962.

Jorrín-Novo JV, Maldonado AM, Echevarría-Zomeño S, et al. Plant proteomics update (2007–2008): Second-generation proteomic techniques, an appropriate experimental design, and data analysis to fulfill MIAPE standards, increase plant proteome coverage and expand biological knowledge. *J Proteomics.* 2009;72:285–314.

Jo H-Y, Jung W-K, Kim S-K. Purification and characterization of a novel anticoagulant peptide from marine echiuroid worm. Urechis unicinctus. *Process Biochem.* 2008;43:179–184.

Karamonova L, Fukal L, Kodicek M, Rauch P, Mills ENC, Morgan MRA. Immunoprobes for thermally-induced alterations in whey protein structure and their application to the analysis of thermally-treated milks. *Food Agric Immunol.* 2003;15:77–91.

Kašička V. Capillary electrophoresis of peptides. *Electrophoresis.* 1999;20:3084–3105.

Kašička V. Recent advances in capillary electrophoresis and capillary electrochromatography of peptides. *Electrophoresis.* 2003;24:4013–4046.

Kašička V. Recent advances in capillary electrophoresis of peptides. *Electrophoresis.* 2001;22:4139–4162.

Kašička V. Recent advances in CE and CEC of peptides (2007–2009). *Electrophoresis.* 2010;31:122–146.

Kašička V. Recent developments in capillary electrophoresis and capillary electrochromatography of peptides. *Electrophoresis.* 2006;27:142–175.

Kašička V. Recent developments in CE and CEC of peptides (2009–2011). *Electrophoresis.* 2012;33:48–73.

Kašička V. Recent developments in CE and CEC of peptides. *Electrophoresis.* 2008;29:179–206.

Kersten B, Agrawal GK, Iwahashi H, Rakwal R. Plant phosphoproteomics: A long road ahead. *Proteomics.* 2006;6:5517–5528.

Kersten B, Agrawal GK, Durek P, et al. Plant phosphoproteomics: An update. *Proteomics.* 2009;9:964–988.

Khalf M, Goulet C, Vorster J, et al. Tubers from potato lines expressing a tomato Kunitz protease inhibitor are substantially equivalent to parental and transgenic controls. *Plant Biotechnol J.* 2010;8:155–169.

Koidis A, Boskou D. The contents of proteins and phospholipids in cloudy (veiled) virgin olive oils. *Eur J Lipid Sci Technol.* 2006;108:323–328.

Kwon SW. Profiling of soluble proteins in wine by nano-high-performance liquid chromatography/tandem mass spectrometry. *J Agric Food Chem.* 2004;52:7258–7263.

Larsen MR, Thingholm TE, Jensen ON, Roepstorff P, Jorgensen TJ. Highly selective enrichment of phosphorylated peptides from peptide mixtures using titanium dioxide microcolumns. *Mol Cell Proteomics.* 2005;4:873–886.

Lavud F, Prevost A, Cossart C, Guerin L, Bernard J, Kochman S. Allergy to latex, avocado pear, and banana: evidence for a 30 kd antigen in immunoblotting. *J Allergy Clin Immunol.* 1995;95:557–564.

Lee SH, Qian Z-J, Kim SK. A novel angiotensin I converting enzyme inhibitory peptide from tuna frame protein hydrolysate and its antihypertensive effect in spontaneously hypertensive rats. *Food Chem.* 2010;118:96–102.

Leone P, Menu-Bouaouiche L, Peumans WJ, et al. Resolution of the structure of the allergenic and antifungal banana fruit thaumatin-like protein at 1.7-Å. *Biochimie.* 2006;88:45–52.

Lopez-Ledesma R, Frati-Munari AC, Hernandez-Dominguez BC, et al. Monosaturated fatty acid (avocado) rich diet for mild hypercolesterolemia. *Arch Med Res.* 1996;27:519–523.

Lu QY, Arteaga JR, Zhang Q, Huerta S, Go VLW, Heber D. Inhibition of protease cancer cell growth by an avocado extract: role of lipid-soluble bioactive substances. *J Nutr Biochem.* 2005;16:23–30.

Luo J, Ning T, Sun Y, et al. Proteomic analysis of rice endosperm cells in response to expression of hGM-CSF. *Proteome Res.* 2009;8:829–837.

Mahdavi F, Sariah M, Maziah M. Expression of rice thaumatin-like protein gene in transgenic banana plants enhances resistance to Fusarium wilt. *Appl Biochem Biotechnol.* 2012;166:1008–1019.

Mann K. The chicken egg white proteome. *Proteomics.* 2007;7:3558–3568.

Mann K, Mann M. In-depth analysis of the chicken egg white proteome using an LTQ Orbitrap Velos. *Proteome Science.* 2011;9:7.

Méchin V, Damerval C, Zivy M. Total protein extraction with TCA-acetone. *Methods Mol Biol.* 2007;355:1–8.

Menke W, Koenig F. Isolation of thylakoid proteins. *Methods Enzymol.* 1980;69:446–452.

Mikkola JH, Alenius H, Kalkkinen N, Turjan-maa K, Palosuo T, Reunala T. Hevein-like protein domain as a possible cause for allergen cross-reactivity between latex and banana. *J Allergy Clin Immunol.* 1998;102:1005–1012.

Moller M, Kayma M, Vieluf D, Paschke A, Steinhart H. Determination and characterization of cross-reacting allergens in latex, avocado, banana, and kiwi fruit. *Allergy.* 1998;53:289–296.

Monaci L, Losito I, Palmisano F, Godula M, Visconti A. Towards the quantification of residual milk allergens in caseinate-fined white wines using HPLC coupled with single-stage Orbitrap mass spectrometry. *Food Addit Contam Part A.* 2011;28:1304–1314.

Monaci L, Losito I, Palmisano F, Visconti A. Identification of allergenic milk proteins markers in fined white wines by capillary liquid chromatography-electrospray ionization-tandem mass spectrometry. *J Chromatogr A.* 2010;1217:4300–4305.

Montealegre C, Esteve C, García MC, Garcia-Ruiz C, Marina ML. Proteins in olive and olive oil. *Crit Rev Food Sci Nutr.* In press. doi:10.1080/10408398.2011.598639 (posted on line 14 January 2013).

Moreno-Arribas MV, Pueyo E, Polo MC. Analytical methods for the characterization of proteins and peptides in wines. *Anal Chim Acta.* 2002;458:63–75.

Natarajan SS, Xu C, Bae H, Caperna TJ, Garrett WM. Characterization of storage proteins in wild (Glycine soja) and cultivated (Glycine max) soybean seeds using proteomic analysis. *J Agric Food Chem.* 2006;54:3114–3120.

Ningappa MB, Srinivas L. Purification and characterization of approximately 35 kDa antioxidant protein from curry leaves (Murraya koenigii L.). *Toxicol In Vitro.* 2008;22:699–709.

Palacin A, Quirce S, Sanche-Monge R, et al. Sensitization profiles to purified plant food allergens among pediatric patients with allergy to banana. *Pediatr Allergy Immunol.* 2011;22:186–195.

Palmisano G, Antonacci D, Larsen MR. Glycoproteomic profile in wine: A "sweet" molecular renaissance. *J Proteome Res.* 2010;9:6148–6159.

Peumans WJ, Barre A, Derycke V, et al. Purification, characterization and structural analysis of an abundant β-1,3-glucanase from banana fruit. *Eur J Biochem.* 2000;267:1188–1195.

Peumans WJ, Proost P, Swennen RL, Van Damme EJM. The abundant class III chitinase homolog in young developing banana fruits behaves as a transient vegetative storage protein and most probably serves as an important supply of amino acids for the synthesis of ripening-associated proteins. *Plant Physiol.* 2002;130:1063–1072.

Pihlström T, Isaac G, Waldebäck M, Osterdahl BG, Markides KE. Pressurised fluid extraction (PFE) as an alternative general method for the determination of pesticide residues in rape seed. *Analyst.* 2002;127:554–559.

Plaza L, Sanchez-Moreno C, de Pascual-Teresa S, de Ancos B, Cano MP. Fatty acids, sterols, and antioxidant activity in minimally processed avocados during refrigerated storage. *J Agric Food Chem.* 2009;57:3204–3209.

Polaskova V, Kapur A, Khan A, Molloy MP, Baker MS. High-abundance protein depletion: Comparison of methods for human plasma biomarker discovery. *Electrophoresis.* 2010;31:471–482.

Radin NS. Extraction of tissue lipids with a solvent of low toxicity. *Methods Enzymol.* 1981;72:5–7.

Rakwal R, Agrawal GK. Rice proteomics: current status and future perspectives. *Electrophoresis.* 2003;24:3378–3389.

Rallo L, Barranco D, Caballero JM, et al. *Variedades del olivo en España.* Madrid: Junta de Andalucía, MAPA and Ediciones Mundi-Prensa; 2005.

Reindl J, Rihs HP, Scheurer S, Wangorsch A, Haustein D, Vieths S. IgE reactivity to profiling in pollen-sensitized subjects with adverse reactions to banana and pineapple. *Int Arch Allergy Immunol.* 2002;128:105–114.

Rey M, Mrázek H, Pompach P, et al. Effective removal of nonionic detergents in protein mass spectrometry, hydrogen/deuterium exchange, and proteomics. *Anal Chem.* 2010;82:5107–5116.

Righetti PG. Happy bicentennial, Electrophoresis! *J Proteomics.* 2009;73:181–187.

Righetti PG, Boschetti E, Kravchuk A, Fasoli E. The proteome buccaneers: how to unearth your treasure chest via combinatorial peptide ligand libraries. *Expert Rev Proteomics.* 2010;7:373–385.

Righetti PG, Boschetti E, Lomas L, Citterio A. Protein Equalizer technology: The quest for a "democratic proteome". *Proteomics.* 2006;6:3980–3992.

Rist MJ, Wenzel U, Daniel H. Nutrition and food science go genomic. *Trends Biotechnol.* 2006;24:1–7.

Rodríguez G, Lama A, Rodríguez R, Jiménez A, Guillén R, Fernández-Bolaños J. Olive Stone as an attractive source of bioactive and valuable compounds. *Biores Tech.* 2008;99:5261–5269.

Rodriguez J, Crespo JF, Burks W, et al. Randomized, double-blind, crossover challenge study in 53 subjects reporting adverse reactions to melon (Cucmis melo). *J Allergy Clin Immunol.* 2000;106:968–972.

Rose JK, Bashir S, Giovannoni JJ, Jahn MM, Saravanan RS. Tackling the plant proteome: practical approaches, hurdles and experimental tools. *Plant J.* 2004;39:715–733.

Rossignol M, Peltier JB, Mock HP, Matros A, Maldonado AM, Jorrín JV. Plant proteome analysis: A 2004–2006 update. *Proteomics.* 2006;6:5529–5548.

Roux-Dalvai F, Gonzalez de Peredo A, Simó C, et al. Extensive analysis of the cytoplasmic proteome of human erythrocytes using the peptide ligand library technology and advanced mass spectrometry. *Mol Cell Proteomics.* 2008;7:2254–2269.

Saez V, Fasoli E, D'Amato A, Simó-Alfonso E, Righetti PG. Artichoke and Cynar liqueur: two (not quite) entangled proteomes. *Biochim Biophys Acta.* 2013;1834:119–126.

Salazar MJ, El Hafidi M, Pastelin G, Ramirez-Ortega MC, Sanchez-Mendoza MA. Effect of an avocado oil-rich diet over an angiotensin II-induced blood pressure response. *J Ethnopharmacol.* 2005;98:335–338.

Samyn B, Sergeant K, Carpentier S, et al. Functional proteome analysis of the banana plant (*Musa ssp.*) using de novo sequence analysis of derivatized peptides. *J Proteome Res.* 2007;6:70–80.

Sanchez-Monge R, Blanco C, Diaz-Perales A, et al. Isolation and characterization of major banana allergens: Identification as fruit class I chitinases. *Clin Exp Allergy.* 1999;29:673–680.

Sanchez-Romero C, Peran-Quesada R, Barcelo-Muñoz A, Pliego-Alfaro F. Variations in storage protein and carbohydrate levels during development of avocado zygotic embryos. *Plant Physiol Biochem.* 2002;40:1043–1049.

Savarese M, De Marco E, Sacchi R. Characterization of phenolic extracts from olives (*Olea europaea* cv. Pisciottana) by electrospray ionization mass spectrometry. *Food Chem.* 2007;105:761–770.

Shahali Y, Sutra JP, Fasoli E, et al. Allergomic study of cypress pollen via combinatorial peptide ligand libraries. *J Proteomics.* 2012;77:101–110.

Seiber JN, Kleinschmidt L. From detrimental to beneficial constituents in foods: tracking the publication trends in JAFC. *J Agric Food Chem.* 2012;60:6644–6647.

Shen Y, Kim J, Strittmatter EF, et al. Characterization of the human blood plasma proteome. *Proteomics.* 2005;5:4034–4045.

Shi T, Zhou JY, Gritsenko MA, et al. IgY14 and SuperMix immunoaffinity separations coupled with liquid chromatography-mass spectrometry for human plasma proteomics biomarker discovery. *Methods.* 2012;56:246–253.

Siciliano RA, Mazzeo MF, Arena S, Renzone G, Scaloni A. Mass spectrometry for the analysis of protein lactosylation in milk products. *Food Res Int.* 2012; in press.

Simó C, Domínguez-Vega E, Marina ML, García MC, Dinelli G, Cifuentes A. CE-TOF MS analysis of complex protein hydrolyzates from genetically modified soybeans–a tool for foodomics. *Electrophoresis.* 2010;31:1175–1183.

Sowka S, Hsieh LS, Krebitz M, et al. Identification and cloning of Prs a 1, a 32-kDa endochitinase and major allergen of avocado, and its expression in the yeast Pichia pastoris. *J Biol Chem.* 1998;273:28091–28097.

Stempfer R, Kubicek M, Lang IM, Christa N, Gerner C. Quantitative assessment of human serum high-abundance protein depletion. *Electrophoresis.* 2008;29:4316–4323.

Sun J, Chu YF, Wu X, Liu RH. Antioxidant and antiproliferative activities of common fruits. *J Agric Food Chem.* 2002;50:7449–7454.

Swisher HE. Avocado oil: from food use to skin care. *J Am Oil Chem Soc.* 1988;65:1704–1706.

Thulasiraman V, Lin S, Gheorghiu L, et al. Reduction of the concentration difference of proteins in biological liquids using a library of combinatorial ligands. *Electrophoresis.* 2005;26:3561–3571.

Tolin S, Pasini G, Curioni A, et al. Mass spectrometry detection of egg proteins in red wines treated with egg white B. *Food Control.* 2012;23:87–94.

Tomás-Barberán FA, Somoza V, Finley J. Food bioactives research and the Journal of Agricultural and Food Chemistry. *J Agric Food Chem.* 2012;60:6641–6643.

van Ree R, Voitenko V, van Leeuwen WA, Aalberse RC. Profilin is a cross-reactive allergen in pollen and vegetable foods. *Int Arch Allergy Immunol.* 1992;98:97–104.

Van Renswoude J, Kemps C. Purification of integral membrane proteins. *Methods Enzymol.* 1984;104:329–339.

Vertommen A, Moller ALB, Cordewener JHG, et al. A workflow for peptide-based proteomics in a poorly sequenced plant: a case study on the plasma membrane proteome of banana. *J Proteomics.* 2011;74:1218–1229.

Vieths S, Scheurer S, Ballmer-Weber B. Current understanding of cross-reactivity of food allergens and pollen. *Ann N Y Acad Sci.* 2002;964:47–68.

Waserman S, Watson W. Food allergy. *Allergy, Asthma & Clinical Immunol.* 2011;7(suppl 1):S7.

Weber P, Steinhart H, Paschke AJ. Determination of the bovine food allergen casein in white wines by quantitative indirect ELISA, SDS-PAGE, Western Blot and Immunostaining. *Agric Food Chem.* 2009;57:8399–8405.

Wessel D, Flugge UI. A method for the quantitative recovery of protein in dilute solution in the presence of detergents and lipids. *Anal Biochem.* 1984;138:141–143.

Wiesner J, Vilcinskas A. Antimicrobial peptides: The ancient arm of the human immune system. *Virulence.* 2010;1:440–464.

Wigand P, Tenzer S, Schild H, Decker HJ. Analysis of protein composition of red wine in comparison with rosé and white wines by electrophoresis and high-pressure liquid chromatography-mass spectrometry (HPLC-MS). *Agric Food Chem.* 2009;57:4328–4333.

Won SR, Lee DC, Ko SH, Kim JW, Rhee HI. Honey major protein characterization and its application to adulteration detection. *Food Res Int.* 2008;41:952–956.

Wu X, Schauss AG. Mitigation of inflammation with foods. *J Agric Food Chem.* 2012;60:6703–6717.

Xu Y, Wang BC, Zhu YX. Identification of proteins expressed at extremely low level in Arabidopsis leaves. *Biochem Biophys Res Commun.* 2007;358:808–812.

Yadav AK, Bhardwaj G, Basak T, et al. A systematic analysis of eluted fraction of plasma post immunoaffinity depletion: implications in biomarker discovery. *PLoS ONE.* 2011;6:e24442.

Yu J, Hu S, Wang J, et al. A draft sequence of the rice genome (*Oryza sativa L. ssp. Indica*). *Science.* 2002;296:79–92.

You S-J, Udenigwe CC, Aluko RE, Wu J. Multifunctional peptides from egg white lysozyme. *Food Res Int.* 2010;43:848–855.

Zimecki M, Kruzel ML. Milk-derived proteins and peptides of potential therapeutic and nutritive value. *J Exp Ther Oncol.* 2007;6:89–106.

Zolotarjova N, Martosella J, Nicol G, Bailey J, Boyes BE, Barrett WC. Reversed-phase high-performance liquid chromatographic prefractionation of immunodepleted human serum proteins to enhance mass spectrometry identification of lower-abundance proteins. *Proteomics.* 2005;5:3304–3313.

CHAPTER 6

Biomedical Involvements of Low-Abundance Proteins

CHAPTER OUTLINE

6.1   INTRODUCTION

This chapter, together with Chapter 4, represents one of the two pillars of the book, since biomedical applications of proteomics are by far the field of most intense research and the one implying the largest majority of labs involved in proteomics. One of the major reasons for that is the fact that most labs, already at the beginning of the new millennium, got deeply involved in biomarker discovery in the hope of finding novel panels of biomarkers able to substitute the old ones currently in use in clinical chemistry labs. All of the old markers were approved during the last century, and none of them display the much needed 100% sensitivity and specificity. Such novel biomarkers, if found and approved for robustness, could represent a quantum jump in human health care, especially in regard to early biomarkers alerting the onset of a given pathology, thus permitting early medical intervention. Unfortunately, as described in the present chapter, the field has not lived up to expectations, and, already starting in the year 2010, it has been under "friendly" fire from highly visible journals. The editorials of these journals have heavily criticized what had been done in a decade and the much too modest results achieved. This criticism has been helpful and has forced the proteomics community to redirect its efforts and devise proper research protocols that should lead to correct biomarker discovery. Major emphasis is of course placed on the proper application of the CPLL methodology. Unlike immune-subtraction, the CPLL methodology appears to be about the only (or major) method that allows access to low- and very low-abundance proteins, those present in sera well below 1 ng/mL, probably representing the treasure chest where new relevant biomarkers might be found. We hope our readers will find here the clues to solving their research problems.

6.2 BIOMARKERS: SOME DEFINITIONS AND GENERAL ASPECTS

Since the field of biomarker discovery has massively colonized proteomics research in the past ten years, it would be appropriate to start this chapter with some definitions and general aspects pertaining to this field. According to the National Cancer Institute (NCI), a biomarker is "a biological molecule found in blood, other body fluids or tissues that is a sign of normal or abnormal processes, or of a condition or disease." Although the NCI's emphasis was likely on cancer, this definition, of course, applies to any pathology. A biomarker could be a molecule normally present in normal and pathological populations, but with strongly altered levels in affected patients, or it could be a molecule newly expressed in pathological tissues and body fluids and thus not found in normal ones. (Paradoxically, it could also be a molecule that disappears in the affected patients, but it would be very difficult to quantify something that is simply not there.)

These alterations could be due to several factors, such as germ-line or somatic mutations, transcriptional changes, and/or post-translational modifications. There is a large variety of biomarkers, not just those restricted to proteins (including enzymes, receptors, antibodies, peptides, as the proteomics field is accustomed to) but including nucleic acids (such as micro-RNA or other noncoding RNAs) and also metabolites. Thus the field of metabolomics is now actively being investigated in order to find panels or differentially expressed metabolites that could well be exploited as biomarkers of pathologies. We can distinguish different classes of biomarkers, based on their function (Henry and Hayes, 2012; Falasca, 2012). A brief list would encompass the following:

- Screening and predisposition biomarkers: those used for screening the general population or individuals at risk
- Diagnostic biomarkers: those for detection of the presence of a particular disease
- Prognostic biomarkers: those used to monitor the progression of a disease and predict its outcome
- Predictive biomarkers: those that could help to show whether a patient will benefit from a specific drug treatment
- Pharmacodynamic biomarkers: an interesting class comprising those biomarkers that could be used to evaluate a given drug efficacy and thus optimize a patient's treatments

Where do we find biomarkers? They are found in tissue biopsies (when available) and, as a much better option, in any biological fluid—either those circulating, such as serum/plasma, follicular fluid, saliva, cerebrospinal fluid (CSF), articular fluid, pleural fluid, or excretory (discharge) mediums, such as urine and bile. Biomarkers should meet several criteria, in order to enable unbiased diagnosis: (i) they have to be highly specific to a given disease (i.e., low false positives) and highly sensitive (i.e., low false negatives); (ii) their assays should be easily performed and standardized by trained health care professionals; and (iii) the readability of their results should be transparent and clear for clinicians. In body fluids, a class of biomarkers of particular interest is constituted by tissue leakage proteins, that is, proteins that are liberated in fluids by diseased tissues differently from healthy tissues. This is how such biomarkers originated in the past, such as liver transaminases (Amacher, 1998), whose increased levels in sera signal destruction of liver cells, or the prostate-specific antigen (PSA; Wang et al., 1981) for prostate cancer or the troponin I (Antman et al., 1996) and troponin T (Mair et al., 1992) for acute myocardial infarction. In today's proteomic world, though, it is widely accepted that, rather than a single biomarker (whose sensitivity and specificity is rarely better than 65–70%), a panel of biomarkers (comprising either proteins or peptides, or both) should be defined, since its clinical value would be much higher, possibly reaching and even exceeding 90% in sensitivity and specificity (Borrebaeck and Wingren, 2009; Rodriguez et al., 2010; Boja et al., 2010).

Some of the listed biological fluids might represent distinct advantages over others, especially those bathing a particular organ or anatomic structure, such as cerebrospinal fluid (CSF), articular or pleural fluids, despite their relatively high complexity similar to plasma. Due to their organ proximity and smaller dilution effects, there could be an increased probability of finding biomarkers, though for only a limited spectrum of diseases. The disadvantage of such fluids would be that they have to be collected in a somewhat invasive manner that is not so different from a needle tissue biopsy.

Yet, some success stories can be attributed to proteomic analysis of such fluids, notably CSF. In this last case, the 14–3–3 proteins have been taken as markers of brain damage and neurodegenerative disorders, including Creutzfeldt-Jakob disease (Hsich et al., 1996), especially when coupled with elevated Tau protein levels (>1300 pg/mL; Zanusso et al., 2011). Such a success with CSF could be attributed to the fact that it collects proteins mostly form one organ, the central nervous system and the spinal cord. Additionally, the mass ratio between this organ and the fluid volume is much more favorable in CSF than in plasma. Yet, one notable disadvantage is the fact that CSF contains much fewer proteins than plasma, typically 1/200, so that the low-abundance species are substantially more dilute than in sera.

The category of excretory body fluids, such as urine and bile, could also represent interesting media for biomarker discovery. Urine especially is rather easy to collect and available in large volumes. Yet, these fluids do not have a homogeneous composition. The protein concentration in urine can change from milligrams to grams, depending on the disease state; it can also change in composition during the day, hence impacting the analytical results. In addition, both fluids contain various (usually high) concentrations of excreted compounds, such as salts, metabolites, complex organic compounds, urea, or solubilizing agents. Thus their analysis by differential, comparative proteomics might be quite difficult. For details on normalization of samples before analysis, see Sections 8.3. and 8.3.2 for urine samples.

6.3 BIOMARKER DISCOVERY: A GLOOMY LANDSCAPE

In the decade 2001–2010 an avalanche of reports appeared claiming biomarker discovery of just about any possible disease. It was a quest for the Holy Grail, pursued by so many labs around the world that it would be impossible to list all of them with their respective claims. Curiously, all proteomics journals accepted such papers without much questioning. In fact, it almost seemed that they welcomed any such study. It is also quite peculiar that the alarm bell was rung not by the scientific community at large, but by free-lance journalists. The strong wind that blew away the fog that covered the entire field of biomarker discovery like a mantle came from Mitchell (2010). In an editorial published in *Nature Biotechnology* he clearly stated that "biomarkers have been the biggest disappointment of the decade; they turned out to be so non-specific as to be next to useless, far from the Holy Grail envisaged….Biomarkers discovered so far are mainly the same proteins that pop up in all kinds of diseases, indicating that the organism is under some kind of stress (p. 665)." He added, ">1250 presumptive biomarkers have been listed so far in the literature, but no one has yet reached clinical trials or been approved. This billion-dollar biomarker fiasco is also largely due to lack of validation (p. 665)."

This challenge was soon picked up and amplified by other authors (Diamandis, 2010; Veenstra, 2011; Issaq et al., 2011). According to Poste (2011), the panorama is gloomier: "by using proteomics and DNA microarrays >150000 papers claiming several thousands biomarkers have been published, but just a few have been validated for routine clinical practice." There is even more, as shown in another editorial by Buchen (2011) on the gloomy story of Gil Mor and his panel of six ovarian cancer biomarkers (OvaSure test) that turned out to have a detection capability of only 34% in women who were then diagnosed with the disease within one year! Worse than the old CA-125 test, which remained the "best of a bad lot," this latter test had a detection ability of 63%.

In reality, Chechlinska et al. (2010) and Kowalewska et al. (2010) also rang the bell on the field, clearly stating that systemic inflammation could be a confounding factor in cancer biomarker discovery and validation. In the words of Chechlinska et al.: "Cancer and inflammation are inextricably linked and cancer patients have local and systemic changes in inflammatory parameters. However, this crucial aspect of tumour biology is often overlooked in biomarker studies and needs to be urgently addressed." Kowalewska et al. (2010) added: "We analyze several modern strategies of tumour marker discovery, namely, proteomics, metabolomics, studies on circulating tumour cells and circulating free nucleic acids, or their methylation degree, and provide examples of scarce, methodologically correct biomarker studies as opposed to numerous methodologically flawed biomarker studies, that examine cancer patients' samples against those of healthy, inflammation-free persons and present many inflammation-related biomarker alterations in cancer patients as cancer-specific. Inflammation as a cancer-associated condition should always be considered in cancer biomarker studies, and biomarkers should be validated against their expression in inflammatory conditions."

More recently, Albrethsen et al. (2012) reached similar conclusions when they critically assessed candidate biomarkers for human colorectal cancer. This bleak situation for plasma-derived biomarker discovery had indeed already been denounced in 2007 by Lescuyer et al. when they stated that "most studies involving the comparison of plasmas obtained from healthy donors with plasma obtained from cancer-diagnosed donors have shown essentially inflammation-related proteins (e.g., haptoglobins or serum amyloid protein) as differential markers, which shows in turn that these markers are indeed not disease-specific" (Ahmed et al., 2004; Le et al., 2005).

What caused this near-disaster situation? One could list many reasons, but one of the most obvious reasons is the tremendous dynamic range of protein levels in plasma, ranging from below the ng/mL to tens of mg/mL, which is 12 orders of magnitude (from millimolar concentration of albumin to femtomolar concentration of tumor necrosis factor, or TNF). As stated, 22 plasma proteins constitute 99% of the protein mass. It is therefore tempting to try to deplete these proteins to enhance the visibility of minor proteins and thus to bring them above the threshold of detection by proteomics.

Immunodepletion methods are described in Chapter 3 along with their difficulties to improve the detectability of the low-abundance species. There are quite a few reasons for that failure. First, because of the enormous dynamic range of plasma proteins, the removal process must be close to 100% efficient to be useful. Even a perfect process removing the 22 most abundant proteins (which is far from being achieved) will leave 8-10 orders of magnitude of protein levels in plasma, which also exceeds by far the current power of proteomics technologies that covers 3 to 5 orders of magnitude. In addition, less abundant interesting proteins may also be eliminated during the depletion process, resulting in loss of reliable information and potential artefactual discrepancies among samples. Albumin alone represents about two-thirds of the entire protein content of plasma. In the top three decades of protein concentration (between 60 mg/mL and 60 µg/mL) likely fewer than 100 different genes are represented. Moreover, it should be understood that tissue-derived proteins (i.e., the clinically used biomarkers) get highly diluted in systemic circulation (i.e., in plasma/sera) to a concentration range of ng/mL and below, which only represents marginal levels of the total plasma protein content. Therefore, to capture relevant targets as biomarkers, the technology must be sensitive enough to reliably identify and accurately quantify the low-abundance species in a highly complex plasma background.

Another important cause of failure has been the fact that most research groups plunged in the deep ocean of plasma, fishing for biomarkers, "blindfolded," that is, without any clue as to what they were looking for. In the vast majority of cases, differential 2D maps were exploited in comparing normal versus pathological sera; any modulated protein would then be listed as a "potential biomarker". Let's face it: Even in the old books of pirate stories, prior to the treasure hunt, there was a hunt for the map to the treasure. Thinking of finding the "biomarker cove" in this vast plasma ocean, where several hundred thousands proteins might be swimming around, is merely wishful thinking. It would hardly materialize in real life.

Prior to closing this section, some more comments are in order. Notwithstanding the reports on the poor performance of immunodepletion (Tu et al., 2010; Zhi et al., 2010), papers staunchly defending it regularly appear (which is quite legitimate, of course) (Roche et al., 2009; Juhasz et al., 2011; Faca et al., 2007). Since the merits and limits of depletion versus enrichment protocols have been amply discussed in Chapters 3 and 4, we will not deal with such topics any longer here. However, we would like to discuss a couple of papers that have reported alternative strategies, both of them related to plasma proteomics, to see if these last ones represent an improvement in the biomarker discovery scenario.

One scenario has been proposed by Kovács et al. (2011). These authors reported the use of combined chromatographic and precipitation techniques for generating a large set of fractions representing the human plasma proteome, referred to as the Analyte Library, with the goal of using the relevant library fractions for antigen identification in conjunction with monoclonal antibody proteomics. Starting from 500 mL normal pooled human plasma, this process generated 783 fractions with an average protein concentration of 1 mg/mL. First, the serum albumin and immunoglobulins were depleted followed by prefractionation of ammonium sulphate precipitation steps. Each precipitate was then separated by size-exclusion, cation and anion exchange chromatographic steps in sequence. The 20 most concentrated ion-exchange chromatography fractions were further separated by hydrophobic interaction chromatography. All chromatography and precipitation steps were carefully designed to maintain the native forms of the intact proteins throughout the fractionation process. In order to follow the efficiency of the process, vitamin D-binding protein was monitored in all major fractionation levels. This last protein is a medium-abundance species in sera; thus, being able to detect just that throughout this very complex fractionation scheme does not look very promising for biomarker discovery. We suspect that, as a consequence of the use of so many adsorption columns, all of them manipulated under native conditions, massive losses of low-abundance species, possibly irreversibly adsorbed to the various resins, might have occurred. Unfortunately, the authors did not examine what remained bound to the various resins via highly performing eluants, such as the boiling SDS-DTT protocol suggested in recent reports (Di Girolamo et al., 2011a,b,c; Di Girolamo and Righetti, 2011), as detailed in Section 8.15. They might have found there all the species missing in action!

The second one was proposed by Hagiwara et al. (2011). These authors sequentially fractionated plasma proteins by using depletion columns for albumin (Cibacron Blue resin) and immunoglobulin (HiTrap Protein G HP column), further separated with an anion-exchange (Resource Q) column. Proteins in each fraction were treated with a solid-phase hexapeptide ligand library (CPLL) and compared to those without treatment. Two-dimensional difference-gel electrophoresis demonstrated an increased number of protein spots in the treated samples. The authors concluded that "the use of CPLL in combination with conventional proteomic modalities such as depletion and anion-exchange columns significantly enhanced trace proteins on SDS-PAGE and increased the number of protein spots on 2D-DIGE, suggesting that the use of the peptide library methodology has great potential for intact plasma proteomics." This statement was immediately denied by the next one, stating: "mass spectrometric protein identification revealed that high- and middle-abundance proteins were enriched by ProteoMiner." The aims and mission of CPLLs are exactly the opposite, however: that is, not to enrich high- and middle- abundance species (who could possibly care about that?), but to enhance the visibility of the low- to very low-abundance species, as demonstrated in all papers in which the technology was properly used. Mission accomplished? Not at all! So, what went wrong? Just about everything: to start with, 5 mL of Cibacron Blue resin were not enough to properly remove albumin, so the resin was oversaturated and had to be loaded and discharged in at least ten cycles for processing the 30 mL of sera. It is known that such a resin adsorbs plenty of low-abundance proteins (Di Girolamo and Righetti, 2011), which are not at all released by the standard elution buffers (50 mM KHPO$_4$, 1.5 M KCl, pH 7.0, a buffer which, if anything, would further favor hydrophobic interaction of such low-abundance species) but only by 2–3% boiling SDS in the presence of 30 mM DTT.

So, the repeated binding/elution cycles might have massively depleted a rather large population of low-abundance species, thus totally subtracting them to the final 2D-DIGE analysis, which in fact revealed only high- to medium-abundance proteins. Were this not enough, the proteins bound to the CPLL beads were eluted only with 8 M urea, 2% CHAPS, and 5% acetic acid, known to release at best 75% of the bound species, and possibly not at all those low-abundance ones that were bound to high-affinity sites. These two examples show that alternative routes, adopting very complex pre-fractionation schemes, might not be valid alternatives to a simple but well-conducted straightforward CPLL sample treatment.

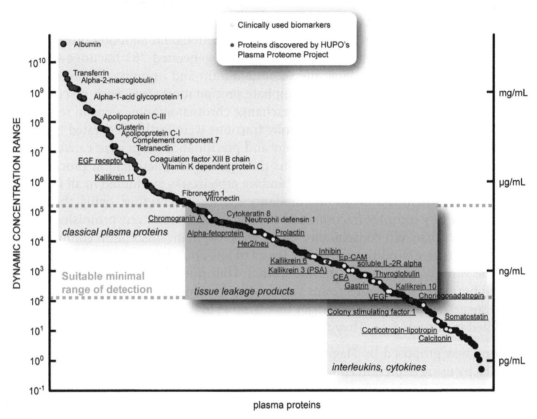

FIGURE 6.1

Plot of the dynamic concentration range of plasma proteins, spanning 11 orders of magnitude. Yellow: clinically used biomarkers; red: proteins discovered by the HUPO Plasma Protein Project. *(From Surinova et al., 2011, by permission).*

We can draw some major conclusions at this point. First, that things did not go so well was also underlined in a recent review by Surinova et al. (2011) and shown in an illuminating figure (Figure 6.1). In this graph, which spans eleven orders of magnitude in dynamic range of plasma protein concentrations, it can be appreciated that the suitable minimal range of detection for clinically useful biomarkers should span a range of 100 ng/mL to at the very least 100 pg/mL. Curiously, though, very few of these biomarkers have been discovered via the HUPO Plasma Proteome Project, and only one (choriogonadotropin) has been found at a level of 100 pg/mL. All the other ones discovered, down to as low as 10 pg/mL, had been discovered during the last century, starting from the 1960s, via immunological assays but without the inmost powerful tool of mass spectrometry available today.

Second, can we change this gloomy landscape into a brighter one? We hope we will be able to show this dramatic change in the following sections, where we will describe what has been accomplished in this field through the CPLL technology. Meanwhile, in anticipation of the incoming attractions, Figure 6.2 shows that the panorama changes dramatically when the use of peptide libraries is

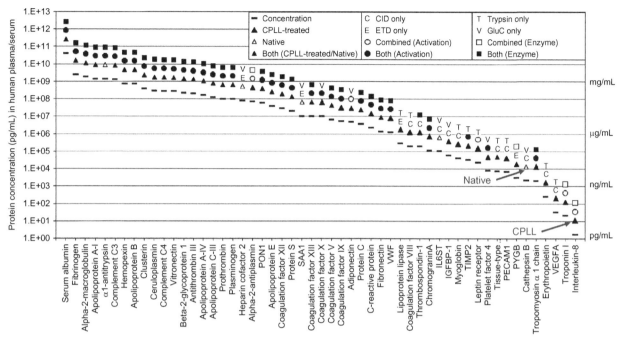

FIGURE 6.2

Plot of 50y swine plasma proteins (spanning 12 orders of magnitude) as detected in native (untreated plasma, open triangles) and after CPLL treatment (black triangles). Note how, in the first case, the lower limit of detection is 10 ng/mL whereas after CPLL capture the lowest sensitivity limit is 10 pg/mL. (Abbreviations: CID: collision-induced dissociation; ETD: electron-transfer dissociation) *(Adapted from Tu et al., 2011).*

adopted as an "enhancing" tool for the low- to very low-abundance proteome (Tu et al., 2011). When using plain, unfractionated plasma, the lowest detection limit appears to be 10 ng/mL. However, if plasma is pretreated with CPLL, species can be detected down to as low as 10 pg protein per mL, with an increment of sensitivity of 3 orders of magnitude. It is in this very low-abundance protein region that biomarkers will likely be hidden, especially when looking for markers of the onset of a disease. The recognition of the merits of combinatorial peptide ligand libraries has known a slow start due to the hammering promotion of immunodepletion methods, claimed as the best way to allow detection of low-abundance proteins. However, in the last two to three years the CPLL method also started to take-off within the controversial domain of the discovery of gene-expression differences when biological disorders take place (see dedicated bibliographic list at the end of this book).

While the enrichment method was progressively improved, its application was extended to many biological extracts and for a variety of diseases. In this domain, a statistical overview on published papers using CPLL is summarized in Figure 6.3, where it is observed that the preferred target is cancer followed by vessel-related pathologies and then infections. Nevertheless, many papers deal with other domains such as pregnancy-related pathologies, diabetes, obesity, and brain disorders. A quite complete review of the detection of biomarkers with samples pretreated with combinatorial hexapeptide libraries was published in 2012 with numerous examples and most powerful technological association possibilities (Boschetti et al., 2012).

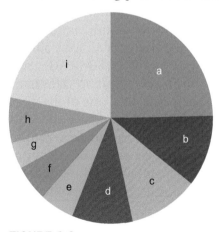

FIGURE 6.3

Statistical review of CPLL-based publications dealing with biomarker discovery studies. a: cancer (various types of cancer); b: vessel and blood pathologies; c: infections; d: pregnancy related pathologies; e: diabetes; f: obesity; g: brain disorders; h: various; i: general reviews.

6.4 PROTEOMICS OF HUMAN BODY FLUIDS

In this section we will review recent advances in the proteomics of body fluids, with particular emphasis on those where CPLLs have been applied for deeper exploration and coverage. An extensive review on this topic was published by Hu et al. (2006) listing 465 references, probably covering just about all that had been published in this field.

6.4.1 Bile Proteomics

Bile is a body fluid produced by the liver and drained by biliary ducts into the duodenum. It has two major functions: first, it contains bile acids, which are critical for the digestion of fats, and second, it is an excretory pathway for many endogenous and exogenous compounds. Proteomic analysis of bile is particularly difficult since this fluid contains high concentrations of various substances that strongly interfere with protein separation and identification techniques. Furthermore, owing to its deep location in the body, bile must be collected by surgical or endoscopic procedures. However, as was speculated for other body fluids, bile appears to be a promising sample for the discovery of disease biomarkers leaking from proximal tissues: the liver, pancreas, or biliary tree.

One of the first major studies on this body fluid was published by Kristiansen et al. in 2004. It involved fractionation of bile by one-dimensional gel electrophoresis and lectin affinity chromatography followed by liquid chromatography tandem mass spectrometry. Overall, they identified 87 unique proteins, including several novel proteins, as well as known proteins whose functions are unknown. By using lectin affinity chromatography and enzymatic labeling of asparagine residues carrying glycan moieties by ^{18}O, they also identified a total of 33 glycosylation sites. A quantum jump came in 2007 when Guerrier et al. treated the bile fluid with CPLLs, in order to concentrate dilute and very dilute species while concomitantly diluting the high-abundance proteins. Analysis of the resulting protein mixture was then performed using LC-MS/MS after having classically separated proteins by a mini-preparative gel electrophoresis. Overall 222 gene products were found, of which 143 were not reported before in proteomics studies. This was an increment of about 300% over the previous study of Kristiansen et al. (2004).

The validity of this approach was recently reviewed by Farina et al. (2010), who highlighted studies that identified bile potential biomarkers for two deadly and difficult to diagnose neoplasms, namely, pancreatic cancer and cholangiocarcinoma. The potential of bile as a source of biomarkers was also underlined by Nagana Gowda (2010), who, however, reported data on metabolomic, rather than proteomic, biomarkers. The latest quantum jump came in 2011, when Barbhuiya et al. were able to identify no less than 2552 proteins in bile fluid, thanks to a multipronged approach (SDS-PAGE, SCX chromatography and Off-Gel fractionation) followed by MS analysis on an LTQ-Orbitrap Velos mass spectrometer using high resolution at both MS and MS/MS levels. This is definitely one of the most extensive catalogs of any body fluid proteome in a single study. These authors also underline the fact that such a catalog should serve as a baseline for future studies aimed at discovering biomarkers from bile in gallbladder, hepatic, biliary cancers, and in general in hepatobiliary disorders. Although this field seems to have made substantial progress, nevertheless one should remember that the collection of bile is not easily achievable and in addition such fluid contains many salts and metabolites that interfere with proteomic analyses.

6.4.2 Saliva Proteomics

Among human body fluids with health diagnostic/monitoring potential, whole saliva has been gaining increasing attention in the last few years. Saliva is produced by the parotid, submandibular and sublingual glands into the oral cavity. It maintains oral health, protects from foreign microorganisms, and facilitates chewing/swallowing of food. Interestingly, molecular constituents in plasma (the current gold standard in biomarker studies) pass into whole saliva, resulting in significant overlap between proteins found in both fluids. Thus, in addition to containing biomarkers of oral health, the salivary proteome holds the potential for predicting systemic diseases such as breast cancer, diabetes, and autoimmune disorders. These features, combined with the fact that whole saliva is easily collected in a noninvasive manner, make it an attractive fluid for biomarker discovery.

Although not as severe as in plasma, proteomic characterization of saliva is also complicated by its broad dynamic range of protein abundance. Coomassie staining identifies approximately ten proteins in whole saliva, which account for nearly 98% of total salivary protein. These high-abundance proteins obscure the detection of proteins present at much lower concentrations. The barrier was broken in 2005 when Xie et al., using a newly developed approach coupling peptide separation in free-flow electrophoresis with linear ion trap tandem mass spectrometry were able to identify 437 proteins with high confidence (false positive rate below 1%), producing the largest catalog of proteins from a single saliva sample and providing new information on the composition and potential diagnostic utility of this fluid. Based on this and other studies, indeed Hu et al. (2007), when reviewing the field, stressed the importance of saliva for biomarker discovery, in particular for oral cancer and Sjögren's syndrome. Soon papers on the discovery of potential biomarkers began to appear. Thus Dowling et al. (2008) when screening the saliva from patients with head and neck squamous cell carcinoma reported a set of potential biomarkers, namely, beta fibrin, S100 calcium binding protein, transferrin, immunoglobulin heavy chain constant region gamma, and cofilin-1, which were found to be significantly increased, and transthyretin, which was significantly decreased.

By the same token, Hu et al. (2008), when profiling whole saliva via C4 reversed-phase liquid chromatography for prefractionation and capillary reversed-phase liquid chromatography with quadruple time-of-flight mass spectrometry, in search for biomarkers for oral squamous cell carcinoma, reported a set of five proteins, namely, M2BP, MRP14, CD59, profilin, and catalase, which were stated to offer a sensitivity of 90% and a specificity of 83% in detecting this pathology. In another investigation, Rao et al. (2009) analyzed the human salivary proteome in search for markers of type 2 diabetes. Although they found no less than 65 proteins (out of a total of 487 identified) modulated in this pathology, they focused on a set of five candidates, whose modulation was further assessed via ELISA, found to be: alpha-2 macroglobulin, alpha-1-antitrypsin, cystatin C, transthyretin and salivary alpha-amylase.

In turn, Cabras et al. (2010) profiled the salivary secretory peptidome in children affected by type 1 diabetes in search for potential biomarkers. They found that statherin, proline-rich peptide P-B, P-C peptide, and histatins, were significantly less concentrated in saliva of diabetic subjects than in controls, while the concentration of α-defensins 1, 2, and 4 and S100A9* was higher. This same group of authors (Castagnola et al., 2011, 2012) also published a critical overview of the results obtained by different proteomic platforms adopted in exploring the salivary proteome. As a further application, De Jong et al. (2010) explored the saliva proteome in order to reliably distinguish premalignant oral lesions from those already transitioned to malignancy. They first prioritized such candidate biomarkers via bioinformatics and validated the selected proteins by western blotting. The end results were an increased abundance of myosin and actin in patients with malignant lesions. The sensitivity/specificity values for distinguishing between the two types of lesions were 100%/75% for actin and 67%/83% for myosin in soluble saliva.

As stated above, there is a non-negligible overlap between proteins found in saliva and plasma. This study came from Yan et al. (2009) who compiled a catalog of the human ductal salivary proteome comprising a grand total of 1166 proteins. This list was compared with a core dataset of the human plasma proteome with 3020 protein identifications, and it resulted in a total of 597 proteins common to the two biological fluids. Gene ontology analysis showed similarities in the distributions of the saliva and plasma proteomes with regard to cellular localization, biological processes, and molecular function, but also revealed differences that may be related to the different physiological functions of saliva and plasma. This list of 1166 proteins in saliva came from a compilation of all data published by several labs. No single study, prior to this one, had reported more than 400 to 500 species in whole saliva.

A major breakthrough came when the CPLL methodology was applied to explore in depth the salivary proteome. Bandhakavi et al. (2009) treated up to 75 mg of saliva proteins with CPLLs and maximized the gains of the CPLL sample handling via a three-dimensional peptide fractionation involving sequential steps of preparative isoelectric focusing, strong cation exchange, and capillary reversed-phase liquid chromatography. The results were quite unique, since the total species identified amounted to 2340 proteins in whole saliva, that is, 400 to 500% more than any other individual discovery up to their report. In a more recent report, Bandhakavi et al. (2011) combined CPLL treatment of saliva with covalent glycopeptide enrichment and MS/MS. By this technology, they identified two times more N-linked glycoproteins and their glycosylation sites than in the untreated sample, thus dramatically increasing the known salivary glycoprotein catalog. Additionally, they compared differentially stable isotope-labeled saliva samples pooled from healthy and metastatic breast cancer women using a multidimensional peptide fractionation-based workflow, analyzing in parallel one sample portion CPLL treated and one portion untreated. Their workflow categorized proteins with higher absolute abundance, whose relative abundance ratios were altered by DRC (dynamic range compression), from proteins of lower absolute abundance detected only after CPLLs. Within each of these salivary protein categories, they identified novel abundance changes putatively associated with breast cancer, demonstrating feasibility and benefits of CPLLs for relative abundance profiling. Collectively, their results showed once more the unique potential of CPLLs for proteomic studies and biomarker discovery. Additional hints on the salivary proteome, with emphasis on the major challenges in protein identification and quantitation, have been offered by Siqueira and Dawes (2011).

6.4.3 Cerebrospinal Fluid Proteomics

Cerebrospinal fluid (CSF) is a colorless body fluid surrounding the brain and the spinal chord in vertebrates. Its main functions are mechanical protection of the brain and transport of metabolically active substances and waste products. Owing to its close contact with the brain, CSF is often investigated when examining disorders related to the central nervous system (CNS); its analysis, for diagnostic purposes, has been performed for over 100 years. This body fluid is secreted at a rate of 0.35 mL/min from the choroid plexus in the brain. The total volume of CSF in adults is around 100 to 150 mL, which means that the fluid is replaced three to four times every day. Most of the chemical compounds found in CSF originate from the blood. The levels of electrolytes, such as Na^+ and Cl^-, are about the same in both body fluids. Water, gases, and lipid soluble compounds move freely from the blood into CSF, whereas glucose, amino acids, and cations are transported by carrier-mediated processes. The protein content in CSF is typically 350 mg/L, which is about 200 times less than in blood. CSF would be an excellent fluid for finding biomarkers of CNS pathologies; thus the knowledge of its proteome would be highly beneficial to clinicians working with neurological disorders.

In one of the major efforts to explore the CSF proteome, Maccarrone et al. (2004) reported a list of 115 unique gene products, as revealed after immunodepletion and shotgun mass spectrometry. This number was substantially incremented when Noben et al. (2006) identified 146 species, while Burgess et al. (2006) listed 299 proteins, but in postmortem CSF, in which it could not be excluded that many species had been released from damaged cells. This number was quite close to the >300 unique proteins in CSF described by Zhang et al. (2005). Waller et al. (2008) reported 156 proteins in CSF, mined by Off-Gel electrophoresis as a first-dimension separation of peptide mixtures from trypsin digests from 2.5 mL of CSF. A further discovery came in the same year, when Shores et al. (2008) listed a total of 200 unique proteins in CSF, found by exploiting a novel technique, a hexameric peptoid diversity library used as baits for capturing proteins.

Apparently, hexameric peptoid could not have the upper hand over hexameric peptides such as those forming CPLLs because a major breakthrough on CSF proteomics came in 2010, when Mouton-Barbosa et al. reported for the first time, on a large pool of CSF from different sources, no less than 1212 unique proteins, of which 745 were only detected after peptide library treatment. However, since for clinical studies small volumes are typically obtained after lumbar punctures, precluding the conventional use of CPLLs with large volume columns, the method was further optimized to be compatible with low-volume samples (as little as 5 μL beads loaded with 1 mL CSF). They could show that the treatment was still efficient with this miniaturized protocol and that the dynamic range of protein concentration was actually reduced even with such small amounts of beads, leading to an increase of more than 100% of the number of identified proteins in one LC-MS/MS run. Moreover, using a dedicated bioinformatics analytical workflow, they found that the method was reproducible and applicable for label-free quantification of series of samples processed in parallel.

Figure 6.4 shows 2D PAGE maps of control CSF (A) and one of the six eluates from the CPLL beads (B): It can be appreciated that this second gel is carpeted with spots, in comparison with the paucity of spots in the control gel, which shows the usual high-abundance proteins in CSF, mostly albumin and IgGs. A second quantum jump occurred in the same year, when Schutzer et al. (2010), by immunoaffinity separation and high-sensitivity and resolution liquid chromatography-mass spectrometry identified 2630 proteins in CSF from normal subjects, of which 56% were CSF-specific, not found in the much larger set of 3654 proteins known in plasma. These two sets of data now provide the basis for proper biomarker discovery in CSF, as already anticipated by Romeo et al. (2005).

There have been several published data of such putative biomarkers in CSF, of which we will report here some of the main findings. Zanusso et al. (2011) reported that, if the surrogate marker such as the 14–3–3 protein, which is found in different CNS pathologies, is coupled with elevated Tau protein levels (>1300 pg/mL), this combined team reaches a specificity of around 100% for sporadic Creutzfeldt–Jakob disease diagnosis. By the same token, Sjodin et al. (2010) by utilizing CPLLs, described a panel of potential markers in CSF of traumatic brain injury, comprising neuron

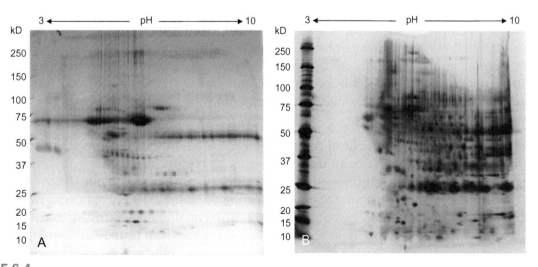

FIGURE 6.4

Two-dimensional maps of cerebrospinal fluid. A: control, untreated CSF; B: represents one of the six eluates (with 7 M urea, 2 M thiourea, 4% CHAPS) from two peptide library treatments. Other eluates—not shown—(three from each peptide library) were equally largely populated of new proteins throughout the entire pH and mass dimensions (see Mouton-Barbosa et al., 2010). Note how the control map is heavily dominated by albumin, transferrin, and heavy and light chains of immunoglobulins. Silver staining.

specific enolase, glial fibrillary acidic protein, myelin basic protein, creatine kinase B-type, and S-100-beta. Martins-de-Souza et al. (2010a, b) reported on a set of potential biomarkers for schizophrenia, comprising apolipoprotein E, apolipoprotein A1 and prostaglandin-H2 D-isomerase (upregulated) and transthyretin, TGF-beta receptor type-1, and coiled-coil domain-containing protein 3 precursor (downregulated). These findings may help to elucidate the disease mechanisms and confirmed the hypothesis of disturbed cholesterol and phospholipid metabolism in schizophrenia.

Another important brain pathology is Alzheimer's disease (AD), of which several familial forms exist, most of them being caused by mutations in the genes that encode the presenilin enzymes involved in the production of amyloid-β (Aβ) from the amyloid precursor protein (APP). In AD, Aβ forms fibrils that are deposited in the brain as plaques. Much of the fibrillar Aβ found in the plaques consists of the 42 amino acid form of Aβ (Aβ1-42), and it is now widely accepted that Aβ is related to the pathogenesis of AD and that Aβ may both impair memory and be neurotoxic. Portelius et al. (2012a,b) demonstrated that carriers of the familial AD (FAD)-associated PSEN1 A431E mutation have low CSF levels of C-terminally truncated Aβ isoforms shorter than Aβ1-40. The same was also observed in symptomatic carriers of the FAD-causing PSEN1 L286P mutation. Furthermore, they showed that preclinical carriers of the PSEN1 M139T mutation overexpressed Aβ1-42, suggesting that this particular mutation may cause AD by stimulating γ-secretase-mediated cleavage at amino acid 42 in the Aβ sequence. In addition, Mattsson et al. (2012) reported that measurements of amyloid-β42 (Aβ42), total-tau (T-tau), and phosphorylated tau (P-tau), as well as the soluble amyloid-β protein precursor protein fragments sAβPPα and sAβPPβ and chromogranin B, may be used to predict future Alzheimer's disease (AD) dementia in patients with mild cognitive impairment. Such novel amyloid precursor protein fragments in the CSF of AD patients had in fact been previously reported by Portelius et al. (2010). These few, selected examples, by all means not exhaustive, show that indeed substantial progress has been made in biomarker analysis in CSF, although a lot still remains to be accomplished. Especially needed is validation via clinical analysis of large cohorts of patients coupled to robust statistical analysis and bioinformatic tools as also advocated, for instance, by Martins-de-Souza (2010).

6.4.4 Urine Proteomics

Compared with other body fluids, urine has several characteristics rendering it a preferred choice for biomarker discovery. First, urine can be obtained in large quantities by using noninvasive procedures. This allows repeated sampling of the same individual for disease surveillance. The availability of urine also allows easy assessment of reproducibility or improvement in sample preparation protocols. Second, urinary peptides and proteins are generally soluble. Third, in general, the urinary protein content is relatively stable, probably because urine "stagnates" for hours in the bladder; hence proteolytic degradation by endogenous proteases may be essentially complete by the time of collection.

Disregarding the work done in the last century, when just a handful of proteins in urines could be identified (e.g., in 1982 Edwards et al. described just 12 proteins, representing the major species in urines) due to lack of MS equipment, intense work on the urinary proteome began in the new millennium. This work, too, concentrated on establishing standard 2D electrophoresis maps against which pathological samples could be confronted via differential display in order to detect disease biomarkers. However, the total booty in terms of discovering the deep urinary proteome was not so exciting. For example, in 2002 Pang et al. described 2D gel maps in which, by excision of each individual spot, a grand total of 104 unique species could be classified. Even when disregarding 2D PAGE mapping and using direct LC-MS/MS analyses of tryptic digests, the catch was only slightly improved to 124 unique proteins (Spahr et al., 2001). Even when resorting to extensive sample treatment (ultrafiltration, size-exclusion chromatography, and immunoaffinity subtraction), although 1400 spots could be displayed in 2D maps, Pieper et al. (2004) could identify just 150 unique proteins. Also, Oh et al. (2004), when trying to establish a near-standard 2D map of urines, could finally identify just 113 proteins. Other labs did not fare much better. Thus Smith et al. (2005) in developing a high-throughput method for urine analysis, could barely list 48 nonredundant proteins and Zerefos et al. (2006) via preparative electrophoresis followed by 2D gel mapping reported just 141 species. This rather shallow horizon of discoveries was summarized in the same period by Thongboonkerd and Malasit (2005).

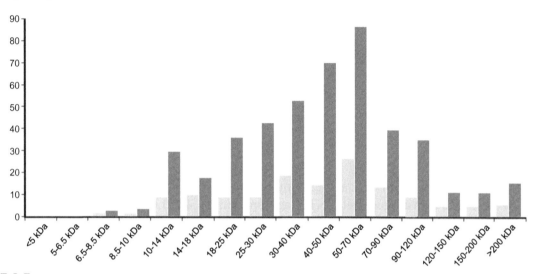

FIGURE 6.5
Number of urinary protein species detected in the two combined eluates (dark gray) from CPLLs as compared with control urines (light gray), as a function of their respective Mr values in the 5 kDa >200 kDa range. Note how, for the first time, proteins with an Mr value well above that of albumin (69 kDa) were detected in urines. *Adapted from Castagna et al. (2005).*

We can confidently state that a major breakthrough came in 2005, when Castagna et al., by means of a simple CPLL treatment, described in a single sweep no less than 471 unique gene products (versus 95 in the control, untreated sample). The trick? Massive sample overload, as is customary with the CPLL methodology: 1.6 liters of urine were concentrated to 45 mL (corresponding to 150 mg total protein) and processed with CPLLs (Figure 6.5). This figure gives an interesting view of the increment in species obtained after treatment with CPLL, while simultaneously showing the molecular mass distribution of identified gene products. The above-mentioned authors, by compiling their data with those already present in previous literature, arrived at a total discovery of about 800 proteins in urine. After this, the trend moved rapidly upward. In a review, Decramer et al. (2008) stated that at least 1500 proteins existed in urine, and at a recent count, this list has more than doubled (Zerefos et al., 2012; Candiano et al., 2012; Santucci et al., 2012). In addition to that, a standard urinary proteome 2D electrophoresis map, developed with the eluates of CPLL beads, now comprises more than 3000 spots (see Figure 6.6). The control (left panel) displays about 600 spots, among which the high-abundance proteins are clearly visible; conversely, the CPLL-treated sample exhibits >3000 spots.

Given this large body of knowledge on the urinary proteome, it is no surprise that several studies have been devoted to extract novel biomarkers for different pathologies from this body fluid, not just limited to kidney disease but to several other body districts (Kentsis et al., 2009; Kentsis, 2011; Zürbig et al., 2011). Here we list some examples of these applications. For instance Petri et al. (2009) tried to assess whether urine could be used to measure specific ovarian cancer proteomic profiles and whether one peak alone or in combination with other peaks or CA125 had the sensitivity and specificity to discriminate between ovarian cancer pelvic mass and benign pelvic mass. After treating urine with CPLLs and examining the various fractions obtained, they found three

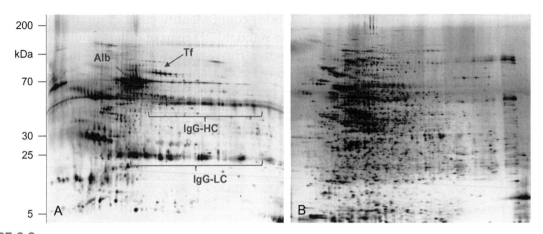

FIGURE 6.6

Two-dimensional electrophoresis pattern comparison of total urinary proteins before (A) and after (B) treatment with CPLL (all combined eluates). The nonlinear IPG gradient was between pH 3 and 10. Protein detection was with silver stain. Abbreviations: Alb, albumin; Tf, transferrin; IgG-HC: immuno-globulins G, heavy chains; IgG-LC, immunoglobulins G, light chains (modified from Candiano et al., 2012).

protein fragments—fibrinogen alpha fragment ($m/z = 2570.21$), collagen alpha 1 (III) fragment ($m/z = 2707.32$), and fibrinogen beta NT fragment ($m/z = 4425.09$)—that could be efficiently exploited (ROC-AUC value of 0.96) for this purpose.

Chen et al. (2011) examined the urine of women with preeclampsia, gestational hypertension, and normal pregnancy in order to screen potential biomarkers for the early diagnosis of preeclampsia and claimed that low urinary angiotensinogen levels could be proposed for screening for this pathology. Efforts have also been focused on finding biomarkers in diabetic nephropathy and nephrotic syndrome (Rao et al., 2007; Candiano et al., 2008; Khurana et al., 2006). To that aim, Fisher et al. (2011) have proposed an extensive fractionation scheme (yielding 700 fractions!) that is stated to greatly facilitate the discovery of potential biomarkers in urine. In closing this short survey of urinary proteomics, we want to mention, too, the field of exosomes (alias nanovesicles, exovesicles), which at present is the focus of intense research (Conde-Vancells and Falcon-Perez, 2012; van der Lubble et al., 2012; Musante et al., 2012; Alvarez et al., 2012; Ramirez-Alvarado et al., 2012; Pisitkun et al., 2012; Raj et al., 2012; Wang et al., 2012).

6.4.5 Serum/Plasma Proteomics

Serum/plasma proteomics has been (and still is, of course) the focus of intense research involving several hundred groups around the world. It is of such widespread interest because it should contain markers of any possible pathology, since it bathes all body tissues and thus receives proteins and peptides exuded from any organ. Its importance has been recognized since the very beginning of the new millennium, when, in 2001, the HUPO (Human Proteome Organization) was funded. Soon after, in April 2002 in Bethesda, Maryland, the PPP (Plasma Proteome Project) was launched, supported by 35 collaborating laboratories with the aim of setting up stringent standards on sera/plasma handling and storage and creating references specimens. The major results of these efforts were presented at the HUPO World Congress held in Munich in August 2005 and were published as a special issue of *Proteomics* (vol. 5, No. 13, 2005). Seventeen of these reports were then re-published in a book edited by Omenn (2006).

It is of interest here to recall some of the major steps taken for digging deeper into the low-abundance proteome, considering that serum could span up to 12 orders of magnitude in dynamic range and that the first 20 high-abundance proteins colonize as much as 99% of the territory, leaving little or no room for sampling the thousands or perhaps hundreds of thousands of low- to very low-abundance species. One major effort was undertaken by Rose et al. (2004), who devised an "industrial-scale proteomics approach" involving the handling of 2.5 liters of plasma, with the aim of subfractionating it to allow access to low-abundance proteins and peptides. This was a mammoth effort whose end result was the generation of no less than 12,690 fractions using chromatographic techniques. Yet, notwithstanding the generation of more than 1.5 million MS/MS spectra and the manual testing and validating of 33,000 of them, the overall discovery was not terribly exciting, since they could list from the various plasma pools a grand total of 700 proteins and polypeptides in plasma, some of them definitely recognized as low-abundance species, such as leptin, ghrelin, and bradykinin.

Things did not fare much better with the introduction of immunodepletion, with commercial resins containing mixed antibody population to the seven, twelve or twenty higher-abundance proteins. As repeatedly stated in this book, according to Tu et al. (2010) and Zhi et al. (2010), this procedure hardly permits access to the low-abundance proteome and cannot be expected to efficiently lead to biomarker discovery. Thus it is not an exaggeration to state that a major breakthrough came in 2007 when Sennels et al. reported treatment of sera with CPLL technology. They started with a rather large sample pool (300 mL) and treated it sequentially with 1 mL of two CPLL libraries, an $-NH_2$ terminus one and a carboxylated one. Yet, when all eluates were analyzed and the data pooled, a grand total of 3896 unique gene products could be listed, by far the most extensive investigation of sera with a single protocol and quite simple experimental setup.

Is this the correct approach for finding biomarkers in plasma/sera? There is a large body of reports dealing precisely with the CPLL approach in biomarker discovery, which is overwhelmingly positive indeed. Since describing each one individually would occupy too much space in the book, we are forced here to just list them and let the readers, by reading their respective titles, sort out the ones of direct interest for their research. Here they are: Au et al. 2007; Sihlbom et al., 2008; Freeby et al., 2008; Hartwig et al., 2009; Callesen et al., 2009; Dwivedi et al., 2010; Ernoult et al., 2010; Drabovich et al., 2010; Fertin et al., 2010; Beseme et al., 2010; Overgaard et al., 2010; Marrocco et al., 2010; Ye et al., 2010; Fröbel et al., 2010; Zhi et al. 2010; Fakelman et al., 2010; Liu et al., 2011; Léger et al., 2011; Fertin et al., 2011; Selvaraju and El Rassi, 2011; Egidi et al., 2011; Wang et al., 2011; Meng et al., 2011; Monari et al., 2011; Hagiwara et al., 2011; Shetty et al., 2011; Huhn et al., 2012; Liang et al., 2012; Hartwig and Lehr, 2012; Cumová et al., 2012.

One wonders if the integration of complementary sample treatments and analysis would offer a more complete view of the composition of an investigated proteome with confirmation of the found data. Different technique associations might be advantageous at the sample pretreatment stage. While some of them could be associated two by two (subtraction could be associated with fractionation or with group separation and so on); some others are incompatible. For instance immunosubtraction and enrichment with peptide libraries might be quite interesting, due to the fact that during CPPL treatment of sera plenty of albumin sticks to the beads (i.e., it does not follow the rule of binding just to its own specific ligand, but colonizes via low-specificity interaction plenty of other sites), thus lowering the capture capability of this method. However, it must be emphasized that the unique advantage of CPLLs is their ability to process large amounts and volumes of sample. On the contrary, the amount of sample that can be treated with immunodepletion is too small (in general 20–100 µL) for a subsequent peptide library treatment where the sample volume is critical for low-abundance species amplification. Streamlining subtraction with fractionation is also difficult because of the low amount of proteins and because fractionation may generate too many fractions to manage within a protein marker discovery.

Indeed, the approach of coupling depletion with CPLL treatment was attempted by Hagiwara et al. (2011) but with less than satisfactory results, as discussed at the beginning of this chapter and here briefly reevaluated. First, in order to circumvent the limits of classic immunodepletion, they treated large volumes of plasma (30 mL, diluted to 90 mL) not with an anti-albumin column, but with a

Cibacron Blue resin (5 mL, repeatedly loaded and eluted with the 90 mL plasma for 10 cycles). The entire albumin depleted plasma was further processed with HiTrap Protein G HP column, known to subtract immunoglobulins (for 20 cycles). In turn, the twice-processed plasma sample was further subfractionated by adsorption on an anion-exchange column, which was eluted stepwise with a NaCl gradient into six fractions, ranging from 0 to 1 M NaCl. The entire process yielded 10 fractions (since in the first two steps not only the eluate was analyzed, but also whatever material had remained bound to the Cibacron Blue and Protein G columns). At the end of this massive protocol, and of running an impressive amount of 2D-DIGE gels, the results were meager, since the work on the treated samples "resulted in the identification of high- and medium-abundance proteins (Hagiwara et al., 2011)."

This excursus thus suggests a note of caution: It is not guaranteed that extensive prefractionation and generation of many plasma/sera fractions would necessarily give access to the low-abundance proteome. It might have the opposite effect, as is also illustrated in the case of Rose et al. (2004) and of Kovács et al. (2011). Results with the enrichment effect of many low-abundance proteins are, however, consistently obtained with serum or plasma treated with combinatorial peptide ligand libraries as depicted by Leger et al (2011). In this case, the number of fractions was limited to four—which is a manageable number for the discovery phase—with the detection of an extremely large number of species that did not show much overlapping (see Figure 6.7). A well-established protocol made on the basis of extreme accuracy in collecting, treating, and storing blood sample with very good reproducibility was defined prior to an organized campaign of blood samples collection as a way to

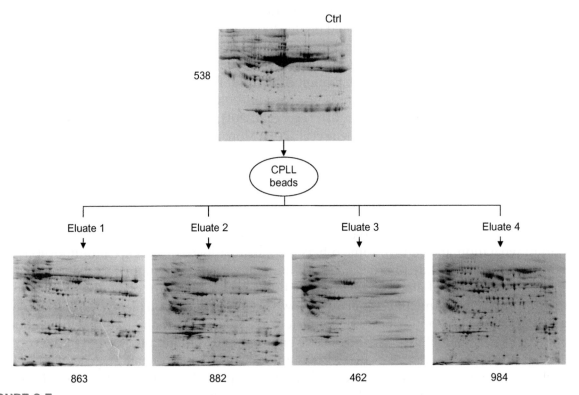

FIGURE 6.7

Two-dimensional gel electrophoresis of four different fractions collected from two peptide libraries. The upper image is from crude plasma control (Ctrl). The aligned maps below represent three eluates from a primary amine hexapeptide library (Eluate 1, 2, and 3) and the eluate of the second library (succinylated primary amine library). The libraries were used sequentially. The numbers underneath each gel plate represent the corresponding number of counted spots. First-dimensional separation was between pH 3 and 10. Staining: colloidal Coomassie Blue. Five mL of crude plasma were incubated with 250 μL of the first library for 1 hour and 30 minutes at 4 °C. The flow-through was eliminated and used to load the second library. Elutions 1, 2, and 3 were carried out using 1 M sodium chloride, 2 M thiourea— 7.7 M urea—4% CHAPS and 8 M urea—2% CHAPS—0.1 M citric acid. The elution from the second library was performed by using 8 M urea—2% CHAPS—0.1 M citric acid *(Adapted from Leger et al., 2011).*

reduce issues related to potential differences originated by contamination or degradation during the various stages. This important foundation work was made after having observed the strong potential of plasma treatment with CPLL for the discovery and identification of circulating biomarkers of atherosclerosis within the frame of a study of searching markers related to coronary events (Meilhac et al. 2009).

6.4.6 Other Body Fluids Proteomics

We summarize here some information on the proteomics of other body fluids, with the proviso that some of them have not been analyzed via CPLL treatment, so that their global proteomic description might be far from exhaustive.

6.4.6.1 TEAR PROTEOMICS

The tear film is a thin layer of fluid that covers the ocular surface and is involved in lubrication and protection of the eye. Its deregulation is associated with disease states, such as diabetic dry eyes; therefore, this makes this body fluid an interesting candidate for in-depth proteomic analysis. Although in 2005 Li et al. identified only 54 proteins in tears, this number was incremented to 491 species the following year (de Souza et al., 2006). However, at the latest count, this number has increased to as many as 1543 proteins (Zhou et al., 2012), a non-negligible accomplishment considering that, in general, barely 5 µL of this fluid are harvested and analyzed. Applications to biomarker discovery soon followed. Thus Versura et al. (2010), in exploring evaporative eye disease, proposed a panel of potential biomarkers, lactoferrin, lipocalin-1, and lipophilin, whose levels are markedly decreased in this pathology. Additionally, Lebrecht et al. (2009) suggested that tear proteomics could even be exploited for diagnosis of breast cancer, via a panel of 20 potential biomarkers enucleated in their studies.

6.4.6.2 AMNIOTIC FLUID PROTEOMICS

The amniotic fluid (AF) is initially formed from maternal plasma that later crosses fetal membranes from 10 to 20 weeks of gestation. It contains large amounts of proteins produced by the amnion epithelial cells, fetal tissues, fetal excretions and placental tissues. The biochemical composition of AF is modified throughout pregnancy, and its protein profile reflects the genotypic constitution of the fetus and regulates feto-maternal physiological interactions. By looking at the composition of the amniotic fluid, scientists can provide valuable information about the health of the fetus and may indicate potential pathological conditions.

Already in 2007 Cho et al., while profiling the human AF proteome of a 16- to 18-week normal pregnancy, identified 1026 unique gene matches corresponding to 842 nonredundant proteins. This list included most of the currently used biomarkers for pregnancy-associated pathologic conditions such as preterm delivery, intra-amniotic infection, and chromosomal anomalies of the fetus. These authors also analyzed by various bioinformatic tools the subcellular localization, tissue expression, functions, and networks of the AF proteome. Subsequently, Buhimschi et al. (2008a) profiled AF to provide insight into the mechanisms of idiopathic preterm birth and enucleated five SELDI peaks as potential markers of this condition. The same authors (Buhimschi et al., 2008b) used AF proteomic analysis for assessing histologic chorioamnionitis and listed a panel of four potential biomarkers (defensins 2 and 1, calgranulins C and A). In turn, Kolialexi et al. (2011) exploited AF proteomic analysis for diagnosis of fetal aneuploidies, while Anagnostopoulos et al. (2010) used the same approach for fetuses with Klinefelter syndrome. In this last case, three proteins (Ceruloplasmin, Alpha-1-antitrypsin, and Zinc-alpha-2-glycoprotein) were found to be upregulated, and four (Apolipoprotein A-I, Plasma retinol-binding protein, Gelsolin, and Vitamin D-binding protein) downregulated. The importance of AF studies in fetal pathologies has also been stressed by Buhimschi and Buhimschi (2008) and by Perluigi et al. (2009).

6.4.6.3 SYNOVIAL FLUID PROTEOMICS

Synovial fluid (SF) is a thin, viscous fluid present in the cavities of synovial joints. It can reduce friction between the articular cartilage and other tissues in joints via lubrication during movement due to the presence of hyaluronic acid. SF may be classified into normal, noninflammatory, inflammatory, septic, and hemorrhagic, with each group associated with certain diagnoses. Regular SF analysis is commonly performed to determine the cause of acute arthritis. SF contains a large number of proteins originating from synovial tissue, cartilage, and serum. The protein composition in SF may reflect the pathophysiological conditions affecting the synovial tissue and articular cartilage. In fact, most of the papers published so far deal with the possibility of biomarker discovery in this fluid. Thus Gobezie et al. (2007) reported that SF profiling could distinguish between healthy individuals and those affected by osteoarthritis. In turn, Baillet et al. (2010) via SF proteomic fingerprinting, could discriminate rheumatoid arthritis from other inflammatory joint diseases through three biomarkers of 10839, 10445, and 13338 Da, characterized as S100A8, S100A12, and S100A9 proteins, respectively. Additionally Gibson et al. (2009) could perform, via SF analysis, stratification and monitoring of juvenile idiopathic arthritis patients. Also, endogenous peptides in human synovial fluid could be successfully profiled by Kamphorst et al. (2007). Synovial fluid proteomics has been reviewed by Cano and Arkfeld (2009) and by Gibson and Rooney (2007).

6.4.6.4 NIPPLE ASPIRATE FLUID PROTEOMICS

Nipple aspirate fluid (NAF) is an interesting body fluid for evaluating biomarkers of breast disease, since it is part of the microenvironment where more than 95% of breast cancers arise. Moreover, nipple aspiration is simple, quick, reliable, and reproducible. In addition, this process is completely noninvasive and requires a device with only minimal cost. Nipple aspiration provides concentrated secreted proteins not diluted with the irrigation fluid that is required to perform procedures such as ductoscopy or ductal lavage. Because of this, NAF is a rich source of proteins, thereby making it a good potential candidate as a source of biomarkers for early diagnosis of breast cancer. To that aim, an extensive database on the NAF proteome was created by Pavlou et al. (2010), listing 800 unique proteins in this body fluid. Although Coombes et al. (2003) via SELDI profiling, were unable to obtain a reliable list of biomarkers for breast cancer, other authors apparently succeeded on that. Thus Pawlik et al. (2006) by differentially screening healthy women versus breast cancer patients were able to enucleate five potential biomarkers, namely: α_2HS-glycoprotein, lipophilin B, β-globin, hemopexin, and vitamin D-binding protein precursor. Also Alexander et al. (2004) performed a similar analysis and proposed three potential biomarkers of breast cancer, namely: gross cystic disease fluid protein-15, apolipoprotein D, and alpha1-acid glycoprotein. There is no agreement between these two last sets of data, and we would like to underline that, as highlighted in the introduction, unfortunately most of these presumptive biomarkers reported in innumerable manuscripts, have not gone through validation, and the vast majority did not even give some basic parameters for their evaluation, namely, sensitivity, specificity, and ROC curve.

6.4.6.5 BRONCHOALVEOLAR LAVAGE FLUID PROTEOMICS

Bronchoalveolar lavage (BAL) conducted with fiber-optic bronchoscopy has been widely used to collect cells and other soluble components from the thin layer of epithelial lining fluid that covers the airway and the alveoli. Bronchoalveolar lavage fluid (BALF) thus collected contains different cell types as well as a wide variety of proteins that either originate from the blood stream or are released locally by epithelial and inflammatory cells. Due to the diverse origin of BALF proteins, analysis of BALF may reveal important pathological mediators and enable more accurate characterization of many lung diseases at the molecular level (Wattiez and Falmagne, 2005; Chen et al., 2008; Govender et al., 2009). There have been numerous reports on the characterization of BALF proteins by proteomic approaches in attempts to uncover biomarkers and pathological mediators for various pulmonary disorders such as idiopathic pulmonary fibrosis, cystic fibrosis, acute lung injury, nonallergic asthma, and coronary obstructive pulmonary disease, notwithstanding the fact that identification and semi-quantitation of interesting proteins that are present at nanogram per milliliter levels in body fluids such as BALF remain challenging.

Thus Fietta et al. (2006) have studied the BALF proteomics in patients suffering from lung fibrosis, in search for suitable biomarkers. Whereas levels of glutathione S-transferase P (GSTP), Cu-Zn superoxide dismutase (SOD) and cystatin SN were downregulated in these patients, significant upregulation of alpha1-acid glycoprotein, haptoglobin-alpha chain, calgranulin (Cal) B, cytohesin-2, calumenin, and mtDNA TOP1 was observed. In turn Wu et al. (2005) when studying BALF in asthmatics patients were able to identify more than 1500 distinct proteins, of which about 10% displayed significant upregulation specific to the asthmatic patients after segmental allergen challenge. In turn, Plymoth et al. (2006) when studying BALF of lifelong smokers and never-smokers identified 481 high- to low-abundance proteins and suggested the potential use of this proteomic approach for assessing the risk of smokers of developing chronic obstructive pulmonary disease—although no specific biomarker was reported in this study.

6.4.6.6 FOLLICULAR FLUID PROTEOMICS

Human follicular fluid (hFF) is the in vivo environment of oocytes during follicular maturation in the ovaries. It contains a large variety of substances implicated in oocyte meiosis, rupture of the follicular wall (ovulation), differentiation of the ovarian cells into the functional corpus luteum, and finally fertilization. Thus, the composition of follicular fluid reflects stages of oocyte development and the degree of follicle maturation and thus the quality of the follicle. Because hFF is a product of both the transfer of blood plasma constituents that cross the blood follicular barrier and the metabolism of the granulosa cells, the concentration of specific peptides and proteins in hFF can be seen as a reflection of this transfer and is influenced by the metabolism of such follicular structures. With regard to the transfer of plasma proteins, the blood–follicle barrier was found to be permeable for proteins with molecular masses <500 kDa. In addition to the size, the charge of the proteins affects the plasma protein transfer into follicular fluid.

Hanrieder et al. (2008) have performed indepth protein analysis of human follicular fluid samples of patients undergoing controlled ovarian hyperstimulation (COH) for in vitro fertilization therapy (IVF). Although a grand total of 69 proteins could be tabulated, at least two relevant compounds essentially involved in hormone secretion regulation during the folliculogenetic process were identified: sex hormone binding globulin (SHBG) and inhibin A (INHA). As in serum, one major difficulty of follicular fluid proteomic analysis is the presence of massive amounts of high-abundance proteins. To simplify the analysis and the detectability of low-abundance proteins, CPLLs were used associated with 2D-PAGE analysis not only for human, but also for porcine and equine FF (Fahiminiya et al., 2011a). After this investigational work, the same authors applied the COH process to analyze low-abundance protein expression of follicular fluid during late follicle development in mares (Fahiminiya et al., 2011b). The enrichment by the peptide library method was considered the most powerful one for the detection and identification of novel low-abundance proteins from follicular fluid emerging upon development.

Ovarian hyperstimulation for fertility treatment of women undergoing in vitro fertilization can induce complications. In order to distinguishi patients developing ovarian hyperstimulatin syndrome (OHS) from others, an in-depth comparative proteome analysis was performed after treatment of follicular fluid with peptide library (Martinkova et al., 2009). In OHS samples, the activity of complement cascade was found to be underexpressed, while few other proteins were upregulated (e.g., ITI-H4 precursor, pigment epithelium-derived factor, ceruloplasmin). Jarkovska et al. (2010) mined hFF for a better understanding of the major problems by which in vitro fertilization (IVF) is fraught. A better understanding of the reproductive process may help increase the IVF birth rate per embryo transfer and at the same time avoid spontaneous miscarriages or life-threatening conditions such as ovarian hyperstimulation syndrome. Their study showed involvement of innate immune function of complement cascade in HFF. Complement inhibition and the presence of C-terminal fragment of perlecan suggested possible links to angiogenesis, which is a vital process in folliculogenesis and placental development. Differences in proteins associated with blood coagulation were also found in the follicular milieu. Schweigert et al. (2006), Estes et al. (2009), and Angelucci et al. (2006) also performed hFF proteomic studies aimed at understanding in deeper detail the problems of IVF. In turn, Kim et al. (2006) adopted the same proteomics tool for a better understanding of the phenomenon of recurrent spontaneous abortion (RSA) in pregnant women. They could assess five aberrantly expressed proteins, namely, complement component C3c chain E, fibrinogen gamma, antithrombin, angiotensinogen, and hemopexin precursor in the follicular fluid from RSA patients.

6.5 BEYOND HUMAN BIOLOGICAL FLUIDS

Before leaving our section on biological fluids, a note should be taken of at least two other examples that are quite common and even if they are not of human origin, they both play a crucial role in the life of vertebrates. These fluids are milk and snake venom.

Milk is one of the most important liquid nutrients for early-age mammals and for adult humans. The protein content of bovine milk can be summarized as comprising about 3.3% protein, of which about 80% are caseins (α_{S1}-, α_{S2}-, β- and κ-caseins); the remaining 20% are serum albumin, β-lactoglobulin, α-lactalbumin, and a large number of other less-abundant proteins. One of the first proteomics studies of milk proteins (Galvani et al., 2000) allowed detecting only seven gene products along with several post-translational modifications. The issue here, as for most of biological fluids, was the presence of massive amounts of high-abundance proteins. With this in mind, the milk proteome was investigated after having removed casein and immunoglobulin G by immunodepletion (Yamada et al., 2002) with the result of finding 15 unique gene products. A few years later, D'Ambrosio et al. (2008) counted a total of 72 unique gene products using 2D MALDI-TOF MS and electrospray-MS/MS analysis of trypsinized slices of SDS-PAGE protein separation.

Perhaps the most extensive findings reported up to 2006 were those from Palmer et al. (2006), who listed 151 unique gene products, of which 83 had not been previously reported.

Although still in its early stage of development, CPLL treatment methodology was applied to milk whey proteins in an attempt to explore this proteome more deeply (D'Amato et al., 2009). This experimental work allowed identifying a total of not less than 149 unique gene products, 100 of them exclusively found thanks to the use of CPLLs. This was by far the most extensive protein list delivered at that time and allowed bringing the number of milk gene products to over 250 when added to what was discovered earlier. More recently, through an extensive ion-exchange fractionation and probably with more modern sophisticated analytical instruments, Le et al. (2011) found 293 unique gene products, of which 176 were newly identified. At that stage, the list of milk proteins could be considered to total more than 500 gene products. However, a year later, using CPLL,

Molinari et al. (2012) identified 415 proteins at once from a pooled milk sample and 161 for the first time, thus bringing the number of known proteins in milk close to 700. Today with the extensive experience with CPLLs and the optimized method of use (more than one library, different pHs, adapted ionic strength, and stringent elutions to collect absolutely all proteins), it should be easy to get many more proteins.

Milk proteomics investigations have not been the only studies to detect an even larger number of gene products. More interestingly, this biological material is also being investigated within the field of nutrition and also, by the composition modification as a function of lactation periods, to understand physiological and nutritional functions. One major question in the nutrition field involves the presence of allergens that could come from native proteins or from modified proteins produced by industrial processes. In this regard Arena et al. (2010) indicated that thermal treatments can modify proteins through the formation of lactose adducts altering protein functions. These modifications were observed not only in major milk proteins, but also in proteins classified as very low-abundance and hence could influence infant nutrition, development, and defense against external agents. Moreover, lactosylated proteins that are resistant to digestion can, at the level of the intestinal tract, enhance allergic responses, as has already been reported for other lactosylated proteins to act as hapten-like antigens (Karamanova et al., 2003).

In the veterinary and medical fields, one should at least mention a few proteomics studies performed after CPLL milk treatment in the domain of bovine Coliform mastitis, a disease of primary importance because of its deleterious effects (Boehmer et al., 2010). It is postulated that with the identification of low-abundance protein markers capable of characterizing the disease, it should be possible to facilitate drug discovery and approval for appropriate treatments.

Milk proteomics is such a large domain that it is impossible to summarize it in few sentences. The reader is urged to examine a recent review of the domain published by Roncada et al. (2012) who report the latest progress in proteomics, peptidomics, and bioinformatics, along with initial sample treatments.

Snake venom is a biochemical poison for subduing, killing, and digesting a prey. This fluid is a complex mixture of peptides and proteins that act on vital systems of the victims. Understanding the composition of venom helps to deconvolute the biochemical mechanisms at the basis of the lethal processes and to design more adapted antidotes especially for human protection. Although very well established, the composition of venom has been improved with the use of two combinatorial peptide ligand libraries, especially with respect to the discovery of low-abundance species (Calvete et al., 2009). Upon 2D gel electrophoresis analysis and mass spectrometry, the approach revealed a full toxin profile with cytotoxic, myotoxic, hemotoxic, and hemorrhagic effects. From the two major protein families (metalloproteinases and serine proteinases), 24 distinct proteins were identified. Four other proteins from medium abundance families were also found (L-amino acid oxidase, cysteine-rich secretory protein, medium-size disintegrin, and PLA_2). Three minor protein families were found, such as vasoactive peptides, endogenous inhibitor of SVMP, and C-type lectin. Moreover, 2-cys peroxiredoxin (found exclusively from regular library eluate) and glutaminyl cyclase (found from both library eluates) were identified. The two latter species, which had never previously been detected in snake venoms, are very interesting because they may contribute in redox processes to the functional diversification of toxins and the maturation of other bioactive polypeptides. Similarly, another venom proteome was analyzed using CPLL (Fasoli et al., 2010) in another snake species, with significant additional proteins found.

6.6 RED BLOOD CELL PROTEOMICS

Probably one of the most crucial examples of low-abundance proteins enhancement is represented by the red blood cell (RBC) lysate, where hemoglobin represents more than 97% of the overall protein mass. Direct analysis of a total hemolysate by 2D electrophoresis gel leads to a major, smearing spot of hemoglobin, with a very small number of other detectable spots. The first in-depth work on the RBC cytoplasmic proteome was performed by Kakhniashvili et al. (2004), who discovered 94 proteins in addition to hemoglobin (Hb). A major breakthrough came in 2006 when Pasini et al. identified 252 cytoplasmic species, in addition to 252 proteins found in the membrane fraction. Interestingly, the following year, Bhattacharya et al. (2007), after Hb depletion via cation-exchange chromatography, detected no fewer than 59 proteins other than Hb! A major improvement came from the work of Ringrose et al. (2008) who, after a double depletion of Hb, identified 700 cytoplasmic proteins in RBCs.

It was only when CPLLs were adopted, however, that a bountiful harvest could be obtained. Roux-Dalvai et al. (2008), when starting from a large amount of cytoplasmic lysate (ca. 6000 mg and using 1 mL of peptide library beads), identified an impressive number of gene products—1578. This was achieved by using a sequence of two libraries and collecting the captured proteins through three elution steps. An SDS-PAGE profiling and 2D-PAGE analysis of the sample after treatment showed a large number of zones and spots, respectively, populating the entire surface of molecular masses and isoelectric points (see Figures 6.8 and 6.9). When investigating the protein list, it appeared that a majority of proteins—72%—were categorized as cytoplasmic; however, a relatively large number of proteins were also classified as nuclear. Although red blood cells are enucleated when they enter the blood circulation, the presence of proteins classified as "nuclear" probably reflects the fact that a certain number that normally stays in the nucleus in the erythroblasts is still present in the mature erythrocytes. Very interestingly, eight different hemoglobin chains were detected; among them were not only the well-known α, β, γ, and δ chains, but also two embryonic ε and ζ chains from genes that are incompletely silenced during development and for the first time μ and θ chains.

FIGURER 6.8

SDS-PAGE profiling of control RBC lysate (Ctrl.) and after treatment with peptide libraries (CPLL). CA: carbonic anhydrase; Hb: α and β globin chains of hemoglobin.

In the same work, Roux-Dalvai et al. (2008) described the addition of 100, 300, or 1000 pmol of yeast alcohol dehydrogenase (ADH) to 680 mg of red blood cell lysate proteins, prior to processing with a 0.5 mL of peptide library beads. Eluates from different spikings and three replicates of each were analyzed by nanoLC-MS/MS after SDS-PAGE separation and digestion by trypsin. They demonstrated a linear increase of signal as the amount of protein spiked in the lysate increased, as shown either on an individual parent ion assigned to ADH or on the global average value calculated from all the different peptides and MS injections. Little variation among three different replicate experiments was shown, with therefore good reproducibility; the relative abundance ratio appeared well-conserved after peptide library treatment, at least for ADH. This showed, for the first time, that differential proteomic analyses could be performed, in search for biomarkers, even after CPLL treatment, provided that the much sought species had to be found in the low-abundance proteome region—that is, below 1 micro-molar concentration.

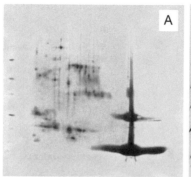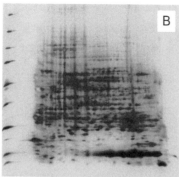

FIGURE 6.9

2D mapping of control RBC lysate (left plaque) versus the six combined eluates from CPLL beads (right slab). Note, in the first case, the massive presence of α and β globin chain with a smear of β-oligomers and only about 80 spots of nonglobin polypeptides versus the ca. 1000 spots in the CPLL combined eluates, notwithstanding the loading of one-half sample amount.

This research had two immediate consequences. The first one concerns the list of eight globin chains, as shown in Table 6.1: Four of them, developed at the embryonic stage and in principle already suppressed at the fetal stage, are indeed found in a 65-year-old normal human adult. According to geneticists, such genes should have been silenced already in the maternal womb, so they should have simply disappeared in adult life, where the respective genes should have been silent. Geneticists also believe that, for proper cell life and growth, at any given time up to 50% of the respective gene products do not need to be expressed. But our data seem to tell a different story—that is, such "silent genes" are not quite silent but perhaps express minute amounts of their protein product, for no other reason than to be ready in case of emergencies. Thus, perhaps such silent genes are rather Stakhanovite genes.

TABLE 6.1 **Hemoglobin Chains Expressed Throughout the Development and Found Before or After Treatment with Combinatorial Peptide Library From Red Cell Lysate**

Globin Chain	Control RBC Lysate	CPLL-Treated Lysate	Comments
α-globin	Massively present	Massively reduced	Most abundant in adult.
β-globin	Massively present	Massively reduced	Most abundant in adult.
γ-globin	Found	Found	Medium abundance in adult.
δ-globin	Found	Found	Partially replaced by α and β upon fetal development.
ε-globin	Found (traces)	Found	Silent in adult. Essentially present in embryos and replaced by α and γ in fetus.
ζ-globin	Not found	Found	Silent in adult. Essentially present in embryos and replaced by α and γ in fetus.
μ-globin	Not found	Found	Function unknown.
θ-globin	Not found	Found	Expressed in fetal erythroid tissue.

The second consequence has been the discovery of the defective gene in a rare RBC disease called congenital dyserythropoietic anemia type II (CDAII). CDAII is an autosomal recessive disease characterized by ineffective erythropoiesis, peripheral hemolysis, erythroblasts' morphological abnormalities, and hypoglycosylation of some RBC membrane proteins. Recent studies indicate that CDAII is caused by a defect that disturbs Golgi processing in erythroblasts; linkage analysis located the CDAII gene in a 5 cM region on chromosome 20, but the molecular basis was still unknown. By using the data of the CPLL RBC treatment that resulted in the detection of 1578 unique gene products in the RBC cytoplasm, Bianchi et al. (2009) enucleated 17 proteins codified by genes mapping onto the CDAN2 locus, which could be potential candidates for this genetic disease, as

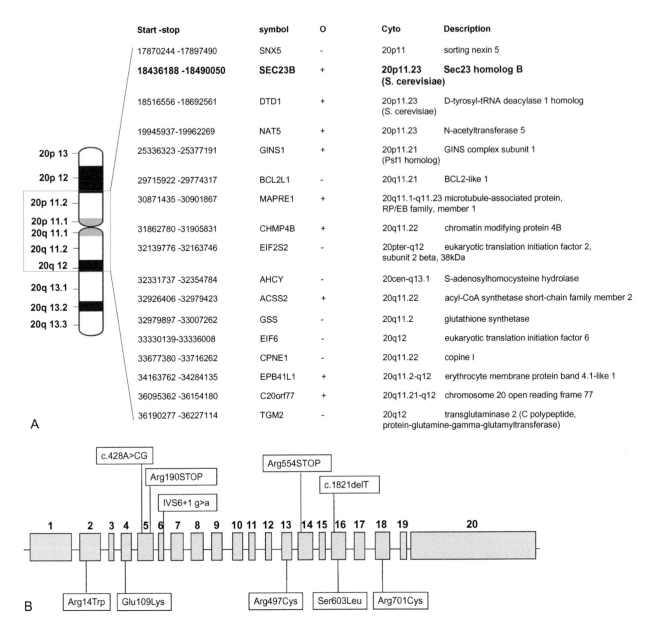

Start -stop	symbol	O	Cyto	Description
17870244 -17897490	SNX5	-	20p11	sorting nexin 5
18436188 - 18490050	**SEC23B**	**+**	**20p11.23 (S. cerevisiae)**	**Sec23 homolog B**
18516556 -18692561	DTD1	+	20p11.23 (S. cerevisiae)	D-tyrosyl-tRNA deacylase 1 homolog
19945937-19962269	NAT5	+	20p11.23	N-acetyltransferase 5
25336323 -25377191	GINS1	+	20p11.21 (Psf1 homolog)	GINS complex subunit 1
29715922 -29774317	BCL2L1	-	20q11.21	BCL2-like 1
30871435 -30901867	MAPRE1	+	20q11.1-q11.23	microtubule-associated protein, RP/EB family, member 1
31862780 -31905831	CHMP4B	+	20q11.22	chromatin modifying protein 4B
32139776 -32163746	EIF2S2	-	20pter-q12	eukaryotic translation initiation factor 2, subunit 2 beta, 38kDa
32331737 -32354784	AHCY	-	20cen-q13.1	S-adenosylhomocysteine hydrolase
32926406 -32979423	ACSS2	+	20q11.22	acyl-CoA synthetase short-chain family member 2
32979897 -33007262	GSS	-	20q11.2	glutathione synthetase
33330139-33336008	EIF6	-	20q12	eukaryotic translation initiation factor 6
33677380 -33716262	CPNE1	-	20q11.22	copine I
34163762 -34284135	EPB41L1	+	20q11.2-q12	erythrocyte membrane protein band 4.1-like 1
36095362 -36154180	C20orf77	+	20q11.21-q12	chromosome 20 open reading frame 77
36190277 -36227114	TGM2	-	20q12	transglutaminase 2 (C polypeptide, protein-glutamine-gamma-glutamyltransferase)

A

B

FIGURE 6.10

A: Drawing of the analyzed region on chromosome 20 with the 17 proteins found to be expressed in RBC cytoplasm. B: Schematic representation of *SEC23B* gene with the mutations identified in CDAII patients. Region displayed: 17,800 K-37,100 K bp; Total genes on chromosome: 824; Total genes in region: 281 (white rectangles: the ten identified mutations). *(From Bianchi et al., 2009, by permission).*

illustrated in Figure 6.10A. Among these species, the researchers' attention was focused on a protein called SEC23B as a likely candidate for expression of the defective gene. *SEC23B* is a member of the SEC23/SEC24 family, a component of the COPII coat protein complex which is involved in protein trafficking through membrane vesicles. Even if the exact function of human *SEC23B* is not completely clear, abnormalities in this gene are likely to disturb ER-to Golgi trafficking, affecting different glycosylation pathways and ultimately accounting for the cellular phenotype observed in CDAII.

When the 20 exons and intronic flanking regions of the *SEC23B* gene were analyzed by direct sequencing in 12 CDA II patients from 10 families, 10- different mutations were detected among the 21 mutated alleles identified: Five of them were missense, two were frame shifts, one splicing, and two stop codons. All the missense mutations affected highly conserved amino acids and were not found in 100 normal alleles examined (Bianchi et al., 2009) (see Figure 6.10B). This was no small achievement, considering that, in modern times, all genetic defects have been discovered through gene analysis and thus genomic tools, whereas here they have been discovered through proteomic tools, a most unusual procedure. Additionally, this finding could be the answer to the various criticisms received from Mitchell (2010) and from other editorials in *Nature*, decrying the fact that nothing had been discovered in 10 years of proteomics. At least here it has been proven that something has been achieved. Once more, these data demonstrate that CPLLs could explore to a remarkable depth any proteome and lead to quite unique and unexpected results.

6.7 TISSUE PROTEOMICS

CPLL technology has been applied predominantly to blood plasma and many other biological fluids, as described in this chapter, but it has not been used extensively to address the issue of dynamic range in tissue samples. This emphasis on secreted bio-fluids, and in particular on blood plasma, has tended to focus on the specific challenges associated with these biological samples, due to the overwhelming presence of albumin, immunoglobulins, and other major plasma proteins. In tissues, the protein dynamic range is often not as exaggerated as in plasma. However, one tissue, in particular the skeletal muscle, exhibits an asymmetry in the dynamic range of protein expression probably as extreme as in blood plasma. In the soluble protein fraction of this tissue, a few tens of proteins, largely comprising the glycolytic enzymes responsible for fueling muscle contraction, are present at a very high level of expression, rendering the analysis of this proteome as much as a challenge as it does in blood plasma. Aware of that, Rivers et al. (2011) have applied the CPLL technology to see to which depth they could explore the proteome of this tissue.

Prior to presenting their data, however, we want to mention another myth about CPLLs that persists in the current literature. Most scientists who have published comparative evaluations of CPLL versus immunodepletion, when listing their advantages and disadvantages state unanimously that the major advantage of immunodepletion is its ability to handle only minute sample volumes (typically 100 μL), whereas the major disadvantage of CPLLs is that one needs to handle larger sample volumes, at least 1 mL or more. Well, according to our reckoning, this does not make much sense. If one can handle only such tiny sample volumes in a biological fluid whose proteome is 99% colonized by barely the first 20 proteins of major abundance, one would never have a chance to find those low- to very low-abundance proteins that might constitute the reservoir of any possible biomarker and that are present in sera at the pico- or sub-pico grams/mL level (as is amply discussed in this chapter). So, CPLL has the unique and distinct advantage of being able to handle not just 1, but 10, 50, and even 100 mL of sera (or more, as needed), where enough copies of such rare species can be captured and rendered visible.

It is with this concept in mind that Rivers et al. (2011) set out to study the proteome of the skeletal muscle: In their remarkable paper, they explored the outcome of an exaggerated oversaturation of CPLLs to input protein pool. It turned out that this was a winning strategy, as illustrated in Figure 6.11. When constant amounts of CPLL beads were exposed to increasing levels of skeletal muscle proteins, from 25 mg up to 1000 mg, more and more proteins could be identified, from barely 35 in the starting material (SM) up to 140 species at the high sample load, plateauing at around 250 mg sample loading. When the data are plotted on a linear scale (panel B), the compression of the dynamic range can be appreciated, with the distribution of protein abundances becoming shallower and extending much further on the protein index axis.

This behavior can be better appreciated by looking at the heat map of Figure 6.12 from the same authors, which shows the appearance of low-abundance species at increasing sample loads. In the control (SM), the group C proteins (in orange color) represent the high-abundance species, which

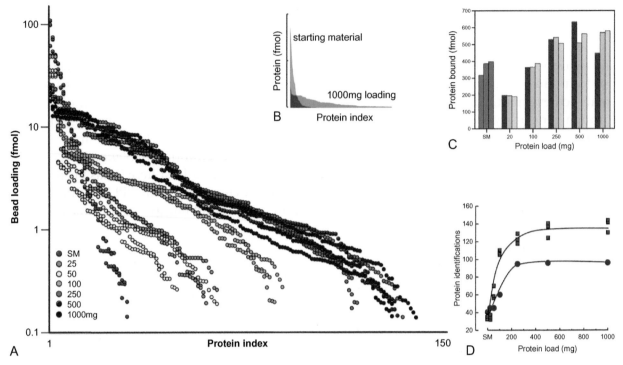

FIGURE 6.11

Representation of the "compression" effect of protein concentration as a function of load. Combinatorial peptide ligand library beads were exposed to very high levels (up to 1000 mg protein/20 mg beads) of chick skeletal muscle soluble proteins in a final volume of 1.0 mL of 20 mM sodium phosphate buffer, pH 7.0. After 2 h, the beads were washed exhaustively (eight successive washes in the same buffer), and the suspension of beads was incubated with trypsin for digestion of bound proteins. The resultant peptides were analyzed by LC-MS/MS by data-dependent gel-LC-MS/MS and data-independent LC-MS/MS. Panel A shows the individually sorted abundance profiles as the beads were loaded from 25 to 1000 mg of starting protein (SM = starting material) and data are expressed as fmol on column. Reduction of dynamic range was evident from the altered profile of the top 100 proteins on a linear abundance scale (panel B). The total protein bound is presented in panel C ($n = 3$) and number of proteins identified by label-free quantification (squares, $n = 3$) or gel-LCMS/MS (circles, $n = 1$) is given in panel D. *(From Rivers et al., 2011, by permission).*

Hierarchical clustering

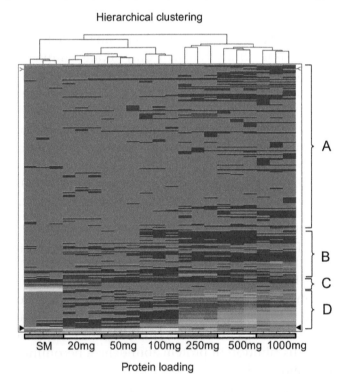

FIGURE 6.12

Heat map showing the reduction of dynamic concentration range as a function of protein load. CPLL beads were exposed to very high levels (up to 1000 mg protein/20 mg beads), and the bound proteins were analyzed by quantitative label-free LC-MS. All quantitative data are expressed as heat diagrams (from red, low abundance to yellow, high abundance; gray: protein absent), arranged according to hierarchical (unweighed average) clustering of samples and K-means (data centroid based search initialized) clustering of the individual protein behaviors. Both clustering methods utilized Euclidian distance methods to assess similarity. The classes of behaviors (labeled A, B, C and D) are discussed in the text. *(From Rivers et al., 2011, by permission).*

overpopulate this region while leaving largely void (the gray area) the region of the low-abundance compounds. However, when analyzing the CPLL eluates at progressively higher sample loads, two phenomena become apparent: The gray area becomes readily populated by the low-abundance proteins, first by group B (probably present at somewhat higher levels in the lysate) and then by group A (probably the very low-abundance compounds). At a sample loading of 250 mg, maximum appearance of the low-abundance species takes place. A curious phenomenon occurs at 500 and 1000 mg loads: The class D region becomes heavily populated, and novel high-abundance species become visible, which were not present in the original lysate. The authors give an interesting interpretation of this phenomenon (which will surely merit further investigation). Apparently, some proteins present in the sample, notably heat shock proteins 90 and 70, α-actinin, calpains, and calmodulins, at high sample loads, might accumulate into successive layers onto the bead surfaces and act as scaffolds for development of large assemblies of protein complexes. Yet, the mission of bringing to the limelight low-abundance species had been well accomplished.

These data are consistent with what we had reported earlier and have discussed throughout this book. Although readers of papers from our laboratories might not have paid attention to it, the very reason why we reported the discovery of very low-abundance proteins was the fact of massively overloading the CPLL beads. Thus, when in 2008 we described, for the first time, no less than 1578 unique gene products in a red blood cell lysate (RBC, with an enormous imbalance in protein concentrations, considering that hemoglobin alone represents 98% of the total cytoplasmic proteins) (Roux-Dalvai et al., 2008), we had loaded onto the CPLL beads no less than about 6 grams total protein! The next best was a paper published in 2006 in which, in the absence of any prefractionation or reduction of dynamic range, the total discovery (believed to be the entire proteome of human RBCs) amounted to a grand total of 252 proteins (Pasini et al., 2006). And when, in a single sweep, we described no less than 3895 proteins in human sera, we had processed onto the CPLL beads something like 300 mL of serum (Sennels et al., 2007). Finally, in an investigation on human CSF, which resulted in the description of 1212 unique gene products, the total amount of CSF processed was 1290 mL, that is rather an exaggeration (Mouton-Barbosa et al., 2010). So, one clear message emerges: If you want to see more, you have to load more (no less, no more!).

Another illuminating example of low-abundance proteins from tissue extract is given by Malaud et al. (2012). These authors investigated how low-abundance-protein markers were capable of discriminating between stable and unstable atherosclerotic plaques from carotid artery tissue in view of a possible early-stage diagnosis of atherosclerosis disease. It was clearly demonstrated that CPLL treatment of these extracts significantly improved the detection of novel species without altering the differential expression of low-abundant proteins such as HSP-27. Although the evaluation of data was complex by the use of two peptide libraries, for two sets of protein extracts (treated and nontreated - control) it was unambiguously found that the expression of HSP-27 was modified especially for one isoform separated by 2D electrophoresis and identified by LC-MS/MS. Thereby, even in this case the use of peptide libraries allowed distinguishing increased species diversity from biological tissue extracts with different molecular mass or different isoelectric point features that could not be detected directly from native extracts.

6.8 LOW-ABUNDANCE ALLERGENS ACCESS

Although not directly related to the composition of human proteomes, allergens modify the expression of human IgE and IgG and generate very serious metabolic disturbances that can be life threatening. The study of allergens, defined as allergomics, increases in importance with proteomics tools due to the possible discovery of proteinaceous allergens that were unexpected. Allergic reactions are peculiar biological processes of defense against external agents that are frequently protein antigens. These allergens are found in large varieties of natural sources, including plant pollens, animal epithelia, insect venoms, various plant proteins, and proteins of animal origin such as eggs and milk. One of the first reviews on the proteomics of allergens was made by Ticha et al. (2002), who described methodologies adapted for these studies. As common tools to separate allergens for their identification, affinity chromatography ligands have been reviewed. Among them there are Poly-proline ligands, lectins ligands such as Concanavalin A, Cibacron Blue, and antibodies. Since then, a number of papers have been published, some of which have reported the discovery of novel allergens.

Ideally, a combinatorial peptide ligand library could constitute a very large source of immobilized affinity ligands also for protein allergens. Most known allergens are generally of an average abundance and can quite easily be identified without the help of enrichment tools; however, the presence of high-abundance proteins in the same extract may preclude proper detection. There are also low- and very low-abundance allergens that in spite of their low concentration can generate strong immunogenic reactions.

With the use of peptide libraries, novel allergens have been discovered from various extracts. One of the first biological materials examined was bovine milk whey, which is known to generate allergic reactions in some individuals. Beyond the current knowledge, it has been found that a number of other proteins were evidenced after CPLL treatment; they produced positive identifications when exposed to selected patients. Among these new allergic species, a polymorphic immunoglobulin not detectable in the control sample was largely amplified (D'Amato et al., 2009). The same authors, using a similar technical approach, investigated *Hevea brasiliensis* exudates in which novel allergens have been found in addition to already known, well-classified allergens. Among the unambiguously detected proteins were a proteasome subunit, a protease inhibitor, a heat shock protein, and glyceraldehyde-3-phosphate dehydrogenase (D'Amato et al., 2010). Here also immunoblots evidenced different allergen patterns when allergic patients sera were tested against the treated extract.

Perhaps the most impressive report came from Shahali et al. (2012) where the examined biological extract was cypress pollen, one of the widespread and highly invasive allergenic sources worldwide. Extracts treated with peptide libraries at different pHs revealed two important things: (i) there were many more IgE-positive proteins than one could expect and (ii) the immune system of an allergic patient is able to raise a diversified IgE response against the found allergenic proteins. Unique responses for each patient might have a medical significance and a role in the diversity and severity of symptoms observed in affected individuals. It was found that some patients allergic to cypress pollen proteins did not respond positively when currently tested against commercial biological assays, whereas they showed an unambiguous positive response to cypress pollen extracts where low-abundance species were amplified. Exclusive allergens found were, for instance, rab-like protein, glyoxalase I, malate dehydrogenase, triosephosphate isomerase, glucanase-like protein and several others. To complete this short review about the interest of using CPLL in allergen discovery, a note is to be made about other published results of fungal allergen discovered in blood subsequent to invasive aspergillosis (Fekkar et al., 2012); peanut extracts (Pedreschi et al., 2012); and hen's egg allergens (Martos et al., 2013).

6.9 CONCLUSIONS

This chapter has been essentially devoted to human biological fluids and tissues, so as to show what proteomics can offer in such an important field. This area of study covers especially the search for a new generation of biomarkers able to substitute those described in the 1960s and 1970s, which quite often do not offer the much-needed high sensitivity and specificity (ideally reaching 100%). Modern biomarkers should be constituted not just by a single protein or polypeptide, as is typical of the old ones, but by a restricted panel of proteins or peptides, judiciously selected so as to provide close to 100% sensitivity and specificity. Unfortunately, as we discussed at the beginning of this chapter, this so far remains a dream and very little has been achieved in 10 years of intense research, a failure that has been strongly criticized in a few editorials in *Nature* and in *Nature Biotechnology*. Yet, as amply described in previous sections, it would appear that a judicious use of the CPLL technique should allow reaching this goal in the not too distant future, since CPLL seems to be the only methodology that offers genuine access to the low- and very low-abundance proteins—precisely those species that should mark the onset of any pathology. The next few years will indeed tell if such a goal has been achieved.

Moreover, it is the research strategy that should change in parallel with the adoption of the proper methodologies for digging deep into any proteome. Thus there will be no more plunging blindfolded into the *mare magnum* of plasma/sera or other biological fields, not having any clue of what to look for. Ideally, one should first analyze, in case of cancer research, the pathological tissue against the normal one, by obtaining, for example, 3 to 4000 thin slices and confronting them with a suitable panel of monoclonal antibodies against the same number of proteins as the number of slices. This will allow detecting which proteins, in the tumor tissue, are overexpressed as compared to normal tissue. At this point, one should go to *in vitro* culture of the same tumor cells and screen the supernatant to see if such potential biomarkers are exuded in the supernatant. If so, chances are that they will also be released into the blood stream. At this point, one will know which proteins one should look for, thus having a "map of the treasure". Such tissue leakage proteins, though, might be present at extremely low levels in the bloodstream, so here, too, capture and amplification will be achieved via CPLLs. Once the goal is achieved—probably after various trials using CPLL under different conditions and increasing loading—and once the proper panel of biomarkers has been found, all that one would have to do in a clinical chemistry environment would be to execute an MRM assay via mass spectrometry analysis. This will mean that the scenario in hospitals will be dramatically changing in the next few decades, to the point that the main instrument for analysis will indeed be a mass spectrometer. It is hoped that we authors and readers of these pages will live long enough to see this new horizon in medical care.

6.10 References

Ahmed N, Barker G, Oliva KT, et al. Proteomic based identification of haptoglobin-1 precursor as a novel circulating biomarker of ovarian cancer. *Br J Cancer*. 2004;91:129–140.

Albrethsen J, Bøgebo R, Møller CH, et al. Candidate biomarker verification: Critical examination of a serum protein pattern for human colorectal cancer. *Proteomics Clin Appl*. 2012;6:182–189.

Alexander H, Stegner AL, Wagner-Mann C, et al. Proteomic analysis to identify breast cancer biomarkers in nipple aspirate fluid. *Clin Cancer Res*. 2004;10:7500–7510.

Alvarez ML, Khosroheidari M, Kanchi Ravi R, Distefano JK. Comparison of protein, microRNA, and mRNA yields using different methods of urinary exosome isolation for the discovery of kidney disease biomarkers. *Kidney Int*. 2012;82:1024–1032.

Amacher DE. Serum transaminase elevations as indicators of hepatic injury following the administration of drugs. *Regul Toxicol Pharmacol*. 1998;27:119–130.

Anagnostopoulos AK, Kolialexi A, Mavrou A, et al. Proteomic analysis of amniotic fluid in pregnancies with Klinefelter syndrome foetuses. *J Proteomics*. 2010;73:943–950.

Angelucci S, Ciavardelli D, Di Giuseppe F, et al. Proteome analysis of human follicular fluid. *Biochim Biophys Acta*. 2006;1764:1775–1785.

Antman EM, Tanasijevic MJ, Thompson B, et al. Cardiac-specific troponin I levels to predict the risk of mortality in patients with acute coronary syndromes. *N Engl J Med*. 1996;335:1342–1349.

Arena S, Renzone G, Novi G, et al. Modern proteomic methodologies for the characterization of lactosylation protein targets in milk. *Proteomics*. 2010;10:3414–3434.

Au JS, Cho WC, Yip TT, et al. Deep proteome profiling of sera from never-smoked lung cancer patients. *Biomed Pharmacother*. 2007;61:570–577.

Baillet A, Trocmé C, Berthier S, et al. Synovial fluid proteomic fingerprint: S100A8, S100A9 and S100A12 proteins discriminate rheumatoid arthritis from other inflammatory joint diseases. *Rheumatology (Oxford)*. 2010;49:671–682.

Bandhakavi S, Stone MD, Onsongo G, Van Riper SK, Griffin TJ. A dynamic range compression and three-dimensional peptide fractionation analysis platform expands proteome coverage and the diagnostic potential of whole saliva. *J Proteome Res*. 2009;8:5590–5600.

Bandhakavi S, Van Riper SK, Tawfik PN, et al. Hexapeptide libraries for enhanced protein PTM identification and relative abundance profiling in whole human saliva. *J Proteome Res.* 2011;10:1052–1061.

Barbhuiya MA, Sahasrabuddhe NA, Pinto SM, et al. Comprehensive proteomic analysis of human bile. *Proteomics.* 2011;11:4443–4453.

Bhattacharya D, Mukhopadhyay D, Chakrabarti A. Hemoglobin depletion from red blood cell cytosol reveals new proteins in 2-D gel-based proteomics study. *Proteomics Clin Appl.* 2007;1:561–564.

Beseme O, Fertin M, Drobecq H, Amouyel P, Pinet F. Combinatorial peptide ligand library plasma treatment: advantages for accessing low-abundance proteins. *Electrophoresis.* 2010;31:2697–2704.

Bianchi P, Fermo E, Vercellati C, et al. Congenital dyserythropoietic anemia type II (CDAII) is caused by mutations in the SEC23B gene. *Hum Mutat.* 2009;30:1292–1298.

Boehmer JL, DeGrasse JA, McFarland MA, et al. The proteomic advantage: Label-free quantification of proteins expressed in bovine milk during experimentally induced coliform mastitis. *Vet Immunol Immunopathol.* 2010;138:252–266.

Boja E, Rivers R, Kinsinger C, et al. Restructuring proteomics through verification. *Biomark Med.* 2010;4:799–803.

Borrebaeck CA, Wingren C. Transferring proteomic discoveries into clinical practice. *Expert Rev Proteomics.* 2009;6:11–13.

Boschetti E, Chung MCM, Righetti PG. "The quest for biomarkers": are we on the right technical track? *Proteomics Clin Applic.* 2012;6:22–41.

Buchen L. Missing the mark. *Nature.* 2011;471:428–432.

Buhimschi IA, Buhimschi CS. Proteomics of the amniotic fluid in assessment of the placenta. Relevance for preterm birth. *Placenta* 2008;29(suppl A):S95–S101.

Buhimschi IA, Zambrano E, Pettker CM, et al. Using proteomic analysis of the human amniotic fluid to identify histologic chorioamnionitis. *Obstet Gynecol.* 2008a;111:403–412.

Buhimschi IA, Zhao G, Rosenberg VA, et al. Multidimensional proteomics analysis of amniotic fluid to provide insight into the mechanisms of idiopathic preterm birth. *PLoS One.* 2008b;3:e2049.

Burgess JA, Lescuyer P, Hainard A, et al. Identification of brain cell death associated proteins in human post-mortem cerebrospinal fluid. *J Proteome Res.* 2006;5:1674–1681.

Cabras T, Pisano E, Mastinu A, et al. Alterations of the salivary secretory peptidome profile in children affected by type 1 diabetes. *Mol Cell Proteomics.* 2010;9:2099–2108.

Callesen AK, Madsen JS, Vach W, et al. Serum protein profiling by solid phase extraction and mass spectrometry: A future diagnostics tool? *Proteomics.* 2009;9:1428–1441.

Calvete JJ, Fasoli E, Sanz L, Boschetti E, Righetti PG. Exploring the venom proteome of the western diamondback rattlesnake, *Crotalus atrox*, via snake venomics and combinatorial peptide ligand library approaches. *J Proteome Res.* 2009;8:3055–3067.

Candiano G, Bruschi M, Petretto A, et al. Proteins and protein fragments in nephrotic syndrome: Clusters, specificity and mechanisms. *Proteomics Clin Appl.* 2008;2:956–963.

Candiano G, Santucci L, Bruschi M, et al. "Cheek-to-cheek" urinary proteome profiling via combinatorial peptide ligand libraries: A novel, unexpected elution system. *J Proteomics.* 2012;75:796–805.

Cano L, Arkfeld DG. Targeted synovial fluid proteomics for biomarker discovery in rheumatoid arthritis. *Clin Proteomics.* 2009;5:75–102.

Castagna A, Cecconi D, Sennels L, et al. Exploring the hidden human urinary proteome via ligand library beads. *J Prot Res.* 2005;4:1917–1930.

Castagnola M, Cabras T, Iavarone F, et al. The human salivary proteome: A critical overview of the results obtained by different proteomic platforms. *Expert Rev Proteomics.* 2012;9:33–46.

Castagnola M, Cabras T, Vitali A, Sanna MT, Messana I. Biotechnological implications of the salivary proteome. *Trends Biotechnol.* 2011;29:409–418.

Chechlinska M, Kowalewska M, Nowak R. Systemic inflammation as a confounding factor in cancer biomarker discovery and validation. *Nat Rev Cancer.* 2010;10:2–3.

Chen G, Zhang Y, Jin X, et al. Urinary proteomics analysis for renal injury in hypertensive disorders of pregnancy with iTRAQ labeling and LC-MS/MS. *Proteomics Clin Appl.* 2011;5:300–310.

Chen J, Ryu S, Gharib SA, Goodlett DR, Schnapp LM. Exploration of the normal human bronchoalveolar lavage fluid proteome. *Proteomics Clin Appl.* 2008;2:585–595.

Cho CK, Shan SJ, Winsor EJ, Diamandis EP. Proteomics analysis of human amniotic fluid. *Mol Cell Proteomics.* 2007;6:1406–1415.

Conde-Vancells J, Falcon-Perez JM. Isolation of urinary exosomes from animal models to unravel noninvasive disease biomarkers. *Methods Mol Biol.* 2012;909:321–340.

Coombes KR, Fritsche Jr. HA, Clarke C, et al. Quality control and peak finding for proteomics data collected from nipple aspirate fluid by surface-enhanced laser desorption and ionization. *Clin Chem.* 2003;49:1615–1623.

Cumová J, Jedličková L, Potěšil D, et al. Comparative plasma proteomic analysis of patients with multiple myeloma treated with Bortezomib-based regimens. *Klin Onkol.* 2012;25:17–25.

D'Amato A, Bachi A, Fasoli E, et al. In-depth exploration of cow's whey proteome via combinatorial peptide ligand libraries. *J Proteome Res.* 2009;8:3925–3936.

D'Amato A, Bachi A, Fasoli E, et al. In-depth exploration of *Hevea brasiliensis* latex proteome and "hidden allergens" via combinatorial peptide ligand libraries. *J Proteomics.* 2010;73:1368–1380.

D'Ambrosio C, Arena S, Salzano AM, et al. A proteomic characterization of water buffalo milk fractions describing PTM of major species and the identification of minor components involved in nutrient delivery and defense against pathogens. *Proteomics.* 2008;8:3657–3666.

De Jong EP, Xie H, Onsongo G, et al. Quantitative proteomics reveals myosin and actin as promising saliva biomarkers for distinguishing pre-malignant and malignant oral lesions. *PLoS One.* 2010;5:e11148.

De Souza GA, Godoy LM, Mann M. Identification of 491 proteins in the tear fluid proteome reveals a large number of proteases and protease inhibitors. *Genome Biol.* 2006;7:R72.

Decramer S, Gonzalez de Peredo A, Breuil B, et al. Urine in clinical proteomics. *Mol Cell Proteomics.* 2008;7:1850–1862.

Di Girolamo F, Bala K, Chung MC, Righetti PG. "Proteomineering" serum biomarkers. *A study in scarlet. Electrophoresis.* 2011b;32:976–980.

Di Girolamo F, Boschetti E, Chung MC, Guadagni F, Righetti PG. "Proteomineering" or not? The debate on biomarker discovery in sera continues. *J Proteomics.* 2011a;74:589–594.

Di Girolamo F, Righetti PG. "Proteomineering" serum biomarkers. *A study in blue. Electrophoresis.* 2011;32:3638–3644.

Di Girolamo F, Righetti PG, D'Amato A, Chung MC. Cibacron Blue and proteomics: The mystery of the platoon missing in action. *J Proteomics.* 2011c;74:2856–2865.

Diamandis EP. Cancer biomarkers: Can we turn recent failures into success? *J Natl Cancer Inst.* 2010;102:1462–1467.

Dowling P, Wormald R, Meleady P, et al. Analysis of the saliva proteome from patients with head and neck squamous cell carcinoma reveals differences in abundance levels of proteins associated with tumour progression and metastasis. *J Proteomics.* 2008;71:168–175.

Drabovich AP, Diamandis EP. Combinatorial peptide libraries facilitate development of multiple reaction monitoring assays for low-abundance proteins. *J Proteome Res.* 2010;9:1236–1245.

Dwivedi RC, Krokhin OV, Cortens JP, Wilkins JA. An assessment of the reproducibility of random hexapeptide peptide library based protein normalization. *J Proteome Res.* 2010;9:1144–1149.

Edwards JJ, Tollaksen SL, Anderson NG. Proteins of human urine. III: Identification and two dimensional electrophoretic map positions of some major urinary proteins. *Clin Chem.* 1982;28:941–948.

Egidi MG, Rinalducci S, Marrocco C, Vaglio S, Zolla L. Proteomic analysis of plasma derived from platelet buffy coats during storage at room temperature. An application of ProteoMiner™ technology. *Platelets* 2011;22:252–269.

Ernoult E, Bourreau A, Gamelin E, Guette C. A proteomic approach for plasma biomarker discovery with iTRAQ labelling and OFFGEL fractionation. *J Biomed Biotechnol.* 2010;2010:927917.

Estes SJ, Ye B, Qiu W, et al. A proteomic analysis of IVF follicular fluid in women ≤ 32 years old. *Fertil Steril.* 2009;92:1569–1578.

Faca V, Pitteri SJ, Newcomb L, et al. Contribution of protein fractionation to depth of analysis of the serum and plasma proteomes. *J Proteome Res.* 2007;6:3558–3565.

Fahiminiya S, Labas V, Dacheux J, Gérard N. Improvement of 2D-PAGE resolution of human, porcine and equine follicular fluid by means of hexapeptide ligand library. *Reprod Domest Anim.* 2011a;46:561–563.

Fahiminiya S, Labas V, Roche S, Dacheux J-L, Gérard N. Proteomic analysis of mare follicular fluid during late follicle development. *Proteome Science.* 2011b;9:54–56.

Fakelman F, Felix K, Büchler MW, Werner J. New pre-analytical approach for the deep proteome analysis of sera from pancreatitis and pancreas cancer patients. *Arch Physiol Biochem.* 2010;116:208–217.

Falasca M. Cancer biomarkers: the future challenge of cancer. *J Mol Biomark Diagn.* 2012;(S):2.

Farina A, Dumonceau JM, Lecuyer P. Proteomics analysis of human bile and potential applications in cancer diagnosis. *Expert Rev Proteomics.* 2010;6:285–301.

Fasoli E, Sanz L, Wagstaff S, et al. Exploring the venom proteome of the African puff adder, *Bitis arietans*, using a combinatorial peptide ligand library approach at different pHs. *J Proteomics.* 2010;73:932–942.

Fekkar A, Pionneau C, Brossas JY, et al. DIGE enables the detection of a putative serum biomarker of fungal origin in a mouse model of invasive aspergillosis. *J Proteomics.* 2012;75:2536–2549.

Fertin M, Beseme O, Duban S, et al. Deep plasma proteomic analysis of patients with left ventricular remodeling after a first myocardial infarction. *Proteomics Clin Appl.* 2010;4:654–673.

Fertin M, Burdese J, Beseme O, et al. Strategy for purification and mass spectrometry identification of SELDI peaks corresponding to low-abundance plasma and serum proteins. *J Proteomics.* 2011;74:420–430.

Fietta A, Bardoni A, Salvini R, et al. Analysis of bronchoalveolar lavage fluid proteome from systemic sclerosis patients with or without functional, clinical and radiological signs of lung fibrosis. *Arthritis Res Ther.* 2006;8:R160.

Fisher WG, Lucas JE, Mehdi UF, et al. A method for isolation and identification of urinary biomarkers in patients with diabetic nephropathy. *Proteomics Clin Appl.* 2011;5:603–612.

Freeby S, Academia K, Wehr T, Liu N, Paulus A. Enrichment of interleukins and low abundance proteins from tissue leakage in serum proteome studies using ProteoMiner beads. *J Proteomics & Bioinformatics.* 2008;S2:171.

Fröbel J, Hartwig S, Paßlack W, et al. ProteoMiner and SELDI-TOF-MS: A robust and highly reproducible combination for biomarker discovery from whole blood serum. *Arch Physiol Biochem.* 2010;116:174–180.

Galvani M, Hamdan M, Righetti PG. Two-dimensional gel electrophoresis/matrix assisted laser desorption/ionization mass spectrometry of a milk powder. *Rapid Commun Mass Spectrom.* 2000;14:1889–1897.

Gibson DS, Finnegan S, Jordan G, et al. Stratification and monitoring of juvenile idiopathic arthritis patients by synovial proteome analysis. *J Proteome Res.* 2009;8:5601–5609.

Gibson DS, Rooney ME. The human synovial fluid proteome: A key factor in the pathology of joint disease. *Proteomics Clin Appl.* 2007;1:889–899.

Gobezie R, Kho A, Krastins B, et al. High abundance synovial fluid proteome: Distinct profiles in health and osteoarthritis. *Arthritis Res Ther.* 2007;9:R36.

Govender P, Dunn MJ, Donnelly SC. Proteomics and the lung: Analysis of bronchoalveolar lavage fluid. *Proteomics Clin Appl.* 2009;3:1044–1051.

Guerrier L, Claverol S, Finzi L, et al. Contribution of solid-phase hexapeptide ligand libraries to the repertoire of human bile proteins. *J Chromatogr A.* 2007;1176:192–205.

Hagiwara T, Saito Y, Nakamura Y, et al. Combined use of a solid-phase hexapeptide ligand library with liquid chromatography and two-dimensional difference gel electrophoresis for intact plasma proteomics. *Int J Proteomics.* 2011;2011:1–11.

Hanrieder J, Nyakas A, Naessén T, Bergquist J. Proteomic analysis of human follicular fluid using an alternative bottom-up approach. *J Proteome Res.* 2008;7:443–449.

Hartwig S, Czibere A, Kotzka J, et al. Combinatorial hexapeptide ligand libraries (ProteoMiner): An innovative fractionation tool for differential quantitative clinical proteomics. *Arch Physiol Biochem.* 2009;115:155–160.

Hartwig S, Lehr S. Combination of highly efficient hexapeptide ligand library-based sample preparation with 2D DIGE for the analysis of the hidden human serum/plasma proteome. *Methods Mol Biol.* 2012;854:169–180.

Henry NL, Hayes DF. Cancer biomarkers. *Mol Oncol.* 2012;6:140–146.

Hsich G, Kenney K, Gibbs CJ, Lee KH, Harrington MG. The 14-3-3 brain protein in cerebrospinal fluid as a marker for transmissible spongiform encephalopathies. *N Engl J Med.* 1996;335:924–930.

Hu S, Arellano M, Boontheung P, et al. Salivary proteomics for oral cancer biomarker discovery. *Clin Cancer Res.* 2008;14:6246–6252.

Hu S, Loo JA, Wong DT. Human body fluid proteome analysis. *Proteomics.* 2006;6:6326–6353.

Hu S, Loo JA, Wong DT. Human saliva proteome analysis. *Ann N Y Acad Sci.* 2007;1098:323–329.

Huhn C, Ruhaak LR, Wuhrer M, Deelder AM. Hexapeptide library as a universal tool for sample preparation in protein glycosylation analysis. *Proteomics.* 2012;75:1515–1528.

Issaq JH, Waybright TJ, Veenstra TD. Cancer biomarker discovery: Opportunities and pitfalls in analytical methods. *Electrophoresis.* 2011;32:967–975.

Jarkovska K, Martinkova J, Liskova L, et al. Proteome mining of human follicular fluid reveals a crucial role of complement cascade and key biological pathways in women undergoing in vitro fertilization. *J Proteome Res.* 2010;9:1289–1301.

Juhasz P, Lynch M, Sethuraman M, et al. Semi-targeted plasma proteomics discovery workflow utilizing two-stage protein depletion and off-line LC-MALDI MS/MS. *J Proteome Res.* 2011;10:34–45.

Kakhniashvili DG, Bulla Jr. LA, Goodman SR. The human erythrocyte proteome: analysis by ion trap mass spectrometry. *Mol Cell Proteomics.* 2004;3:501–509.

Kamphorst JJ, van der Heijden R, DeGroot J, et al. Profiling of endogenous peptides in human synovial fluid by NanoLC-MS: method validation and peptide identification. *J Proteome Res.* 2007;6:4388–4396.

Karamanova L, Fukal L, Kodicek M, et al. Immunoprobes for thermally-induced alterations in whey protein structure and their application to the analysis of thermally-treated milks. *Food Agric Immunol.* 2003;15:77–91.

Kentsis A, Monigatti F, Dorff K, et al. Urine proteomics for profiling of human disease using high accuracy mass spectrometry. *Proteomics Clin Appl.* 2009;3:1052–1061.

Kentsis A. Challenges and opportunities for discovery of disease biomarkers using urine proteomics. *Pediatrics International.* 2011;53:1–6.

Khurana M, Traum AZ, Aivado M, et al. Urine proteomic profiling of pediatric nephrotic syndrome. *Pediatr Nephrol.* 2006;21:1257–1265.

Kim YS, Kim MS, Lee SH, et al. Proteomic analysis of recurrent spontaneous abortion: Identification of an inadequately expressed set of proteins in human follicular fluid. *Proteomics.* 2006;6:3445–3454.

Kolialexi A, Tounta G, Mavrou A, Tsangaris GT. Proteomic analysis of amniotic fluid for the diagnosis of fetal aneuploidies. *Expert Rev Proteomics.* 2011;8:175–185.

Kovács A, Sperling E, Lázár J, et al. Fractionation of the human plasma proteome for monoclonal antibody proteomics-based biomarker discovery. *Electrophoresis.* 2011;32:1916–1925.

Kowalewska M, Nowak R, Chechlinska M. Implications of cancer-associated systemic inflammation for biomarker studies. *Biochim Biophys Acta.* 2010;1806:163–171.

Kristiansen TZ, Bunkenborg J, Gronborg M, et al. A proteomic analysis of human bile. *Mol Cell Proteomics.* 2004;3:715–728.

Le A, Barton LD, Sanders JT, Zhang Q. Exploration of bovine milk proteome in colostral and mature whey using an ion-exchange approach. *J Proteome Res.* 2011;10:692–704.

Le L, Chi K, Tyldesley S, et al. Identification of serum amyloid A as a biomarker to distinguish prostate cancer patients with bone lesions. *Clin Chem.* 2005;51:695–707.

Lebrecht A, Boehm D, Schmidt M, et al. Diagnosis of breast cancer by tear proteomic pattern. *Cancer Genomics Proteomics.* 2009;6:177–182.

Leger T, Lavigne D, Le Caër JP, et al. Solid-phase hexapeptide ligand libraries open up new perspectives in the discovery of biomarkers in human plasma. *Clin Chim Acta.* 2011;412:740–747.

Lescuyer P, Hochstrasser D, Rabilloud T. How shall we use the proteomics toolbox for biomarker discovery? *J Proteome Res.* 2007;6:3371–3376.

Li N, Wang N, Zheng J, et al. Characterization of human tear proteome using multiple proteomic analysis techniques. *J Proteome Res.* 2005;4:2052–2061.

Liang C, Tan GS, Chung MC. 2D DIGE Analysis of serum after fractionation by ProteoMiner™ beads. *Methods Mol Biol.* 2012;854:181–194.

Liu C, Zhang N, Yu H, et al. Proteomic analysis of human serum for finding pathogenic factors and potential biomarkers in preeclampsia. *Placenta.* 2011;32:168–174.

Maccarrone G, Milfay D, Birg I, et al. Mining the human cerebrospinal fluid proteome by immunodepletion and shotgun mass spectrometry. *Electrophoresis.* 2004;25:2402–2412.

Mair J, Dienstl F, Puschendorf B. Cardiac troponin T in the diagnosis of myocardial injury. *Crit Rev Clin Lab Sci.* 1992;29:31–57.

Malaud E, Piquer D, Merle D, et al. Carotid atherosclerotic plaques: Proteomics study after a low-abundance protein enrichment step. *Electrophoresis.* 2012;33:470–482.

Marrocco C, Rinalducci S, Mohamadkhani A, D'Amici GM, Zolla L. Plasma gelsolin protein: a candidate biomarker for hepatitis B-associated liver cirrhosis identified by proteomic approach. *Blood Transfus.* 2010;8:s105–s112.

Martinkova J, Jelinkova L, Jarkovska K, et al. Protein profiling of human follicular fluid: Quest for biomarkers of ovarian hyperstimulation syndrome. *Cancer Genom Proteom.* 2009;6:58–65.

Martins-de-Souza D, Maccarrone G, Wobrock T, et al. Proteome analysis of the thalamus and cerebrospinal fluid reveals glycolysis dysfunction and potential biomarkers candidates for schizophrenia. *J Psychiatr Res.* 2010a;44:1176–1189.

Martins-de-Souza D, Wobrock T, Zerr I, et al. Different apolipoprotein E, apolipoprotein A1 and prostaglandin-H2 D-isomerase levels in cerebrospinal fluid of schizophrenia patients and healthy controls. *World J Biol Psychiatry.* 2010b;11:719–728.

Martins-de-Souza D. Is the word 'biomarker' being properly used by proteomics research in neuroscience? *Eur Arch Psychiatry Clin Neurosci.* 2010;260:561–562.

Martos G, López-Fandiño R, Molina E. Immunoreactivity of hen egg allergens: influence on in-vitro gastrointestinal digestion of the presence of other egg white proteins and of egg yolk. *Food Chemistry* 2013;136:775–781.

Mattsson N, Portelius E, Rolstad S, et al. Longitudinal cerebrospinal fluid biomarkers over four years in mild cognitive impairment. *J Alzheimers Dis.* 2012;30:767–778.

Meilhac O, Leger T, Lavigne D, et al. The biomarkers of coronary events study (biocore): from plasma sampling to discovery of new circulating biomarkers of atherosclerosis using differential proteomic analysis. *Arch Cardiovascular Diseases.* 2009;102:S12.

Meng R, Gormley M, Bhat VB, Rosenberg A, Quong AA. Low abundance protein enrichment for discovery of candidate plasma protein biomarkers for early detection of breast cancer. *J Proteomics.* 2011;75:366–374.

Mitchell P. Proteomics retrenches. *Nature Biotech* 2010;28:665–670.

Molinari CE, Casadio YS, Hartmann BT, et al. Proteome mapping of human skim milk proteins in term and preterm milk. *J Proteome Res.* 2012;11:1696–1714.

Monari E, Casali C, Cuoghi A, et al. Enriched sera protein profiling for detection of non-small cell lung cancer biomarkers. *Proteome Science.* 2011;9:55–65.

Mouton-Barbosa E, Roux-Dalvai F, Bouyssié D, et al. In-depth exploration of cerebrospinal fluid by combining peptide ligand library treatment and label free protein quantification. *Mol Cell Proteomics.* 2010;9:1006–1021.

Musante L, Saraswat M, Duriez E, et al. Biochemical and physical characterisation of urinary nanovesicles following CHAPS treatment. *PLoS One.* 2012;7:e37279.

Nagana Gowda GA. Human bile as a rich source of biomarkers for hepatopancreatobiliary cancers. *Biomarkers in Medicine* 2010;4:299–314.

Noben JP, Dumont D, Kwasnikowska N, et al. Lumbar cerebrospinal fluid proteome in multiple sclerosis: Characterization by ultrafiltration, liquid chromatography, and mass spectrometry. *J Proteome Res.* 2006;5:1647–1657.

Oh J, Pyo JH, Jo EH, et al. Establishment of a near-standard two-dimensional human urine proteomic map. *Proteomics.* 2004;4:3485–3497.

Omenn GS, ed. Exploring the Human Plasma Proteome. Weinheim: Wiley-VCH; 2006:1–372.

Overgaard AJ, Hansen HG, Lajer M, et al. Plasma proteome analysis of patients with type 1 diabetes with diabetic nephropathy. *Proteome Sci.* 2010;8:4.

Palmer DJ, Kelly VC, Smit AM, et al. Human colostrum: Identification of minor proteins in the aqueous phase by proteomics. *Proteomics.* 2006;6:2208–2216.

Pang JX, Ginanni N, Dongre AR, Hefta SA, Opitek GJ. Biomarker discovery in urine by proteomics. *J Proteome Res.* 2002;1:161–169.

Pasini EM, Kirkegaard M, Mortensen P, et al. In-depth analysis of the membrane and cytosolic proteome of red blood cells. *Blood.* 2006;108:791–801.

Pavlou MP, Kulasingam V, Sauter ER, Kliethermes B, Diamandis EP. Nipple aspirate fluid proteome of healthy females and patients with breast cancer. *Clin Chem.* 2010;56:848–855.

Pawlik TM, Hawke DH, Liu Y, et al. Proteomic analysis of nipple aspirate fluid from women with early-stage breast cancer using isotope-coded affinity tags and tandem mass spectrometry reveals differential expression of vitamin D binding protein. *BMC Cancer.* 2006;6:68–75.

Pedreschi R, Nørgaard J, Maquet A. Current challenges in detecting food allergens by shotgun and targeted proteomic approaches: A case study on traces of peanut allergens in baked cookies. *Nutrients.* 2012;4:132–150.

Perluigi M, Di Domenico F, Cini C, et al. Proteomic analysis for the study of amniotic fluid protein composition. *J Prenat Med.* 2009;3:39–41.

Petri AL, Simonsen AH, Yip TT, et al. Three new potential ovarian cancer biomarkers detected in human urine with equalizer bead technology. *Acta Obstet Gynecol Scand.* 2009;88:18–26.

Pieper R, Gatlin CL, McGrath AM, et al. Characterization of the human urinary proteome: A method for high-resolution display of urinary proteins on two-dimensional electrophoresis gels with a yield of nearly 1400 distinct protein spots. *Proteomics.* 2004;4:1159–1174.

Pisitkun T, Gandolfo MT, Das S, Knepper MA, Bagnasco SM. Application of systems biology principles to protein biomarker discovery: Urinary exosomal proteome in renal transplantation. *Proteomics Clin Appl.* 2012;6:268–278.

Plymoth A, Yang Z, Löfdahl CG, et al. Rapid proteome analysis of bronchoalveolar lavage samples of lifelong smokers and never-smokers by micro-scale liquid chromatography and mass spectrometry. *Clin Chem.* 2006;52:671–679.

Portelius E, Brinkmalm G, Tran A, et al. Identification of novel N-terminal fragments of amyloid precursor protein in cerebrospinal fluid. *Exp Neurol.* 2010;223:351–358.

Portelius E, Zetterberg H, Dean RA, et al. Amyloid-β1-15/16 as a Marker for γ-Secretase Inhibition in Alzheimer's Disease. *J Alzheimers Dis.* 2012a;31:335–341.

Portelius E, Fortea J, Molinuevo JL, et al. The amyloid-β isoform pattern in cerebrospinal fluid in familial PSEN1 M139T- and L286P-associated Alzheimer's disease. *Mol Med Report.* 2012b;5:1111–1115.

Poste G. Bring on the biomarkers. *Nature.* 2011;469:156–157.

Raj DA, Fiume I, Capasso G, Pocsfalvi G. A multiplex quantitative proteomics strategy for protein biomarker studies in urinary exosomes. *Kidney Int.* 2012;81:1263–1272.

Ramirez-Alvarado M, Ward CJ, Huang BQ, et al. Differences in immunoglobulin light chain species found in urinary exosomes in light chain amyloidosis (Al). *PLoS One.* 2012;7:e38061.

Rao PV, Lu X, Standley M, et al. Proteomic identification of urinary biomarkers of diabetic nephropathy. *Diabetes Care.* 2007;30:629–637.

Rao PV, Reddy AP, Lu X, et al. Proteomic identification of salivary biomarkers of type-2 diabetes. *J Proteome Res.* 2009;8:239–245.

Ringrose JH, van Solinge WW, Mohammed S, et al. Highly efficient depletion strategy for the two most abundant erythrocyte soluble proteins improves proteome coverage dramatically. *J Proteome Res.* 2008;7:3060–3063.

Rivers J, Hughes C, McKenna T, et al. Asymmetric proteome equalization of the skeletal muscle proteome using a combinatorial hexapeptide library. *PLoS One.* 2011;6:e28902.

Roche S, Tiers L, Provansal M, et al. Depletion of one, six, twelve or twenty major blood proteins before proteomic analysis: the more the better? *J Proteomics.* 2009;72:945–951.

Rodriguez H, Rivers R, Kinsinger C, et al. Reconstructing the pipeline by introducing multiplexed multiple reaction monitoring mass spectrometry for cancer biomarker verification: an NCI-CPTC initiative perspective. *Proteomics Clin Appl.* 2010;4:904–914.

Roncada P, Piras C, Soggiu A, et al. Farm animal milk proteins. *J Proteomics.* 2012;75:4259–4274.

Romeo MJ, Espina V, Lowenthal M, et al. CSF proteome: A protein repository for potential biomarker identification. *Expert Rev Proteomics.* 2005;2:57–70.

Rose K, Bougueleret L, Baussant T, et al. Industrial scale proteomics: From liters of plasma to chemically synthesized proteins. *Proteomics.* 2004;4:2125–2150.

Roux-Dalvai F, Gonzalez de Peredo A, Simó C, et al. Extensive analysis of the cytoplasmic proteome of human erythrocytes using the peptide ligand library technology and advanced mass spectrometry. *Mol Cell Proteomics.* 2008;7:2254–2269.

Santucci L, Candiano G, Bruschi M, et al. Combinatorial peptide ligand libraries for the analysis of low-expression proteins. validation for normal urine and definition of a first protein map. *Proteomics.* 2012;12:509–515.

Schutzer SE, Liu T, Natelson BH, et al. Establishing the proteome of normal human cerebrospinal fluid. *PLoS One.* 2010;5:e10980.

Schweigert FJ, Gericke B, Wolfram W, Kaisers U, Dudenhausen JW. Peptide and protein profiles in serum and follicular fluid of women undergoing IVF. *Hum Reprod.* 2006;21:2960–2968.

Selvaraju S, El Rassi Z. Reduction of protein concentration range difference followed by multicolumn fractionation prior to 2-DE and LC-MS/MS profiling of serum proteins. *Electrophoresis* 2011;32:674–685.

Sennels L, Salek M, Lomas L, et al. Proteomic analysis of human blood serum using peptide library beads. *J Proteome Res.* 2007;6:4055–4062.

Shahali Y, Sutra JP, Fasoli E, et al. Allergomic study of cypress pollen via combinatorial peptide ligand libraries. *J Proteomics.* 2012;77:101–110.

Shetty V, Jain P, Nickens Z, et al. Investigation of plasma biomarkers in HIV-1/HCV mono- and coinfected individuals by multiplex iTRAQ quantitative proteomics. *OMICS: A J Integr Biol.* 2011;15:705–717.

Shores KS, Udugamasooriya DG, Kodadek T, Knapp DR. Use of peptide analogue diversity library beads for increased depth of proteomic analysis: Application to cerebrospinal fluid. *J Proteome Res.* 2008;7:1922–1931.

Sihlbom C, Kanmert I, Bahr H, Davidsson P. Evaluation of the combination of bead technology with SELDI-TOF-MS and 2-D DIGE for detection of plasma proteins. *J Proteome Res.* 2008;7:4191–4198.

Siqueira WL, Dawes C. The salivary proteome: Challenges and perspectives. *Proteomics Clin Appl.* 2011;5:575–579.

Sjodin MO, Bergquist J, Wetterhall M. Mining ventricular cerebrospinal fluid from patients with traumatic brain injury using hexapeptide ligand libraries to search for trauma biomarkers. *J Chromatogr B.* 2010;878:2003–2012.

Smith G, Barratt D, Rowlinson R, Nickson J, Tonge R. Development of a high-throughput method for preparing human urine for two-dimensional electrophoresis. *Proteomics.* 2005;5:2315–2318.

Spahr CS, Davis MT, McGinley MD, et al. Towards defining the urinary proteome using liquid chromatography-tandem mass spectrometry. *I. Profiling an unfractionated tryptic digest. Proteomics.* 2001;1:93–107.

Surinova S, Schiess R, Hüttenhain R, et al. On the development of plasma protein biomarkers. *J Proteome Res.* 2011;10:5–16.

Thongboonkerd V, Malasit P. Renal and urinary proteomics: current applications and challenges. *Proteomics.* 2005;5:1033–1042.

Ticha M, Pacakova V, Stulik K. Proteomics of allergens. *J Chromatogr B.* 2002;771:343–353.

Tu C, Li J, Young R, et al. Combinatorial peptide ligand library treatment followed by a dual-enzyme, dual-activation approach on a nano-flow liquid chromatography/orbitrap/electron transfer dissociation system for comprehensive analysis of swine plasma proteome. *Anal Chem.* 2011;83:4802–4813.

Tu C, Rudnick PA, Martinez MY, et al. Depletion of abundant plasma proteins and limitations of plasma proteomics. *J Proteome Res.* 2010;9:4982–4991.

van der Lubbe N, Jansen PM, Salih M, et al. The phosphorylated sodium chloride cotransporter in urinary exosomes is superior to prostasin as a marker for aldosteronism. *Hypertension.* 2012;60:741–748.

Veenstra TD. Where are all the biomarkers? *Expert Rev Proteomics.* 2011;8:681–683.

Versura P, Nanni P, Bavelloni A, et al. Tear proteomics in evaporative dry eye disease. *Eye.* 2010;24:1396–1402.

Waller LN, Shores K, Knapp DR. Shotgun proteomic analysis of cerebrospinal fluid using off-gel electrophoresis as the first-dimension separation. *J Proteome Res.* 2008;7:4577–4584.

Wang MC, Papsidero LD, Kuriyama M, et al. Prostate antigen: a new potential marker for prostatic cancer. *Prostate.* 1981;2:89–96.

Wang Y-S, Cao R, Jin H, et al. Altered protein expression in serum from endometrial hyperplasia and carcinoma patients. *J Hematol Oncol.* 2011;4:15–23.

Wang Z, Hill S, Luther JM, Hachey DL, Schey KL. Proteomic analysis of urine exosomes by multidimensional protein identification technology (MudPIT). *Proteomics.* 2012;12(2):329–338.

Wattiez R, Falmagne P. Proteomics of bronchoalveolar lavage fluid. *J Chromatogr B.* 2005;815:169–178.

Wu J, Kobayashi M, Sousa EA, et al. Differential proteomic analysis of bronchoalveolar lavage fluid in asthmatics following segmental antigen challenge. *Mol Cell Proteomics.* 2005;4:1251–1264.

Xie H, Rhodus NL, Griffin RJ, Carlis JV, Griffin TJ. A catalogue of human saliva proteins identified by free flow electrophoresis-based peptide separation and tandem mass spectrometry. *Mol Cell Proteomics.* 2005;4:1826–1830.

Yamada M, Murakami K, Wallingford JC, Yuki Y. Identification of low-abundance proteins of bovine colostral and mature milk using two-dimensional electrophoresis followed by microsequencing and mass spectrometry. *Electrophoresis.* 2002;23:1153–1160.

Yan W, Apweiler R, Balgley BM, et al. Systematic comparison of the human saliva and plasma proteomes. *Proteomics Clin Appl.* 2009;3:116–134.

Ye H, Sun L, Huang X, Zhang P, Zhao X. A proteomic approach for plasma biomarker discovery with 8-plex iTRAQ labeling and SCX-LC-MS/MS. *Mol Cell Biochem.* 2010;343:91–99.

Zanusso G, Fiorini M, Righetti PG, Monaco S. Specific and Surrogate cerebrospinal fluid markers in Creutzfeldt–Jakob disease. *Adv Neurobiol.* 2011;2:455–467.

Zerefos PG, Aivaliotis M, Baumann M, Vlahou A. Analysis of the urine proteome via a combination of multi-dimensional approaches. *Proteomics.* 2012;12:391–400.

Zerefos PG, Vougas K, Dimitraki P, et al. Characterization of the human urine proteome by preparative electrophoresis in combination with 2-DE. *Proteomics.* 2006;6:4346–4355.

Zhang J, Goodlett DR, Peskind ER, et al. Quantitative proteomic analysis of age-related changes in human cerebrospinal fluid. *Neurobiol Aging.* 2005;26:207–227.

Zhi W, Purohit S, Carey C, Wang M, She JX. Proteomic technologies for the discovery of type 1 diabetes biomarkers. *J Diabetes Sci Technol.* 2010;4:993–1002.

Zhou L, Zhao SZ, Koh SK, et al. In-depth analysis of the human tear proteome. *J Proteomics.* 2012;75:3877–3885.

Zürbig P, Dihazi H, Metzger J, Thongboonkerd V, Vlahou A. Urine proteomics in kidney and urogenital diseases: Moving towards clinical applications. *Proteomics Clin Appl.* 2011;5:256–268.

CHAPTER 7

Other Applications of Combinatorial Peptide Libraries

7.1 SCREENING FOR AFFINITY LIGAND SELECTION

From the beginning of the 1970s, affinity chromatography met with immense success because this was a unique approach to isolate given proteins in a single step. In practice, this great promise has been counterbalanced by a large number of practical concerns, the main one being the selection or preparation of selective ligands for specific purposes. Some approaches have been given in Chapter 3 where various affinity technologies are described for separating proteins or groups of proteins.

The case of affinity chromatography is complicated by the availability of specific ligands. Many of them have been described, but proteins are so numerous that the majority of this field is yet to be discovered. It is within this context that libraries of ligands have been conceived from where ligands could be fished out and then used for protein purification purposes. Aside from specific antibodies for immunoaffinity chromatography (Moser and Hage, 2010) (polyclonal, monoclonal, camelides, ribosome display ligands), which are not necessarily good examples of a ligand library, there are various approaches. Some ligands are of proteinaceous origin, such as affibodies (mentioned in Section 3.2.1) and phage display libraries generating a large number of protein ligand candidates that need to be screened to detect the one having the appropriate affinity dissociation constant and specificity for the target protein. Some others are based on polynucleotides such as DNA- and RNA-aptamers, with an extraordinary diversity guaranteeing the existence of an appropriate affinity ligand for each single application, but that has to be selectively sequenced and then synthesized prior to being grafted on solid chromatographic phases (Zhao and Chris, 2009; Mairal et al., 2008; Javaherian et al. 2009). Finally, there are small molecules such as chemical libraries (Leimbacher et al., 2012; Linhult et al., 2005; Roque et al., 2005), peptides (Lam et al., 1997; Tozzi and Giraudi, 2006;

TABLE 7.1 Examples of Peptide Identified as Ligands for Affinity Chromatography

Protein	Selected Affinity Peptides	References*
Streptavidin	TPHPQ	Lam et al., 1991
p60(c-src) tyrosine kinase	YIYGSFK	Lam et al., 1995
Melanoma MSH receptors antagonist	WRL	Quillan et al., 1995
Glycosomalphosphoglycerate kinase	NWMMF	Samson et al, 1995
Ribonuclease S	YNFEVL	Buettner et al., 1996
Von Willebrand factor	RVRSFY	Huang et al., 1996
Alpha-6-beta-1-integrin	LNIVS-VNGRHX	Pennington et al., 1996
Coagulation factor IX	YANKGY	Buettner et al., 1996
Alpha-1-proteinase inhibitor	RAFWYI	Bastek et al., 2000
Fibrinogen	FLLVPL	Kaufman et al., 2002
Staphylococcal enterotoxin B	YYWLHH	Wang et al., 2004
Fc region of human IgG	HWRGWV	Yang et al., 2005
Anti-insulin Antibodies	EFDWNH	Lehman et al., 2006
Prion protein	GLERPE	Lathrop et al., 2007
Interleukin 2	GVASED	Sarkar et al., 2007
Porcine insulin	HWWWPAS	Dong et al., 2007
Alpha-amylase	FHENWS	Liu et al., 2007
Beta actin	Various	Miyamoto et al., 2008
αvβ3 integrin	RGD	Xiao et al., 2010
Anti-hemophilic Factor VIII	Cyclic peptide	Kelley et al., 2010
Bovine Carbonicanhydrase II	IHRYWF	Lee et al., 2010

References do not necessarily report only one peptide sequence.

Pande et al., 2010), and peptoids (Kodadek and Bachhawat-Sikder, 2006) that can be used as a source of affinity ligands. To stay within the scope of this book, only peptide libraries will be discussed as a potential source of chromatographic baits.

A major breakthrough in this domain came from Lam et al. in 1991. Since then, many other applications have been reported in the literature, and they still continue (see examples in Table 7.1). The principle of peptide ligand discovery is to incubate the bead peptide library with the protein of interest and then identify beads that are positive once probed with a specific antibody labeled with a dye or an enzyme.

7.1.1 Strategy Definition

Before entering the process of selecting a peptide ligand out of a library comprising millions of possible structures, it is advisable to state the final objective in exact terms. In Chapter 4 (Sections 4.4.2 and 4.4.3), it has been extensively explained how many possible molecular interactions are generated and, due to the complexity of the structures, that they are present under different forms in each hexapeptide ligand. Most of these interactions are dependent on the physicochemical conditions, especially pH, ionic strength, and hydrophobic-hydrophilic balance of the buffer constituents. Therefore, to select a proper ligand, the initial conditions must be perfectly known. For example if the proteins are to be extracted from a crude protein solution in a physiological buffer, the selection process must be exactly done in this same buffer. If the protein for which a specific ligand is searched belongs to a tissue extract where detergents or urea or other chemical agents are present, the screening conditions must comply with these requirements.

In theory, affinity peptide ligands could be sorted out for many types of environmental conditions, including the capture of proteins at acidic pH with elution at neutral pH. This could be the case when the feedstock is an acidic eluate from a previous chromatographic column and needs to be directly injected into the affinity column without any preliminary modification of pH and/or ionic strength. Moreover, among potential peptide ligands that are chosen, a second selection is operated for elution at preestablished conditions: acidic or alkaline or neutral conditions or even physiological conditions (e.g., phosphate buffered saline). Under these conditions, the affinity interaction must be completely lost to have the protein desorbed from the column. The capture of a given protein with its elution with neutral buffers is a very special and rare situation but could be very convenient for simplifying the sample treatment before and after affinity separation.

It is also advisable to select the temperature that will be the one used at the exploitation stage. In fact, hydrophobic interactions are dependent on temperature, and if the ligand selection is operated at room temperature but the final affinity column is used at 4 °C, the obtained results could be very different from the expected ones.

As a reminder, a combinatorial peptide library is made by solid-phase synthesis directly on the beads. Once the peptide ligand is selected, it is strongly recommended that the affinity column be prepared by solid-phase synthesis on the same solid support. The chemical immobilization of the same peptide even on a similar chromatographic media could produce different results in terms of specificity for the target protein.

7.1.2 Technical Approaches with Unmodified Peptide Libraries

A number of variants have been reported for screening solid-phase combinatorial peptide ligand libraries with the objective of identifying a specific ligand for a given protein. In practice, a protein extract is incubated under predefined conditions with the peptide ligand library for the time necessary to reach the equilibrium. Then the solid phase is separated by filtration or centrifugation and washed with the buffer used for incubation until the excess of proteins is eliminated. Then the peptide library beads on which protein partners are adsorbed are incubated with an antibody against the target protein under the same buffering conditions, followed by a wash to eliminate the excess of antibody. Finally, the beads carrying the target protein with its specific antibody are revealed by immunostaining with fluorescent dyes or with enzymes. Beads that are thus highlighted are subtracted from the mixture using appropriate tools and the proteins desorbed using stringent elution agents selected among those suggested in Section 8.15. These beads are then submitted to analysis of the sequence of peptides that are covalently attached and thus get the structure of the hexapeptide capable of recognizing the target protein. Figure 7.1 depicts a schematic view of the various steps.

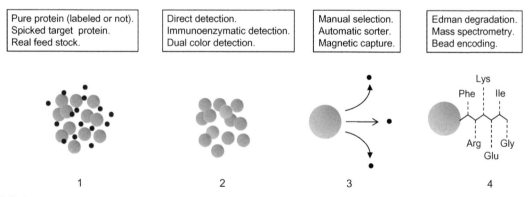

FIGURE 7.1

Schematic representation of an a affinity peptide identification from a large library in four main steps.

1: The library beads (large gray bowls) are incubated with the biological sample (small black dots; see box above) under predefined conditions of pH and ionic strength. The sample could be one of those indicated in the above box. Then the protein excess is eliminated by simple washings.

2: Positive beads (beads that contain the target protein) are detected in the entire library by various specific staining procedures (see box above).

3. Positive beads are then isolated from the ligand library by means of various possible mechanical approaches as described in the insert above and the captured protein(s) desorbed completely.

4: Each positive depleted bead is sequenced to obtain the peptide composition. Sequencing is performed according to several methods, as those described in the insert above. *Adapted from Boschetti and Righetti, (2012).*

The procedure can easily be carried out using the purified target protein with a cleaner process. However, the process is more effective when the crude extract containing the target protein is used instead. The most important advantage here is to have the process of screening similar to the conditions of separation where the target protein competes with other species for the same or similar ligand but with different affinity dissociation constants. In spite of taking these precautions, a quite large number of false positives needs to be circumvented to end up to an optimized process. In fact, at the end of the selection many beads appear as positive, and from each of them the carried peptide is to be sequenced, thus producing a quite large useless workload. False positives can come from (i) the adsorption of the target protein by a low-affinity constant, (ii) the adsorption of antibodies alone on beads that do not have the target protein but are simply adsorbed because of the recognition process between the peptide and the antibody itself, and (iii) the co-adsorption of other proteins on beads that are thus not very specific for the target. All these false positives constitute the most important technical issue that needs to be resolved by various technical approaches. They could include a second screening process of positive beads from a first selection, thus reducing the level of false positives. Special washings could be adopted to reduce the phenomenon.

Another technique, as detailed below, consists in a double staining (Lam et al., 1995; Buettner et al., 1996). Nonetheless, it is very difficult to eliminate all possible false-positive situations. In spite of false-positive issues, short peptide chains screened from combinatorial banks probably represent one of the most attractive approaches to identify affinity ligands for protein isolation by affinity chromatography.

Over time, numerous methods of ligand bead selection have been reported. Therefore, the technology has been progressively improved in view of reducing the number of false-positive responses. Purified targets as well as crude protein extracts containing or not the target protein have been used so as to retrieve or deduct rational responses. The target protein could be labeled or not. Even when using stringent washing conditions of captured species the reduction of false positive can still be large depending on the complexity of the initial protein extract. The elution conditions of the target protein also need to be checked to comply with normal exploitation situations. The interaction of a protein with a given ligand could be so strong and specific that the elution becomes a real challenge without the denaturation of the target protein. It is within the context of reducing false positives and of obtaining good exploitation conditions that variations of the main screening principle have been devised. In addition, owing to the very large number of ligand combinations, improvements have also been suggested to increase the throughput of screening and ligand identification methods. The following sections list the efficient, most well-known peptide ligand identification approaches from solid-phase libraries that have chronologically been reported.

7.1.2.1 DIRECT METHOD

Described by Lam et al. (1991), the direct method proposes that, prior to incubation with the peptide beads, the target protein is grafted with an enzyme that can be easily detected (e.g., alkaline phosphatase). Other possibilities are to label the target protein with fluorescent dyes. Whatever the labeling procedure, the target protein is mixed with the peptide library and incubated under gentle stirring. After exhaustive removal of protein excess with repeated washings, the enzymatic activity carried by the captured target protein is revealed with a colored substrate. Positive colored beads are then removed from the bead bulk and analyzed. As an alternative to the enzymatic reaction, a light of appropriate wavelength could be used when the target protein is labeled with fluorescent dyes. Selected beads are then treated with a stripping agent (e.g., 6 M guanidine-HCl, pH 6) to desorb the target protein and then are submitted to the peptide sequencing. Since this process still comprises a number of false positives, Sepetov et al. (1995) described variants and optimizations of this method.

The direct method was used to identify an octapeptide for separation or recombinant erythropoietin (Martinez-Ceron et al., 2011). Erythropoietin was previously labeled with either Texas Red or biotin, and, after incubation with the peptide library, the positive beads were evidenced by either the color or the reaction with avidin. About 70 positive peptides were identified, and two of them were used for the purification of the target protein from a cell culture supernatant with a purity between 95 and 97% and 90% yield.

The use of red fluorescent dyes as a direct method for sorting out positive beads was further investigated by Marani et al. (2009). Streptavidin was used as a probe protein conjugated with either Atto 590 or Texas Red dyes. Once the bead library was incubated with the labeled target protein, the positive beads were separated by flow-sorting equipment. While the sorter does not distinguish the true from false positive, it was found that an in-depth observation of separated beads made it possible to make the distinction. The false-positive beads showed bright homogeneous fluorescence, while real positive beads showed a sort of fluorescence gradient from the surface to the center, with the greatest fluorescence intensity at the bead surface (this is a phenomenon known in protein adsorption on chromatographic beads). From this difference it was possible to separate the two populations of beads and then make a peptide sequence by MALDI TOF mass spectrometry. As a final note, the authors indicated that best results are obtained with Texas Red.

7.1.2.2 DUAL COLOR

With the objective of reducing the number of false positives, a quite similar process based on a dual-color screening was described by Lam et al. (1995). Basically, the beads were labeled and screened first with one color and then by a second orthogonal reagent producing a second color. Orthogonality of staining allowed a better selection of beads that were also in a much smaller number for a faster decoding.

7.1.2.3 EVOLUTION OF DUAL COLOR

A year later, in 1996, another two-color approach was reported by Buettner et al. (1996). The reported method allowed screening a library for the discovery of a peptide having affinity and specificity for human coagulation Factor IX. The method consisted in multiple sequenced injections into HPLC library packed columns alternated by immunostaining, releasing two different colors. The first phase consisted in injecting just albumin, followed by the antibody against the target protein, and finally the secondary antibody alkaline phosphatase conjugate. Then all beads were submitted to the enzymatic detection using a first substrate releasing a blue color (5-bromo-4-choro-3′-indolyphosphate-p-toluidine salt/nitrobluetetrazolium chloride). The second phase consisted in contacting the resin with the target protein (human Factor IX) in the presence of the same buffered albumin and then followed by the sequence of the same two antibodies. This time the enzymatic activity was determined using a different substrate: naphtol AS-MX phosphate releasing a red color. Red color unambiguously indicated reactivity for the target protein, although some beads took some of this red color up, turning beads from blue to brown depending on the degree of uptake. Although quite elaborate and necessitating programmed HPLC, this approach significantly reduced the number of false positives. Then the extracted beads were analyzed to sequence the affinity peptide, and the latter was validated for the separation of coagulation Factor IX.

7.1.2.4 IMAGE SUBTRACTION

Lehman et al. (2006) have proposed another strategy to reduce the number of false positives based on the use of two protein mixtures as screening probes and described as "image subtraction." The objective was to rapidly identify few true-positive beads out of a still large number of false positives. The method is based on a two-stage bead selection process and uses an image subtraction technique. The first step (made twice) involves the detection of all possible nonspecific bindings by incubating the peptide library beads with a blocking mixture comprising albumin and Tween-20 in phosphate buffered saline, followed by the addition of biotinylated antibodies against insulin target protein, and finally with the secondary streptavidin-alkaline phosphatase conjugate. Between additions, the solid phase was always washed to remove the excess of reagents. Finally, the phosphatase activity was revealed with 5-bromo-4-chloro-3′-indoly-phosphate as a substrate. The false-positive beads thus turned to blue. After the addition of reference red-colored beads, the whole bead mixture was evenly dispersed in a low-melting agarose to form a thin layer of transparent agarose gel where beads were optically distinguishable. Then the enzyme substrate was added over the exposed surface of the agarose gel, and a first image was immediately taken with a scanner. Then a second image was taken after a two-hour incubation. Differences between images (image subtraction) allowed distinguishing the positive beads for the specific adsorption of antibodies against insulin. This method appeared simpler than the tag encoding, with the advantage of using regular laboratory equipment.

7.1.2.5 BEAD-BLOT

Lathrop et al. (2007) described a procedure named "bead-blot," which was radically different from all the methods described above. After incubation with the crude sample containing the target protein and the elimination of protein excess by repeated washings, the peptide beads were embedded within a gel layer of low-melting agarose, forming a sort of bead array. Beads were thus dispersed and immobilized as a monolayer within an agarose gel layer. The proteins captured by the beads were then transferred in a manner similar to DNA transfer in a Southern blot and seized by a protein-binding membrane placed on top of the agarose gel layer. When the diffusion of proteins from the beads resulting from progressive desorption of proteins from the beads (mostly active, but also passive) took place, each protein was captured by a blot membrane where the proteins were immediately immobilized and then probed for the target protein presence. The positioning of beads from where the positive spot protein was revealed was retrieved by superposition, and the identified beads were taken out of the agarose gel. (A detailed procedure is described in Section 8.22.)

The peptide ligand was usually sequenced by Edman degradation technology using an automatic polypeptide sequencer. In the present technology, false positives were largely reduced because the proteins were probed upon diffusion on the membrane as reported, but were not completely cleared by the concomitant co-capture of other proteins by the same bead as described in Section 4.4.1. This procedure allowed identifying peptide ligands for several proteins during the same experiment; hence this approach is very useful for the design of cascade affinity chromatography where various proteins are to be isolated from the same feedstock. Pseudo-false positives here are only due to the co-adsorption of proteins on the same bead. This phenomenon is almost unavoidable and can be managed by optimizing conditions of adsorption and desorption at the separation process level. The authors demonstrated the ability of the method to sequence peptides usable for separating vWF/FVIII complex from undiluted human plasma. The captured and eluted fraction was considerably enriched. It has also been successfully adapted for finding peptide ligands specific to fibrinogen (Kaufman et al., 2002) and more importantly peptides capable of binding prion proteins from whole blood and plasma (Lathrop et al., 2007).

7.1.2.6 LOW-LOADING BEAD APPROACH

The presence of false positives by nonspecific binding was hypothesized as being the result of high-ligand densities on the beads. In fact, the high-ligand density may induce simultaneous interactions with more than one peptide ligand at different amino acid locations. In order to resolve this question, Chen et al. (2009) proposed to screen a library where the peptide ligand density was reduced by a factor of 10 and tested against a high-density control library. The work was constructed around the identification of a specific ligand against several SH2 domains of Csk and Fyn kinases using a phosphotyrosyl-containing peptide library. While with the high-density peptide library there were a number of false-positive sequences that did not allow measuring affinity values for the target proteins, with the low-density library it was possible to retrieve specific recognition sequences for each of the target protein domains. The negative aspect of this approach is that the small amount of peptides in each bead requires a very sensitive analytical method to make a proper sequencing and also the low amount of the target protein obtained makes its identification challenging.

To the question of peptide bead screening, Maillard et al. (2009) have provided additional insights into peptide libraries. Table 7.2 summarizes the advantages and disadvantages of bead identification.

All in all, to keep the process simple but effective, it is advised that the above-summarized methods be combined and that stringent washings be made in between. Basically, once the incubation of the protein extract is completed and protein excess is totally washed out, the peptide library should be washed with the same incubation buffer containing 1 M sodium chloride or potassium chloride. This increase of ionic strength eliminates a large number of nonspecific binding due to electrostatic interactions. Then the beads are incubated with the selected labeled antibody for the identification of positive beads. The latter are separated from all other beads, and then the captured proteins are completely desorbed by washing with the desorbing agents described in Section 8.14. The washed peptide beads are equilibrated with a 0.1–0.2 M glycine HCl, pH 2.6, and then incubated with the target protein previously labeled with a fluorescent dye. Upon washing, only the negative beads (nonfluorescent) are taken for determining the peptide sequence. This second step ensures that the target protein adsorbs with phosphate buffered saline and that it does not move while washing with 1 M sodium chloride, but very probably elutes in the presence of an acidic buffer. A third incubation with the initial crude extract followed by a stringent wash could be performed and single beads loaded in a few µL of Laemmli buffer followed by electrophoresis separation. Compared with the pure target protein used as a control, the pattern can reveal the purity of the target protein that could be additionally formally identified by immunoblot.

As described in Chapter 4, one bead out of a large peptide library generally captures more than one protein at a time, even in physiological conditions where the buffer comprises 150 mM sodium

TABLE 7.2 Main Bead Identification Procedures: Comparative Features

Screening Method	Advantages	Disadvantages
Direct methods	Easy procedure for direct contact and detection	Nonspecific binding with a number of false positives.
Dual-color methods	Reduction of a number of false positives.	Complex procedure with still non-specific binding.
Image subtraction	Further reduction of false positives. Reduction of the number of sequencings.	Complex approach with sophisticated instruments.
Bead blot	High specificity. Reduction of false positives to the minimum.	Long process. Manual approach with low possibility of automation.
Low-loading beads	Takes advantage of thermodynamics. Very low level of non-specific binding.	Low amounts of peptide to decode and low amounts of protein collected for purity analysis.

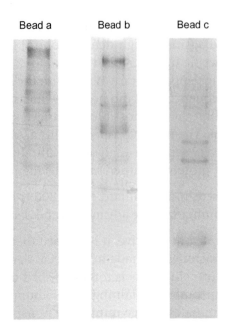

FIGURE 7.2

SDS-polyacrylamide gel electrophoresis of eluates from individual beads that were part of a hexapeptide library contacted with a large excess of human serum in physiological conditions. Once isolated, individual beads were boiled in 10 µL of Laemmli buffer and directly loaded for electrophoresis separation. Protein staining was performed with a silver procedure. More than one protein was found per bead. Proteins are not the same from bead to bead. *Adapted from Righetti and Boschetti, (2009).*

chloride, which reduces the electrostatic involvement. Figure 7.2 shows the protein content of three single beads taken randomly from a library incubated with a human serum sample. A selected screening process to reduce false positives should not only involve the use of double staining and other tricks as described above, but should also take into consideration the probability that a single ligand could be specific for two or more proteins, with probably slightly different dissociation constants. If the specificity is defined as being the result of multiple types of interactions due to the existence of different side chains of amino acids composing the peptide, a way to proceed could be to apply washings so that each of them individually targets single types of interactions. For example, electrostatic interactions can be reduced by means of salt washing, as recommended above. Hydrophobic associations should be weakened by a wash with about 20% ethylene glycol or 0.1% of a non-ionic detergent. Finally, hydrogen-bonding interactions should be destroyed by a wash with urea 4–8 M concentration. Proteins that remain still anchored to the peptide bead should thus interact by multiple forces that act all together to form the affinity docking. Figure 7.3 depicts the process. If all these washings are necessary for a specific binding of the target protein on the selected peptide, they are also to be operated when the affinity chromatography process is designed.

Libraries used to reduce the dynamic concentration range are generally designed with a quite high density of peptides, with, for instance, about 80 µmoles of peptide grafted to 1 mL of wet beads. This high density favors the docking of proteins by mechanisms already described and probably influences what is qualified as nonspecific for the process of selecting ligand for affinity separations. As said above, to reduce this phenomenon, Chen et al. (2009) found first that the major reason for false positives comes from high-ligand density; they consequently proposed to reduce the ligand density by a factor of 10. However, when reducing the ligand density, the peptide sequencing becomes more challenging due to the very low availability of material. This work was perhaps the confirmation of a previous observation made by Wang et al. (2005) where the authors exploited beads composed of a bilayer of two different ligand densities: a low concentration of random peptides on the outer layer where the high screening stringency could be operated for better

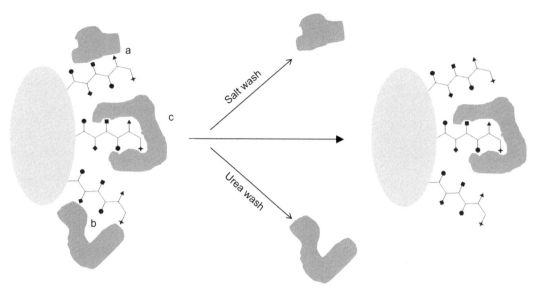

FIGURE 7.3

Schematic representation of multiple protein docking on the same bead (same peptide sequence) with true affinity and nonspecific binding. "a" is a protein that is supposed to be captured by electrostatic interaction with a few amino acids of the peptide. "b" is another protein that interacts by means of hydrogen bonding with other amino acids of the peptide ligand. Protein "c" is the one that interacts by several molecular forces and is qualified for true affinity. To get a better specificity "a" and "b" are washed out, using weakeners of either electrostatic interactions (e.g., salt wash) or hydrogen bonding (e.g., urea), leaving in place only the protein captured more specifically.

specificity, and a full substitution inner layer comprising enough of the same peptide to facilitate its identification by microsequencing. All these suggested precautions should also take into consideration the oversaturation of the beads.

This phenomenon described for the reduction of dynamic concentration range at Chapter 4 is not considered in the investigations for the detection of affinity ligands. Nevertheless, due to a very intense protein competition when overloading the beads, false positives may increase upon overloading. Moreover, whatever the overloading extent, it does not mimic what really happens during the following affinity chromatography process when purifying the target protein. In fact, the exploitation of affinity chromatography always requires staying below the saturation of the binding capacity, which contrasts with a screening under an oversized sample.

The bead decoding operation (peptide sequence)—another important aspect of the ligand discovery process—can be performed by Edman degradation (Lam et al., 1991) or by mass spectrometry (Redman et al., 2003). A side-by-side comparison of both methods has recently been performed demonstrating that they are consensual at least when using these methods for determining the identity of solid-phase peptide ligands for bovine carbonic anhydrase II (Lee et al., 2010).

7.1.3 Bead Encoding as a Way to Identify the Positive Peptides

As indicated above, there are two tasks that are labor-intensive after having evidenced positive beads: (i) the separation of the positive beads, an operation that is generally performed either manually under a microscope or with sorting automatic systems when the tag is a fluorescent dye; and (ii) the sequencing of grafted peptides on the beads after identification of positive signals and bead separation. This second operation may be difficult, especially when the number of beads to analyze is large. To simplify these tasks, some technical approaches have been proposed.

To facilitate the separation of positive beads from the bead bulk, a specific investigative work has been made by Astle et al. (2010) with an interesting proposal. After incubation with the protein mixture containing the target protein, the bulk beads are incubated with the antibody against the

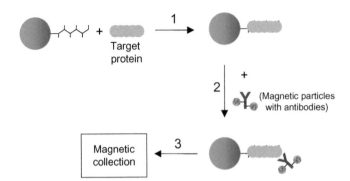

FIGURE 7.4

Schematic representation of the separation of positive beads by a magnetic field. First, the library of peptides is incubated with the crude extract comprising the protein to purify. After capture of the target protein (1) and elimination of the excess of protein load followed by stringent washings to eliminate additional nonspecifically adsorbed species, the bead mixture is incubated with magnetic particles (2) supporting specific antibodies against the target protein. After appropriate incubation, the bead mixture is submitted to a magnetic field, and the attracted proteins are easily separated (3).

target previously covalently attached on very small particles of magnetic iron oxide. Beads that bind the target protein also bind the magnetic beads; this recovered property is advantageously used to separate the population of constructs on magnetic surfaces from all other beads. It is from this separated population that the peptide sequence is performed on each bead. Figure 7.4 represents the principle.

The question of directly linking the labeled bead with the structure of the peptide has been approached in several different ways; for instance, the beads themselves are precoded during or before the preparation of the ligand library so that a different code is associated with each peptide sequence.

Since solid-phase beads are not positionally encoded as they would be in a microarray, various technologies have been devised for identifying the ligand on the beads without having to proceed for peptide sequencing by Edman degradation. The main approaches use molecular tags of identifiable chemicals or physical signatures. In spite of numerous possibilities, the tagging process must take into consideration the compatibility of tags with the synthesis of peptides, on the one hand, and the easy deconvolution without having to use complex procedures of both synthesis and analysis, on the other hand. Another important feature is the possibility of covering a very large number of combinations without overlapping.

Without mentioning pioneering investigations in this area, Affleck (2001) described creative encoding concepts, with interesting options to select encoding techniques with respect to matching library design and compound management criteria.

In 2002 Liu et al. described another concept based on the use of "topologically segregated bifunctional beads." Beads were first made in a biphasic solvent and were then used to build a ligand library. While the coding tag area is segregated within the interior of the beads, the external part of the beads was used for affinity screening therefore the two zones of each bead did not interfere with each other.

In 2004 Hwang et al. reported a mass-tag encoding strategy for a small-molecule combinatorial library with a Quantum Dot/COPAS assay. Although this approach offers certain advantages over previous methods, it appears relatively complicated to be reduced into practice by common biochemistry laboratories.

In 2010 Marcon et al. developed a so-called colloidal bar coding involving the grafting of fluorescent microparticles to the beads during the split-and-mix synthesis process. The beads could then be separated by a sorter capable of identifying the ligand sequence. The process involves the use of colloidal silica particles containing identifiable combinations of fluorescent dyes. In this way a relatively limited amount of dyes is enough to encode millions of compounds. In an example, the authors demonstrated that by using a small peptide library encoded during the peptide synthesis, it was possible to identify peptide sequences with cleavage sites for trypsin.

In another approach, Kim et al. (2011) proposed encoding with surface-enhanced Raman spectroscopic dots adsorbed on the beads' surface during the synthesis of the library, thereby encoding each corresponding amino acid. Consisting of silver nanoparticle-coated silica, they allow large encoding possibilities covering millions of combinations and easily readable by Raman spectroscopy.

In a recent review, Mannocci et al. (2011) described the merits of DNA-encoded libraries as representing an effective and cheap tool for identifying ligands from a library. Each ligand is linked to a unique DNA tag usable as an amplifiable bar code by PCR. Thus this type of encoding allows identifying structures at subpicomolar concentrations. Several strategies of DNA encoding have been reported for preparing libraries comprising millions of compounds without overlapping.

While the ideal bead encoding is yet to be discovered, progress has been made in a field where this question is crucial for reducing labor and achieving better management of the library screening processes.

7.1.4 Examples of Successful Affinity Peptide Ligands Discovery

A number of published reports describe the use of affinity ligands for separating a given protein after ligand selection from a peptide library. Table 7.1 summarizes the situation of most known protein purifications issued from peptide library screening. Following are several examples related to the specificity and difficulties associated to the ligand selection.

Alpha-1-proteinase inhibitor was purified from human plasma using a grafted hexapeptide on chromatography beads that was previously identified from a peptide library (Bastek et al., 2000). The total number of peptidomers of the library was 34 million. However, only 2% of the library was tested, representing around 680,000 hexapeptide sequences. Analysis of positive beads revealed a relatively low level of consensus sequences, even if significant similarities were shown. For example, all sequences comprise at least an aromatic ring (Phe, Trp, Tyr) and a cationic amino acid (Arg, His, Lys). After having observed the best conditions of most complete alpha-1-proteinase inhibitor capture and the best elution in the presence of 2% acetic acid, several sequences were selected to validate the capability of these chosen peptides to separate the target product by affinity chromatography using first protein mixture models and then a real feedstock from the Cohn plasma fractionation process. The best peptides were VIFLVR and RAFWYI, with yields of 79% and 80%, respectively, and producing alpha-1-proteinase inhibitor at a purity of 52% and 69%, with a major contaminant that was apolipoprotein A1.

By using an immunostaining screening technique, a hexapeptide specific for fibrinogen was isolated and adopted for its purification (Kaufman et al., 2002). After peptide sequencing of the positive beads, the most viable seemed the sequence FLLVPL, which was grafted to chromatographic beads with which the purification of fibrinogen was tested directly from crude plasma. The loading was performed with a 20 mM HEPES buffer pH 6.8 followed by a wash with 0.1 M sodium chloride; the protein elution was operated by means of 2.5% acetic acid. SDS-polyacrylamide gel electrophoresis and Western blot analysis confirmed that the separated protein was effectively fibrinogen. After having explored the influence of various physicochemical parameters, it was possible to optimize the separation with an almost homogeneous band as attested by sodium dodecyl sulfate (SDS)-polyacrylamide gel electrophoresis.

The one-bead–one-compound screening was also used to identify small molecule ligands targeting proteins from cytoplasm cell extracts. This was the case of small ligands active against actin present in Ramos B-lymphoma cells (Miyamoto et al., 2008). The screening procedure was performed by two distinct steps. The first was a screening by immunostaining using the whole-cell extract followed by incubation with an anti-actin antibody. The latter was then recognized by a secondary antibody conjugate supporting alkaline phosphatase activity that was evidenced with an appropriate substrate. As a second step, the beads were submitted to a second screening based on image subtraction (see above). A novel peptide synthesis of the identified sequences was then prepared on hydrophilic beads (beads with polyethylene glycol) and exposed to the cell extract. This *de novo* solid phase demonstrated its ability to capture Ramos proteins, and formal identification was confirmed by immunoblotting. The bioactivity effect of these ligands was also examined on the proliferation of lymphoma cells, with a clear reduction of the cell population indicating that these molecules could affect different protein targets. This suggests that these libraries could also constitute a great mine for molecules of biological importance.

With a detection of positive beads using either Texas Red or biotin conjugated target protein, an octapeptide library was screened in view of discovering an affinity ligand for recombinant erythropoietin, a factor used for anemia therapy (Martinez-Ceron et al., 2011). After cleavage of the five dozen selected peptides from the positive beads, their sequencing was performed by mass spectrometry and their ability to bind erythropoietin was determined by Plasmon resonance. Leads with dissociation constants of 1.8–2.7 μmol were then grafted on agarose beads for chromatography, and separation trials started using a CHO cell culture supernatant containing human erythropoietin. The authors demonstrated that the two ligands with the largest sequence consensus allowed separating the target protein with a yield of about 90% and a purity higher than 95–97%.

In recent years, a pressing need has arisen to find small high-affinity IgG-binding ligands from various species and from different biological fluids. Among a number of attempts for ligand discovery, libraries of peptides are perhaps the most promising. From a combinatorial hexapeptide library, Yang et al. (2005) described for the first time a few working sequences. The most interesting were HYFKFD, HFRRHL, and HWRGWV, all of which sorted out from the solid-phase library using a three-step screening.

Antibodies are large proteins with a substantial number of potential areas of interaction with peptides from large libraries. The difficulty consists in finding the one that is able to interact with the Fc region of immunoglobulins, particularly the site that is common for many different types of IgG. The reported peptides have been shown to interact with a selected targeted site, thus opening up a variety of possibilities for IgG separation and purification.

Preliminary investigations showed the capability of a HWRGWV bead conjugate to capture antibodies from a mammalian cell culture medium containing 10% fetal calf serum, with purity comparable to commercially available resins such as protein A and A2P agarose sorbents. Investigations performed on influencing factors such as acetylation of the ligand, its density on the resin, and temperature, supported the impression that this ligand is probably the preferred candidate for large-scale affinity separation against current best performing resins (Yang et al., 2009). Encouraged by these data, the investigation was pushed for a better understanding of the performance and docking mechanism (Yang et al., 2010), where the interaction with the Fc portion of IgG was confirmed. In-depth studies have been performed in order to identify the docking portion of the antibody where the hexapeptide interacts and also to determine by comparison if this is similar to the well-known interaction epitopes between protein A and the Fc region of antibodies. From these investigations it has been hypothesized that HWRGWV binds to amino acids of the loop Ser383-Asn389 of the C_H3 domain.

Three different approaches have been used for this experimental work: (i) verification of whether the hexapeptide was still effectively interacting with IgG after complete deglycosylation of the Fc fragment, (ii) competitive binding against proteinA, and (iii) binding with fragments of IgG resulting from protease degradation. Interestingly, this peptide ligand is not affected in the binding properties once the antibodies are deglycosylated. This means that the carbohydrate portion is not involved in binding site recognition in a lectin-like interaction manner. This is an important point teaching that the interaction does not happen on the hinge proximal region of the C_H2 domain. Moreover, it was found that neither protein A nor protein G interferes with the hexapeptide and IgG binding; this finding suggests that the interaction takes place in a different docking site. The absence of displacement effects is clearly not due to a peculiar thermodynamic situation since the affinity of protein A for IgG ($10^8 M^{-1}$) is higher than the affinity between the hexapeptide and IgG ($10^5 M^{-1}$). The formal identification of the interaction site was then deciphered by studies using protease fragments instead of whole IgG. It was found that the portion Gly371–Lys392 that contains a loop structure (Ser383–Asn389) exposed to the environmental solvent is probably the best option for the interplay with HWRGWV peptide.

As expected, this quite specific binding involves various molecular concomitant forces such as hydrophobic associations, hydrogen bonding, and electrostatic interactions, as is also suggested by the composition of the peptide. In spite of these interesting properties, it was shown that the modifications of physicochemical conditions have an influence on the specificity for immunoglobulin classes (Liu et al., 2011b). When investigating the comparative properties of the three selected hexapeptides (HWRGWV, HYFKFD, and HFRRHL), it appeared that the first had the strongest binding to IgM, followed by IgA and IgG. Affinity results could, however, differ, depending not only on environmental conditions but also on the peptide grafting density on the beads and the nature of the feedstock with variations in recovery and purity. Overall, it appeared that HWRGWV was the most promising peptide for preparative-scale purification of immunoglobulins.

As a confirmation of these studies, a HWRGWV peptide sequence was successfully used for the preparative purification of humanized IgG1 and IgG4, as well as chimeric monoclonal antibodies from the CHO cell culture (Naik et al., 2011). The capturing buffer used contained sodium chloride and sodium caprylate to reduce the level of nonspecific binding and hence maximizing the final purity and recovery of collected antibodies that were estimated to be at least 94% and 85%, respectively. With a dynamic binding capacity of around 20 mg/mL this ligand compared pretty well with protein A affinity columns. It was also demonstrated that, for potential large-scale purification purposes, this affinity column could be cleaned repeatedly with stringent chemical agents such as 2 M guanidine-HCl or combinations of phosphoric acid and urea used sequentially, without affecting the binding of antibodies. Further analysis related to regulatory compliance was the residual level of host DNA that was around four logs, comparable to the effect of protein A resins.

More recently, the same team reported the effective use of the HWRGWV peptide ligand for the capture of human polyclonal antibodies from the Cohn fraction II+III and from bovine milk (Menegatti et al., 2012). Here, too, sodium caprylate and sodium chloride were added to the binding buffer to minimize the level of parasite protein binding. Under these conditions, immunoglobulins from human plasma were recovered, with a yield of about 84% and purity around 95%. This application is of industrial interest because protein A resins are too costly for this application, while regular ion-exchange chromatography does not allow obtaining very high yields and purity and is not necessarily easier to perform. The fact that the peptide is a small ligand compared to the molecular mass of IgG authorizes high productivity when combining binding capacity and probably higher velocities. In other words, it should be possible to use smaller columns or other filtration devices, with high-frequency cycling justified by the extremely large volumes of plasma or plasma fractions from where massive amounts of IgG are extracted, compared to cell culture supernatants.

Another antibody binder usable for chromatographic purposes has been the result of combinatorial peptide library screening, further refined by some chemical modifications (Lund et al., 2012). It is composed of natural amino acids such as arginine and glycine, as well as a synthetic, aromatic acid,

2,6-di-t-butyl-4-hydroxybenzyl derivative. The best reported binding capacity for IgG was up to about 40 mg IgG/mL of resin, depending on the type of chromatographic material where the ligand was grafted and also very much depending on the isoelectric point of the antibody to be separated. IgG recovery and purity was estimated to be close to 80% and 90%, respectively. Beyond thermodynamic data and various comparisons with other competitive IgG binders, the authors did not give details on the molecular interaction interplay between this peptide and the Fc amino acid sequence where the peptide ligand would interact. The comparison of their respective molecular structure does not illuminate the understanding because there are too many differences in composition, shape, and interaction association mechanisms (see Figure 7.5). The only common features are represented by the presence of arginine in both structures and some hydrophobic degree. However, comparative side-by-side studies of these IgG binders could be of interest to see if they are complementary with the possibility to be used, for instance, as a column sequence.

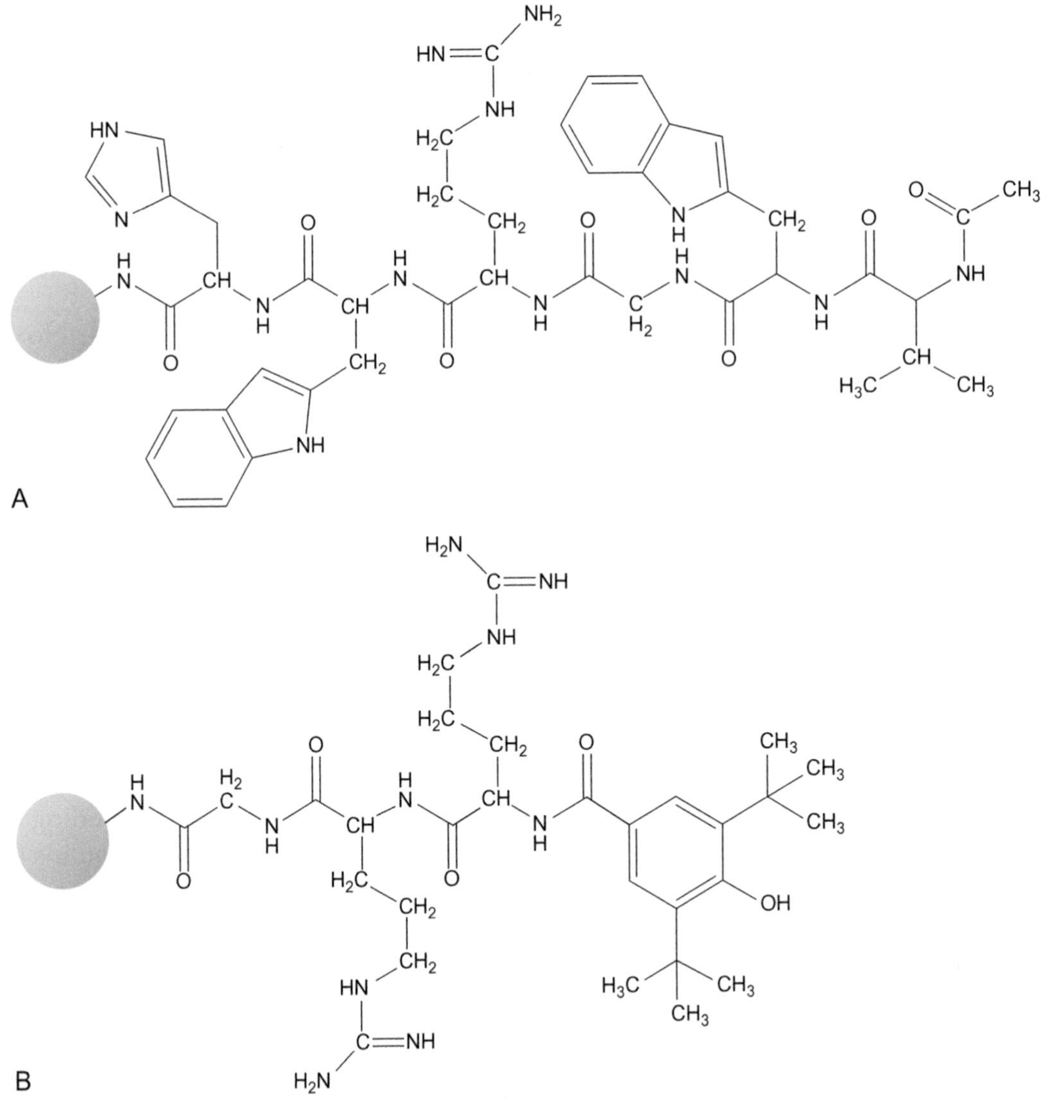

FIGURE 7.5

Structural comparison of two peptides selected from ligand libraries and presenting a similar interaction with human immunoglobulins G. Peptide of panel A is from Yang et al. (2005); peptide from panel B is from Lund et al. (2012). These peptide sequences are very different; however, there are common features such as arginine residues and terminal hydrophobic moieties. In one case, the alkaline character of arginine (panel B) is replaced by a histidine residue (panel A).

7.2 POLISHING OUT PROTEIN IMPURITIES FROM PURE BIOLOGICALS

An important objective in the production of recombinant proteins for human therapy is to reach a high degree of purity from very crude feedstocks containing large amounts of host-cell proteins. Recombinant proteins are generally expressed either in eukaryotic systems such as CHO cells or in microorganisms like yeasts (e.g., *Pichia pastoris, Saccharomyces cerevisiae)* and prokaryotes (e.g., *Escherichia coli)*. The resulting purified injectable proteins must comply with very stringent rules of purity defined as "homogeneity." Nevertheless, trace proteins could still be present and could generate undesirable side effects when injected repeatedly such as immunogenic reactions when impurities are host-cell proteins of nonhuman origin. Several strategies have been described to remove the final protein impurities whose efficiency depends on their nature and amount. This process is called *polishing*. When polishing is intended to remove DNA-related molecules and endotoxins, most generally anion exchangers are used (Petsch and Anspach, 2000; Kepka et al., 2004). The success of this operation is contingent upon the net charge of the target protein; if the protein has an acidic isoelectric point, then the separation process may be much less efficient in terms of overall yield. The protein of interest may in fact be partially co-adsorbed on the chromatographic media, thus significantly reducing the recovery.

In current real practice, even after having deployed considerable efforts in downstream processing, it is hard to achieve purity levels better than 99% (Janson et al., 1998; Simpson, 2004; Janson and Jönsson, 2011). The elimination of the last remaining impurities is generally prohibitively expensive and leads to significant losses of the valuable target biopharmaceutical protein.

Purification processes of recombinant biopharmaceuticals are composed of a number of steps, including chromatographic separations under sequences that have orthogonal complementary properties to increase the efficiency of the separation process. The first chromatographic separation step is designed to concentrate the protein of interest while eliminating the maximum amount and number of protein impurities. When economically and technically possible, an affinity chromatography column is used with a high degree of selectivity. A concrete example is the use of protein A columns for the initial extraction of antibodies that was illustrated by Östlund et al. as early as 1987. Since then, thousands of applications have been developed, especially for monoclonal antibodies (see, for instance, a review by Darcy et al., 2011). But even in this case the purity of the target protein only rarely reaches levels higher than 95%. Most contaminants are host-cell proteins from the culture broths that are co-adsorbed on solid-phase sorbents during the chromatographic capturing step. Although it is still unclear if these contaminating proteins are co-captured by non-specific binding on the antibody itself or on the solid-phase sorbent, it has been reported that a substantial fraction of contaminating proteins could be removed by selected washes. In fact, non-negligible amounts of these host-cell proteins adsorb on the beads and co-elute with the antibody, with consequent contamination. In spite of washing treatments before the elution of antibodies, a number of impurities remain, and their amount seems to be dependent on the type of antibody produced (Nogal et al., 2012), suggesting that mAbs might represent the preferential location where subsets of HCPs (host cell proteins) bind. This is why intermediate separation processes capable of reducing the impurity presence are added, such as ion-exchange chromatography or hydrophobic chromatography (Liu et al., 2010).

Because 100% purity is extremely difficult to achieve (the target product contains traces of a large number of proteins coming from the expression system or from the biological crude extract, as well from as protease fragments of the main protein), a polishing step needs to be added. This last task is accomplished according to a few available technologies. One of them is gel permeation chromatography based on molecular size discrimination (Zanette et al., 2003; Dieryck et al., 2003), a fully orthogonal separation mode compared to the previous steps. The problem with gel filtration is its poor productivity resulting from a small load for large columns and the slow separation process. To these issues is to be added the large dilution effect of the initial loaded sample. All these reasons largely limit the applicability of gel filtration that otherwise could be effective for removing the last impurity traces except when they have a similar molecular mass compared to the protein to be polished.

Other more popular approaches are the use of special ion exchangers (Mrabet, 1992; Herzer et al., 2003); hydrophobic chromatography packings (Lauer et al., 2000; Roder et al., 2000); or mixed mode sorbents (Chen et al., 2008). Many marketing communications have stressed the concept of polishing as being able to simultaneously eliminate protein impurities, DNA products, and viruses. In reality, the rationale behind this approach is quite poor because, if not specifically adapted to each situation, these sorbents can easily also adsorb the target protein, especially when it has a complementary charge to the sorbent or has the appropriate hydrophobic balance. For ion exchangers, quaternary amines are generally used; if DNA products are indeed preferentially captured and thus removed from the target product, this is not necessarily true for all present protein traces. The flow-through from these sorbents could still contain cationic protein traces and, more importantly, could reduce the overall recovery of the target biopharmaceutical partially co-adsorbed on the resin. As an example, these quaternary resins cannot be applied to proteins with an acidic isoelectric point even if some salt could be added to the buffer, thus reducing the electrostatic interaction (Liu et al., 2010).

In spite of their reported efficacy, ion exchange- and hydrophobic-based polishing methods remain of narrow specificity. They either apply to preselected categories of proteins or when the target protein has a different hydrophobic and/or electrostatic behavior compared to the impurities. Current polishing alternatives are illustrated in Figure 7.6.

To accomplish a real polishing task, the ideal chromatographic support should be able to capture specifically all impurities and leave all or almost all the target protein in the flow-through. Due to the trace level of protein impurities to be removed, this ideal column would be of small size compared to the volume of the biopharmaceutical to treat and could also be accomplished under fast flow rates. To add to this ideal situation, the polishing sorbent should be as generic as possible. In other words, it should be applicable to all polishing situations without modifying or even adjusting the conditions

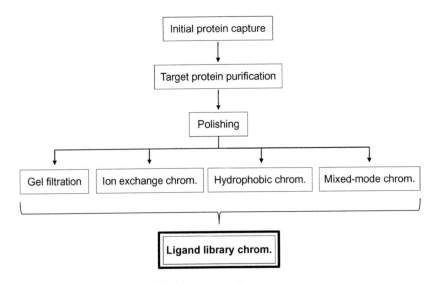

FIGURE 7.6
Schematic representation of current flow-through polishing alternatives for the final purification of proteins, especially monoclonal antibodies (the first column being in this case an affinity sorbent made for example with protein A). At the polishing stage are represented well-known possible alternatives, depending on the properties of the target protein and the nature of protein impurities present. These alternatives could be replaced by a sort of universal polishing, with a mixed bed of combinatorial ligands that is independent of the sample nature and its composition.

of work. Although this approach appears idealistic, for single interacting multimode ligands it can be turned quite elegantly into reality by the use of solid-phase combinatorial ligand libraries as described in the following paragraphs.

7.2.1 Reversing the Effect of Dynamic Concentration Range

Chapter 4 explains the mechanism of action of combinatorial peptide ligand libraries in detail, along with all possible rules for capturing and eluting proteins from complex mixtures. Briefly, each bead acts as a single-affinity adsorbent and is thus specific for one or a group of proteins. It is also indicated that such a mixed-bed adsorption is used under very large overloading conditions exactly on a mode that is opposite to a regular affinity chromatography separation. Affinity adsorption, the most selective way for capturing single proteins from very crude dilute feedstocks, is in fact used under loading conditions that are always below the saturation. Once the binding capacity of the solid phase is approached, the process is stopped, the column is washed, the adsorbed protein is recovered, and a second cycle can be initiated.

If this simple but performant mechanism is extended to a large number of different affinity solid-phase ligands mixed together, many proteins can be captured at a time by their own complementary ligand. Each unique affinity ligand bead within the pool would bind its specific protein partner(s) up to the point of ligand saturation. If in such a mixed bed of affinity beads, a mixture of proteins—one of them being largely dominant (e.g., 98+ % purity) and many others at very low concentrations or trace amounts—is loaded onto the mixed-affinity bed, the dominant protein will very rapidly saturate its corresponding bead(s) while impurity traces will not. Using loading conditions where the impurities do not saturate the solid-phase ligands, the flow-through of the column will be composed of the dominant protein, with purity substantially higher than before the process as a result of the capture of all or almost all impurity traces by their corresponding bead partners. Under optimized conditions with a solid-phase pool containing a large diversity of ligands, the process can be used to remove impurity traces at the end of many downstream separation processes.

This separation principle could be reduced to practice using the same combinatorial peptide ligand libraries described for the reduction of dynamic concentration ranges (Fortis et al., 2006). Naturally, other types of libraries could alternatively be used when the mixed bed of beads contains a large number of diverse affinity ligands. Figure 7.7 shows the mechanism by which an almost pure protein is polished out of the impurity traces that are present.

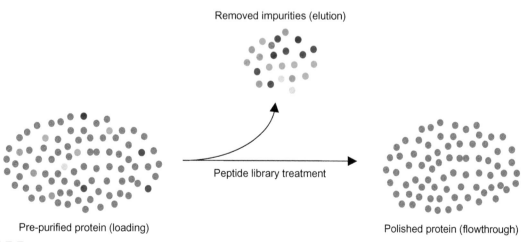

Removed impurities (elution)

Peptide library treatment

Pre-purified protein (loading)

Polished protein (flowthrough)

FIGURE 7.7

Representative cartoon of a polishing process using peptide libraries. The initial purified sample (gray dots still containing traces of protein impurities in relatively large number—colored dots) is contacted with a peptide library under predetermined conditions (could be physiological buffers). The process is operated under nonsaturating conditions: Protein impurities are captured along with very minimal amounts of the main protein the latter being collected in the flowthrough (right). The protein impurities captured by the beads are completely desorbed to start another polishing run. *Adapted from Boschetti and Righetti, (2012).*

The following section reports how the same peptide ligand library has been used so that the saturation is not reached with potential application to a final universal polishing step, in particular for removing the last protein traces from purified biopharmaceuticals.

7.2.2 "Polishing" Purified Biologicals

To demonstrate the ability to remove protein traces from a given purified protein, the choice of a model is a convenient approach, as described by Fortis et al. (2006). The selected pure protein of one of the examples reported was myoglobin from sperm whale. Since this commercially available protein is not of absolute purity, foreign proteins present were first removed by a mini-preparative ion-exchange chromatography to obtain a final myoglobin purity close to 100%, with the presence of a small amount of dimer and isoforms. Its mass is 17.8 kDa, and its isoelectric point is around 8.3. The purity of this protein was attested by SDS-polyacrylamide gel electrophoresis as well as by two-dimensional electrophoresis. This obtained pure material was purposely contaminated by a protein extract from *E. coli*, the total amount of protein added representing around 5% of the amount of myoglobin.

This "contaminated" myoglobin (95% purity) was then treated with a combinatorial hexapeptide ligand library under conditions described in Figure 7.8. The flow-through of the peptide ligands column was

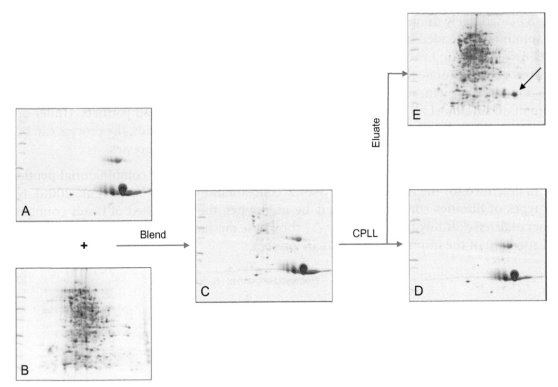

FIGURE 7.8

Two-dimensional electrophoresis of a sequence of polishing samples. Myoglobin (panel A) was mixed with 5% of an *E. coli* protein extract (panel B). The resulting sample (panel C) was submitted to a polishing process with a hexapeptide combinatorial ligand library from which two fractions were collected: the polished myoglobin (panel D) and the captured *E. coli* proteins (panel E). The latter comprises all initial protein spots, with some amount of myoglobin represented by the arrow.

Myoglobin from sperm whale was first purified by anion-exchange chromatography to remove impurities that are naturally present in the commercial preparation. Pure myoglobin was then added with an *E. coli* protein whole extract at a rate of 5 mg of protein per 100 mg of pure myoglobin. The buffer used was 25 mM phosphate, pH 7.0.

400 μL of contaminated myoglobin solution was mixed with 80 μL of peptide library beads for 20 minutes at room temperature. The mixture was centrifuged to collect the supernatant containing polished myoglobin.

Analysis experiments performed by using a 12% polyacrylamide gel in the SDS-dimension in a pH gradient between 3 and 10 in the focusing dimension. Coomassie blue staining.

collected and then analyzed by electrophoretic means. In addition, the proteins captured by the beads were desorbed using a solution of 9 M urea-citric acid, pH 3.5 and also checked by electrophoretic analysis in parallel. Before the addition of *E. coli* proteins, the purified myoglobin shows a high level of purity (panel A). Few isoforms are present, and three main traces of impurities are visible at isoelectric points of 7.8–8.3, 7.1, and 5.8 and molecular mass values of, respectively, 35–40 kDa, 6–7 kDa, and 12 kDa. After the addition of 5% of *E. coli* proteins (panel B), a much larger number of protein impurities is detectable throughout a large range of isoelectric point and masses extending from 10 to 200 kDa (panel C). Upon treatment with the peptide library, myoglobin appears to be almost as pure as before contamination: Only minor spots are detectable (panel D) all of them coming from *E. coli* protein extract. The contaminating protein capture is demonstrated by analysis of the eluted fraction (panel E) where many bacterial proteins are detected, along with a small amount of myoglobin concomitantly captured by the ligand library, as expected (see arrow). Therefore, the experimental data clearly illustrate the capability of the described process to remove a large number of very diverse proteins in terms of isoelectric point and molecular mass.

Direct mass spectrometry analyses of samples (Figure 7.9) in a mass window between 1 and 20 kDa (this is the zone where two-dimensional electrophoresis shows low performance, especially below 10 kDa) were performed using a copper ions-chelating chromatographic surface (SELDI TOF). It was found that the contaminated myoglobin (panel A) comprises a large number of low-molecular-mass protein impurities, most of them having higher masses compared to myoglobin, as shown by the SDS-PAGE pattern. Mass spectrometry data do not show the presence of low-molecular-mass contaminants because they are at a concentration below the detectability of the analysis. High- and low-mass molecules that are captured by the ligand library column are shown in panel C, respectively,

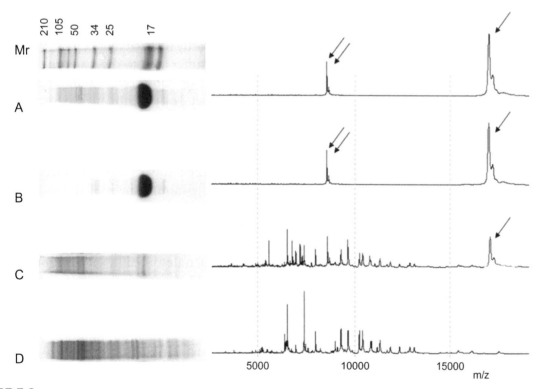

FIGURE 7.9

Analytical determinations (SDS-PAGE and mass spectrometry) of fractions from polished myoglobin.

On the left, SDS-PAGE patterns show the *E. coli*-contaminated myoglobin (A); the polished myoglobin (B); and the eluate of captured proteins (C) and its comparison with the initial *E. coli* extract (D). Mr represents the protein mass ladder.

On the right are the corresponding mass spectrometry data focusing on the molecular mass range between 3 and 18 kDa. The single arrow indicates the signal of myoglobin, and the double arrow is the signal of double-charged myoglobin. All other signals are polypeptides from the *E. coli* extract.

by SDS-PAGE analysis and mass spectrometry, thus yielding a polished myoglobin as shown in panel B (17.8 kDa, single arrow and double charged, double arrow in the mass spectrometry pattern). These data fully confirm and complement the two-dimensional electrophoresis analysis. For comparison purposes, the *E. coli* protein extract is represented in panel D, SDS-PAGE (left), and mass spectrometry (right).

Within the context of protein polishing, the capture of *E. coli* proteins is of great practical interest since many recombinant proteins are expressed in this microorganism. Therefore such similar impurities may potentially be part of the purified recombinant proteins and can represent a medical issue for patients repeatedly receiving the purified recombinant protein with bacterial polypeptide impurities present even in trace amounts.

In summary, this process model demonstrated that the purity of contaminated myoglobin after treatment was similar to the initial uncontaminated product except that very few impurities remained present (probably those that either saturated their bead partner or did not find a peptide partner to be captured). The amount of myoglobin lost in the process was estimated to be lower than 2%, a value well below the standard losses from current chromatographic separations. This is essentially because the column was largely overloaded with myoglobin and statistically only few beads of the mixed bed were present to capture myoglobin. The operation performed in the presence of 25 mM phosphate buffer pH 7.0 could probably be even further improved by using a lower or higher ionic strength; by adopting a different pH (for details on the use of peptide library, see Chapter 8); or by enlarging the composition of the library. The operation was performed by using a spin column that is far less effective than a real chromatography column for a proteins separation; therefore, the expected results from real chromatography could be further improved.

Indeed, in general, a polishing process is performed in a chromatographic column. The protein solution to be polished is prepared in an appropriate buffer and directly injected onto a packed solid-phase combinatorial peptide library pre-equilibrated with the same buffer used to dissolve the proteins. The chromatographic column is connected to a chromatographic setup comprising a pumping system and UV/pH detection unit for recording events at the column outlet. For preliminary experiments and in order to determine the amount of loading, a frontal analysis is performed and the collected fractions is analyzed to check the absence of contaminants. Thus the column is loaded continuously with the protein solution to be polished at an average linear flow rate of 30–50 cm/hour. The flow-through is collected under fractions of a few mL each to analyze the capability of the solid phase to remove protein impurities. Likewise a frontal analysis is applied for determining the protein-binding capacity (Bjorklund and Hearn, 1996; Perez-Almodovar and Carta, 2009). The breaking point can be calculated according to the degree of desired impurity removal. At this stage the loading is stopped and is replaced by the same buffer used for the separation, until reaching the UV baseline. The captured protein impurities, along with small amounts of the target protein, are then desorbed by an appropriate elution mixture among those presented in Chapter 8, Sections 8.14–8.17. A good compromise is to elute with an aqueous solution of 9 M urea-citric acid, pH 3.5. This stripping procedure ensures good regeneration of the polishing column that can be easily re-equilibrated with the buffer and reused for another cycle. A detailed protocol of a polishing preliminary experiment to determine the volume of combinatorial peptide ligand library for a given amount of purified protein can be found in Section 8.23.

The most laborious phase is the optimization of the loading to get the desired polishing effect. Once this is defined, the loading amount (or volume of feedstock) is reduced to about 10% to ensure good reproducibility as well as to compensate for possible reduction of binding capacity for impurities over cycles.

A real-life example of this polishing procedure is exemplified here by recombinant human albumin expressed in *Pichia pastoris* after purification by anion-exchange chromatography up to a purity estimated at 90–95%, as attested by SDS-polyacrylamide gel electrophoresis. In this case, the recombinant human albumin concentration prior to polishing was 10 mg/mL. The detectable impurities present, as determined by electrophoresis, were relatively limited in number: Four main bands and three less concentrated ones of various masses were visible. For the purpose of polishing, using a ligand library in this situation (limited number of impurities but present at relatively high concentration) represents a difficult case because it is postulated that only a limited number of ligands will be used for removing the impurities and hence the volume of beads would have to be quite large. As a consequence, a large volume of beaded ligand library statistically signifies that a large number of beads with affinity for the target protein will also be present, contributing to reducing the final recovery.

As mentioned above, to determine the necessary volume of solid-state peptide libraries to polish a given amount of sample, a frontal analysis followed by an elution step of captured impurities was performed. The flow-through fraction-by-fraction was analyzed by SDS-polyacrylamide gel electrophoresis and compared to the initial sample. Once the purity was considered insufficient, the frontal analysis was stopped; the column was washed with the initial buffer; the captured impurities were eluted with a phosphate buffered saline (25 mM phosphate buffer, 150 mM sodium chloride, pH 7.2); and finally the adsorbed proteins were stripped out with a solution of 9 M urea-citric acid, pH 3.5.

In the first part of the flow-through (see Figure 7.10, insert 1), recombinant albumin showed better purity compared to both the initial sample (Ctrl) and the fraction collected at the last part of the

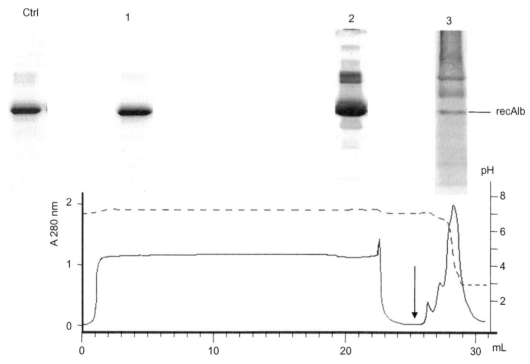

FIGURE 7.10

Polishing experiment corresponding to a frontal analysis made using a pre-purified recombinant albumin (expressed in *P. pastoris*) and a column of hexa-peptide ligand library. The volume of the column was 1.4 mL pre-equilibrated with 0.02 M phosphate buffer, 150 mM sodium chloride, pH 7.2. Twenty mL of sample in the same buffer was injected at a flow rate of 39 mL/hour at room temperature. At the end of the loading, the column was washed with a few mL of buffer followed by an elution of the captured protein impurities with a solution of 9 M urea-citric acid, pH 3.5. The composition of the protein solution at the column outlet was followed by SDS-polyacrylamide gel electrophoresis, as illustrated above the chromatogram trace. The continuous line represents the UV trace, and the broken line is the pH detected at the column outlet. The arrow indicates the injection into the column of the desorption solution of acidic urea. "Ctrl" is the analysis of the initial sample where the major band is albumin with some amount of dimer. Image inserts 1, 2, and 3 represent the SDS-PAGE analysis of the collected samples as the chromatogram was developed. They are positioned at their collection level: Sample 1 was collected at the beginning of the chromatogram, while sample 3 was collected at the last part of the chromatogram. 3 represents the collected impurities removed from the initial sample. *Adapted from Fortis et al. (2006).*

frontal analysis (insert 2). The eluted protein fraction contained a large number of protein impurities that were not visible in the crude initial sample along with some amount of albumin, as expected and shown in the eluate (see insert 3). This experiment again demonstrated the ability of a peptide ligand library to capture protein impurities until the limit of binding capacity dictated by the available beads specific to given impurities. In this example, it appeared that the column volume for the removal of *Pichia pastoris* impurities and albumin fragments starting from 500 mg of total proteins would have been about 2 mL. Larger volumes would allow obtaining greater purity of recombinant albumin with, however, higher losses.

This polishing procedure was also applied to a commercial purified transferrin from human plasma contaminated by traces of serum protein, as described by Fortis et al. (2006). The initial purity of this protein was estimated to be close to 95%; after polishing under the described conditions, the purity was significantly enhanced. The purity was nevertheless a function of the column load as mentioned above. When the load was limited to 50 mg of proteins per mL of ligand library beads, the purity reached was close to 99%. Conversely, when the load was increased to 1000 mg of transferrin per mL of ligand library beads, the impurities could not be substantially reduced. This relatively limited column loading compared to the myoglobin case doped with *E. coli* proteins is explainable by the small number of impurities present in the estimated 5% impurity bulk.

The limited binding capacity of peptide libraries for a single protein means that the described process is very effective when a large number of protein impurities are present, each one in minute amounts, as is almost always the case for host-cell proteins (Eaton, 1995; Shukla et al., 2008). Conversely, if, instead of many species, only one or a few protein impurities are present for a similar level of impurity amount, they may exceed the binding capacity of partner beads and thus would not be completely removed unless the volume of beads was large, with consequently larger losses of the main protein. Because the situation of protein polishing is different from case to case, it is necessary to proceed by a preliminary set of experiments to determine the ratio between the sample and the column volume using a sort of frontal analysis as described above for albumin polishing and in Section 8.23.

The qualitative and quantitative composition of impurities is thus of critical importance to the success of the procedure because each bead-ligand present in the library has a binding capacity estimated close to 1-3 ng of protein per bead. Calculated on the basis of mg/mL of packed bed beads, this corresponds to an average of about 10 mg of proteins per mL of packed bed when the diversity of protein impurities present is very large. In theory, when 1000 mg of a protein contains 2% of a multitude of polypeptidic impurities (or 20 mg), the minimum amount of beads for their removal should be close to 1.5 mL of beads. In practice, a calculated experiment design, instead of a real experiment to remove a given amount of unknown impurities, is relatively complex and highly speculative because it involves probability calculations as a consequence of both the ligand diversity and the number of protein impurities, along with the amount of each of them. This is why a polishing process design based on combinatorial ligands has to be approached by a trial-and-error fashion in order to define the best capturing configuration in terms of bead-sample volume ratio as provided by frontal chromatography. The relatively large sample volume loaded to a small column and the collection of "polished" protein in the flow-through add the advantage of a very limited dilution factor compared to current chromatographic separation processes: The larger the sample to column volume ratio, the smaller the dilution factor.

The described method is applicable to a variety of situations, rendering the polishing step generally applicable. The diversity of ligands makes the library quite universal for the adsorption of protein impurities regardless of their composition and origin. Adjustments of separation conditions may have to be made, as for instance verifying the best pH for impurity removal as well as the optimized ionic strength.

When making performance comparisons to gel filtration polishing, one could argue about polishing's ability to formulate the protein solution in the final storage buffer. Actually, this feature is an attribute of isocratic chromatography separations, which is the case for this polishing process. The composition of the buffer for polishing can in fact be selected in order to meet formulation requirements in terms of pH, ionic strength, and additives, provided these latter do not compete with the capture of impurities to be eliminated. Table 7.3 summarizes the advantages and disadvantages of current polishing methods compared to the use of combinatorial peptide ligand libraries.

For preparative polishing applications where it is common to run the sorbent for many repetitive cycles, few additional points are to be considered. It is important to check that the binding capacity of the ligand library does not decrease over cycles. Therefore, all the protein impurities captured by the previous cycle should be stripped out, using stringent desorption agents as described in Chapters 4 and 8. Another point to consider is related to peptide leakages that might contaminate the polished protein that is always harvested in the flow-through with no change in the physicochemical conditions of the solid phase. This mechanism of polishing without the modification of physicochemical separation conditions and the fact that the polished protein is collected in the flow-through largely reduces the risk of ligand leakage. In the event of problems or doubts, the difference in molecular mass between the polished protein and the leached ligands renders the elimination very easy with standard methods. Finally, as noted earlier, proteases such as exopeptidases cannot hydrolyze the peptide ligand as a consequence of its structural design.

TABLE 7.3 Comparative Features of Current Polishing Methods

Polishing Chromatography Methods	Advantages	Disadvantages
Gel filtration	Generic method with no optimization of separation conditions. Easily removes protein fragments.	Slow process. Dilution of target protein. Requires large columns.
Ion exchange	Small columns. Rapid process. Low cost. Easy implementation.	Requires optimization of separation conditions. Unspecific for HCP. Co-adsorption of the target. Ineffective with impurities of similar isoelectric point.
Hydrophobic	Small columns. Rapid process. Low cost. Easy implementation.	Requires optimization of separation conditions. Unspecific for HCP. Co-adsorption of the target. Ineffective with impurities of similar hydrophobic degree.
Mixed beds of combinatorial ligands	Small columns. Generic method with no optimization of separation conditions. Almost no losses of target protein. Rapid process. Easy implementation.	High sorbent cost* Effective with high number of impurities present in low and very low amount.

*This is to counterbalance the small size of the column, the repeatability of the impurity depletion cycles, and the "universality of the process"

7.3 INTERACTION STUDIES

Before proceeding, a note of caution is in order. Protein interaction with other protein partners involves epitopes in a very complex manner. Conformational epitopes from both partners are most generally at the basis of associations that are difficult to occur with just short linear hexapeptides such as those used to reduce the dynamic concentration range as reported in this book. Interacting peptides with proteins to decipher natural interplays or to help investigating the interaction itself are probably much longer with natural folding possibilities. This being said, a number of papers have reported the use of relatively short peptides for protein–protein interaction studies.

Protein–protein complex formation, which is at the basis of molecular communications for the control of various mechanisms of living cells to induce all the cell systems to cooperate synergistically, as for instance in signal transmission, has been approached by various methods. One method and probably the best known, is the so-called yeast two-hybrid (Fields and Song, 1989) capable of identifying binary interactions. Affinity separations associated with mass spectrometry are another interesting way to detect interactions between proteins and protein complexes. Genomics also helps elucidate these interactions (Bhardwaj and Lu, 2005), which are extremely complex. In addition, they can have a transient character as in signaling. One of the important obstacles in these investigations is the availability of protein partners that belong to the group of low-abundance species. Moreover, recombinant heterologous expression of such species is hampered by technical difficulties, such as poor solubility, toxicity for the host, low expression, and the problem of purification, and thus is very difficult to obtain in the amounts required for analytical and structural studies.

In their review, Benyamini and Friedler (2010) highlighted the advantages of using synthetic peptides for investigating protein–protein interactions. Easy synthesis even for long peptides, systematic synthesis such as combinatorial constructions with high probabilities to cover exactly protein interaction domains, post-translational modifications and labeling are all good reasons to use peptides as models for these investigational experiments.

In 2000, Kay et al. postulated that short peptides from combinatorial synthesis could be used for protein–protein interaction studies; they showed that certain peptides interact specifically with some proteins. Thus it was suggested that peptide ligands would first be identified from libraries and would then be isolated with the objective of accelerating not only protein functional analysis of proteomes but also drug discovery.

With the use of phosphotyrosyl (pY) solid-phase combinatorial peptide libraries, protein–protein interaction decoding resulted in the identification of four Src homology 2 (SH2) domains derived from protein tyrosine phosphatase and inositol phosphatase (Sweeney et al., 2005). Binding competition studies indicated that the selected pY peptides bind to the same site on the SH2 domain surface, suggesting that peptide library approaches could be used for the discovery of other protein domains of interaction.

However, the interaction domain mapping of a protein is more efficiently determined by using peptide arrays compared to beaded peptide libraries. This approach can almost simultaneously evidence the identity of several peptides. Depending on the complexity and the length of the peptides in an array, overlapping sequences can be present and a higher precision can be reached. In spite of that, methods involving combinatorial peptides for the discovery of protein–protein interaction are increasingly being used and are associated with computer-assisted models to enhance the power of the identification of binding sites (Guntas et al., 2010; Katz et al., 2011).

The discovered interacting peptides can be used in protein–protein interactions in competitive or noncompetitive ways in an agonistic manner to restore a missed mechanism or as inhibitors of undesired interactions.

7.4 PRESENT TRENDS AND FUTURE PROSPECTS OF COMBINATORIAL PEPTIDE LIBRARIES

For years, combinatorial peptide ligand libraries have proven their potential and their reality for a number of applications in biochemistry and biology. Two current large applications can be cited: The first is their use in reducing the dynamic concentration range of polypeptides present in biological extracts (the main object of this book); the second is their use as a source of affinity ligands for the purification of proteins. Both applications are expected to be developed further, with potential variants in their mode of exploitation. In the first case, more focused peptide libraries can easily be synthesized for the treatment of groups of proteins. They can also be used under different conditions so as to elicit the capture and/or the elution of categories of proteins. These two approaches could open the way to enhance detection conditions for targeted proteins of diagnostic interest and therefore the possibility to prepare ready-to-use kit assays.

With regard to identification of a peptide from the entire ligand library for affinity chromatography purposes, further efforts are needed on the one hand to find methods for reducing nonspecific binding or false positives, and on the other hand to more easily separate the positive beads or even to decode the sequence of the found specific peptide. With these developments, peptide libraries will no doubt become the mine for affinity ligands that affinity chromatography has been seeking for decades. Hybrid peptide ligands can also be envisioned as, for instance, associated with chelating moieties, hydrophobic tails, and other possible structures (see the examples in Section 8.2). These additions could be introduced either following the peptide library synthesis or during the split-and-pool process with amino acid derivatives. As a suggestion, a chelating moiety could be introduced at the end of the library synthesis by reacting bromoacetic acid at the distal free terminal amine, thus creating an iminodiacetic acid capable of forming complexes with transition metal ions which are currently used for metal chelated affinity chromatography. The specificity will thus not only be played on the chelate with the metal ion, but interestingly also with both the chelated metal ion and the peptide sequence.

Outside these two major areas, peptide libraries can easily be modified chemically to develop other applications. For instance, the search for enzyme inhibitors is a domain that has already been explored. An interesting attempt is reported by Liu et al. (2011a), where a cyclic peptide library was prepared and evaluated to evidence properties of docking at the substrate site of activated T cell nuclear factors. Using the most recent technological advancements (spatially segregated beads, magnetic bead sorting, bead encoding at their interior, etc.), several compounds have been identified as good substrate inhibitors, thus highlighting the interest in finding new classes of selective and less toxic immunosuppressive agents.

A more specific aspect is the research for phosphorylated peptides as substrates or inhibitors for kinases. On the model used for deciphering protein–protein interaction studies for identifying Src homology domains (Sweeney et al., 2005), phosphotyrosyl, phosphoseryl and phosphothreosyl peptide libraries could be made and tested to identify specific substrates or inhibitors. Protease inhibitor design and/or discovery are other fields of investigation. Already in 2000, Buchardt et al. described a solid-phase library made with phosphinic peptides for the discovery of metalloproteinase inhibitors.

The discovery of substrates for peptidases could therefore become a possible application of peptide libraries. For example, Comellas et al. (2009) reported the evaluation of solid-state peptide libraries for a cytosolic serine peptidase called prolyloligopeptidase, which is associated with schizophrenia. The objective was to learn more about the action of these enzymes and possibly discover substrates allowing a direct control.

Among possible molecules with potential therapeutic application are peptides that can be retrieved from a peptide library as reported by Roof et al. (2009). The library screening allowed identifying an inhibitor called RGS4 (a regulator of G protein signaling). During the study, the mechanism of action of the discovered peptide inhibitors was also elucidated.

Another interesting emerging application could be the discovery of novel biological activities from complex protein extracts or, better, from circulating and noncirculating body fluids. Sarkar et al. (2007) described an interesting use of combinatorial peptide libraries to detect factors promoting cell growth from the whole human blood. In this respect, whole blood was contacted with a peptide library where proteins are captured according to the principle described in Section 4.4.1. Blood was preferred to plasma or serum because it is free of any degradation subsequent to manipulation of coagulation reactions, thus preserving intact most, if not all, biological properties of all components. Red and white cells were not removed since they do not interfere with the protein capture by the beads and are washed out after the protein adsorption process. The diversity of peptide ligands creates conditions of protein docking that are governed by different dissociation constants. When the affinity is not extremely strong, the dissociation of the captured protein occurs quite slowly and redissolves progressively in the liquid medium once the excess of blood proteins is eliminated.

This interesting mechanism related to the mass action law has been exploited for the detection of cell-promoting growth when the beads are mixed with cells in culture. Beads associated with clusters of cells are then isolated and the activating factor is identified. Concurrently, the same library loaded with proteins can be checked for specific biological activities once distributed individually in well plates comprising a very large number of wells (up to 100,000). The published results are very convincing and open a large number of possibilities to detect enzymatic activities and other important cell-interacting factors. This general method also offers the possibility of variants with, for instance, the design of targeted libraries after first results from a generic one. It also represents an interesting approach to discover peptides of therapeutic interest.

7.5 References

Affleck RL. Solutions for library encoding to create collections of discrete compounds. *Curr Opin Chem Biol.* 2001;5:257–263.

Astle JM, Simpson LS, Huang Y, et al. Seamless bead to microarray screening: Rapid identification of the highest affinity protein ligands from large combinatorial libraries. *Chem Biol.* 2010;17:38–45.

Bastek PD, Land JM, Baumbach GA, Hammond DH, Carbonell RG. Discovery of alpha-1-proteinase inhibitor binding peptide from the screening of a solid phase combinatorial library. *Sep Sci Technol.* 2000;35:1681–1706.

Benyamini H, Friedler A. Using peptides to study protein–protein interactions. *Future Med Chem.* 2010;2:989–1003.

Bhardwaj N, Lu H. Correlation between gene expression profiles and protein–protein interactions within and across genomes. *Bioinformatics.* 2005;21:2730–2738.

Bjorklund M, Hearn MT. Characterisation of silica-based heparin-affinity adsorbents through column chromatography of plasma fractions containing thrombin. *J Chromatogr A.* 1996;743:145–162.

Boschetti E, Righetti PG. Mixed beds: beyond the frontiers of classical chromatography of proteins. In: Grushka E, Grinberg N, eds. Advances in Chromatography. vol. 50. Boca Raton, FL: CRC Press; 2012:1–46.

Buchardt J, Schiødt CB, Krog-Jensen C, Delaissé JM, Foged NT, Meldal M. Solid phase combinatorial library of phosphinic peptides for discovery of matrix metalloproteinase inhibitors. *J Comb Chem.* 2000;2:624–638.

Buettner JA, Dadd CA, Baumbach GA, Masecar BL, Hammond DJ. Chemically derived peptide libraries: A new resin and methodology for lead identification. *Int J Pept Protein Res.* 1996;47:70–83.

Chen J, Tetrault J, Ley A. Comparison of standard and new generation hydrophobic interaction chromatography resins in the monoclonal antibody purification process. *J Chromatogr A.* 2008;1177:272–281.

Chen X, Tan PH, Zhang Y, Pei D. On-bead screening of combinatorial libraries: Reduction of nonspecific binding by decreasing surface ligand density. *J Comb Chem.* 2009;11:604–611.

Comellas G, Kaczmarska Z, Tarrago T, Teixido M, Giralt E. Exploration of the one-bead one-compound methodology for the design of prolyloligopeptidase substrates. *PLoS One.* 2009;4:e6222.

Darcy E, Leonard P, Fitzgerald J, Danaher M, O'Kennedy R. Purification of antibodies using affinity chromatography. *Methods Mol Biol.* 2011;681:369–382.

Dieryck W, Noubhani AM, Coulon D, Santarelli X. Cloning, expression and two-step purification of recombinant His-tag enhanced green fluorescent protein over-expressed in Escherichia coli. *J Chromatogr B*. 2003;786:153–159.

Dong X, Fu L, Yu H. Affinity purification of insulin by peptide ligand affinity chromatography. *Trans Tianjin Univers*. 2007;13:313–317.

Eaton LC. Host cell contaminant protein assay development for recombinant biopharmaceuticals. *J Chromatogr*. 1995;705:105–114.

Fields S, Song O. A novel genetic system to detect protein-protein interactions. *Nature*. 1989;340:245–246.

Fortis F, Guerrier L, Righetti PG, Antonioli P, Boschetti E. A new approach for the removal of protein impurities from purified biologicals using combinatorial solid phase ligand libraries. *Electrophoresis*. 2006;27:3018–3027.

Guntas G, Purbeck C, Kuhlman B. Engineering a protein-protein interface using a computationally designed library. *Proc Natl Acad Sci U S A*. 2010;107:19296–19301.

Herzer S, Kinealy K, Asbury R, Beckett P, Eriksson K, Moore P. Purification of native dehydrin from Glycine Max cv., *Pisumsativum*, and *Rosmarinum officinalis* by affinity chromatography. *Protein Expr Purif*. 2003;28:232–240.

Huang PY, Baumbach GA, Dadd CA, et al. Affinity purification of von Willebrand factor using ligands derived from peptide libraries. *Bioorg Med Chem*. 1996;4:699–708.

Hwang SH, Lehman A, Cong X, et al. OBOC small molecule combinatorial library Encoded by halogenated mass-tags. *Org Lett*. 2004;6:3830–3832.

Janson JC. *Protein Purification* [Janson JC, Rydén L, eds.]. New York: Wiley-VCH; 1998:3–40.

Janson JC. Jönsson JA. Introduction to chromatography. *Methods Biochem Anal*. 2011;54:25–50.

Javaherian S, Musheev MU, Kanoatov M, Berezovski MV, Krylov SN. Selection of aptamers for a protein target in cell lysate and their application to protein purification. *Nucleic Acids Res*. 2009;37:2–10.

Kay BK, Kasanov J, Knight S, Kurakin A. Convergent evolution with combinatorial peptides. *FEBS Lett*. 2000;480:55–62.

Katz C, Levy-Beladev L, Rotem-Bamberger S, Rito T, Rüdiger SG, Friedler A. Studying protein-protein interactions using peptide arrays. *Chem Soc Rev*. 2011;40:2131–2145.

Kaufman DB, Hentsch ME, Baumbach GA, et al. Affinity purification of fibrinogen using a ligand from a peptide library. *Biotechnol Bioeng*. 2002;77:278–289.

Kelley B, Jankowski M, Booth J. An improved manufacturing process for Xyntha/ReFacto AF. *Haemophilia*. 2010;16:717–725.

Kepka C, Lemmens R, Vas J, Nyhammar T, Gustavsson PE. Integrated process for purification of plasmid DNA using aqueous two-phase systems combined with membrane filtration and lid bead chromatography. *J Chromatogr A*. 2004;1057:115–124.

Kim JH, Kang H, Kim S, et al. Encoding peptide sequences with surface-enhanced Raman spectroscopic nanoparticles. *ChemCommun (Camb)*. 2011;47:2306–2308.

Kodadek T, Bachhawat-Sikder K. Optimized protocols for the isolation of specific protein-binding peptides or peptoids from combinatorial libraries displayed on beads. *Mol Biosyst*. 2006;2:25–35.

Lam KS, Lebl M, Krchnak V. The "One-Bead-One-Compound" combinatorial library method. *Chem Rev*. 1997;97:411–448.

Lam KS, Salmon SE, Hersh EM, Hruby VJ, Kazmierski WM, Knapp RJ. A new type of synthetic peptide library for identifying ligand-binding activity. *Nature*. 1991;354:82–84.

Lam KS, Wade S, Abdul-latif F, Lebl M. Application of a dual color detection scheme in the screening of a random combinatorial peptide library. *J Immunol Methods*. 1995;180:219–223.

Lathrop JT, Fijalkowska I, Hammond D. The bead blot: a method for identifying ligand–protein and protein–protein interactions using combinatorial libraries of peptide ligands. *Anal Biochem*. 2007;361:65–76.

Lauer I, Bonnewitz B, Meunier A, Beverini M. New approach for separating *Bacillus subtilis metalloprotease* and alpha-amylase by affinity chromatography and for purifying neutral protease by hydrophobic chromatography. *J Chromatogr B*. 2000;737:277–284.

Lee SS, Lim J, Tan S, et al. Accurate MALDI-TOF/TOF sequencing of one-bead-one-compound peptide libraries with application to the identification of multiligand protein affinity agents using in situ click chemistry screening. *Anal Chem*. 2010;82:672–679.

Lehman A, Gholami S, Hahn M, Lam KS. Image subtraction approach to screening one-bead-one-compound combinatorial libraries with complex protein mixtures. *J Comb Chem*. 2006;8:562–570.

Leimbacher M, Zhang Y, Mannocci L, et al. Discovery of small-molecule interleukin-2 inhibitors from a DNA-encoded chemical library. *Chemistry*. 2012;18:7729–7737.

Linhult M, Gülich S, Hober S. Affinity ligands for industrial protein purification. *Protein Pept Lett*. 2005;12:305–310.

Liu FF, Wang T, Dong XY, Sun Y. Rational design of affinity peptide ligand by flexible docking simulation. *J Chromatogr A*. 2007;1146:41–50.

Liu HF, Ma J, Winter C, Bayer R. Recovery and purification process development for monoclonal antibody production. *MAbs*. 2010;2:480–499.

Liu R, Marik J, Lam KS. A novel peptide-based encoding system for "one-bead one-compound" peptidomimetic and small molecule combinatorial libraries. *J Am Chem Soc*. 2002;124:7678–7680.

Liu T, Qian Z, Xiao Q, Pei D. High-throughput screening of one-bead-one-compound libraries: Identification of cyclic peptidyl inhibitors against calcineurin/NFAT interaction. *ACS Comb Sci*. 2011a;13:537–546.

Liu Z, Gurgel PV, Carbonell RG. Effects of peptide density and elution pH on affinity chromatographic purification of human immunoglobulins A and M. *J Chromatogr A*. 2011b;1218:8344–8352.

Lund LN, Gustavsson PE, Michael R, et al. Novel peptide ligand with high binding capacity for antibody purification. *J Chromatogr A*. 2012;1225:158–167.

Maillard N, Clouet A, Darbre T, Reymond J-L. Combinatorial libraries of peptide dendrimers: Design, synthesis, on-bead high-throughput screening, bead decoding and characterization. *Nature Protocols*. 2009;4:132–142.

Mairal T, Ozalp VC, Lozano-Sanchez P, Mir M, Katakis I, O'Sullivan CK. Aptamers: molecular tools for analytical applications. *Anal Bioanal Chem*. 2008;390:989–1007.

Mannocci L, Leimbacher M, Wichert M, Scheuermann J, Neri D. 20 years of DNA-encoded chemical libraries. *Chem Commun (Camb)*. 2011;47:12747–12753.

Marani MM, MartínezCeron MC, Giudicessi SL, et al. Screening of one-bead-one-peptide combinatorial library using red fluorescent dyes. Presence of positive and false positive beads. *J Comb Chem*. 2009;11:146–150.

Marcon L, Battersby BJ, Rühmann A, et al. "On-the-fly" optical encoding of combinatorial peptide libraries for profiling of protease specificity. *Mol Biosyst*. 2010;6:225–233.

Martínez-Ceron MC, Marani MM, Taulés M, et al. Affinity chromatography based on a combinatorial strategy for rerythropoietin purification. *ACS Comb Sci*. 2011;13:251–258.

Menegatti S, Naik AD, Gurgel PV, Carbonell RG. Purification of polyclonal antibodies from Cohn fraction II + III, skim milk, and whey by affinity chromatography using a hexamerpeptide ligand. *J Sep Sci*. 2012;35:3139–3148.

Miyamoto S, Liu R, Hung S, Wang X, Lam KS. Screening of a one bead-one compound combinatorial library for beta-actin identifies molecules active toward Ramos Blymphoma cells. *Anal Biochem*. 2008;374:112–120.

Moser AC, Hage DS. Immunoaffinity chromatography: An introduction to applications and recent developments. *Bioanalysis*. 2010;2:769–790.

Mrabet NT. One-step purification of Actinoplanesmissouriensis D-xylose isomerase by high-performance immobilized copper-affinity chromatography: Functional analysis of surface histidine residues by site-directed mutagenesis. *Biochemistry*. 1992;31:2690–2702.

Naik AD, Menegatti S, Gurgel PV, Carbonell RG. Performance of hexamer peptide ligands for affinity purification of immunoglobulin G from commercial cell culture media. *J Chromatogr A*. 2011;1218:1691–1700.

Nogal B, Chhiba K, Emery JC. Select host cell proteins coelute with monoclonal antibodies in protein A chromatography. *Biotechnol Prog*. 2012;28:454–458.

Östlund C, Borwell P, Malm B. Process-scale purification from cell culture supernatants: monoclonal antibodies. *Dev Biol Stand*. 1987;66:367–375.

Pande J, Szewczyk MM, Grover AK. Phage display: Concept, innovations, applications and future. *Biotechnol Adv*. 2010;28:849–858.

Pennington ME, Lam KS, Cress AE. The use of a combinatorial library method to isolate human tumor cell adhesion peptides. *Mol Divers*. 1996;2:19–28.

Perez-Almodovar EX, Carta G. IgG adsorption on a new protein A adsorbent based on macroporous hydrophilic polymers. I. Adsorption equilibrium and kinetics. *J Chromatogr A*. 2009;1216:8339–8347.

Petsch D, Anspach FB. Endotoxin removal from protein solutions. *J Biotechnol*. 2000;76:97–119.

Quillan JM, Jayawickreme CK, Lerner MR. Combinatorial diffusion assay used to identify topically active melanocyte-stimulating hormone receptor antagonists. *Proc Natl Acad Sci U S A*. 1995;92:2894–2898.

Redman JE, Wilcoxen KM, Ghadiri MR. Automated mass spectrometric sequence determination of cyclic peptide library members. *J Comb Chem*. 2003;5:33–40.

Righetti PG, Boschetti E. The art of observing rare protein species with peptide libraries. *Proteomics*. 2009;9:1492–1510.

Roder C, Krusat T, Reimers K, Werchau H. Purification of respiratory syncytial virus F and G proteins. *J Chromatogr B*. 2000;737:97–101.

Roof RA, Roman DL, Clements ST, et al. A covalent peptide inhibitor of RGS4 identified in a focused one-bead-one-compound library screen. *BMC Pharmacol*. 2009;9:9.

Roque AC, Gupta G, Lowe CR. Design, synthesis, and screening of biomimetic ligands for affinity chromatography. *Methods Mol Biol*. 2005;310:43–62.

Samson I, Kerremans L, Rozenski J, et al. Identification of a peptide inhibitor against glycosomalphosphoglycerate kinase of Trypanosomabrucei by a synthetic peptide library approach. *Bioorg Med Chem*. 1995;3:257–265.

Sarkar J, Soukharev S, Lathrop J, Hammond DJ. Functional identification of novel activities: Activity-based selection of proteins from complete proteomes. *Anal Biochem*. 2007;365:91–102.

Sepetov NF, Krchnák V, Stanková M, Wade S, Lam KS, Lebl M. Library of libraries: approach to synthetic combinatorial library design and screening of "pharmacophore" motifs. *Proc Natl Acad Sci U S A*. 1995;92:5426–5430.

Shukla AA, Jiang C, Ma J, Rubacha M, Flansburg L, Lee SS. Demonstration of robust host cell protein clearance in biopharmaceutical downstream processes. *Biotechnol Prog*. 2008;24:615–622.

Simpson RJ. In: Simpson RJ, ed. Purifying Proteins for Proteomics. Cold Spring Harbor: Cold Spring Harbor Laboratory Press; 2004:17–40.

Sweeney MC, Wavreille AS, Park J, Butchar JP, Tridandapani S, Pei D. Decoding protein-protein interactions through combinatorial chemistry: sequence specificity of SHP-1, SHP-2, and SHIP SH2 domains. *Biochemistry*. 2005;44:14932–14947.

Tozzi C, Giraudi G. Antibody-like peptides as a novel purification tool for drugs design. *Curr Pharm Des*. 2006;12:191–203.

Wang G, De J, Schoeniger JS, Roe DC, Carbonell RG. A hexamer peptide ligand that binds selectively to staphylococcal enterotoxin B: Isolation from a solid phase combinatorial library. *J Pept Res*. 2004;64:51–64.

Wang X, Peng L, Liu R, Xu B, Lam KS. Applications of topologically segregated bilayer beads in "one-bead one-compound" combinatorial libraries. *J Pept Res.* 2005;65:130–138.

Xiao W, Wang Y, Lau EY, et al. The use of one-bead one-compound combinatorial library technology to discover high-affinity αvβ3 integrin and cancer targeting arginine-glycine-aspartic acid ligands with a built-in handle. *Mol Cancer Ther.* 2010;9:2714–2723.

Yang H, Gurgel PV, Carbonell RG. Hexamer peptide affinity resins that bind the Fc region of human immunoglobulin G. *J Pept Res.* 2005;66:120–137.

Yang H, Gurgel PV, Carbonell RG. Purification of human immunoglobulin G via Fc-specific small peptide ligand affinity chromatography. *J Chromatogr A.* 2009;1216:910–918.

Yang H, Gurgel PV, Williams Jr. DK, et al. Binding site on human immunoglobulin G for the affinity ligand HWRGWV. *J Mol Recognit.* 2010;23:271–282.

Zanette D, Soffientini A, Sottani C, Sarubbi E. Evaluation of phenylboronateagarose for industrial-scale purification of erythropoietin from mammalian cell cultures. *J Biotechnol.* 2003;101:275–287.

Zhao Q, Chris X. Recent advances in aptamer affinity chromatography. *Se Pu.* 2009;27:556–565.

Detailed Methodologies and Protocols

8.1 PREPARATION OF COMBINATORIAL PEPTIDE LIGAND LIBRARIES

8.1.1 Main Objective

The combinatorial peptide ligand library adopted for reducing the dynamic concentration range in protein extracts and biological fluids follows specific rules that are not necessarily the same for other applications. For simplicity, peptides are linear and must cover the broadest possible diversity. To reach this goal, the number of building blocks should be large. In this respect, when using the collection of natural amino acids, the total number of obtainable diverse ligands is very large. However, several restrictions need to be considered, such as secondary reactions of cysteine, thereby reducing the number of possible building blocks. This is also the case if only very stable amino acids are selected (for instance, asparagine and tryptophan may not be extremely stable and could be eliminated or replaced by other non-natural amino acids). The selection of building blocks is thus dictated by practical considerations, price, longevity, and number of diversomers. The selection also impacts the final behavior of the library. For instance, if only ionic amino acids are selected, ionic interaction for the capture of proteins will dominate. Conversely, if hydrophobic amino acids are used, the hydrophobic associations will be at the basis of protein docking. In general, the selection will have to balance all these considerations to obtain a large-spectrum library for generic protein capture. Typically, 10 to 18 amino acids are enough to obtain a good diversity of hexapeptides.

The solid support also has to be selected carefully. As a first objective, solid beads must withstand all chemistry conditions of the peptide synthesis process. Another important consideration is the hydrophilicity of the polymer matrix in order to comply with the requirements of protein solution and compatibility with aqueous buffers. Non-specific binding should be as small as possible (ideally without traces of this noxious phenomenon). The porosity of the beads is another important aspect that is directly related to the protein-binding capacity: the larger the pore size and the surface area, the larger the binding capacity. Nevertheless, when the surface area increases, the size of the pores generally decreases; a best compromise is thus to be sought. Hydrophilic methacrylates are good candidates for preparation of the combinatorial peptide libraries.

Finally, an important consideration is the particle size of the polymeric beads. Since the definition of the library is that each bead carries a single peptide structure, the size of the beads dictates the number of diversomers for a given volume of settled beads. With large beads the total volume necessary to cover all the diversity is very large and may be incompatible with the volume of the biological protein extract. Therefore to decrease the volume of the gel beads while maintaining the diversity, it is necessary to use small particle sizes. Table 8.1 indicates the various issues in terms of bead volume and diversity as a function of the particle size.

8.1.2 Principle of the Synthesis of a Peptide Library

As described by Furka (1998), reaction vessels with a frit at the bottom for filtration are mounted on a vacuum manifold. The construct is placed over a shaker so that all reaction vessels are gently agitated concomitantly. The number of reaction chambers is equal to the number of building blocks (selected amino acids). Each vessel receives an identical volume of beads, the latter comprising a number of beads largely exceeding the number of diversomers targeted at the end of the synthesis. For instance, if the number of diverse amino acids is 12 and the peptide is a hexamer, the number of diversomers will be a little less than 3 million. In this case the number of beads should be at least 50 times larger or about 150 million. From this large number of beads and the bead diameter, it is possible to calculate the volume of beads required. Once the setup is in place, the first step is to attach a linker to the beads. The second step is to graft an amino-protected amino acid different from one reaction vessel to another. At the end of the operation, the beads are washed and mixed together and finally deprotected. At this stage the total volume of beads is split again into the same number of aliquots as

TABLE 8.1 Library Diversity and Coverage as a Function of Bead Size. The Considered Hexapeptide Library here is Composed of 20 Amino Acids and therefore the Total Number of Diverse Peptides is 64×10^6

Bead Diameter (µm)	Number of Beads Per mL Packed Bed	Hexapeptide Library Coverage (%)
5	8×10^9	12,500
10	1×10^9	1500
20	125×10^6	200
30	35×10^6	54
65	3.5×10^6	5

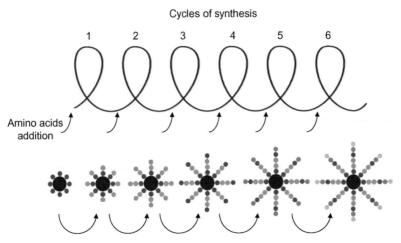

Cycles of synthesis

Amino acids addition

FIGURE 8.1
Schematic representation of the preparation of a peptide library cycle after cycle by using the so-called split-and-mix approach as also described in Section 4.3.2.

in the first step and introduced into each reaction vessel. Under similar conditions, the second series of amino acids is attached (one per reaction vessel as for the first grafting step). The beads are then washed again and another cycle starts. Figure 8.1 illustrates schematically the cycling process. For the preparation of hexapeptides, the number of cycles is six. The synthesis of combinatorial peptide libraries can advantageously be obtained by using automatic systems that ensure accuracy and speed.

8.1.3 Recipes

Since the possibilities of such a synthesis are very numerous, just a few indications and references are given here. What is considered a peptide library for the reduction of dynamic concentration range is to generate an equimolecular peptide mixture, this being obtainable only if the number of beads is significantly larger (at least 50–100 times) than the number of diverse peptides synthesized. For this solid-phase peptide synthesis, Fmoc-α-amine-protected amino acids (Fmoc = Fluorenylmethyloxycarbonyl) are generally used due to their easy deprotection (Lloyd-Williams et al., 1997). In this type of combinatorial synthesis, each bead may carry a different peptide, and the amino acid to be grafted may react at different speeds according to the terminal amino acid of the peptide bead. This situation is very peculiar because it is not of interest to know whether the reaction is complete on average but rather if it is complete for all beads. To this end it is necessary to monitor almost bead-by-bead by staining with bromophenol blue and continue the reaction until all beads are clear (no primary amine still free from the solid-phase previous reaction). With these libraries the number of amino acids is less critical than the proper selection of amino acids.

Actually the absence of some amino acids such as glycine or alanine does not significantly affect the functional behavior of the library for the uses described in this book. On the contrary, there are critical amino acids that are very important for the final function as demonstrated by Bachi et al. (2008). As described by Lam et al. (1997), if the motif results from an arrangement of three amino acids, different arrangements could be made within the hexapeptide framework and the final resulting number of diversomers would be substantially lower than for a complete hexapeptide combinatorial library. Maillard et al. (2009) describe the split and mix detailed procedure, along with the proper follow-up of the synthesis achievement and the deprotection of Fmoc after each synthesis step. Critical steps and troubleshooting are also indicated.

8.1.4 Expected Results

Only analytical data and their interpretation can verify the success of the synthesis. First of all, since the library is equimolar, the individual analysis of amino acids should give a similar number of moles of each amino acid engaged in the synthesis. Determining the length of each peptide is also a means to check if the reaction sequence was complete. This can be performed after cleavage of peptides from the beads and then by setting up an appropriate analysis of the resulting soluble peptides (e.g., mass spectrometry and/or HPLC).

8.1.5 Variants

A large number of variants can be considered. The first is the use of a selection of amino acids (natural and unnatural). Another is to use *D*-amino acids instead of *L*, or even both. The libraries discussed in the context of this book are hexapeptide and linear; however, they could be very different in length and concept. Libraries of variable length can be made as well as the so-called library of libraries. For instance, after each randomization step (there could be three, four, or more steps), the library could be separated into three of four parts followed by a coupling of different mixtures of amino acids. Since combinatorial libraries can also be used as a source of ligands for specific interaction investigations, several encoding technologies have been devised (see Section 7.1.3).

Already in 2002, Liu et al. described the mass-tag encoding strategy for a small-molecule combinatorial library. The mass spectrometry isotope pattern of each tag could define the component building blocks of each selected bead within a quite limited library screened to find a ligand partner to streptavidin. The identification was performed by means of both enzyme-linked colorimetric assay and quantum dot/COPAS assays. Hwang et al. (2004) published a second method of encoding with quantum dot/COPAS assay. More recently, a fluorescent encoding was described, with the goal of using and recognizing each bead by cytofluorimetry (Surawski et al., 2008). In another example, isobaric encoding was adopted for applications in mass spectrometry (Hu et al., 2007).

8.2 CHEMICAL MODIFICATION OF COMBINATORIAL PEPTIDES (E.G., SUCCINYLATION)

8.2.1 Objective of the Chemical Modification

Chemical modifications of peptide libraries with primary amino terminal ends can be easily implemented in order to create another library with slightly different properties, depending on the type of chemical structure that is brought to react with the terminal primary amine of the peptide. The resulting modified library will deliver different results compared to the initial one, thus enlarging the possibility of capturing different proteins or the same proteins with different dissociation constants. The present example deals with the introduction of carboxylic groups by a simple succinylation reaction.

8.2.2 Recipe

After deprotection, the peptide library beads carry peptides with free primary amines (terminal and side chains from lysine). These amines can be endcapped or reacted with a number of chemicals such as succinic anhydride to transform primary amines and introduce carboxylic acids. This transformation either changes the affinity interaction for proteins by, for instance, adsorbing proteins that are normally not recognized by regular peptide libraries, or it changes the affinity constant for given proteins as described in Section 4.5.7. Here two recipes are described for the succinylation of peptide library beads (see Figure 8.2).

8.2.2.1 METHOD 1

Rehydrate 300 mg dry peptide library beads with 5 mL of a 50% methanol-50% distilled water mixture for one hour at room temperature. Then wash three times with 5 mL of distilled water for a few minutes and remove the supernatant by centrifugation. The resulting wet peptide library beads are equilibrated with saturated sodium acetate solution (15–20 gram sodium acetate in 50 mL of distilled water, pH about 8.5–9.2) by successive addition of 5 mL solutions, agitation for few minutes, and elimination of the supernatant by decantation or centrifugation. This operation is performed three times.

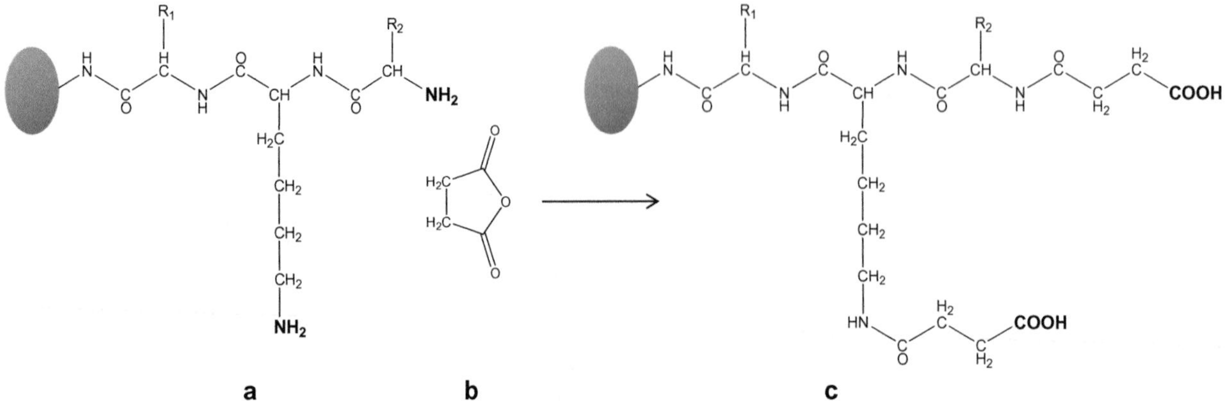

FIGURE 8.2
Succinylation scheme of free primary amines of combinatorial peptide library by means of a succinylation reaction. The reaction takes place on all primary amines, including those from lysine side chains. The terminal carboxylic acid turns the weak cationic character of the library into a weak anionic one, with changes in affinity for bait proteins.

To the suspension containing the bead pellets and 4 mL of sodium acetate, 400 mg of solid succinic anhydride are added, and the mixture is maintained under agitation for 3–5 hours. At the end of the process, the anhydride dissolves completely. The pH of the suspension is monitored and should not go below 8; otherwise it is advisable to adjust with 1–6 M sodium hydroxide dropwise. However, it is advisable not to exceed pH 9.2 for stability purposes. Then the supernatant is discarded, replaced by fresh sodium acetate solution, and another 400 mg of dry succinic anhydride is added. The mixture is then maintained under agitation overnight at room temperature. The solution of sodium acetate and all by-products are eliminated by ten repeated washes with a seven-volume excess of distilled water. The separation of supernatants is performed by centrifugation. The resulting succinylated beads are stored in the cold, preferably after equilibration with a phosphate physiological buffer.

8.2.2.2 METHOD 2

The bead rehydration process is to be performed as described above. One mL of rehydrated beads is washed with distilled water to eliminate salts and preservatives and then washed twice with pure ethanol. Centrifugations follow each time to eliminate the supernatant. A mixture of 9 mL of DMF (dimethyl formamide) and 1 mL of pyridine is then prepared. To the bead pellet 4 mL of previous solvent mixture are added and, under gentle shaking, 400 mg of dry succinic anhydride are added. The suspension is gently stirred for 30–60 minutes or more at room temperature and washed extensively with water and with a phosphate physiological buffer. The succinylated beads are stored at 4 °C.

8.2.3 Variants

Instead of using sodium acetate, other possibilities are available to maintain the pH between 8 and 9.2—for example, one can add sodium/potassium hydroxide, sodium carbonate.

Carboxylation of terminal amines or lysine side chain can be obtained using other anhydrides from dicarboxylic acids such as glutaric anhydride. The process remains the same.

Another possibility is to use chloracetic or bromoacetic acid as well as other analogs instead of anhydrides. In this case, the reaction conditions are different. Stronger alkaline conditions are needed using with 1 M sodium hydroxide, with the risk of some level of degradation when weak amino acids are part of the building blocks such as asparagine. The reaction would be performed at temperatures between 20 and 60 °C, depending on the time allowed for completion of the reaction. Acyl halides may also react on other amino acid sites, rendering the final structure less predictable. Besides carboxylic groups such as distal substituents of the peptide library, many other molecules can be used, generating libraries with different properties. Examples of chemical endcapping are to graft chelating agents (e.g., EDTA anhydride resulting

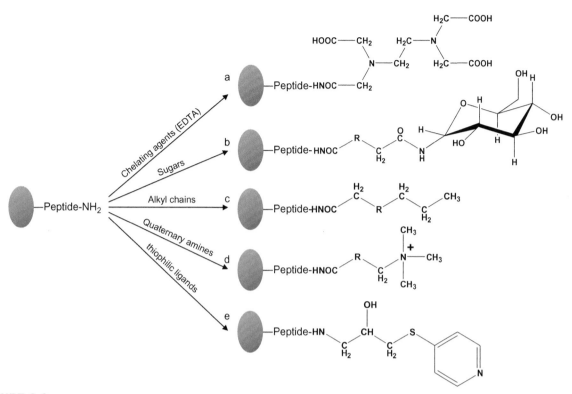

FIGURE 8.3

Examples of chemical modification of terminal amines of a peptide library with various substituents to confer a dominant global character to the ligands, with consequent changes in specificities for protein capture.

a: Introduction of chelating agents such as EDTA (by means of a reaction with EDTA di-anhydride). The inclusion of transition metal ions will introduce a chelating chromatography variant to the library behavior.
b: Sugars can be grafted at the primary amine distal moiety by a variety of chemical approaches.
c: Alkyl chains (aliphatic or aromatic) attached at the primary amino groups extend some level of hydrophobic properties to all beads beyond those already comprising a dominant number of hydrophobic amino acids.
d: Reinforcement of cationic character by grafting quaternary amines; guanidine groups could also be added instead.
e: Addition of a sort of thiophilic moiety comprising a sulfur group and a pyridine ring. Other possibilities are available. *(For a review see Boschetti, 2001).*

in a combined mode of chelating interaction, depending on the metal ion used and the presence of the peptide sequence), to make covalent attachment of sugars such as a reaction with gluconolactone, and to introduce hydrocarbon chains to add strong hydrophobicity (see examples of schemes in Figure 8.3).

8.2.4 Comments and Remarks

Succinic anhydride solubility in water is quite poor. It solubilizes progressively while the succinylation reaction takes place. Under conditions of large excess, the reaction solution is thus kept saturated. When larger amounts of succinic anhydride are used, the solution may become acidic, with consequent decrease of reaction speed. In this case it is recommended that the pH be maintained between 8 and 9.2 by dropwise addition of 2M sodium hydroxide. The final wash is critical for the use of modified peptide libraries. All reagents and by-products must be completely eliminated. The modified library should be under sodium salt and stored in the cold. For long storages it may be necessary to add current preservatives used in chromatography, such as 20% ethanol in water or 0.02% sodium azide.

8.3 PREPARATION OF BIOLOGICAL EXTRACTS

Biological extracts, including biological fluids, can be directly treated with peptide libraries, with no modification of ionic strength, pH, and without additives. To this general rule there are a number of exceptions. For instance, a warning is necessary when the extracts are made using highly concentrated chaotropic agents, such as urea and detergents. Specific recommendations are given below. Nonetheless, for consistent studies it is recommended to check the protein solution for pH and ionic strength or even to proceed for buffer equilibration (diafiltration, desalting, dialysis). Since protein extracts frequently comprise active proteases that hydrolyze progressively proteins, adding one or two tablets of protease inhibitor cocktails per gram of protein in the extract is always recommended. This preliminary treatment ensures the best reproducibility of results since the capture of protein by the library is dependent on physicochemical conditions. Specific cases are explained below.

8.3.1 Biological Fluids

This category comprises blood plasma or serum, cerebrospinal, follicular and amniotic fluids, as well as urine, saliva, and bile. These fluids all contain proteins at physiological pH and ionic strength. They do not need any chemical adjustment. If insoluble matter is present, it is to be removed by centrifugation. Only clear solutions should be put in contact with the beaded peptide library. The latter is pre-equilibrated in physiological saline composed of a 10–25 mM phosphate buffer containing 150 mM sodium chloride (PBS). Attention is to be paid with the manipulation of blood plasma since it contains all components for the coagulation process (clotting factors, fibrinogen, etc.) and the risk of forming a clot is high. To prevent this phenomenon when contacting the fluid with the beads, the addition of 100 mM sodium citrate (final concentration) is strongly recommended (Leger et al., 2011). Sodium citrate could be replaced by 20–50 mM EDTA.

In cerebrospinal fluid the protein concentration is 100–200 times lower than in serum, and for in-depth studies where the volume of CSF needed is large, a preliminary concentration is advised. This is made first by a dialysis against a volatile salt solution (e.g., 50–200 mM ammonium carbonate), followed by a lyophilization operation. The cutoff of the dialysis membrane should be 3500 Da. The lyophilized powder is then diluted in the buffer selected for the treatment with CPLLs. For more details see Mouton-Barbosa et al. (2010).

Another specific case of biological fluid is bile. Besides proteins, bile contains a lot of hydrophobic material and aggregates such as bile salt, phospholipids, cholesterol, bilirubin, and mucins, all of which are deleterious for CPLL protein capture. All this material must therefore be eliminated prior to the enrichment phase. It is advised first to eliminate material in suspension by high-speed centrifugation (e.g., 20,000 g for 20 minutes at 4 °C) and then to treat with Cleanascite as described by Kristiansen et al. (2004) to eliminate lipidic material. The resulting clear supernatant is dialyzed or desalted exhaustively against the selected buffer prior to enrichment for low-abundance proteins (Guerrier et al., 2007a).

8.3.2 Urine

Urine composition may change even within the same day for a same individual (protein concentration, metabolites, pH, etc.) depending on a number of factors. This variation induces modifications of the affinity constants of proteins for peptides of the library and as a consequence the reproducibility might be affected. In addition comparisons between individuals (e.g. biomarker discovery studies) cannot be made without a normalization step. To reduce this issue, it is recommended to dialyze against a buffer of choice (generally, PBS or regular phosphate buffer without salts). Dialysis could be replaced by any other physical procedure such as diafiltration of desalting chromatography. Urine solutions might also need to be centrifuged to eliminate possible particles in suspension. (Detailed information can be found in Castagna et al., 2005, Decramer et al., 2008, Candiano et al., 2009a, 2009b, 2012).

8.3.3 Cell Culture Supernatants

The first precaution to be considered is to eliminate any possible presence of cells. Cell culture supernatants (eukaryotic cells, yeasts, bacteria) also need to be clear prior to protein capture. Generally, cell culture supernatants comprise proteins secreted by cells or delivered by cell death. However, for eukaryotic cell culture, foreign proteins can be present. They come from the culture medium such as albumin, transferrin, and insulin for some media, or even fetal bovine serum up to 10% for other media. In these examples, a massive amount of albumin is present, compared to the secreted proteins by cells in culture. When albumin is added as supplement, the proportion of albumin against other proteins can be far higher than in human serum where it already represents roughly 50–60% of the global protein mass. The recommendation is to use large volumes of cell culture supernatant for a small volume of peptide library beads so as to saturate very rapidly the beads with albumin and leave all other very low-abundance proteins captured by the beads. When fetal bovine serum is added as supplement to standard cell culture basal medium, the CPLL treatment is performed as it is usually applied to human serum (Colzani et al., 2009). A possible preliminary albumin immunodepletion step could also be performed with the risk of losing some low-abundance proteins. A summary scheme is illustrated on Figure 8.4. Generally, the expressed proteins are very dilute, and only with increasingly larger volumes of cell culture supernatant for a given small volume of beads is it possible to progressively evidence low-abundance proteins (see also Figure 4.30).

8.3.4 Microbial Extracts

When dealing with proteomes of microbial cells and yeasts, the first operation is to make lyses of cells and then recover cell proteins. Unfortunately, not only proteins will be released but also nucleic acids. The latter interfere with CPLL and must be eliminated prior to use. One way to proceed is to make extraction of proteins under conditions where nucleic acids are extracted. Otherwise it is advised to make a protein precipitation using one of the methods described in Sections 8.4 and 8.5. The elimination of nucleic acids is also critical to obtain clean results from two-dimensional electrophoresis.

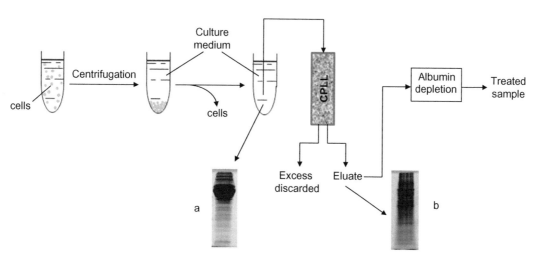

FIGURE 8.4

Schematic representation for the treatment of cell culture supernatants. The final treatment (depletion with anti-albumin antibodies) is optional. It should, however, take place after the enrichment phase with CPLLs because culture supernatants are generally dilute in expression proteins, and the albumin presence is too high to comply with the low binding capacity of immunodepletion media. Thus the first phase enriches in low-abundance species while decreasing the albumin levels. The second phase eliminates the residual albumin.

Between phases the eluate is to be equilibrated in a physiological buffer as, for instance, PBS.

Inserts represent SDS-polyacrylamide gel electrophoresis of a cell culture supernatant before a and after b treatment by CPLL. *(Adapted from Colzani et al., 2009).*

8.3.5 Plant Extracts

Plants contain relatively low amounts of proteins within a mixture of large amounts of various materials. They are generally represented by pigments, polysaccharides, polyphenols, nucleic acids, and metabolites. These macromolecular compounds disturb the normal behavior of peptide ligand libraries to capture proteins. In order to prevent this parasite phenomenon, pretreating the initial crude extract is advised. Various approaches are proposed; among them are extraction of contaminants and protein precipitation. Figure 8.5 illustrates alternatives to plant sample extract pretreatments. The plant extract is first neutralized and then if necessary delipidated (see Section 8.3.10). Examples of plant extract and their pretreatments before CPLL protein adsorption are given in Section 8.13.2.

8.3.6 Viscous Fluids

Biological fluids or extracts may contain mucins, nucleic acids, or various other viscous materials. This is the case, for example, for egg white and plant exudates. Depending on the type of extract, it is first recommeneded to dilute with the buffer selected for the adsorption of proteins on CPLL. As a second option, these fluids are vigorously agitated up to fluidification. If need be, they can also be precipitated with saturated ammonium sulfate in water at room temperature. After a few hours of stirring, the precipitate is collected by centrifugation and solubilized in the buffer used for protein capture with CPLLs. Precipitation options are described in Sections 8.4 and 8.5. Once redissolved, it is advised to dialyze the protein solution against the same buffer overnight to ensure elimination of the excess of ammonium sulfate. Dialysis can be replaced by chromatography desalting or appropriate membrane filtration, with cutoff of about 3500 Da. Due to residual viscosity of the protein solution, it is recommended to operate the protein capture by the CPLL beads in suspended bulk and not in packed columns. For example, egg white is manually separated from the yolk and homogenized with a magnetic stirrer until there is strong reduction of viscosity while preventing foam generation (see, for example, D'Ambrosio et al., 2008). This operation, performed at room temperature, might induce some proteolysis due to the action of proteases. This is prevented by the addition of complete protease inhibitor cocktails (generally one or two tablets per gram of protein in the extract). It is this extract that is equilibrated with PBS and contacted with CPLLs overnight in suspension.

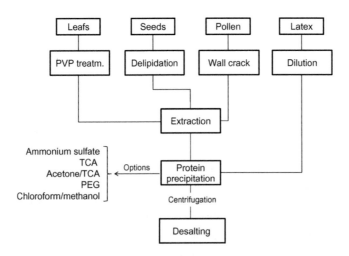

FIGURE 8.5

Proposals of plant sample treatment as a function of organ origin concluding protein extraction in view of further treatments, such as fractionation, depletion, or enrichment with CPLL. *(Adapted from Boschetti et al., 2009) and recently completed by Boschetti and Righetti, 2012.*

8.3.7 Animal Tissues

Proteins are usually extracted using physiological buffered solutions. The pieces of tissue are first finely homogenized. During homogenization or protein extraction phases, antiprotease inhibitors need to be added and the extraction preferably be operated in the cold (e.g., 4 °C). Once the clear supernatants are separated from the pellets by centrifugation, no additional steps are needed unless the protein capture is operated under a different buffer. In this case the sample must be equilibrated in this selected buffer, by dialysis or by buffer exchange or diafiltration.

8.3.8 Cell Extracts

Whole-cell extracts can be used to "amplify" the signal of low-abundance proteins and thus detect and identify rare species. Eukaryotic cells including yeasts are considered here. Major problems to be avoided are the activation of proteases present in quite massive amounts, which could hydrolyze native proteins. Thus it is always advised to add antiprotease cocktails before proceeding for cell homogenization. Nucleic acids could also be released in large quantities. They are not harmful per se, but represent an obstacle for a proper capture of proteins by the peptide libraries, especially libraries that have a dominant cationic charge. Nucleic acids present in the extract of mammalian cells as well as in yeasts need to be eliminated by protein precipitation under conditions where nucleic acids remain in solution. Ammonium sulfate precipitation (Section 8.4) and PEG fractionation (Section 8.5) are recommended procedures. Nucleases can also be used to hydrolyze nucleic acids into simple bases with concomitant reduction of viscosity. For more detailed recommendations, see Bachi et al. (2008) and Guerrier et al. (2007b).

8.3.9 Beverages

There are very different categories of beverages. Those that are of plant origin may contain proteins; however, their detection and analyses are relatively difficult first because of the low amount of residual proteins and second because the presence of polyphenols is deleterious to enrichment processes. Examples of the presence of polyphenols are red wines. Although they do not always contain polyphenols, their removal is critical. This is operated by a treatment with polyvinylpyrrolidone as described by Wigand et al. (2009). Briefly, to remove polyphenols and possibly other pigments, red wine (equivalent to about 100 mg of dry powder) is mixed with 3 mL of 0.1 M sodium phosphate buffer, pH 7.0 containing 300 mg of polyvinylpyrrolidone. The mixture is stirred overnight at room temperature and then centrifuged at 3000 g for about for 10 min. This procedure is repeated once, and the remaining polyvinylpyrrolidone is separated by filtering through a 0.22 μm filter. For a general review reporting a number of beverage examples, see Righetti and Boschetti (2012).

8.3.10 Delipidation

Lipid-rich biological material needs to be treated for lipid removal prior to protein extraction. This is the case for most plant seeds (see, for example, Sussulini et al., 2007, Xu et al., 2011). First, the initial material is powdered by any physical means without heating. Then the delipidation is performed by using nonpolar organic solvents. For example, for 50 gr of powdered corn seeds, 50 mL of a mixture of chloroform-ethanol (70%–30%, v/v) is added, and the suspension is shaken for one or two hours at room temperature. The solid phase is then separated by centrifugation, and the supernatant containing extracted lipids is discarded. An extensive wash with the same organic mixture of solvent is recommended prior to collecting the solid material. The latter is then dried, for example, under vacuum. Proteins are then extracted from the dry powder using classical procedures with an aqueous buffered solution. It is this extract that is then ready for protein capture with CPLL. See, for example, Fasoli et al. (2009).

8.4 Protein Precipitation by Means of Ammonium Sulfate

Protein precipitation prior to CPLL treatment is justified when a protein extract contains molecules incompatible with a proper capture. Nucleic acids, polysaccharides, as well as pigments and polyphenols are the most important molecules to eliminate. Also, the presence of ionic surfactants, especially cationic surfactants, is a reason for protein precipitation since they adsorb onto CPLL beads.

Proteins are generally soluble in dilute salts; this phenomenon is known as salting-in. However, when the salt concentration increases, the solubility decreases, and with certain salts proteins become insoluble (salting-out). The concentration must be very high to precipitate proteins; therefore very soluble salts are generally used for protein precipitation (Hofmeister series). In addition, some proteins do not precipitate even in the presence of extremely high concentration of salts, especially proteins of low molecular mass.

The most common salt for protein precipitation is ammonium sulfate, as described by Burgess (2009). It is one of the first of the series of lyotropic salts and probably the most effective for protein precipitation (Bull and Breese, 1980). Protein behavior in the presence of ammonium sulfate differs as a function of not only molecular mass but also of its composition. Very ionic proteins behave differently from hydrophobic proteins. In this context, ammonium sulfate is used not only for protein precipitation to separate other molecules of non-proteinaceous nature, but also for fractionation purposes. Some proteins in fact precipitate at relatively low concentration of ammonium sulfate, whereas others need massive amounts of this salt. Ammonium precipitation has no adverse effects on enzyme activity and does not denature protein structures. The protein functionality is fully recovered upon resolubilization in appropriate physiological buffers.

When full precipitation of proteins is required, the easiest approach is to use a saturated solution of ammonium sulfate. Conversely, for fractionated precipitation, the ammonium sulfate concentration is progressively increased step by step, and after each step the precipitated proteins are removed by centrifugation. Considering that the proteins to be precipitated are usually already in solution in an aqueous buffer, ammonium sulfate is added dry so as to know perfectly the final concentration. Adding solid ammonium sulfate to an initial given volume of protein solution is easy to calculate; however, the addition of large amounts of salt also increases the volume of the solution itself. Therefore, the final concentration might be incorrect. This is why specific tables are made to know each time the exact amount of this salt that is required to pass from one concentration to another, taking into account the volume of the solution. All these calculations are based on the percentage of saturation of ammonium sulfate and are valid at a given temperature. Table 8.2 gives the amount of ammonium sulfate to add to a 10 mL of protein solution at 0 °C.

8.4.1 Recipe

First, the protein extract solution is clarified by centrifugation and elimination of the pellets. Then the protein solution is maintained in the cold (generally ice bath) and the volume of the solution exactly measured. The protein solution is placed over on a magnetic stirrer and added with the amount of solid ammonium sulfate according to the precalculated Table 8.2. For instance, if the volume of the solution is 10 mL and it does not contain ammonium sulfate and the objective is to bring it to 80% saturation, 5.16 grams of ammonium sulfate are added. Instead, if the protein solution already contains 10% saturation ammonium sulfate, the amount needed to reach 80% saturation is 4.52 grams. As a third example, if the solution does not contain ammonium sulfate and the objective is to separate two fractions, the first at 60% saturation and the second at 90% saturation, 3.61 grams of this salt are first added, the mixture stirred for two hours, and then centrifuged at 10,000 g for 15 minutes at 4 °C to separate the protein precipitate. Then another 2.01 grams are added, and a second precipitation is formed and separated by centrifugation. The precipitated proteins can be dissolved in

TABLE 8.2 Precalculated Table of Ammonium Sulfate Needed for the Precipitation of 10 mL Proteins Solution (From a Given Saturation to Another)

Initial Concentration (% Saturation)	Percentage Saturation at 0°C								
	20	30	40	50	60	70	80	90	100
	Solid Ammonium Sulfate (Grams) per 10 mL of Protein Solution								
0	1.06	1.64	2.26	2.91	3.61	4.36	5.16	6.03	6.97
10	0.53	1.09	1.69	2.33	3.01	3.74	4.52	5.36	6.27
20	0	0.55	1.13	1.75	2.41	3.12	3.87	4.69	5.57
30		0	0.56	1.17	1.81	2.49	3.23	4.02	4.88
40			0	0.58	1.20	1.87	2.58	3.35	4.18
50				0	0.60	1.25	1.94	2.68	3.48
60					0	0.62	1.29	2.01	2.79
70						0	0.65	1.34	2.09
80							0	0.67	1.39
90								0	0.7
100									0

physiological buffers. Before use with CPLL beads, the protein solution is dialyzed overnight (cutoff 3500 Da) to eliminate the excess of the precipitating salt. (Equivalent desalting methods can also be used such as desalting chromatography or diafiltration.)

8.5 ALTERNATIVE METHODS FOR PROTEIN PRECIPITATION

Precipitation of proteins in view of the elimination of agents incompatible with CPLL treatment or concentration of dilute extracts can be performed by several other methods than ammonium sulfate. Polyethylene glycol, trichloroacetic acid, acetone, chloroform, and mixtures of them are used for this purpose. These agents are more denaturing than ammonium sulfate precipitation and cannot be used for fractionated precipitations. The following described protocols may involve some degree of protein losses and possible proteins that do not precipitate, thus escaping the collection. Since all subsequent precipitation protocols involve very harsh chemical agents, they must immediately be dialyzed or submitted to equivalent procedures such as buffer exchange to re-dissolve them and to re-equilibrate with a biologically compatible buffer. Figure 8.6 illustrates the possible precipitation alternatives.

8.5.1 Precipitation with Trichloroacetic Acid (TCA)

Protein precipitation with TCA is very common and largely used for plant and animal proteins (Yeang et al., 1995). It is an easy operation to perform, as described below. First, a stock solution is made by dissolving 500 g of pure TCA into 350 mL of distilled water at room temperature. This solution is stable for months.

The protein extract solution to be precipitated must be perfectly clear; if not, it is recommended to centrifuge for at least 10 minutes at 10,000 g at 4 °C to remove all possible particles in suspension.

One volume of stock solution of TCA is mixed with four volumes of clear protein solution while stirring. This mixture is left for 15 minutes at 4 °C to complete the precipitation process. A centrifugation follows at 10,000 g for 5–10 minutes at 4 °C, and the supernatant is eliminated. The pellet is then washed with 0.5 volumes of cold acetone to eliminate TCA and recovered also by centrifugation. The acetone wash is repeated twice or three times to eliminate all TCA and to dry the protein precipitate. The final drying operation is made by simple evaporation that can be accelerated by a mild heating. The protein pellet should not be left drying completely because its re-solubilization could become difficult. The protein powder is then dissolved in an appropriate

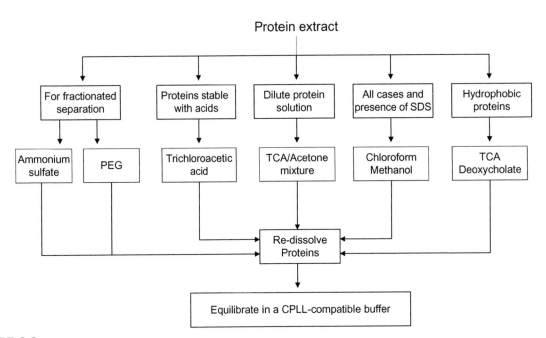

FIGURE 8.6

Options of protein precipitation according to the context. For fractionated precipitation of proteins, it is advisable to adopt either ammonium sulfate or polyethylene glycol. In both options, the precipitation agents are to be eliminated by diafiltration, desalting, or dialysis using membranes with cutoff of 3.5 kDa.

buffer compatible with CPLL treatments and possibly dialyzed (cutoff 3500 Da) to be sure that all by-products are eliminated. Equivalent desalting operations can also be performed instead of dialysis. TCA precipitation is not recommended for further glycoprotein investigations. Although some authors (Sakuma et al., 1987) claim 99.4% serum protein removal in 10% TCA, others (Zellner et al., 2005) suggest, in the case of blood platelets, adding as much as 25% TCA, such a level ensuring only 91% protein precipitation. (*Warning*: Any level of TCA above 50% will, instead of precipitating, re-dissolve proteins!)

8.5.2 Precipitation with an Acetone/TCA Mixture

This protocol is generally applied to the precipitation of dilute protein solutions. This is also a quite common precipitation for plant protein isolation from other non-proteinaceous material (Mechin et al., 2007). A general method is as follows.

A fresh 10% TCA solution in acetone is first prepared. The protein extract solution to be precipitated must be perfectly clear. It is recommended to centrifuge for at least 10 minutes at 10,000 g at 4 °C to remove all possible particles in suspension.

Ten volumes of TCA/acetone are mixed with one volume of protein solution sample and stirred vigorously prior to storing at −20 °C for a minimum of three hours to overnight. The mixture is then centrifuged at 10,000 g for 10–15 minutes at 4 °C and the supernatant is removed. The pellets are washed with pure cold acetone two or three times to eliminate traces of TCA. Once the final washed pellet is recovered, it is dried and the protein powder is re-dissolved immediately with an appropriate buffer compatible with the subsequent CPLL treatment or for other analytical determinations. The pellet should not be left completely dry because the re-solubilization of proteins could become difficult.

8.5.3 Precipitation with Polyethylene Glycol

Proteins can be precipitated by means of polyethylene glycol (PEG) (Kim et al., 2001). By using various concentrations and types of PEG of different molecular mass, it is possible to obtain precipitated fractionations just as one can do with ammonium sulfate (Section 8.4). This approach is interesting not only for protein cleaning, but also for simplifying the analyses of complex proteomes when the precipitation is fractionated (Xi et al., 2006).

The clear protein solution is added with a 50% (w/v) PEG 4000 stock aqueous solution up to a final concentration of 8% PEG. The mixture is gently stirred for 30 min in an ice bath to ensure protein precipitation and then centrifuged at 2500 g for 10 minutes at 4 °C. The pellet is the first protein fraction, and the supernatant still containing proteins can be precipitated again by adding another aliquot of stock solution of PEG up to the final concentration of 16%. This operation of fractionation can be extended to other additions of PEG stock solution up to a final concentration of 25%. Each time the precipitated proteins are collected by centrifugation and stored in the cold. At the end, the protein fractions are dissolved in the selected buffer, and the residual PEG is eliminated by buffer exchange or any other convenient means prior to further analysis or treatment with CPLL. A simple way to eliminate PEG is to make an additional precipitation of each protein fraction by the acetone/TCA method (see Section 8.5.2).

8.5.4 Precipitation with Chloroform/Methanol

This method is also largely used for protein precipitation from animal and plant extracts. It generally delivers high-protein recovery as described by Fic et al. (2010) when comparing various precipitation techniques for brain tissue. The protein solution to be precipitated must be perfectly clear. It is recommended to centrifuge for at least 10 minutes at 10,000 g at 4 °C to remove all possible particles in suspension.

To the clear protein solution, four volumes of cold pure methanol are added, and the mixture is stirred vigorously for a few minutes. Then three volumes of pure cold chloroform are added, and the mixture is mixed vigorously again for a few additional minutes. The concoction is then diluted with three volumes of water and the precipitation process is left to completion for 10–15 minutes at room temperature. The precipitated proteins are recovered by centrifugation at 15,000 g for about 5 minutes at 4 °C, but instead of being collected as solid pellets they should be located at the liquid interface. At this stage, the top aqueous layer is pipetted out. Four volumes of pure methanol are then added and the blend is stirred again. A second centrifugation follows at 15,000 g for about 5 minutes to remove the maximum volume of liquid without disturbing the protein precipitate. The latter is then dried, for instance, by speed-vac or other comparable means. Finally, the protein is re-dissolved with an appropriate buffer compatible with the subsequent CPLL treatments or for analytical determinations. This precipitation method is recommended when the sample contains sodium dodecyl sulfate.

8.5.5 Precipitation with TCA/Deoxycholate

For the precipitation of slightly hydrophobic proteins, the addition of detergent may help the recovery while improving the following analysis results. This case was described by Wu et al. (2004) to improve the precipitation of proteins from bronchial epithelial tissue by adding deoxycholate to TCA. A general method is as follows.

The protein solution to be precipitated must be perfectly clear. It is recommended to centrifuge for at least 10 minutes at 10,000 g at 4 °C to remove all possible particles in suspension.

Two % aqueous deoxycholate (DOC) solution is added to the protein solution (1 volume per 100 volumes), and the mixture is gently stirred for a few minutes to get a homogeneous solution without producing foam and is left in an ice bath for about 30 minutes. Then pure TCA is added upon stirring to a final concentration of 15%. The agitation is continued for one minute, and then the blend is left to complete the protein precipitation for at least one hour or better overnight. The suspension is centrifuged at 10,000 g for 10–15 minutes and the supernatant is removed. The pellet is washed in an ice bath with cold acetone or cold ethanol, and the supernatant is recovered again by centrifugation. This operation is repeated twice to ensure the complete elimination of DOC and TCA traces.

The protein precipitate is dried under a mild stream of nitrogen or speed-vac procedure. However, it is advised not to dry the protein precipitate too much; otherwise the re-solubilization in aqueous buffers could become difficult. Finally, the proteins are re-dissolved with an appropriate buffer compatible with the subsequent CPLL treatment or analytical determination. Other surfactants can be used instead of deoxycholate such as sodium dodecyl sulfate, Triton X-100, N-octylglucoside, Nonidet NP40, Tween-20, and a few others (Chang, 1992).

8.6 BEADS CONDITIONING PRIOR TO USE

8.6.1 Objective of the Procedure

Combinatorial peptide bead libraries are generally supplied as an aqueous slurry in the presence of preserving agents against potential bacterial growth. This material could also be supplied under powdered dry beads. In both cases, the beads need to be conditioned prior to use according to the purpose.

8.6.2 Recipe

8.6.2.1 AQUEOUS SLURRY

To 100 μL bead suspension, 1 mL of one of the following required buffer is added according to the envisioned application:

- Phosphate buffered saline (25 mM phosphate buffer, pH 7.2 containing 150 mM sodium chloride) if beads are intended to capture proteins under physiological applications.
- 25 mM phosphate buffer, pH 7 in case of protein capture in low-salt conditions.
- 50 mM sodium acetate buffer, pH 4.0 containing or not 150 mM sodium chloride when the capture is to be performed in acidic conditions as part of a whole process where capture would have to be performed at three different pHs (see Section 8.9 and Chapter 4, Section 4.5.2).
- 50 mM Tris-HCl buffer, pH 9.0 containing or not 150 mM sodium chloride when the capture is to be performed in alkaline conditions as part of a whole process where capture would have to be performed at three different pHs (see Section 8.9).
- Any other aqueous solution adapted to the capture of proteins from specific extracts. In this case the composition of the aqueous solution should be the same as the one where proteins are solubilized.

After a few minutes of vortexing, the suspension is centrifuged to eliminate the excess of supernatant along with the bead-preserving agents. This washing procedure is repeated at least three times to be sure that all undesired preserving agents are eliminated. The beads are then centrifuged to remove the excess of solution and used as described in subsequent protocols for the capture of proteins.

8.6.2.2 DRY POWDERED LIBRARY

The library comprises very hydrophilic, ionic, and hydrophobic peptides as a result of different combinations of amino acids. Therefore, their rehydration may not be accomplished at the same rate. To circumvent this question, it is first advised to slurry 100 mg of dry beads in 2 mL methanol for 30 minutes while shaking gently and then add 2 mL of phosphate buffer (e.g., 25 mM pH 7). The rehydration is to be extended overnight at room temperature. The rehydrated beads are then washed extensively, with the buffer selected for the capture of proteins as described above for the aqueous slurry. Rehydrated and buffer-equilibrated beads can be stored in the cold at 4 °C and used within the day. Once the bead library carries adsorbed proteins, it can be stored at 4 °C for a few days. Before desorbing the proteins, the beads are to be equilibrated at the same temperature at which the protein capture has been performed.

8.6.3 Variants

Equilibration of peptide beads can be performed in a variety of buffers containing various agents. Among the latter are detergents (at low concentrations not exceeding 0.1–0.5%), organic or mineral salts, additives, anti-oxidizing agents, and low concentrations of urea (up to 3 M). The presence of chemical agents might change the affinity constants between given proteins and their peptide partners.

8.6.4 Comments and Remarks

After protein desorption, the re-use of the beads for another protein capture is not recommended. Actually, in case of incomplete protein desorption, some carryover is expected with cross contaminations.

8.7 GENERAL METHOD FOR PROTEIN CAPTURE IN PHYSIOLOGICAL CONDITIONS

8.7.1 Objective of the Experiment

Reducing the dynamic concentration range under physiological conditions means targeting only natural molecular recognitions that are not enhanced by artificial means such as the increased ionic interaction in low ionic strength or hydrophobic associations enhanced in the presence of lyotropic salts. Only interactions that take place in normal life conditions are considered. As a corollary, the interactions under physiological conditions do not denature the function of proteins even temporarily. The review by Righetti et al. (2010) summarizes the situation and refers to a number of relevant bibliographic references.

In spite of the fact that protein capture by CPLL is generally operated in physiological conditions, it is here underlined, as repeatedly described, that a decrease of ionic strength or the modification of pH may improve the capture of proteins or categories of them. This is due to the fact that the affinity constant changes and also that the main interaction parameters (electrostatic attraction) are reinforced when the ionic strength is low.

8.7.2 Recipe

Common protein extracts are made with physiological buffers, thus simplifying direct contact with the beads; however if other buffers are used, a buffer conditioning is necessary before CPLL operation. Other parameters to consider are: (i) the protein concentration in the sample (ii) and the overall protein amount. The protein concentration should be between 1 and 10 mg/mL. Lower concentrations may render the capture of very low-abundance proteins challenging when the dissociation constant is too high. The total amount of protein from the sample should be larger than 50 mg when 100 μL of hexapeptide ligand library is used. In the case of human serum, it is recommended to use at least 1 mL of serum for 100 μL of hexapeptide ligand library. However, the larger the protein load, the greater the probability to find very low-abundance proteins. When small samples are available, the only possibility is to decrease the volume of hexapeptide beads. The smallest volume of beads usable is around 10 μL; lower volumes could also be used with the risk of loosing diversity and hence the number of protein captured.

Once all the above-mentioned conditions are optimized, the reduction of the dynamic concentration range can start. First, possible small particles in suspension in the biological extract are to be eliminated by filtration through a 0.8-mm filter membrane. Alternatively, a centrifugation can be performed at 10,000–20,000 g for 20 minutes at 4 °C and the clear supernatant taken for subsequent operations. In case protease traces are present, appropriate inhibitors are to be added such as a protease inhibitor cocktail. In case the protein sample is very dilute and the volume is large, it is advised to concentrate by using current methods adopted in biochemistry. Preferred ways are lyophilization after dialysis, followed by solubilization of proteins in PBS (phosphate-buffered saline). Membrane concentration under centrifugation is an interesting alternative for small sample volumes (e.g., Centricon, centrifugal filters from Millipore).

The hexapeptide ligand library is first equilibrated with the selected physiological buffer (generally PBS). This is obtained by washing the library (e.g., 100 μL) at least three times with 10 volumes of PBS each time under low-speed centrifugation at about 1000–2000 g at 20 °C for 5 minutes. At the end of three washings, the beads are pelleted by centrifugation or vacuum filtration. Then the protein extract is contacted with the solid-phase peptide library. In practice, this operation can be performed at room temperature either in a spin column or in a chromatographic-packed column. With spin columns, it is critical to maintain the hexapeptide ligand library beads as a slurry so as to ensure good contact between the beads and the proteins. The shaking should be gentle to prevent the formation of foam. It is recommended to incubate the sample with the beads for 2 to 3 hours. The suspension can also be left overnight under agitation without negative consequences. Once the protein adsorption phase is completed the supernatant is eliminated and the beads are washed with the selected physiological buffer to eliminate the excess of proteins.

If the operation is processed on a column, the bead library is first packed in a small chromatographic tube (e.g., 1000 μL of beads filled in a column of 6 mm diameter and a height of 36 mm). Then, by means of a peristaltic pump, about 10 volumes of physiological buffer are passed through the column at a flow rate of 0.5 mL/min. Then the clear protein extract is injected into the column at a flow rate ensuring a residence time of 10–20 minutes. The column can also be stopped overnight, providing the temperature does not change. Once the loading is completed, three column volumes of physiological buffer are injected to ensure that the excess of proteins is totally eliminated (baseline signal at 280 nm).

Throughout the process (suspension, spin column, or packed column), do not let the hexapeptide ligand beads dry completely, especially during the elimination of protein excess. Protein excess can be preserved in view of quantitative analysis. The desorption of captured proteins can be obtained by various approaches. Recipes are detailed in Sections 8.14–8.18.

8.7.3   Expected Results

The reduction of the concentration of high-abundance protein is expected, as is the enhancement of many diluted species. It may happen that some proteins do not find their peptide partner. The process has been largely described in a number of publications.

8.7.4   Variants

The protein solution prior to treatment with CPLLs could comprise additives compatible with physiological conditions of the process. They could be small amounts of different salts, sugars, low concentrations of non-ionic detergents (up to 0.1–0.5%), chelating agents such as EDTA and urea (up to 2–3 M), amino acids and vitamins. The protein-capturing phase could be operated at different temperature than ambient. At low temperatures ionic interactions are enhanced, whereas at high temperatures (up to 40 °C) hydrophobic associations are promoted. All operations can indifferently be made in columns or in suspension.

8.7.5   Comments and Remarks

For good reproducibility of the capturing process, precision in sample as well as on bead volumes is necessary. Biological samples may vary from one to another; it is therefore important to be sure that the ionic strength and pH are always the same. This is particularly true with urine where physicochemical parameters may vary quite largely.

8.8 PROTEIN SAMPLE TREATMENT AT LOW IONIC STRENGTH

8.8.1 Objective of the Approach

Compared to the capture of proteins under physiological conditions, a decrease in ionic strength allows a better harvest of rare proteins as a result of enhanced ionic interaction. This happens even when the pH remains neutral. The approach can be used whatever the type of sample; however, it is recommended when the main objective is the capture of rare species rather than reducing the concentration of very-high-abundance proteins. Most generally, when operating under low ionic strength, the amount of proteins captured is higher as a result of improved binding capacity (Guerrier et al., 2006). The number of gene products retrieved is also higher (Di Girolamo et al., 2011).

8.8.2 Recipe

The recipe is exactly the same as described in the previous Section 8.7. What is different is the composition of the neutral buffer that generally contains less salt (e.g., potassium or sodium chloride), or no salt at all. It is also possible to play on the concentration of the buffer itself.

An example of the effect of low ionic strength capture *versus* physiological conditions is presented in Figure 8.7.

8.8.3 Comments and Remarks

The counterpart of the reduction of ionic strength results in an increase of the binding capacity of peptide beads and as a consequence is a lower discrimination for protein capture. It is recommended to reduce first the ionic strength of PBS by, for instance, decreasing the concentration of sodium chloride from 150 mM to 50 mM. Then the salt can even be completely removed if the capture proceeds as expected. Other pHs than neutrality could be used, as indicated in the following section. The latter option is especially recommended when the concentration of the protein in the sample is very low. In all cases such experimental conditions improve the ability of the beads to collect proteins.

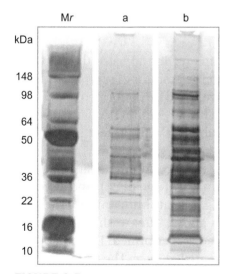

FIGURE 8.7

SDS-PAGE of TUC eluates of human serum proteins treated with CPLL at different ionic strengths. Mr lane: mass ladder of proteins standard. Lane a: serum proteins retrieved in phosphate buffer containing 150 mM sodium chloride. Lane b: serum proteins captured in phosphate buffer without salt. *(Adapted from Di Girolamo et al., 2011).*

8.9 PROTEIN SAMPLE TREATMENT AT DIFFERENT pHS

Since the environmental pH influences quite largely the dynamics of protein capture by CPLLs (influence of dissociation constants between preys—sample proteins—and baits—peptides of the library), it is possible to play on this parameter to target groups of proteins. Capturing proteins from a biological sample at different pHs may also increase the efficacy of the CPLL technology to enlarge the possibility of enrichment. To this end, two options are recommended: use of sequenced operations or use of parallel operations (Figure 8.8 depicts the options).

8.9.1 Objective of the Experiment

It has been clearly stated that the variations of physicochemical conditions modify the affinity of proteins for their respective peptide ligand in virtue of the mass action law (see Chapter 4, Section 4.4.2). The pH is perhaps the most important parameter governing the protein–peptide docking because (i) it changes the ionization of mainly proteins but in some extent also the ionization of peptides and (ii) because electrostatic interactions are the most dominant in this kind of affinity recognition.

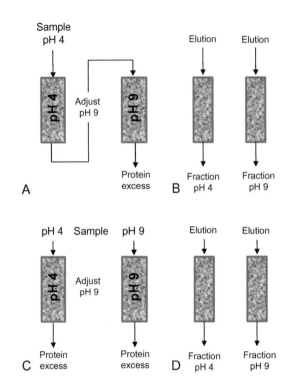

FIGURE 8.8
Schematic representation of protein capture at different pHs. The example deals with adsorptions at pH 4 and pH 9. More pHs could be considered. Upper panel: protein capture performed in series. A: The sample adjusted at the selected pH value (e.g., pH 4) is primarily injected into the first column previously equilibrated at the same pH. The column effluent is adjusted at the pH of the second column (e.g., pH 9) and then injected into the second column previously equilibrated at the same pH. B: Once the adsorption is completed and the excess of proteins is washed out, the columns are separately eluted with selected desorbing agents (see general method of elution and variants in Sections 8.14 to 8.18). Lower panel: protein capture performed in parallel. C: The sample is divided into two parts; the first is adjusted at pH 4 and the second at pH 9, and they are separately injected in their respective CPLL columns previously equilibrated at pH 4 and pH 9, respectively. Once the adsorption is completed and the excess of proteins is washed out, the columns are separately eluted (D) with selected desorbing agents (see general method of elution and variants section 8.14 to 8.18). These operations can be made in suspension using small Eppendorf tubes and separations between solid and liquids operated by centrifugation.

By this principle it is possible to capture proteins with CPLL beads after having adjusted the pH value without modifying all other parameters (temperature, ratio between the amount of sample, and volume of CPLL beads, ionic strength, etc.). For instance, the first capture could be performed at pH 9; in this case, the protein solution must be titrated up to pH 9 and CPLL beads equilibrated with a buffer of pH 9, with the same ionic strength of the biological sample. Incubation with protein proceeds as described in Section 8.7. It could be operated in column or in batch (e.g., spin columns). Once the capture phase is over, the excess of proteins is recovered and used for the second capture at a different pH. In practice, proteins that are not captured are titrated at the second pH (e.g., pH 4), and a second aliquot of CPLLs is adjusted at this pH with an appropriate buffer (e.g., acetate buffer pH 4). Then the second capture can be started according to the regular procedure as above. After washing the CPLLs from both capture operations (pH 9 and pH 4), elution from the beads is performed. This can be operated separately, or beads could be mixed together and elution operated at once. The former option allows identifying two groups of proteins where a number of them are redundant. When the elution is operated after having mixed the CPLLs, a single protein solution is obtained that can be analyzed as a whole (separate elutions followed by elutions blend will result in the same protein mixture; however, common proteins are more concentrated). Generally, the pattern of proteins is larger compared to a capture at a single pH as, for instance, pH 7. This sequenced procedure is recommended when the biological sample availability is limited. These options can be performed in small bulk or in columns, depending on the scale of the experiment (Figure 8.8 represents the options schematically).

Many papers have been published using two or three pH values during capture with interesting data (Fasoli et al., 2010; D'Amato et al., 2010; Fasoli et al., 2011). Capturing proteins from a biological sample at different pHs may also increase the efficacy of the CPLL technology to enlarge the possibility of enrichment.

8.9.2 Recipe

8.9.2.1 SEQUENCED COLUMNS

Conditions of capture are similar to those described in Section 8.7. Nevertheless, the pH of both the protein extract and CPLL must be adjusted before use. For instance, the first capture could be performed at pH 4; in this case, the protein solution must be titrated to pH 4 with acetic acid, and the CPLL beads equilibrated with a buffer of pH 4 with the same ionic strength of the biological sample. Incubation with protein proceeds as described in Section 8.7. It could be operated in column (Figure 8.8) or in batch (e.g., spin columns). Once the capture phase is over, the excess of proteins is recovered and used for the second capture at a different pH. In practice, proteins that are not captured are titrated at the second pH (e.g., pH 9), and a second aliquot of CPLLs is adjusted at this pH with an appropriate buffer (e.g., Tris buffer pH 9). Then the second capture can be started according to the regular procedure as above. After washing the CPLLs from both capture operations (pH 4 and pH 9), the elution is activated. This can be operated separately; beads could also be mixed together and elution operated at once. The former option allows identifying two groups of proteins where a number of them are redundant. When the elution is operated after having mixed the CPLLs, a single protein solution is obtained that can be analyzed as a whole. Generally, the pattern of proteins is larger compared to a capture at a single pH as, for instance, pH 7. This sequenced procedure is suggested when the biological sample availability is limited.

8.9.2.2 PARALLEL COLUMNS

In this case, two simultaneous protein-capturing operations are considered. The initial sample is divided into two equal parts as well as the CPLL volume. The first part of the protein sample and the first part of CPLLs are equilibrated at a desired pH (e.g., pH 4 with acetic acid to comply with the example above), and the second part of the sample as well as the second part of CPLLs equilibrated with the second pH (e.g., pH 9 with Tris(hydroxymethyl)aminomethane). Capturing operations, performed either in batch or in columns, are operated in parallel up to the washing step to eliminate the protein excess. Then the washed CPLLs could be eluted using the same approach as indicated above: separately or after having blended the CPLLs.

8.9.3 Expected Results

Since the protein capture is dominantly governed by electrostatic interactions, the result of capturing at different pH is to find two distinct groups of proteins based on their isoelectric points. Nevertheless all proteins that are essentially adsorbed on the CPLL beads, thanks to a dominant hydrophobic interaction, are basically unchanged and can be found in both groups of eluted proteins. The total number of gene product that can be found by scanning the adsorption pH from 4 to 9 is largely increased. Acidic pHs generally allow adsorbing a larger number of proteins compared to alkaline pHs (see examples in Figure 8.9).

8.9.4 Variants

The pH can be changed from the suggested values above. For instance, pH 2 can also be used with the advantage of having all proteins positively charged since they are titrated to a pH value below their isoelectric points. Ionic strength can also be changed as well as the addition of some chemical agents, as described in the general method (see Section 8.7). Elutions could be operated according to all methods described below (see Sections 8.14–8.18).

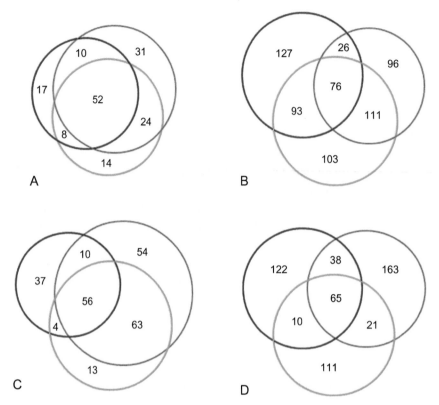

FIGURE 8.9
Venn diagrams representing proteins captured at different pHs and identified by LC-MS/MS after SDS-PAGE separation, lane slicing, and trypsination.

A: *Limulus polyphemus* haemolymph proteins captured sequentially at different pHs and eluted using 4% sodium dodecyl sulfate aqueous solution containing 25 mM DTT at boiling temperature for 5 minutes. The red circle represents the capture at pH 4.0; the blue and green circles depict proteins captured at pH 7.0 and 9.5, respectively. (From D'Amato et al., 2010).

B: Human serum proteins captured sequentially at different pHs and eluted with 9 M urea, 2% CHAPS, and 50 mM citric acid pH 3.3 (Fasoli et al., 2010). The red circle represents the capture at pH 4.0; the blue and green circles regard proteins captured at pH 7.2 and 9.3, respectively. Protein elutions were performed using 9 M urea containing 2% CHAPS and 50 mM citric acid pH 3.3.

C: Gene products from a spinach leaf extract (from Fasoli et al., 2011). Red circle represents unique proteins harvested at pH 4.0; blue circle assembles unique proteins retrieved at pH 7.0; green circle comprises proteins from a capture at pH 9.3. All elutions were performed using 4% SDS solution containing 25 mM DTT at boiling temperature.

D: Unique protein count identified from an artichoke extract at three different pHs: 4.0, 7.2, and 9.3. Color codes are the same as above. *(Calculated from data reported by Saez et al., 2013).*

8.9.5   Comments and Remarks

Between two columns, the sample could be stored, for example, overnight in the cold. However, before being put in contact with the beads at a second pH, the protein solution must be equilibrated at the same temperature of the previous capture. Sequential capturing is preferred when the volume of the available sample is small. Sequential capturing is also preferred when the analysis of eluates has to be performed separately. To obtain reproducible results, it is necessary to adjust carefully the pH of protein solutions and check if it drifts over time. Temperature is also critical since it impacts the affinity constants.

8.10   PROTEIN EXTRACT TREATMENT IN THE PRESENCE OF LYOTROPIC SALTS

8.10.1   Objective of the Experiment

Since the environmental conditions influence quite largely the dynamics of protein capture by CPLLs (dissociation constants and competition), it is possible to play on the presence of lyotropic salts by enhancing the propensity of proteins to associate by hydrophobic interaction. Hydrophobic associations are largely present in protein binding and folding as described in Section 4.4.3 and are demonstrated representing an important contribution to capture proteins by peptide ligand library. To this end two technical options are recommended when using different concentrations of salts: sequenced or parallel operations (Figure 8.8 that represents schematically the options for pH exploration is also applicable in the present case if instead of changing the pH in between columns the addition of lyotropic salts will be added). To enhance this type of interaction it is suggested to add some amount of lyotropic salts in both the protein solution and the buffer used for bead equilibration.

For instance the first capture could be performed under regular conditions, as described in Section 8.7 and the second capture in the presence of the hydrophobic-enhancing salt. The composition comparison of collected captured proteins by any dissociation agents will thus give an idea of how the protein patterns are different with the possibility to segregate hydrophilic from hydrophobic proteins. The experiment will also allow detecting novel species that are normally not well captured under regular conditions enlarging by consequence the vision on proteome compositions. This approach has been described by Santucci et al. (2013) for the analysis of human serum proteins. Depending on the availability of sample volume this operation could be performed either in column or in suspension. The recipe below describes the procedure as a suspension.

8.10.2 Recipe

The clear protein sample is diluted twice with a neutral phosphate buffered solution of 2 M ammonium sulfate in order to reach a final concentration of 1 M. The diluted protein sample is vortexed repeatedly and then left for about one hour at room temperature. In parallel CPLL beads are equilibrated (two-three washes) with a phosphate-buffered solution of 1 M ammonium sulfate and the excess of liquid eliminated by centrifugation. The sample and the CPLL beads are then mixed and left under gentle agitation for about two hours at room temperature. The amount of sample and the volume of CPLL are those discussed in the general method of protein capture Section 8.7.2.

The excess of protein is then eliminated by centrifugation followed by two washes with an equal volume of phosphate-buffered solution of 1 M ammonium sulfate. Proteins captured by the beads are eluted by one of the desorption methods described in Sections 8.14–8.18. In the published report by Santucci et al. (2013) proteins were eluted using 4% SDS solution containing 50 mM DTT under boiling conditions. The collected fraction is then ready for SDS-PAGE followed by LC-MS/MS or for other types of analyses after having eliminated the elution agents. For the elimination of sodium dodecyl sulfate refer to Section 8.5.4.

8.10.3 Variants

Most important variants to the enhancement of hydrophobic species capture are based on the use of other lyotropic salts than ammonium sulfate. To this end the Hofmeister's series provides several suggestions (Hofmeister et al., 1888; Zhang and Cremer, 2006).

The concentration of the selected lyotropic salt may also vary according to the propensity of proteins to precipitate and to the temperature of the experiment.

Small protein samples are better treated in suspension while large samples can easily be performed in column. The latter could be used in series or in parallel with similar results; however, the use of a series of columns will give eluates with lower number of proteins per fraction for easier analyses.

The pH of the lyotropic salt solution used can range from 3 to 11 with direct consequences on its lyophilicity and the precipitation propensity of proteins present in the sample. Generally the highest concentration of lyotropic salt that allows maintaining all proteins is solution is selected.

Protein elution can be operated using single stringent conditions or by using sequential desorption as described in Section 8.16.

8.10.4 Comments and Remarks

One important risk is to have protein precipitation during the adsorption and elution phase within the beads. This can be minimized by lowering the concentration of the lyotropic salt. Since the hydrophobic aggregation of proteins depends on the temperature, it is advised avoid temperature changes during the experiments. Very abundant proteins that are also largely susceptible to aggregate can advantageously be eliminated before the CPLL treatment by selected depletion methods. In the case of serum, albumin can be eliminated (even partially) by the use of Cibacron Blue sorbents as described by Santucci et al., (2013), while immunoglobulins G can be subtracted by using Protein A sorbents.

8.11 PROTEIN SAMPLE TREATMENT IN THE PRESENCE OF SURFACTANTS

8.11.1 Purpose of the Method

The presence of detergents in a biological extract is generally due to the extraction phase where their function is to enhance the extraction process. This is necessary for low-solubility proteins, especially those that are involved in membranes or that are inside spaces protected by membranes. Detergents may contain functional groups incompatible with CPLL use. They are detergents with ionic groups. Zwitterionic detergents are also incompatible for proper protein capture. Other non-ionic and relatively mild surfactants are compatible with the use of CPLL when they are used in low concentrations. They may be less efficient for the extraction of proteins but constitute a good compromise to enhance the extraction.

8.11.2 Recipe of CPLL Treatment

The protocol for these extracts is similar to what is described in Section 8.7. The concentration of the surfactant should not exceed 0.5% and should be below this concentration whenever possible. If the initial concentration during extraction is higher, the sample must be diluted to the maximum tolerated concentration; otherwise a large number of proteins will not be captured. Specific details on the use of detergents for protein extraction followed by CPLL treatment are given for plant proteins in Section 8.13.2. Sample dilution must be performed by using a buffer of the same nature adopted for extraction. Another important preliminary operation is the equilibration of CPLL beads with the same buffer containing the same detergent at the same final concentration. For the purpose of comparison, all experimentation must be performed at the same temperature since hydrophobic interactions are sensitive to temperature changes. The subsequent elution operations are similar to any other CPLL recipe where protein desorption can be performed singularly or sequentially (see Sections 8.14–8.18 below).

8.11.3 Suggested Detergents and Variants

Non-ionic detergents that can be used such as Brij 58, Tween-20, and octylglucoside are preferably aliphatic (see Figure 8.10). Aromatic non-ionic detergents such as Nonidet NP40 and Triton X-100 can also be used (Figure 8.11). However, in all cases, it is advisable to test their compatibility with the maximum tolerated concentration for both protein extraction and protein capture by the peptide library beads. Non-ionic detergents can also be used as a mixture; in this case the final total concentration should not exceed 0.5%. It is suggested, for instance, that for each chosen detergent there should be a little series of preliminary experiments where the concentration of the detergent ranges between 0.1 and 0.5%. Nonidet NP40 has been successfully used for the discovery of many proteins from platelet lysates (Guerrier et al., 2007b) and from cypress pollen lysates (Shahali et al., 2012).

For very complex extracts like eukaryotic cell lysates, Fonslow et al. (2011) found that the addition of 0.1% SDS allows increasing the number of captured proteins and hence the number of spots found by 2D gel analysis. These authors justified this approach based on the assumption that this detergent denatures proteins and further exposes high-affinity epitopes, disrupts protein interactions within protein complexes, and may minimize hydrophobic hexapeptide–protein interactions.

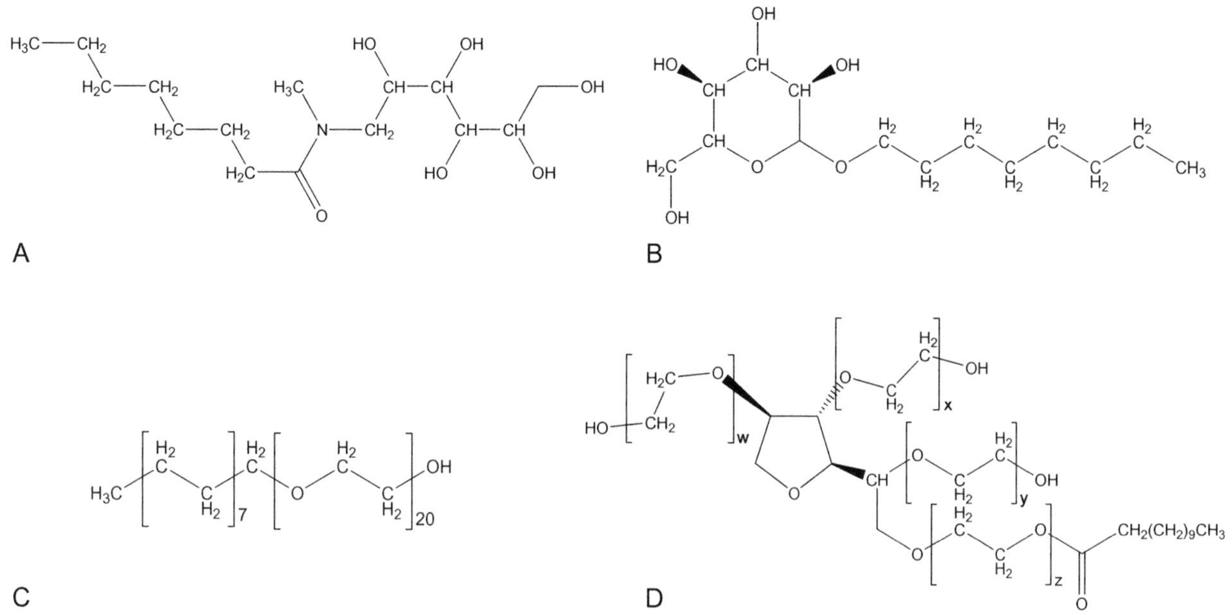

FIGURE 8.10
Examples of non-ionic aliphatic surfactants. A: Octanoyl-N-methylglucamide; B: Octyl-β-glucoside; C: Brij-58 or Polyoxyethylene (20) cetyl ether; D: Tween-20 or Polyethylene glycol sorbitan monolaurate.

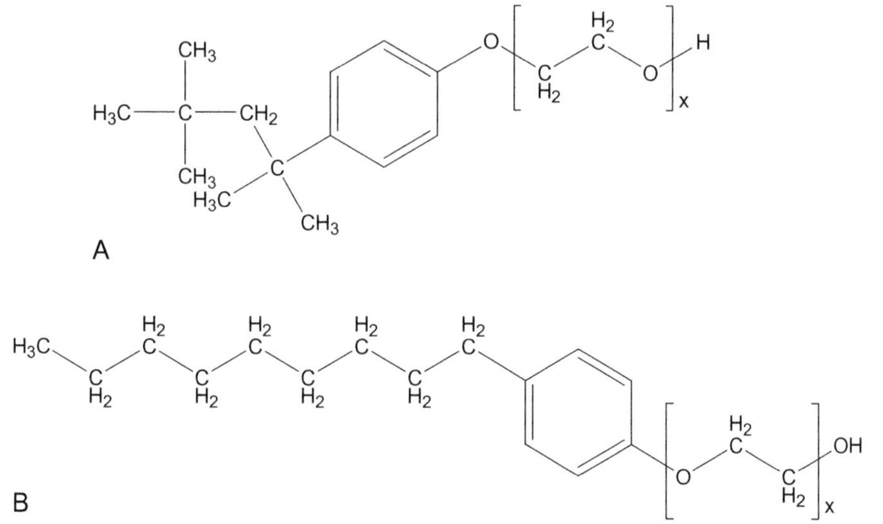

FIGURE 8.11
Examples of non-ionic aromatic surfactants used in biochemistry. A: Triton X-100 or Octylphenolpoly(ethyleneglycolether)$_x$; B: Nonidet NP-40 or Nonylphenyl-polyethylene glycol.

Naturally, using larger amounts of SDS will have an opposite effect since this chemical is used for the elution of proteins after CPLL capture (see Section 8.15).

8.11.4 Suggestions and Remarks

Whatever the surfactant selected, it must be pure for protein applications and especially free of aldehydes. Detergents are molecules that decrease the surface tension and thus promote the formation of foam. Gentle agitation is required during protein capture in bulk to prevent the phenomenon. The free surfactant can also be eliminated before protein elution by a rapid wash with the same buffer without detergent. The protein elution then follows as usual.

8.12 USE OF TWO COMPLEMENTARY PEPTIDE LIBRARIES

Chemically modified libraries have properties that may allow enlarging the protein capture, as for capturing proteins at different pHs (Section 8.9). This is the case when a succinylated peptide library (see Section 8.2) is used instead of a regular primary amino terminal peptide library. Effects resulting from chemical modifications of peptide libraries are discussed in Section 4.5.7.

In the case of unmodified peptides, the overall character is at best neutral or somewhat cationic due to the terminal primary amines, as well as to the presence of primary amines from the side chain of lysine. Conversely, after succinylation, the presence of carboxylic groups confers a global anionic character to the library. Generally, in the first case anionic proteins are preferably captured, while in the second case cationic proteins are better captured instead. Although this is not a generalizable rule, due to the presence of a number of other factors that contribute to protein docking on the respective bead peptide, it has been demonstrated that carboxylation induces a significant ionic effect. With different libraries, such as the complementary ionic versions discussed above, it is possible to proceed following three different practical applications options. The first is to use the libraries as a sequence of protein-capturing steps; the second is to use the two libraries in parallel; and the third is to use them as a mixture. (Figure 8.12 represents schematically the first two options.) The order of libraries within a sequence has no effect on the final results, as is demonstrated in Figure 8.13. This is a logic consequence of a massive overloading.

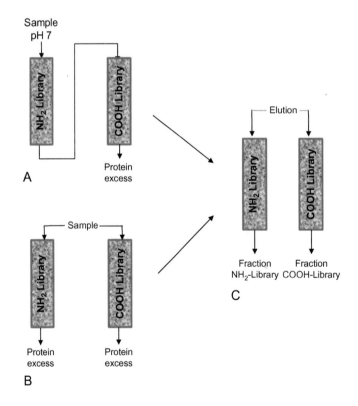

FIGURE 8.12

Schematic representation of protein capture with two different libraries (a primary terminal peptide library and a carboxylated one).

A: protein capture is performed in series. The sample equilibrated with the same buffer as the columns is primarily injected into the first column. The column effluent is directly loaded into the second column.

B: protein capture performed in parallel in both columns independently.

C: once the adsorption is completed (first or second option) and the excess of proteins washed out, the columns are separately eluted with selected desorbing agents (see general method of elution and variants in sections 8.14 to 8.18). These operations can be made in bulk using small Eppendorf tubes and separations between solid and liquids operated by centrifugation. The eluted proteins could also be blended, depending on the objective of the experiment.

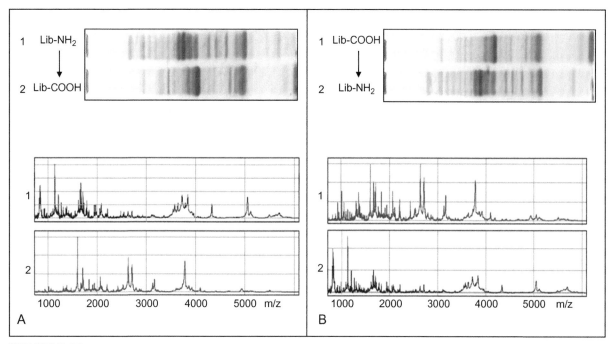

FIGURE 8.13

Combined sequential use of two peptide libraries for reduction of the dynamic range of human serum. A primary amino terminal library was used in combination with the same library after succinylation. "A" represents analytical data of serum proteins after treatment with CPLL using first an amino-terminal library, followed by the carboxylated-library. Analytical determinations are SDS-PAGE (upper panel) and mass spectrometry SELDI using a Q protein chip array (lower panel). "B" is a similar experiment where the order of libraries was reversed (carboxyl-library was used first). Analytical determinations are the same. It is to be noticed that whatever the order of use of peptide libraries, the final results are extremely similar (Lib: library).

8.12.1 Objective of the Experiment

Clearly, due to the complementary effect of these libraries, the general objective is to enhance the capability of the process for capturing more low-abundance species from a protein extract. Interestingly, contrary to what is described when using the same library at different pHs, there is no need to adjust the pH from one column to another. Once the capture phase is completed, the excess of proteins is washed away and the bound proteins are collected by regular elution rules. Eluates can be mixed together or kept separated for the subsequent analytical determinations such as 2D-PAGE or LC-MS/MS. The capturing operations can be performed in small bulk or in columns depending on the scale of the experiment.

8.12.2 Recipe

8.12.2.1 SEQUENCED COLUMNS

After having equilibrated each library with the appropriate buffer (generally, PBS or neutral phosphate buffer without NaCl or even sodium chloride with no buffering molecules), CPLL libraries are used as a sequence in batch (e.g., spin columns) or in columns (see Figure 8.12A). Conditions of capture are similar to those described in Section 8.7. As a reminder, the protein sample should be equilibrated with the same buffer or salt solution as the peptide libraries.

Once the capture phase of the first column is over, the excess of proteins (flow-through) is directly used for the capture on the second column. If small-chromatography columns are used, the outlet of the first column is connected directly to the inlet of the second one. The sample is pushed through both columns and then washed by injection of the selected equilibration buffer up to the elimination of protein excess from the second column. The columns are then disconnected, and elution is operated separately using the same eluting agents (see Sections 8.14–8.18). As a possible option, since the

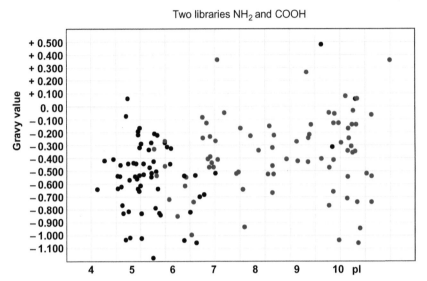

FIGURE 8.14

Two-dimensional positioning of proteins captured from two libraries classified as a function of their isoelectric point and Gravy index (hydrophobicity). Beads were a primary amine hexapeptide combinatorial library (black dots) and its carboxylated version (red dots). The clear segregation by ionic properties of the proteins is notable, whereas segregation by hydrophobicity is irrelevant. *(From Guerrier et al., 2007b, by permission).*

elution is performed under the same conditions, the beads could even be mixed together and elution performed as a single operation. The separate elution yields two groups of proteins with some degree of overlapping, although there is a dominance of anionic and cationic proteins, respectively, from primary amine library and carboxylated library (see Figure 8.14).

8.12.2.2 PARALLEL COLUMNS

An equal amount of the two distinct bead libraries is equilibrated with the same buffer or aqueous solution. The usual buffer is PBS, but it could also be a neutral phosphate buffer with reduced concentration of sodium chloride or total absence of salt or even just sodium chloride as stated earlier. The biological sample is also to be equilibrated with the same buffer and separated into two equal parts. Then the protein load starts in parallel and is operated similarly (same temperature, same incubation time, etc.). The columns are then washed with the same buffer to eliminate all unbound proteins. Finally, the captured proteins are recovered as usual by elution with selected agents, according to the rules indicated in Sections 8.14–8.18. As per the indication given for "in-series" use of libraries, it is optionally possible to mix the libraries after having captured all proteins and elute them simultaneously. However, this process does not add any advantages to a mixed-bed library as described below.

8.12.2.3 MIXED-BED LIBRARIES

Peptide libraries can easily be used as a blend. This is especially the case with complementary libraries that are amino-terminal and carboxyl-terminal peptides. First, they both need to be washed in relatively high ionic strength as, for instance, buffered 1 M sodium chloride. Then they are mixed together in appropriate proportions (50%–50% in volume is the first choice) and mixed thoroughly so as to get a homogeneous suspension and prevent aggregation. The bead mixture is then equilibrated with the selected buffer and loaded with the biological sample followed by an extensive washing with the buffer, and the captured proteins are collected upon elution with one of the desorption methods described in Sections 8.14–8.18.

8.12.3 Expected Results

A number of examples have been published to date that give quite clear picture of results that are expected compared to the use of a single library. First, the use of a primary amino terminal peptide library associated with a carboxyl terminal one, allows separating populations of proteins with a dominant cationic and dominant anionic character, respectively. The sequence of these libraries has only a very limited effect on the segregation of hydrophobic dominant interactions. Figure 8.14 illustrates the case of proteins extracted from human platelets (Guerrier et al., 2007b). Second, the number of gene products that are found when using two libraries has always been larger than the species found from each library, thus enlarging the repertoire of proteins that can be targeted. Although a large number of proteins is redundant (see Figure 8.15), the number of identifiable proteins is always enlarged not only in terms of gene products but also in terms of isoforms. Third, proteins that escape from the capture by peptide beads for reasons that are discussed in Chapter 4 are largely reduced as a consequence of the complementary action of the second library. The eluate from the use of a mixed library is equivalent to the blend of eluates from the first two options of sequential or in parallel modes.

8.12.4 Variants

A most obvious variant is to play on the composition of the buffer used throughout the process of capturing: composition, pH, and ionic strength.

8.12.5 Comments and Remarks

Serial columns are preferred when the amount of sample is small; the effluent of the first column goes directly into the second column. When the enrichment of very dilute species is more important than the high-abundance protein reduction, it is preferred to use the libraries as a blend. In this case and in the absence of experimental indications, 50% blend is the first choice (see also Chapter 4, Sections 4.5.6 and 4.5.7).

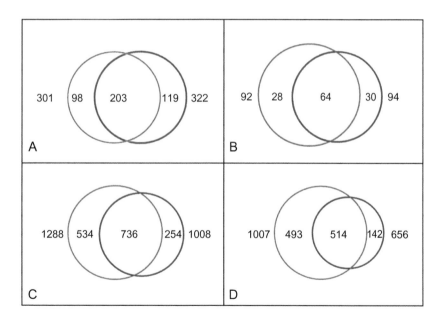

FIGURE 8.15

Venn diagrams representing proteins captured by two sequential libraries. The first library (blue circles) is a primary amino hexapeptide combinatorial library, and the second (red circles) is a carboxylated version of the first.

Protein capture was always performed under physiological conditions, and the collected proteins are a pool of three elution protocols: 2 M thiourea-7 M urea-2% CHAPS; 9 M urea-citric acid pH 3.3; 6% acetonitrile-12% isopropanol-10% ammonia at 20%-72% distilled water.

A: Human platelets extract (adapted from Guerrier et al., 2007b).
B: Bovine lactoserum (adapted from D'Amato et al., 2009).
C: Human red blood cells extract (adapted from Roux-Dalvai et al., 2008).
D: Human cerebrospinal fluid (adapted from Mouton-Barbosa et al., 2010).

What is important to observe is that the libraries capture a large number of same-gene products, but the second library adds a significant number of additional species demonstrating its complementary effect.

Sequential capturing is preferred when the volume of the available sample is small. Sequential or parallel capturing is preferred over a library blend when a better understanding about the behavior of captured proteins is desired; the analysis of eluates needs to be performed separately. Between two columns used as a sequence, the sample could be stored, for example, overnight in the cold. However, before being put in contact with the beads at a second pH, the protein solution must be equilibrated at the same temperature of the previous capture.

8.13 PLANT PROTEIN ENRICHMENT

8.13.1 General Considerations

Plant protein extracts are not always dominated by a single or a few high-abundance proteins such as human serum or red blood cell extracts. However, there are cases where one protein is largely present among traces of many other proteins. This is the case of leaf extracts where RuBisCO (ribulose-1,5-biphosphate carboxylase/oxygenase) accounts for more than 40% of the total protein amount (McCabe et al., 2001). The presence of that protein justifies per se the use of CPLL for the reduction of the dynamic concentration range. This is additionally justified because of the poor protein content of plant extracts (Rose et al., 2004) where very low-abundance gene products need to be enriched. Nevertheless, the question of plant extracts is not easy because during extraction many other undesirable compounds are solubilized. Therefore, before entering the process of proteome studies, several points have to be considered. Plant cells are rich in proteases, which require the presence of inactivating agents to prevent protein hydrolysis during the extraction process. They are rich in acidic polysaccharides as well as in pigments, lipids, nucleic acids, polyphenols, oxidative enzymes, terpenes, organic acids, and secondary metabolites, all of them interacting directly with CPLL beads and thus preventing a proper protein capture or even competing with proteins. These compounds may also interfere with protein separation and analyses (Gengenheimer, 1990). Moreover, plant proteins are heavily glycosylated, thus adding consequent problems of abnormal migrations in two-dimensional electrophoresis and difficult ionization in mass spectrometry.

Several clean-up protocols applicable to plant extracts are available; however, most protocols are not compatible with CPLLs. Thus the following general rules should be adopted:

- When dealing with seeds, it is advisable to extract lipids by means of solvent extraction (see Section 8.3.10).
- The aqueous extraction should be performed in relatively low ionic strength to prevent the solubilization of nucleic acids.
- A treatment with phenol, associated with some amounts of polyvinylpyrrolidone, eliminates general pigments and polyphenols.
- With highly viscous material, such as latex and honey, a dilution with an aqueous buffer is recommended.
- When dealing with proteins that are engaged within the cell wall, such as pollen proteins, some amounts of non-ionic detergent (less than 0.5%) and urea (less than 3 M) should be used; these concentrations are compatible with CPLL treatment.

Several examples of CPLL treatments of plants extracts after preliminary cleaning operations are given in the literature, with detailed technical information (Boschetti et al. 2009; Frohlich and Lindermayr, 2011; Fasoli et al. 2011; Koroleva and Bindschedler, 2011; Righetti and Boschetti, 2012b). A preliminary lipid removal step with solvents is particularly recommended with plant seeds such as soya beans, peanuts, corn, sunflower, and many others. The extracted plant proteins can be precipitated with ammonium sulfate or trichloroacetic acid or acetone, or polyethylene glycol to collect precipitates that are free or almost free of CPLL interfering substances. Acidic precipitation can also be operated just by acidifying the solution with acetic acid at pH 3-4. The use of trichloroacetic acid as a precipitating agent may induce denaturation and therefore the loss of biological properties. Sample pretreatments are important because they contribute enormously to obtaining significantly better analytical results, especially when using two-dimensional electrophoresis and related methods.

8.13.2 Recipe for Plant Protein Extraction Prior to CPLL Treatment

Proteins are only poorly present in plants and fruits; moreover, they are difficult to dissolve. The primary action for protein extraction is to treat the plant tissue with one of the current buffers used for extraction of proteins from animal organs (e.g., 125 mM Tris-HCl, 50 mM NaCl, 1% CHAPS, pH 7.4). This works relatively well; however, a large portion of proteins is not extracted. An alternative method is to use buffers on which some amount of ionic detergent such as sodium or lithium dodecyl sulfate is added. The difference in protein content and protein diversity that can be reached using these two different extractions can be large. This is exemplified by Esteve et al. (2012) from banana extracts. The authors demonstrated that the number of gene products that could be found by replacing CHAPS 2–4% by 3% SDS and 25 mM dithiothreitol associated with high-temperature extraction has been greatly improved. However, such a SDS concentration prevents the capture of extracted proteins to CPLL unless the detergent is eliminated. In the present case, the authors precipitated the extracted proteins by means of methanol/chloroform mixture as described in Section 8.5.4. In another paper, in view of circumventing the time-consuming protein precipitation, the concentration of sodium dodecyl sulfate was reduced to 1%. Then the extracted proteins were diluted 10 times to decrease the SDS from 1 to 0.1%, that is, below the critical concentration for protein capturing with CPLL. Under these conditions, the number of extracted and identified proteins compared to the extraction with Tris-NaCl-CHAPS was substantially increased (Saez et al., 2013). Figure 5.4 depicts possible options when dealing with extraction of plant proteins and enrichment differences resulting from the protein extraction methods.

8.13.3 Preliminary Treatments

Prior to analyses or capture with CPLLs, plant extracts could require a preliminary treatment. Delipidation is, however, not an obligation with aqueous tissues. For instance, this was not the case for banana extracts described earlier. Following are various options depending on the source of the plant extract.

8.13.3.1 DELIPIDATION

A detailed recipe of delipidation is described in Section 8.3.10. This applies mostly to seeds and any other plant organ that are supposed to contain fats or oils.

8.13.3.2 PHENOL EXTRACTION

Phenol extraction is a relatively old method used to remove proteins initially from carbohydrates and then from nucleic acid extracts. Today it still represents the preferred method for removing proteins from nucleic acids, and it is increasingly adopted for protein plant extraction while eliminating undesired material. Phenol is a weak acid that is only partially soluble in water (about 7%), but it interacts quite strongly with proteins by hydrogen bonding (see the explanations of hydrogen bonding in Section 4.4.3, subset c). As a result, proteins become denatured and soluble in the organic phase. Proteins are found in the phenol phase and no longer in the aqueous phase. This method is largely documented for proteins difficult to extract and for elimination of non-proteinaceous material. Fruits (Saravanan and Rose, 2004), tubers (Mihr and Braun, 2003), leaves (Carpentier et al., 2005; Wang et al., 2003), and wood (Mijnsbrugge et al., 2000) are among the best known examples.

In practice, plant tissues are ground to a homogeneous fine powder using appropriate devices or liquid nitrogen. Then 1 g (or multiple) of powdered tissue is mixed with 3 mL of 500 mM Tris-HCl, 50 mM

EDTA, 700 mM sucrose, 100 mM KCl, and pH is adjusted to 8.0, vigorously agitated, and then incubated upon shaking for several minutes overnight in the cold. Ice bath is easily applicable. If the extraction takes longer than a few minutes, the addition of a cocktail of proteases inhibitors is recommended to preserve the proteins' integrity. Afterward, an equal volume of phenol previously saturated with 50 mM Tris-HCl pH 7.5 is added, and the mixture is incubated under shaking for 15 minutes at room temperature.

The insoluble material is separated by centrifugation at 3000–6000 g for a few minutes at 4 °C. The phenolic phase on the top of the tube is collected while taking care of not contacting with the liquid interphase and then mixed with 3 mL of 500 mM Tris-HCl, 50 mM EDTA, 700 mM sucrose, 100 mM KCl, pH to 8.0. The sample is shaken for a few minutes and centrifuged again for phase separation. The phenol phase still on top of the tube is collected and precipitated by adding four volumes of 0.1 M ammonium acetate in methanol. The mixture is then agitated overnight in the cold (preferably at −20 °C). After this precipitation phase, the proteins are collected by centrifugation and washed with the precipitation solution once or twice. The last operation consists of a wash with acetone followed by drying in air or under vacuum.

8.13.3.3 POLYVINYLPYRROLIDONE (PVP) EXTRACTION

The removal of polyphenol compounds is possible by adding 1% solid PVP to the aqueous solution of proteins. The resin adsorbs the undesired material while leaving proteins in solution. PVP can advantageously be added at the stage of protein extraction after having reduced the plant tissue to a fine homogeneous powder (see paragraph above).

8.13.3.4 PROTEIN PRECIPITATION

The collected proteins from phenol extraction and PVP adsorption can also be further cleaned by a precipitation step. First, the proteins are to be dissolved in a physiological buffer, unless they are already in solution. The solution must be clear or clarified by centrifugation to remove all material in suspension. Several approaches are available for protein precipitation, as described in detail in Sections 8.4 and 8.5.

8.13.4 Protein Capture and Dynamic Range Reduction

After removal of undesired nonproteinaceous material, the plant protein extract can be treated with CPLLs under relatively standard conditions. The general protocol is as follows:

The solid-phase hexapeptide ligand libraries, whatever the presentations, dry or wet, need to be equilibrated with the buffer solution selected for protein capture (e.g., PBS) before use (refer to Section 8.6 for equilibration details). The plant extract in the selected buffer used for protein capture and containing the antiprotease cocktail is centrifuged at 12,000 g to remove all solid material in suspension. For an optimal capture, the protein concentration of the sample should be at least 0.1–1 mg/mL. Lower concentrations may render the capture of very low-abundance proteins challenging especially when the affinity constant is too low. The total amount of protein from the sample should be greater than 50 mg when 100 µL of hexapeptide ligand library is used. The larger the protein load, the greater the probability of enriching and thus detecting very low-abundance proteins. In order to concentrate protein solutions, the current methods adopted in biochemistry labs can be used. The preferred ways are lyophilization after dialysis or membrane concentration under centrifugation (Centricon, centrifugal filters from Millipore).

The protein solution with antiproteases is mixed with the drained CPLL beads, and the suspension is shaken gently for at least two hours at room temperature. It is critical to maintain an even slurry of the hexapeptide ligand library beads so as to ensure intimate contact with the proteins. The agitation should not be too strong so as to prevent the formation of foam, especially if the protein solution contains small amounts of non-ionic detergents (capped test tubes rotating on a tilted platform are an excellent solution). The suspension can be left overnight under agitation without negative consequences. The protein capture step can be performed in physiological conditions of ionic strength and pH; however, it can also be made in lower ionic strength to increase the amount of captured proteins. The pH is also critical to modulate the protein capture since it acts on the affinity constant of proteins for their corresponding peptide baits (see Section 8.9). To enlarge the captured protein coverage, it is advised to operate at three different pHs as described by Fasoli et al. (2010). This treatment can be performed under different configurations as described above for other types of proteins. The excess of proteins is washed out from the hexapeptide ligand library by centrifugation at about 2000g, followed by at least three washings with 500 µL of the same buffer used for protein capture. In between centrifugations, the bead slurry is shaken gently for 10 min. All along the washing process, CPLLs should not be left completely dry.

8.14 MOST COMMON SINGLE-ELUTION PROTOCOLS

The proteins captured by the peptide library beads under the described conditions are desorbed and analyzed with the expectation of reducing the concentration of high-abundance species with a concomitant increase (and hence detection) of many other species that are generally below the sensitivity of current analytical methods. As a result of this phenomenon, the collected protein fraction contains an apparent larger number of proteins that can be in excess of 5 times the initial protein count. The pool is thus more complex than it appeared to be before the treatment, with possible issues involving the subsequent analytical determinations. For instance, a 2D gel analysis of a CPLL-treated sample may show a large number of overlapped spots that are not present before the treatment. Such a situation renders challenging the possibility of excising precisely a spot for singular analysis. On the contrary, there are no protein spots that are largely abundant, as this is the case for albumin in serum. An example is the 2D electrophoresis image of human follicular fluid shown in Figure 8.16. It is because of the larger qualitative complexity of the treated sample that fractionated elutions are justified. Single and sequenced elution protocols are described below with several variants, so that the user can easily select the optimal solution for the expected application. However, it should be emphasized that proteins are captured by quite strong interactions and some of them are even difficult to dissociate under harsh conditions. As a consequence, when single elutions are considered, it is important to select those that guarantee the total desorption of all species captured. Figure 8.17 illustrates the options for the recovery of captured proteins.

8.14.1 Recipe

Elutions can be performed using the same column after the capturing phase, or they can be performed as a small batch or spin columns. The proteins are desorbed by three successive washings

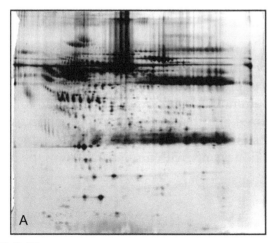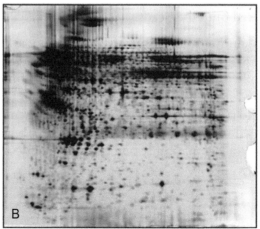

FIGURE 8.16
Two-dimensional electrophoresis of human follicular fluid.

A: initial extract. B: extract treated with CPLL under normal physiological conditions. *(From Dr. Angelucci, by courtesy)*

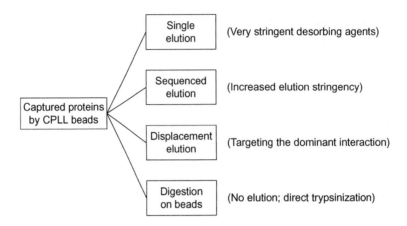

FIGURE 8.17

Schematic representation of options available once proteins are captured by CPLL beads. The most common is to use a single elution by adopting very stringent desorbing agents. Sequenced elutions are generally recommended for in-depth analysis of a proteome. On-bead digestion is also possible for a direct analysis by LC-MS/MS and is recommended when the amount of proteins is insufficient to make a 2D gel analysis. It is also used for multidimensional protein fractionation and identification technology (known under the term *MudPIT*) with advantages when dealing with comparisons of protein expression differences.

with the elution solution. Basically, one volume of selected desorbing solution is mixed with the drained beads where the supernatant is previously eliminated by centrifugation and incubated for about 10 minutes under gentle shaking. The beads are separated from the supernatant by centrifugation at about 1500 g for a few minutes. The filtrate is collected and stored in the cold. Then a second extraction is made using the same procedure, and the collected liquid fraction is mixed with the first one. The operation is performed a third time to be sure that all desorbed proteins are collected. If the elution is made from a small chromatographic column, the desorbing solution is injected into the column inlet by means of a peristaltic pump and proteins are detected in-line with a UV monitor at 280 nm. The elution will end when the UV absorption reaches the base-line.

Warning: In both cases (spin and chromatographic columns), it is worth checking if proteins are fully desorbed at the end of the elution process. This can easily be done by boiling the beads for a few minutes in a small volume of SDS solution that is currently used for solubilization of proteins prior to SDS-PAGE analysis (e.g., Laemmli buffer, Laemmli, 1970). The most common elution protocol is to use a solution composed of dissociating agents, detergents, and acidic pH. Each of these ingredients contributes to dissociation of the interactions between the protein and the hexapeptide bait grafted onto the beads.

A list of eluting solutions is given in Table 8.3. The most common one is composed of 9 M urea, 2-4% CHAPS and 0.5 M acetic acid. The latter can be replaced by citric acid, especially when the initial sample is human plasma because it contributes to preventing possible coagulation reactions. An alternative composition is to replace acetic acid by ammonia at the same concentration to get a pH of 11. The collected sample is immediately neutralized and then dialyzed against a selected buffer or desalted using alternative methods before being stored frozen for further analysis. Table 8.3 also indicates the compatibility of recommended desorption agents, with further analysis of the collected proteins.

8.14.2 Note

In addition to what is presented in Table 8.3, there could be specific situations where the elution solution is to be customized. This is the case when 2D-DIGE analysis is used after sample treatment. Although the chemical agents present in the eluates can quite easily be eliminated by the methods described above, it is also possible to design an elution solution that is directly compatible with the

TABLE 8.3 Composition of Most Common Aqueous Solutions for Single-Protein Desorption From Peptide Ligand Libraries and their Corresponding Benefits

Desorbing Solution for Single Elution	Comments and Compatible Analytical Methods
9 M urea, 2-4% CHAPS, 25 mM cysteic acid	Dissociates most interactions except very strong hydrophobic associations. Fully compatible with isoelectric focusing and 2D electrophoresis.
9 M urea, 2-4% CHAPS, 0.5 M acetic acid	Dissociates most interactions except very strong hydrophobic associations. Fully compatible with isoelectric focusing, 2D-PAGE and MALDI if neutralized and eightfold diluted with water. Otherwise proteins need to be precipitated for further analytical methods.
9 M urea, 2-4% CHAPS, citric acid to pH 3.0-3.5	Dissociates most interactions except very strong hydrophobic associations. Fully compatible with isoelectric focusing, 2D-PAGE and MALDI if neutralized and eightfold diluted with water. Otherwise proteins need to be precipitated for further analytical methods.
9 M urea, 2.4% CHAPS, ammonia to pH 11	Dissociates most interactions except very strong hydrophobic associations. Fully compatible with isoelectric focusing, 2D-PAGE and MALDI if neutralized and eightfold diluted with water. Otherwise, proteins need to be precipitated for further analytical methods.

cyanine dyes labeling. This is the case of the desorption solution composed of 20 mM Tris containing 7 M urea, 2 M thiourea and 4% CHAPS pH 8.5. As an alternative, 20 mM sodium carbonate solution containing 7 M urea, 2 M thiourea, and 4% CHAPS pH 8.5 can be used.

8.15 ABSOLUTE ELUTION PROTOCOLS

8.15.1 Objective of the Experiment

The objective of eluting all proteins together at the highest recovery is to take total advantage of CPLLs, especially to detect those that are of very low-abundance and that are captured vigorously. In fact, the usual eluting agents might not be effective enough for 100% elution, thus leaving gene products on the beads that are of interest. The absolute protein desorption process could be a single or a sequential operation; therefore in the latter case the absolute elution will be operated as last step. Complete protein desorption also contributes for better reproducibility.

8.15.2 Desorbing Agents Producing Absolute Elutions

Perhaps the most stringent desorbing agent is sodium dodecyl sulfate (SDS). The desorption process is attributed to the fact that such a strong anionic detergent adsorbs tightly on the protein structure, thus neutralizing all cationic charges and forming internal hydrophobic associations and thus conferring to each protein the same global net negative charge. This is why this protein treatment is also used in electrophoresis where highly dissociating conditions are required, so that the mobility depends exclusively on the mass (Hashimoto et al., 1983). Sodium (or lithium) dodecyl sulfate dissolved in distilled water at a concentration between 3 and 6% containing 25 mM reducing agents such as dithiothreitol can therefore easily be used to collect all proteins captured by CPLL beads, including those that interact with particular intensity. This solution should be used at a boiling temperature for 5 to 10 minutes. All captured proteins are thus desorbed from the CPLL beads and are found in the supernatant.

When the protein elution is performed by using SDS solutions, the most logical analytical procedure is SDS-PAGE, which is directly applicable with no further sample treatment. On the contrary, if other analytical determinations are adopted, such as two-dimensional electrophoresis, the elimination of SDS is necessary. To this end, a precipitation of proteins is recommended. The best precipitation is the one involving a mixture of chloroform-methanol described in Section 8.5.4. This elution method has proven to be extremely effective with urinary proteomes where a large panel of protein is present. Illustrated in Figure 4.44 (see Chapter 4) is the efficiency of SDS single-desorption (panel D) compared to the sequenced elution using thiourea-urea-CHAPS solution (A), followed by urea-citric acid solution (B) and finally by 6 M guanidine-HCl pH 6 (C). The remaining traces of proteins collected with the last solution represented only a small percentage of all proteins adsorbed and desorbed by SDS.

A second choice for absolute elution from CPLLs is the use of guanidine-HCl (generally at neutrality, with pH close to 6). This solution is also very active in destroying the affinity of proteins for the CPLL beads. Proteins collected after this treatment are denatured and difficult to use for further analysis. For instance, performing the first dimension (isoelectric focusing) of two-dimensional electrophoresis is practically impossible. This protein elution is also incompatible with antibody-based detections. The presence of 6 M guanidine induces a high ionic strength which is also deleterious to electrophoresis analysis. After elution, the proteins should be precipitated once or twice to eliminate the desorbing agent.

8.16 FRACTIONATED (OR SEQUENTIAL) ELUTION PROTOCOLS

One would want to desorb sequentially primarily when the number of gene products after CPLL treatment is so large that it may prevent making an exhaustive analysis of each component. For example two-dimensional electrophoresis would be too much populated of overlapping spots with difficulties to excise them separately. Another example is when the SDS-PAGE analysis used for further slicing prior LC-MS/MS comprises too many bands. Finally, this is also the case when the collected complex protein mixtures are trypsinized prior to two-dimensional chromatography separation of peptides for multidimensional protein identification technology (MudPIT). A fractionated elution from the beads concurs with the simplification of the single fractions in terms of number of proteins present, and hence the following steps are more reliable and more easily interpreted.

8.16.1 Objective

The captured proteins can easily be desorbed sequentially by two, three, or even four fractionation steps. To this end, the objective is first to collect protein groups that are adsorbed by a common dominant physicochemical interaction. Thus electrostatic interactions could be weakened by the addition of salts, while hydrophobic interactions are broken by the presence of a hydrophobic challenger. Hydrogen bondings are weakened by agents such as concentrated urea. Naturally, the multiplication of eluted protein fractions adds more labor since each one needs a two-dimensional electrophoresis step or trypsination, followed by mass spectrometry analysis or other parallel analytical determinations.

8.16.2 Main Recipe

The methodologies that have been described in this chapter can be operated either in column or in suspension. The latter is detailed here. One hundred μL of CPLL beads carrying proteins are mixed with 200 μL of neutral 1 M NaCl (or KCl) and agitated gently to maintain beads in suspension for about 10 minutes at room temperature (0.5 M salt solution could also be used instead of 1 M). Then the beads are separated by centrifugation, and the supernatant is collected. The bead pellets are then treated again with the same volume of the same salt solution, and the process is repeated. The second supernatant is mixed with the first and stored in the cold or frozen for further analysis. This first treatment allows collecting proteins that are dominantly adsorbed by an electrostatic interaction effect and depends very much on the pH.

The bead pellets are then added with 200 μL of 50% ethylene glycol in water to desorb the dominantly captured proteins by hydrophobic associations. In certain cases less than 50% ethylene glycol in water can be used instead with lower efficiency, but also lower risks of protein denaturation. The CPLL beads are gently stirred for about 10 minutes at room temperature, and the supernatant is separated by centrifugation and collected. A second operation with the use of ethylene glycol solution is to be carried out to complete the elution process. The second supernatant is added to the first one and stored in the cold. The bead pellets are then added with 200 μL of 0.2 M glycine-HCl buffer, pH 2.5, for the elution of other proteins that are tightly adsorbed by a variety of concomitant molecular interactions. The suspension is gently agitated for 10 minutes at room temperature, and then the supernatant is separated by centrifugation and collected. A second operation of protein desorption using the same acidic buffer is needed to complete the elution. The second eluted protein solution is added to the previous one and stored in the cold for further analysis. The beads are finally treated with a harsh stripping solution to remove all possible other proteins that are still attached on the affinity peptide beads. This is a solution composed of 6% acetonitrile-12% isopropanol-10% of 17 M ammonia-72% distilled water.

As described above for other eluting solutions, this stripping operation is performed twice. The two eluted protein solutions are also mixed together and stored in the cold. At this stage the CPLL beads are empty of proteins and can be disposed. To check if all proteins are eliminated from the beads, 100 μL of SDS Laemmli buffer are added and the mixture boiled for about 10 minutes; finally the supernatant can be used for a SDS-PAGE analysis. A number of alternative sequential elutions can be performed as shown in Table 8.4.

TABLE 8.4 Composition of Aqueous Solutions for Sequential Protein Desorption From Peptide Ligand Libraries and their Corresponding Benefits

Elution Solution	Comments and Compatible Analytical Methods
0.5-1 M Sodium chloride	Desorbs proteins that are captured by a dominant ion-exchange effect. Compatible with certain types of liquid chromatography such as HIC, RPC, and IMAC.
50% ethylene glycol in water	Dissociates mild hydrophobic interactions. Compatible with most analytical methods except HIC and RPC.
4-8 M urea	Annihilates hydrogen bonding. Compatible with most protein analytical methods, including SELDI except MALDI unless extensively diluted.
2 M thiourea-7 M urea- 4.4% CHAPS	Dissociates mixed modes, hydrophobic associations, and hydrogen bonding. Fully compatible with isoelectric focusing, 2D-PAGE, and SELDI.
9 M urea, 2% CHAPS, citric acid to pH 3.0-3.5	Dissociates most interactions except very strong hydrophobic associations. Fully compatible with isoelectric focusing, 2D-PAGE, and SELDI if neutralized and eightfold diluted. Otherwise proteins need to be precipitated for further analytical methods.
9 M urea, 2.4% CHAPS, ammonia to pH 11	Dissociates most interactions except very strong hydrophobic associations. Fully compatible with isoelectric focusing, 2D-PAGE, and SELDI if neutralized and eightfold diluted. Otherwise proteins need to be precipitated for further analytical methods.
200 mM Glycine-HCl, pH 2.5	Deforming solution for the dissociation of biological affinity interactions (e.g., Ag-Ab). Proteins need to be precipitated for further analytical methods.
6 M Guanidine-HCl, pH 6	Dissociates all types of interactions; proteins are fully denatured. Proteins need to be precipitated for further analytical methods.
Acetonitrile (6%)-isopropanol (12%)-trifluoroacetic acid (1%)-water (81%)	Dissociates strong hydrophobic interactions. Compatible with most analytical methods except HIC and RPC. Solvents can be eliminated by speed-vac.
Acetonitrile (6%)-isopropanol (12%)-17 M ammonia (10%)-water (72%)	Dissociates strong hydrophobic interactions. Compatible with most analytical methods except HIC and RPC. Solvents can be eliminated by speed-vac.

8.16.3 Expected Results

With sequential elutions, proteins are released according to their affinity for the immobilized peptides. Since the same protein may share different peptide structures, their recovery is not necessarily exhaustively operated by one elution, but rather the same protein could be found within more than a single eluate. In fact a given protein interacts first with the highest affinity peptide partner. However, when this is saturated, the same protein could form a complex with another peptide, with possibly a lower affinity constant and so on as long as the protein in question is available.

Nevertheless, the protein fractions largely differ from each other. A classical result is the one illustrated in Figure 8.18. Here the proteome from human red blood cell lysate has been captured by two distinct libraries, and the proteins have been recovered by fractionated elution. In both cases, it is interesting to notice that a large overlap of proteins appears among fractions. This extends over the three elution steps. Another important point is that the behavior of the libraries is very different due to the terminal group (primary amine for library A and carboxylic acid library B). In the first case, the UCA desorbing mixture is the most efficient, while in the second case the most efficient is the TUC mixture.

8.16.4 Variants of the Main Sequential Process

A variant of the main process is to replace 1 M salt solution by 0.5 M neutral sodium or potassium chloride solution.

Other variants of the main sequence are as follows:

- Use 2 M thiourea, 7 M urea, 2% CHAPS (TUC) as the third step.
- Use 9 M urea pH 3.5 by acetic acid as the third step. This essentially desorbs proteins anchored by a dominant hydrogen-bonding interaction.
- Use 6 M guanidine-HCl, pH 6.0, to replace both the third and the fourth steps as a single stripping solution.

Other alternative methods are described in Table 8.4, with corresponding comments.

Technical alternatives are to elute proteins in column instead of small bulk or spin column. In this case, the centrifugation is avoided and replaced by a UV monitoring at the outlet of the chromatographic column.

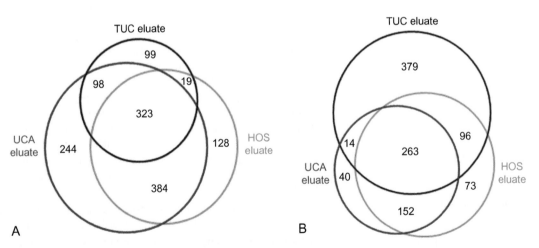

FIGURE 8.18

Venn diagrams of a human red blood cells lysate, independently treated with two combinatorial peptide ligand libraries and the captured proteins eluted according to the same sequence. Protein capture was performed under physiological conditions. Elutions were obtained by using first a mixture of 2 M thiourea, 7 M urea, 2% CHAPS (TUC) followed by a mixture of 9 M urea, citric acid to pH 3.3 (UCA), and completed by a solution composed of 6% acetonitrile, 12% isopropanol, 10% ammonia at 20%, and 72% water (HOS). The desorbed proteins were separated by SDS-PAGE, lanes were cut into 20 slices, and each slice was digested with trypsin and peptides analyzed by nanoLC-MS/MS.

8.16.5 Comments and Remarks

When acidic or alkaline elution solutions are used, the collected protein fractions are immediately neutralized by adding either an aqueous solution of 3 M Tris-(hydroxymethyl)-aminomethane or 3 M acetic acid. If the collected protein solutions are very dilute, a precipitation may be required to concentrate them for further analyses. In this case, refer to Sections 8.4 and 8.5 for the selection of the most appropriate precipitation method. Alternatively, exhaustively dialyze the collected proteins against 20–100 mM ammonium carbonate solution and then lyophilize. The recovered proteins are ready to be analyzed by SDS-PAGE, two-dimensional electrophoresis and mass spectrometry.

8.17 MILD ELUTION METHODS FOR FUNCTIONAL STUDIES

Most desorption methods of proteins captured by CPLLs are strong so as to ensure the total recovery of proteins. However, such an approach frequently denatures proteins at least at the level of their biological integrity. Under these conditions, biological function determinations can be difficult, if not impossible. Enzymatic activities, transport capabilities, antigen properties, recognition of the corresponding receptor, and so on are diminished. After elution, proteins could be immediately equilibrated to the best physiological possible conditions, but it may be insufficient to recover all biological functions. It is within this context that mild elutions have been envisioned and devised.

8.17.1 Recipes for Mild Elution

Following the general procedure described in Section 8.14, (in column or spin columns), the first solution to be used is 1 M sodium chloride in water. Proportions are as usual 200–300 μL of salt solution per 100 μL of beads. The proteins captured by a dominance of electrostatic interactions will thus be desorbed. This operation is performed twice or three times. The second elution process involves using a solution of 0.1–0.5 M acetic acid containing 40% ethylene glycol. The acid has mild deforming properties for desorption, while ethylene glycol contributes to hydrophobic dissociations. It is also performed twice or three times. Depending on the proteins in question, it may be possible to use a single eluting agent resulting from a compromise of the three above-described eluting agents: NP40, 1 M sodium chloride containing 0.5 M acetic acidic, and 2% NP40. The 2% NP40 could be replaced by 2% Tween-20, which is possibly milder than NP40. It is also to be noted that acetic acid is less denaturing than the glycine-HCl buffer described in the sequential elution protocols in Section 8.16.

8.17.2 Remarks and Recommendations

Each of the above-described mild elutions taken singularly may be insufficient to collect quantitatively captured proteins; however they could be complementary and all together comprise all captured proteins. To this end, it is advised to mix eluates immediately after neutralization. An equilibration to the appropriate buffer is also recommended, all operations being performed at 4 °C. If the resulting protein solution is too dilute, concentration by any nondenaturing means (e.g., dialysis and lyophilization, spin concentration) is advised. To be certain that all proteins are desorbed from CPLL beads, boil 100 μL of beads with 200 μL of SDS Laemmli buffer for 5 minutes and perform a SDS-polyacrylamide gel electrophoresis.

8.18 DISPLACEMENT ELUTION

Since proteins are captured by immobilized peptide ligands, it could be possible to desorb them by using competitive peptides in solution. This approach is only of academic interest because the cost of such an operation is too high for the expected benefit. However, since some amino acids are more effective than others in capturing proteins (see Bachi et al., 2008), playing with combinations of amino acids as eluting agents may be possible. The approach is probably too simplistic since the interaction points on the peptides are different and a single amino acid might not be strong enough to release proteins. Nevertheless, if amino acids are associated with agents that weaken most important interactions such as hydrophobic (e.g., with ethylene glycol), ionic (e.g., with sodium of potassium chloride), and hydrogen bonding (e.g., with urea), the probability of protein elution happening is high. It is with this in mind that the first attempts to elute using this general rule were successfully made for urinary proteins by Candiano et al. (2012).

8.18.1 Recipe

As stated earlier, this is an envisioned mode of elution that was demonstrated only with preliminary experiments and should be used with caution. A given volume of CPLL beads where proteins are adsorbed is mixed with a double or triple volume of 100–200 mM L-lysine pH 7.4, and after an incubation of 30 minutes at room temperature under gentle shaking, the supernatant is separated by centrifugation. The operation is repeated to ensure that all proteins are desorbed and collected. L-lysine can be added with weakening agents such as 40% ethylene glycol, and/or 3 M urea and/or 1 M sodium chloride. Instead of L-lysine alone, a mixture of four well-differentiated amino acids could be used such as arginine (50 mM), lysine (50 mM), aspartic acid (25 mM), and glutamic acid (25 mM) (D E R K mixture), as described. From the list of Bachi et al. (2008) it could also be possible to use different mixtures. When the hydrophobic interactions are also intensively present, it is advisable to add amino acids such as isoleucine, valine, leucine, and phenylalanine.

8.18.2 Remarks and Comments

The selection of different amino acids in association with weakening agents could result in the elution of specific groups of proteins. To be certain that all proteins are desorbed from the CPLL beads with amino acids, it is advisable to make a second elution with one of the agents described in Section 8.15 and/or boil 100 μL of beads with 200 μL of SDS Laemmli buffer for 5 minutes, in both cases performing an SDS-polyacrylamide gel electrophoresis. Note that the D E R K mix might only work for urines, since some hydrophobic binding sites on the CPLL beads are blocked by pigments and bile salts normally present in urines.

8.19 A NOTE OF CAUTION ON ELUTION PROTOCOLS

Before closing this long list of elution protocols, we have to dispel some myths that are presently plaguing the field and would confuse the user. One myth of major concern regards some published elution methods, such as the one of Bandow (2010). When prefractionating sera in search of biomarkers and comparing the performance of CPLLs versus immunodepletion (Seppro IgY14 System), Bandow came to the conclusion that there was no difference between the two systems: Both of them performed poorly in discovering biomarkers. Bandow stated that "detectable protein spots in the different plasma fractions contained exclusively high-abundance proteins normally present in plasma at concentrations between 1 μg and 40 mg/mL." There seems to be a general consensus in the current literature that this could be so in the case of immunodepletion. But that this could apply also to CPLLs is incorrect, as is discussed below. So, what went wrong? An explanation is evident from Table 8.5, which lists most of the elution cocktails proposed in the literature. Among them, Bandow's seems to be the poorest one, as it is based on an eluant comprising 4 M urea and 1% CHAPS. It is well known that urea at 4 M levels cannot even denature proteins. Protein unfolding starts at 5 M urea and in most cases is completed in 8 to 9 M urea. One should not live with the impression that the linkage between a given protein and its hexapeptide partner in the CPLL beads is a weak one, given that the peptide bait is rather short. It turns out that such a binding event is quite strong and requires highly denaturing conditions to be split (Candiano et al., 2009b) (see also Sections 4.4.2 and 4.4.3 describing molecular interactions).

TABLE 8.5 Different Elution Systems From ProteoMiner Beads

Reference	Eluant	Results
Sihlbom et al. (2008)	1 M NaCl, 20 mM HEPES, pH 75. 200 mM glycine pH 2.4, 60% ethylene glycol. Hydro-organic IPOH, ACN, TFA	Single elution (4 M urea, 1% CHAPS): 91 peaks; combined 4 elutions: 330 peaks in SELDI.
Ernoult et al. (2010)	4 M urea, 1% CHAPS, 5% acetic acid; 6 M guanidine HCl, pH 6.0.	320 proteins in CPLL-treated sample; 332 after immunodepletion
Fakelman et al. (2010)	4 M urea, 1% CHAPS, 5% acetic acid.	Many more SELDI peaks than control.
Froebel et al. (2010)	8 M urea, 2% CHAPS, 5% acetic acid. 7 M urea, 2 M thiourea, 4% CHAPS, 40 mM Tris.	Identical spots in 2D electrophoresis maps.
Beseme et al. (2010)	4 M urea, 1% CHAPS, 5% acetic acid.	Control: 157 spots; CPLL-treated: 557 spots, IPGs pH 4–7.
De Bock et al. (2010)	8 M urea, 2% CHAPS, 5% acetic acid.	Control: 48 Peaks; CPLL- treated: 136 Peaks in SELDI.
Bandow (2010)	4 M urea, 1% CHAPS.	Only high-abundance proteins 1 µg to 40 mg/mL.
Bandhakavi et al. (2009)	100 mM Tris-HCl pH 6.8, 4% SDS, 10% 2-ME, boil.	Untreated saliva: 251 proteins; CPLL-treated: 693 proteins.
Candiano et al. (2009b)	100 mM Tris-HCl pH 6.8, 4% SDS, 10% 2-ME, boil.	Urine proteins: Three times more spots with CPLL than control.
Dwivedi et al. (2010)	As per Bio-Rad instruction manual	108 proteins unique to CPLL; 100 proteins unique to IgY; 404 total proteins sera.

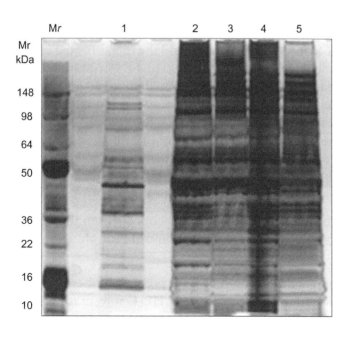

FIGURE 8.19
SDS-PAGE profiling of the various CPLL eluates from human serum. The track 1 represents elution in 4 M urea and 1% CHAPS, according to the published protocol from Bandow (2010). The other four following tracks to the right represent the mixture of the three eluates at the three pH values (4.0 + 7.0 + 9.3) (Lane 2) as well as the three individual eluates (pH 4.0, pH 7.0, and pH 9.3, respectively lanes 3, 4, and 5), all eluted in boiling 4% SDS and 25 mM DTT.

What has occurred has been elegantly explained by Di Girolamo et al. (2011) and is shown in Figure 8.19. The eluant used by Bandow (2010) will barely elute 20–25% of the captured species and certainly not those having high affinity for the hexapeptide ligands—that is, those trace proteins that had to compete hard with the overwhelming presence of the high-abundance species (HAP) that might have had lower affinities for the same baits! As a result of the insufficient elution protocol, only the high- to medium-abundance species were desorbed, thus leaving on the beads the precious booty of low-abundance compounds (LAP). It has been reported that, in order to recover ~99% of the bound species, the beads have to be boiled in 4% SDS containing 30 mM DTT (Candiano et al., 2009b) (6 M guanidine-HCl performing just as well). Thus there is no need to stir up "muddy waters" (with due respect for the great American singer) and publish protocols that do not work!

8.20 ON-BEAD DIGESTION

Elution of proteins from CPLL beads is justified when protein analyses with their integral structure, such as one- or two-dimensional electrophoresis, isoelectric focusing, MALDI mass spectrometry, antigen-antibody reactions and other determinations of biological activity, is desired. However, if the objective is limited to identification of captured proteins by means of the analysis of peptides, it is possible to hydrolyze the captured proteins directly on the beads (Fonslow et al., 2011; Meng et al., 2011). This approach is also recommended when the volume of beads is very small.

8.20.1 Recipe

8.20.1.1 METHOD 1

This protocol is given for 100 μL bead volume (volume estimated after removal of all liquid water by, for instance, low-speed centrifugation). After protein capture on the peptide library beads (whatever the method or the physicochemical conditions), the beads are rapidly washed twice with 200 μL of 100 mM ammonium bicarbonate containing 0.1% Rapigest supplied by Waters (this is not mandatory, but it facilitates the proteolysis process). This mixture is obtained by adding 1 mL of 100 mM ammonium bicarbonate to the 1 mg Rapigest vial lyophilizate and shaking gently for few minutes). The bead suspension is then vortexed for a few minutes.

To this mixture 300 μL of 10 mM DTT are added, and the mix is heated at 65 °C for 1 hour under gentle stirring or occasional shaking. Then another 300 μL of 55 mM iodoacetamide are added, and the mixture is stored in the dark for 60 minutes. After incubation, 60 μL of 0.2 μg/μL of sequencing grade trypsin is added. If the succinylated peptide library beads are used, the amount of trypsin should be four times larger. The bead suspension is vortexed a few seconds and incubated overnight at 37 °C under gentle shaking. The day after 200 μL of 500 mM formic acid are added, the mixture is vortexed for a few seconds and incubated for about 40 minutes at room temperature. At that stage, the supernatant is recovered by filtration (30,000 MWCO) under centrifugation (12,000 rpm for 20 minutes) in order to separate it from insoluble material and beads.

In order to fully extract the remaining peptides embebbed within the bead network a wash is recommended under centrifugation once with 50 μL of 500 mM formic acid. This second filtrate is mixed with the first one.

The stripped beads could then be kept at −20 °C for possible further analysis. The solution of peptides is then dried by speed-vac and re-dissolved in 20 μL HPLC solvent for LC-mass spectrometry analysis.

8.20.1.2 METHOD 2

One hundred μL are added with 600 μL of a solution composed of 8 M urea, 100 mM Tris(2-amino-2-hydroxymethyl-1,3-propanediol) pH 8.5, and 5 mM tris(2-carboxyethyl)phosphine. After a few seconds of vortex stirring, the suspension is agitated gently for 30 minutes at room temperature. Cysteine residues are then acetylated by the addition of 10 mM iodoacetamide and incubated in the dark for at least 15 minutes. The bead suspension is then diluted with 100 mM Tris-(2-amino-2-hydroxymethyl-1,3-propanediol) pH 8.5 to reach a final concentration of 2 M urea. Trypsin is added (20 μg as 0.5 μg/μL; this corresponds roughly to 1:50 protease/protein ratio) along with $CaCl_2$ at 1 mM concentration, and the resulting mixture is incubated overnight at 37 °C. The peptides obtained are separated from the beads by centrifugation at 1000 g and stored at −80 °C until the day of analysis. Before analysis, the peptide solution is acidified to 5% formic acid and centrifuged at 18,000 g to remove possible solid material.

8.20.2 Variants

Protein reduction can be performed with other chemical agents such as 2-mercaptoethanol or its dimer or other reducing agents, according to current methods of protein/peptide treatment prior to mass spectrometry.

Another variant is to make the trypsination after having labeled proteins using stable isotope-labeled internal standards (QconCAT), as described by Rivers et al. (2011). QconCAT is added to the beads carrying the captured proteins after equilibration with 50 mM ammonium bicarbonate (or added to the initial sample prior to treatment with CPLLs). The digestion is performed with trypsin at a protein:enzyme ratio of 20:1 at 37 °C for 24 hours. The absence of intact proteins is to be verified by SDS-PAGE analysis.

8.20.3 Comments and Remarks

If succinylated peptide library beads are used, the amount of trypsin should be four times larger since trypsin itself is partially captured by the carboxylated beads.

8.21 GLYCOPROTEIN ENRICHMENT

Most proteins from eukaryotic cells are glycosylated. Glycosylation covers a large variety of structures that are dependent on the type of sugar and the number of sugars involved in the glycan moiety. The interest in capturing specifically glycoproteins focuses on investigation of this post-translational modification. Glycoprotein species usually include very dilute glycoproteins that may be the objective of an investigation. For glycoprotein enrichment, it is suggested using the CPLL general procedure, followed or not by the glycoprotein fractionation process. The process is performed by adopting a few selected lectins. The use of a first capture with lectins, followed by an enrichment with CPLLs, may not be as efficient at least for very low-abundance species since the affinity constant of glycanes for lectins is generally low. Glycoproteins having a concentration lower than the dissociation constant would therefore not be separated by lectins. The enrichment process is thus composed of two steps: enrichment of all proteins by a combinatorial peptide ligand library followed by glycoprotein specific capture/fractionation.

8.21.1 Enrichment Step

The enrichment step is performed by using a regular capture with CPLL beads or one of the variants described in this series of protocols.

In practice, for instance, 100 μL of beads are equilibrated with the selected buffer and then mixed with the protein sample dissolved in the same buffer. The amount of protein involved in the process should be higher than 50 mg; below this value the enrichment might not be very effective owing to the insufficient amount of low-abundance species. The mixture is vortexed for a few seconds and left for about 60–90 minutes at room temperature while vortexing from time to time. Beads with captured proteins are then separated by centrifugation at 2000 g for 1–2 minutes. They are washed rapidly once or twice for a few minutes with 200 μL of buffer and separated by the same centrifugation process. At this stage few options are possible: (i) the study of glycans from CPLL-captured glycoproteins and (ii) collection of all CPLL-captured proteins for further glycoprotein capture and / or fractionation (see next Section 8.21.2). For direct on-bead glycan release, 40 µL of 2% SDS and 20 µL water are added, and the mixture is shaken for 10 minutes at room temperature, followed by a second incubation for 10 minutes at 60 °C. After cooling down at room temperature, 40 µL of PBS/PNGase F solution (composed of 5 x PBS containing 4% Nonidet NP40 and 0.5 mU of PNGase) are added, and the mixture is sealed and shaken for 10 more minutes. The enzymatic release is then allowed to proceed overnight at 37 °C. For more details, especially about the analysis of released glycans, see Huhn et al. (2012). If the glycan release is performed on proteins in solution, a preliminary desorption of all captured proteins is to be performed. The collected proteins are immediately equilibrated in PBS prior treatment with PNGase.

8.21.2 Separation of Glycoproteins

Once captured, the proteins are desorbed and rapidly equilibrated in a physiological buffer. They are then separated by using multilectin chromatography where three main lectins are adopted. Lectin selection comprises concanavalin A, jacalin, and wheat germ agglutinin, which demonstrated high recovery of glycoproteins and good coverage of glycan spectra (Disni et al., 2008; Kullolli et al., 2010). The enriched protein sample from CPLL elution is equilibrated with 0.1 M Tris-HCl buffer containing 0.1 M sodium chloride, pH 7.4, supplemented with 1 mM calcium chloride, and 1 mM manganese chloride. Then it is loaded onto the sequence of lectin columns at a flow rate of 0.5 mL per minute. After having disconnected the lectin column sequence, the adsorbed glycoproteins are then eluted with 100 mM glycine pH 2.5 and immediately neutralized. The collected fractions are then used for glycan release for further structure analyses. Figure 8.20 illustrates the setup schematically.

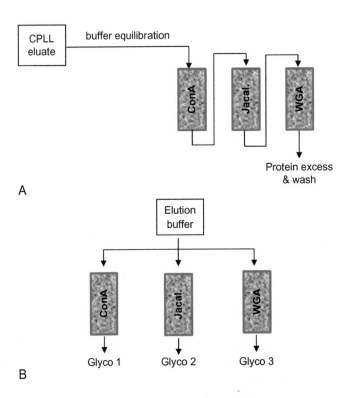

FIGURE 8.20

Separation of glycoproteins and fractionation according to the glycan structure. Proteins are first treated with CPLL under classical conditions to enrich for low-abundance species. The collected proteins are then adsorbed (A) on three different lectin sorbents (Con A: concanavalinA; Jacal: jacaline; WGA: wheat germ agglutinin). Once all unreacted proteins are eliminated, they are desorbed separately from their respective column (B) and collected for further analysis.

8.21.3 Variants

The most obvious variant is the alternative selection of lectins. The selection depends on the type of glycan targeted. Another approach that is most general without differentiating between glycans is the use of an affinity sorbent carrying boronic acid as a sugar binder (see Section 3.3.2).

8.22 IDENTIFICATION OF PEPTIDE LIGANDS BY THE BEAD-BLOT METHOD

As detailed in Section 7.1, a peptide library can be considered a source of peptide ligands for the purification of a given protein. To this end the specific interaction between the targeted proteins and the corresponding beads is first to be evidenced. Then the isolated bead is to be analyzed in order to identify the peptide sequence to build the affinity chromatography media. A number of protocols has been suggested (Sections 7.1.2 and 7.1.3), but most of them produce false positives, generating a lot of work for the identification of the appropriate specific peptide binder. In this methodological section, we selected the one that gives a good chance to reduce nonspecific binding or false positives, at the same time allowing the identification of ligands for more than one protein.

8.22.1 Description of the Principle and the Main Objectives

Within the domain of affinity ligand selection for protein separation, the step of sorting out the right beads corresponding to the best affinity candidate is of crucial importance. Among various technologies, the one selected here is the so-called bead-blot as reported by Lathrop et al. (2007). This choice is justified by the low risk of false positives generated by this method that is one of the major technological obstacles (see Section 7.1.2). The principle is based on a blot analysis of a print from beads that are dispersed as a monolayer throughout an agarose gel surface. After capture and washing, proteins that are adsorbed on their corresponding beads have the property to slowly diffuse out of the bead by an active fluid flow, releasing progressively proteins that are immediately trapped by a blot paper placed just above the agarose surface. The membrane is used to identify the positioning of the target protein and hence the localization of the corresponding bead partner.

8.22.2 Recipe

The combinatorial peptide ligand library is first equilibrated in the selected buffer (e.g., 25 mM phosphate, 150 mM sodium chloride, pH 7.2); this buffer is the same one used for solubilization of the protein sample containing the target protein. One volume of the library (settled bed) is then mixed with 10 volumes of protein sample (the crude extract) at a concentration between 1 and 25 mg/mL. The mixture is incubated under gentle shaking at room temperature for approximately 1 hour. The beads are then washed with the same buffer repeatedly to eliminate noncaptured proteins. The UV absorbance at 280 nm of the washes should be equal to the unused buffer.

A solution of 1% agarose in distilled water is separately prepared using current procedures, and the hot solution is layered on a surface and left until solidification. A second solution of 0.5% of low-melting agarose is prepared, and 900 µL of it is taken and kept at 25–41 °C to prevent solidification. To this solution 10 to 100 µL of washed beads loaded with proteins is mixed, and the mixture is maintained at the same temperature while gently agitating for a few minutes. Then the mixture is layered on the top of the first agarose surface and allowed to solidify horizontally at 4 °C. The gels are then placed (bead side up) on top of a spongy paper or gauze strip, with extensions on the side to be in contact with a transfer buffer tank. Over the gel top layer, a protein binding membrane is placed without trapping air bubbles and is then covered by adsorbing paper towels. Under these conditions, part of the captured proteins vacate the beads and are captured by the binding membrane. After 10–24 hours, the binding membrane is removed for the detection of the target proteins. The gel is stored at 4 °C, waiting for the alignment with the revealed binding membrane.

The detection of the target protein is performed by using an antibody followed by a secondary antibody labeled for appropriate enzymatic colored reactions. The evidenced colored spots represent the location of the positive beads placed on the surface of the gel layer. By superimposing the enzymatically stained membrane on the gel layer, it is possible to locate the corresponding beads carrying the affinity peptide ligand. The alignment is a relatively complicated step because finding the exact initial positioning of the membrane is problematic because the gel layer is very soft and may tend to distort upon manipulations. To overcome this issue, alignment beads can be artificially added to the peptide beads after the capture of proteins. The positioning of alignment beads is detailed in the following section. Once the beads are identified, they are taken out of the gel and the carried peptide is sequenced.

8.22.3 Variants

The process also allows identifying a peptide ligand for more than one protein contained in the crude protein sample treated by the beads. To this end, a sequence of blotting is to be done and positioning is identified by using different antibodies.

A second important variant is the use of various solutions to desorb proteins from the beads. For instance 0.45 M sodium chloride in 20 mM citrate buffer pH 7.4 followed by 1 M sodium chloride in the same buffer and concluded with a 6 M solution of guanidine-HCl can be used. Other types of dissociating agents could be used; they could simplify the identification of optimized washing, desorbing, and cleaning solutions for the future affinity columns.

The alignment of beads within the agarose gel with the blot membrane can also be considered a variant of the main protocol.

The operation is performed as follows: 1–3 μL of alignment beads are added to the 10–100 μL aliquot of protein-loaded peptide beads, and the blend is mixed with low melting point agarose solution. Before mixing with peptide beads, alignment beads, as for instance protein G-Sepharose, are incubated with alkaline phosphatase-labeled immunoglobulin G and adsorbed by affinity. This can alternatively be performed by using first human IgG and then a secondary antibody as, for instance, a labeled mouse antibody against human IgG. The beads are washed and incubated with a chromogenic alkaline phosphatase substrate (e.g., Fast Red colors the positive beads in red). After agarose layering, the color remains on the beads and some IgG diffuse in the binding membrane from which they could be specifically stained with the same substrate. At the end of the process, the red spots can easily be aligned with the red beads to re-locate the initial position of the blot membrane and thus identifying beads.

8.22.4 Comments

In spite of the quite elaborate process, this mode of tracking the right affinity beads for a given protein is relatively clean because it does not produce as many false positives as other methods (see description in Section 7.1.2). However, the collection of beads (around 60–70 μm diameter) remains a delicate operation that could be performed under the microscope. Sequencing of peptides is performed by using Edman degradation technology (Lam et al., 1991) or mass spectrometry (Redman et al., 2003).

8.23 DETERMINATION OF BINDING CAPACITY BY FRONTAL ANALYSIS

This determination is a current operation in liquid chromatography where it is used for the optimized use of ion exchangers or hydrophobic or affinity solid media. Generally, it applies to homogeneous media and also to a single protein at a time. In this section the method is adapted to polishing processes by means of mixed beds (they are not homogeneous) and involving many proteins that are present in trace amounts in a purified biological.

8.23.1 Description of the Principle and Main Objectives

Removing traces of impurity from purified biological products (e.g., recombinant proteins for therapeutic use) by chromatography requires following well-known rules. One of them is to maintain the amount of sample to be polished under the binding capacity of the column against impurities. This question is not trivial because the binding capacity depends on the type of impurity and their number. This is why a systematic experimental determination is critical before starting the exploitation of this process.

What is suggested here is the adoption of the so-called frontal analysis which consists in loading continuously a chromatography column at a very constant flow rate and simultaneously determining the presence of impurities at the outlet of the column. The main protein that is supposed not to be captured (or very little) will appear at the outlet of the column around the void volume of the packing material, while the impurities are delayed until the binding capacity of the column is saturated. Figure 8.21 depicts the procedure schematically. Naturally, since impurities are indistinguishable from the main proteins by just the UV detection at the column outlet, protein pattern analyses are necessary. They are thus taken as a basis for calculating the column volume to sample amount ratio. This calculation remains valid as long as the amount and the diversity of impurities versus the main protein remain constant.

How does the removal of many impurities at a time works? As already explained in Section 4.4.1, a ligand library displays its binding capabilities against a large number of proteins and particularly those species that are present in low amounts. The removal of impurities is here also considered a sort of capture of low-abundance proteins.

However, the main difference of using peptide ligand libraries for dynamic range compression and removal of protein impurities is at the level of the loading. While the decrease of dynamic protein concentration is operated under a large excess of sample, the removal of protein impurities must not exceed the binding capacity of the library. Considering explanations given in Section 4.4, the binding capacity is relatively constant at given physicochemical conditions, which contrasts with the diversity in the number and concentration of protein impurities possibly present in biotech drugs. For this reason, despite the willingness to make complicated calculations, it is always preferable to determine experimentally the volume of the column prior to applying this polishing procedure.

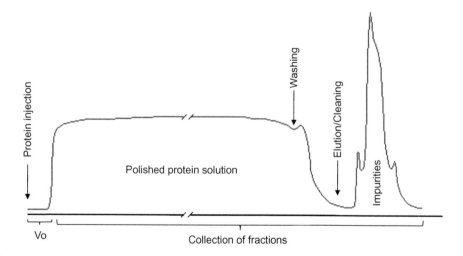

FIGURE 8.21
Schematic representation of a frontal analysis profile for the determination of binding capacity of a mixed bed for a mixture of protein traces present in a purified biological. Upon injection, the protein detection at the column outlet appears close to the void volume of the column (Vo) or at about one-third of the column volume. Upon continuous injection, the fractions are collected and analyzed by, for instance, SDS-PAGE. Once the column is considered saturated of protein impurities (they start to appear in SDS-PAGE analysis), the injection of the sample stops and the column is washed with the buffer. Finally, the adsorbed protein impurities are eluted by stringent agents.

When in such a mixed bed a complex mixture composed of many proteins, one of them being largely dominant (e.g., 99% purity) is loaded, this latter will saturate its corresponding solid-phase ligand very rapidly while impurity traces will not. Using loading volumes so that impurities do not saturate the solid-phase ligands, the flow-through will comprise the dominant protein with a level of purity substantially higher than before the process. Nevertheless, protein impurities are not all present at the same concentration; thus the least concentrated will appear last and the predominant one will appear first. It is when the predominant impurity appears at the column outlets that the binding capacity for the impurities taken as a whole must be considered reached. The ideal situation would be to have a very large number of protein impurities each present in very low concentration. The initial purity of the main proteins is also very critical because it determines the volume of the column for a given amount of sample. For example, if instead of 1% the presence of the same protein impurities is 2%, the polishing column volume would have to be doubled.

8.23.2 Recipe

1 mL of solid-phase combinatorial peptide ligands is packed in a chromatographic column (e.g., 10 mm ID x 13 mm length) and equilibrated with a selected buffer (generally, 25 mM phosphate buffer containing 150 mM sodium chloride, pH 7.2). The column is connected to a chromatographic setup comprising a pumping system and a UV/pH detection unit for recording both events at the column outlet. The solution of protein to be purified from trace protein contaminants is also equilibrated with the same buffer by dialysis or buffer exchange or equivalent techniques. The sample volume should largely exceed the column volume, and the protein concentration could range from a fraction of mg/mL to a dozen of mg/mL. Ideally for 1 mL column, the volume of the sample is 10–20 mL and the protein concentration is 1–2 mg/mL. The column is then loaded continuously with the protein solution at a fixed and accurate flow rate of 20 cm/hour, while the chromatographic events are recorded by UV and pH detection at the column outlet.

This configuration allows some proteins to be captured by the bead partners up to the saturation as it happens in a current frontal analysis for the determination of binding capacity of an adsorption column (Staby et al., 1998) except that fractions from the column outlet are also collected. The volume of each fraction is the same as the column volume—in this case 1 mL. In the present case, since the solid phase is a blend of many affinity beads of different specificity, the UV trace recorded at the column outlet signifies only the evolution of the main protein to be polished. Conversely, the composition of each fraction can reveal the main protein's differences of purity. Once the sample load is completed, a PBS solution is introduced to wash out the excess proteins while continuing to collect fractions. Finally, the captured proteins (generally they are impurities) are desorbed from the beads by a solution of 9 M urea-citric acid, pH 3.5, and the eluate is collected as a single fraction following the UV trace. To check the elimination of protein impurities, each fraction is analyzed by SDS-polyacrylamide gel electrophoresis stained with very sensitive dyes or analyzed by MALDI MS when the masses of proteins to polish is relatively low (e.g., < 30 kDa) against a control (the protein before frontal analysis).

8.23.3 Expected Results

The first collected fractions should contain the proteins to be polished without impurities or a significant decrease of them. The impurity depletion depends on the binding capacity of the column for each protein. Considering that each impurity is not present in the same concentration, the most dilute impurities are better removed than the most concentrated ones. The fraction where the main protein to be polished is contaminated by an amount of impurities that is considered unacceptable for the purpose allows calculating the maximum sample loading for a given column volume. The eluted fraction at the end of the process comprises protein impurities that were not detectable in the initial sample. It also comprises some minute amounts of the main protein whose levels allow calculating the recovery of a real polishing operation (after calculation of the ideal column to sample ratio). The analysis of the captured impurities could be of importance for detecting the nature and origin of the proteins (e.g., they can come from the recombinant host or from the growth medium where the main protein is expressed) (Fortis et al., 2006a and 2006b; Antonioli et al., 2007). A real case of polishing, dealing with recombinant human albumin, is discussed in Section 7.2 and experimental results are illustrated in Figure 7.10.

8.23.4 Variants

The most obvious variable parameter for frontal analysis is the buffer where the determination is performed. Instead of using phosphate-buffered saline, other buffers can be adopted with variations in their composition, ionic strength, and pH. These three variables can substantially modify the capture of protein impurities. Since no constraints are part of the buffer selection, it is recommended that a few different experiments be made and that the one be selected that allows using the best column volume to sample volume ratio for the same final degree of purity of the target protein. Although the flow rate can also be modified, it must remain constant throughout the entire experiment and should not exceed 50 cm/hour (linear flow rate), especially if the target protein is larger than 50 kDa.

8.23.5 Comments and Recommendations

A solid-phase combinatorial peptide pool contains a very large diversity of ligands; therefore it is adapted to clean up virtually all protein traces present in a sample regardless of the origin and the species. Best conditions for determining protein-binding capacity by frontal analysis are to use columns with an aspect ratio (height versus diameter) of at least 2 and preferably larger.

8.24 REUSING COMBINATORIAL LIGAND LIBRARY BEADS?

The reuse of combinatorial hexapeptide beads is not recommended in proteomics analysis because (i) some level of protein carryover from previous sample may appear, with consequent misinterpretation of data, and (ii) some peptides may have been modified as a result of stringent elution conditions from previous operations.

In Section 8.15 (also illustrated in Figure 4.44), the difficulty of desorbing all captured proteins has been demonstrated; therefore, the risk of leaving traces on the beads is high. If so, the proteins exposed to these beads will compete with residual proteins from previous samples, thus causing possible displacement effects that would not occur with unused CPLL beads. The solution capable of desorbing all (or almost all) proteins is SDS, as described. Traces of this detergent may still be present on regenerated beads and would in this case modify the environment where the new capture takes place.

Reuse of CPLL is thus always risky either when trying to decipher a proteome composition or more importantly, when trying to compare proteomes in view of detecting expression differences for identifying protein markers used as possible diagnostic signatures.

8.25 MONODIMENSIONAL SDS-PAGE PRIOR TO LC-MS/MS

Within the scope of this book and the objectives of these protocols, SDS-PAGE is here considered as a preliminary separation method prior to LC-MS/MS analysis of peptides from protein band digestion. The objective of this protocol is therefore to maximize the number of identifiable gene products resulting from a CPLL sample treatment. Globally, the sample is first separated by electrophoresis, and protein bands are analyzed separately after trpsinization. The recipe described below has been used a number of times with great success; however, variants are possible, as is confirmed by published methods.

8.25.1 Recipe

The recipe involves 100–200 µg of proteins eluted from CPLL beads diluted in Laemmli buffer (4% SDS, 20% glycerol, 10% 2-mercaptoethanol, 0.004% bromophenol blue, 0.125 M Tris-HCl, pH approximately 6.8), boiled for 5 minutes and immediately loaded in a 12% polyacrylamide gel (or in a 8–18% gradient gel). The electrophoretic run is performed by setting a voltage of 100 V until the dye front reaches the bottom of the gel. Generally, this separation is made concomitantly with a control sample that is not treated with CPLL. Each gel lane is cut into 20 homogeneous slices that are washed in 100 mM ammonium bicarbonate for 15 minutes at 37 °C followed by a second wash in 100 mM ammonium bicarbonate, acetonitrile (1:1) for 15 minutes at 37 °C. Reduction and alkylation of cysteine residues are performed by mixing the gel pieces in 10 mM DTT for 35 minutes at 56 °C, followed by 55 mM iodoacetamide for 30 minutes at room temperature in the dark. An additional cycle of washes in ammonium bicarbonate and ammonium bicarbonate/acetonitrile is then performed. Proteins are digested by incubating each gel slice with 0.6 µg of modified sequencing-grade trypsin in 50 mM ammonium bicarbonate overnight at 37 °C. The resulting peptides are extracted from the gel in three steps: a first incubation in 50 mM ammonium bicarbonate for 15 minutes at 37 °C and two incubations in 10% formic acid, acetonitrile (1:1) for 15 minutes at 37 °C. The three collected extractions are finally pooled with the initial digestion supernatant, dried by speed-vac, and re-suspended with 14 µL of 5% acetonitrile, 0.05% trifluoroacetic acid. The peptide samples (one per line slice) are directly used for LC-MS/MS analysis. Figure 8.22 depicts the main steps.

8.26 TWO-DIMENSIONAL ELECTROPHORESIS AFTER PEPTIDE LIBRARY ENRICHMENT

The objective of this analysis that follows a protein treatment by combinatorial peptide ligand library can be dual. The first objective is to make a side-by-side comparison of the treated sample with the initial control and perform a qualitative spot count. The second objective is to inspect visually or by means of dedicated software for identifying spots that are differently expressed when comparing, for instance, a control with a diseased sample. This inspection is by far the most efficient, and, since it is global, it allows having a general quick idea of the main differences. There is also a third justification of a two-dimensional gel electrophoresis (2D-PAGE): to identify specific spots reacting with antibodies using immunoblots in view of identifying more specifically targeted proteins. This is the case of isoforms, protein hydrolysates, allergens, and so forth.

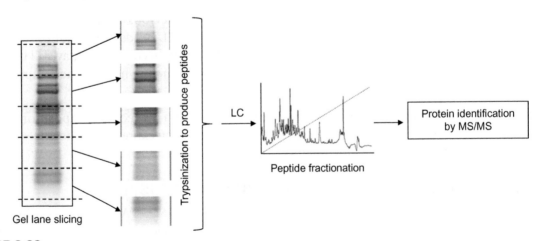

FIGURE 8.22

Global analysis of a proteome after CPLL treatment. The eluted proteins are first separated by SDS-PAGE, and, after light staining, the separation lane is sliced into a predetermined number of gel bands (here the number of slices is 5 for simplicity, but generally it is between 15 and 20). From each gel band, proteins are extracted, hydrolyzed by trypsin, and once the peptides are obtained and cleaned, they are separated by HPLC (one or two orthogonal separations) and the fractions obtained analyzed by MS/MS for protein identification.

8.26.1 Recipe of 2D gel Electrophoresis

CPLL eluates and nontreated samples (or even pools of eluates) are equilibrated in two-dimensional electrophoresis sample buffer generally composed of 2% thiourea, 7% urea, 3% CHAPS and 40 mM Tris. The total protein load is between 80 and 200 μg, which is largely obtainable using 100 μL of CPLL beads. Disulphide bond reduction is performed at room temperature for 60 minutes by the addition of TCEP [Tris (2-carboxyethyl) phosphine hydrochloride] at a final concentration of 5 mM. For alkylating sulphydryl groups De-Streak [Bis-(2-hydroxyethyl)disulphide, $(HOCH_2- CH_2)_2S_2)$] is added to the solution at a final concentration of 150 mM (by dilution from the 8.175 M stock solution, Sigma-Aldrich), followed by 0.5% Ampholine (diluted directly from the stock, 40% Ampholine solution) and a trace amount of bromphenol blue. The sample is then separated by isoelectric focusing (first dimension) after having incubated the IPG strip with the protein solution for at least 5 hours. Isoelectric focusing separation is carried out with a dedicated equipment (e.g., Protean IEF cell from Bio-Rad) in a linear voltage gradient from 100 to 2000 V for 5 hours and 2000 V for 4 hours, followed by an exponential gradient up to 10,000 V until each strip is electrophoresed for at least 25 kV-h.

For the second dimension, the IPG strips are equilibrated for 25 minutes in a solution containing 6 M urea, 2% SDS, 20% glycerol, and 375 mM Tris-HCl (pH 8.8) under gentle shaking. The IPG strips are then laid on an 8–18% acrylamide gradient SDS-PAGE gel with 0.5% agarose in the cathode buffer (192 mM glycine, 0.1% SDS, Tris to pH 8.3). The electrophoretic run is performed by setting a current of 5 mA/gel for 1 hour followed by 10 mA/gel for 1 hour and 15 mA/gel until the dye front reaches the bottom of the gel. Gels are incubated in a fixing solution containing 40% methanol and 7% acetic acid for 1 hour followed by silver staining. Destaining is performed in 7% acetic acid until the background is clear, followed by a rinse in pure water. The two-dimensional electrophoresis gels are scanned with an image analysis system (e.g., Versa-doc from Bio-Rad) by fixing the acquisition time at 10 s; the relative gel images are evaluated using appropriate software (e.g., PDQuest software from Bio-Rad). After filtering the gel images to remove the background, the spots are automatically detected, manually edited, and then counted.

8.26.2 Variants

Major variants of the above-described protocol are the size of the gel plate as well as the concentration of polyacrylamide gel. The latter could also be a wider gradient of concentration (e.g., 6–20% T; T=total amount of monomers). Another variant is the selection of the pH range for the isoelectric focusing. Generally, one should start with the largest range (pH 3–10), but much narrower pH gradients can be used (typically, two pH units or even less for overcrowded acidic pH regions).

Coomassie staining can be replaced by other methods of spot detection with increased sensitivity. For instance, silver staining can be used to detect protein spots (Hanash et al., 1982) with a nanogram or even a subnanogram level of sensitivity. Alternatively, to increase the sensitivity, dyes with fluorescent properties can be employed to advantage (Valdes et al., 2000); some of them are also fully compatible with mass spectrometry analysis when the spot is considered to be excised for trypsination and identification. Fluorescent staining molecules such as Sypro dyes are able to detect protein traces as little as 1–2 ng with a good linear dynamic range (Steinberg et al., 1996; Cong et al., 2008).

8.26.3 Expected Results

Ideally, spots are almost circular in shape and quite distinct from each other on a clear background. In practice, a perfect assembly of regular spot is rare. There are frequent distortions due to the presence of traces of insoluble material and of macromolecules such as mucopolysaccharides, or even traces of nucleic acids. These issues underline the importance of the sample pre-treatment as described in Sections 8.3, 8.4 and 8.5. The positioning of a given gene product within the two-dimensional space is not always exactly the same. Manual interpretations are frequently necessary. Sophisticated software is available to circumvent some migration distortions and make the right interpretation.

8.27 IMMUNOBLOT AND SPOT IDENTIFICATION

8.27.1 Principle

Immunoblots are used for the localization of selected proteins throughout a pattern (mono-dimensional and two-dimensional electrophoresis) exploiting antibodies. This technology is used, for instance, to evidence proteins that react to the same antibody (e.g., isoforms) or to a group of antibodies (e.g., IgE for the detection of allergic antigens). The principle is based on the migration of the protein mixture as usual for a single-dimension electrophoresis, two-dimensional electrophoresis, or even an iso-electric focusing. Separated proteins are visualized by regular staining. To be detectable by specific antibodies, proteins have to be transferred from the polyacrylamide gel to a membrane (generally cellulose nitrate or PVDF, polyvinylidene difluoride). Protein binding is based essentially on nonspecific hydrophobic association. In the case of reactive membranes, the proteins are chemically grafted. The protein transfer can be performed by passive diffusion or, better, by electroblotting while keeping the pattern they had after electrophoretic migration within the gel. The membrane where proteins have migrated is then incubated with appropriate solutions of, for instance, antibodies and immuno-chemical reactions revealed by a second labeled antibody (Kurien and Scofield, 2003).

8.27.2 Recipe

After proper protein migration (1D, 2D and IEF) one-half of the gel plate is incubated in a colloidal Coomassie Blue solution, and de-staining is performed as usual in 7% acetic acid until a clear background is obtained. Then a rinse in pure water follows. The patterns are scanned with an image analysis system (e.g., Versa-Doc from Bio-Rad), while fixing the acquisition time at, for instance, 10 seconds. The gel image is thus captured via the PDQuest software. The other half of the gel plate is blotted on a CNBr-activated nitrocellulose (NCa) sheet performed according to Demeulemester et al. (1987). After rehydration of nitrocellulose membranes, proteins are transferred by using a semi-dry blotting apparatus for 75 minutes at $1\,mA/cm^2$ of gel.

The nitrocellulose sheet is then dried and blocked with PBS containing 0.3% Tween-20. The sheet is covered with the minimum volume of interacting solution containing either antibodies (e.g., against the proteins to be revealed) or serum (e.g., from patients that are supposed to have IgE antibodies against allergens diluted 10 times with PBS containing 0.1% Tween-20) and incubated overnight at 20 °C. After washing, the nitrocellulose sheets are incubated with alkaline phosphatase-conjugated anti-antibodies or goat anti-human IgE for 2 hours at 20 °C, followed by the alkaline phosphatase substrate 5-bromo-4-chloro-3-indolyl phosphate+nitroblue tetrazolium in 0.1 M Tris–HCl buffer, pH 9.5. Positive spots (separated proteins from initial sample having reacted with antibodies) are thus visualized. By superimposing the blot results and the Coomassie stained plate images, it is possible to locate the positioning of positive proteins in the original gel plate. The latter is thus used for spot excising, protein digestion, and identification by nanoLC-MS/MS. A scheme of the overall process is illustrated in Figure 8.23.

8.27.3 Variants

Nitrocellulose membranes can be replaced by PVDF membranes. These sheets are less fragile and easily manipulable. Depending on the type of membrane, the blocking operation can be performed by using 3% bovine serum albumin or 3% de-fatted milk proteins. Tween-20 could be replaced by Triton X-100. The blocking treatment goal is to passivate nonspecific binding sites where the

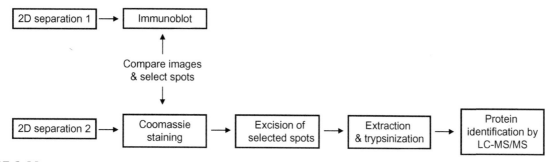

FIGURE 8.23
Schematic representation of immunoblot use after CPLL protein elution and 2D gel electrophoresis separation.

antibodies could be adsorbed and thus create mistakenly labeled zones. The final revelation can be performed by using various enzyme-labeled antibodies. Peroxidase and alkaline phosphatase are the most common labels. For a broad immunochemical reaction polyclonal specific antibodies are used.

8.28 2D-DIGE FOLLOWING PEPTIDE LIBRARY ENRICHMENT

A variant of two-dimensional electrophoresis, so-called two-dimensional differential in-gel electrophoresis (2D-DIGE), was developed around the question of reproducibility of two-dimensional electrophoresis by having two samples concomitantly separated on the same plate. This method was described for the first time by Unlu et al. (1997). This approach with its intrinsic comparison properties also allowed enhancing significantly the detection sensitivity of the protein spots. In practice, the initial protein extracts to be compared are separately labeled with cyanine fluorescent dyes having the same mass but capable of being excited at different wavelengths. These dyes have distinct spectral properties and hence can be imaged with different excitation and emission filters to discriminate the proteins from each sample (Lilley and Friedman, 2004). Then the labeled samples are mixed in equal proportions and submitted to two-dimensional separations. The obtained map has spots composed of proteins from two samples but are differently labeled so as to be evidenced independently.

Differences in signal intensity (compared to a reference) indicate possible modifications of the expression regulation. The reference can be represented by an internal standard labeled with a third fluorescent dye. Typically, Cy3 and Cy5 are used for labeling samples, and the Cy2 dye is used as an internal standard. The detection limit of minimal labeling with Cy dyes is about 2 orders of magnitude over silver staining and even more compared to currently used dyes (e.g., Coomassie Blue). Protein spots are visualized by confocal laser scanning associated with advanced software capable of sorting out relative differences in spot intensity and thus concentration from the samples used on the same plate. Over time, this method has been improved in terms of reproducibility and reliability for differential protein expression (Alban et al., 2003).

8.28.1 Recipe

The following detailed method is adapted from Sihlbom et al. (2008) and is directly usable for samples collected from CPLL treatment. About 50 µg protein from CPLL eluates are first equilibrated or dialyzed against a solution composed of 7 M urea, 2 M thiourea, 4% CHAPS, 30 mM Tris, pH 8.5. The volume of the solution should be around 15 µL to keep the protein concentration quite high and also for good reproducibility. The pH of protein solution is to be checked and, if necessary, adjusted to 8.0–9.0 by adding 1 M Tris base solution (few µL). After reconstitution, Cy dyes (Amersham, GE Healthcare) are mixed with protein solution so that 50 µg of protein solution is labeled with 400 pmol of dye solution.

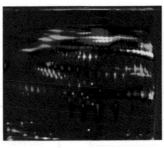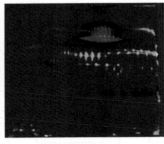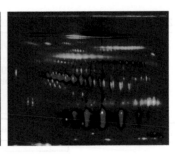

FIGURE 8.24

Three examples of 2D-DIGE images obtained by mixing a protein extract before and after treatment with CPLLs. Left: The sample is human serum. Center: The sample is a serum-bound fraction on Blue agarose column. Right: The sample is a serum-bound fraction on Protein G agarose column. *(Adapted from Hagiwara et al., 2011).*

The reaction is performed in the cold (0–4 °C) and in the dark for about 30 minutes, and then 10 μL of 10 mM lysine aqueous solution is added to stop the dye labeling reaction. The protein samples (the one labeled with Cy3 dye, the other labeled with Cy5 dye, and the possible internal standard labeled with Cy2 dye) are mixed. To the protein-labeled mixture an equal volume of 7 M urea, 2 M thiourea, 4% CHAPS, 30 mM Tris, 20 mM DTT and 0.05% IPG buffer, pH 8.5 is added and rapidly vortexed. This mixture is kept in the dark at 4 °C. Immobiline Dry Strips (pH 3-11 nonlinear or other selected pH ranges from, for example, Amersham Biosciences, Bio-Rad, Serva, or other alternative suppliers) are used for the first-dimensional isoelectric focusing migration. Each gel strip is first rehydrated for 12 hours. and the protein sample is then added by using cup loading. Isoelectric focusing is carried out at 150 V in 2 hours, then 300 V for 3 hours, 600 V for 3 hours, 2000 V for 3 hours, 2000–8000 V for 3 hours, and finally 8000 V for 3 hours. Alternative programs can also be applied according to the suppliers of gels and equipment. Generally, a maximum of 50 μA is applied per gel strip. Once the first-dimension migration is achieved, the strips are equilibrated in two steps: (i) 15 minutes in 65 mM DTT and (ii) 15 minutes in 260 mM iodoacetamide for reducing and alkylating the cysteins in the proteins. Both solutions also contain 6 M urea, 30% glycerol 2% SDS, and 50 mM Tris-HCl pH 8.8 and 0.007% bromophenol blue.

After this treatment the gel strips are sealed to the top of a 12.5% precast Tris-glycine SDS-PAGE gels by means of 0.6% agarose dissolved in a hot bath in running buffer (24 mM Tris base, 192 mM glycine, and 0.1% SDS from Bio-Rad). The gels are finally mounted in a Protean Plus Dodeca cell (Bio-Rad) filled with Tris/glycine/SDS running buffer. The electrophoretic protein migration is performed with continuous mixing and cooling at 60 mA/gel for one hour and then 90 mA/gel until the tracking dye, bromophenol blue, reaches the anodic side of the gel. The total migration time is generally close to 6–7 hours. The maximum voltage should be close to 500 V (and 50 W); the temperature should not exceed 25 °C. The separated protein spots and overall images are then visualized using adapted scanners such as Progenesis SameSpots (Nonlinear Dynamics, Ltd.). The program is used for spot detection and relative quantification. Gel matching is also possible when more than one gel plate is used. Figure 8.24 presents an example of 2D-DIGE results.

8.29 References

Alban A, David SO, Bjorkesten L, et al. A novel experimental design for comparative two-dimensional gel analysis: Two-dimensional difference gel electrophoresis incorporating a pooled internal standard. *Proteomics.* 2003;3:36–44.

Antonioli P, Fortis F, Guerrier L, et al. Capturing and amplifying impurities from purified recombinant monoclonal antibodies via peptide library beads: A proteomic study. *Proteomics.* 2007;7:1624–1633.

Bachi A, Simó C, Restuccia U, et al. Performance of combinatorial peptide libraries in capturing the low-abundance proteome of red blood cells. II: Behavior of resins containing individual amino acids. *Anal Chem.* 2008;80:3557–3565.

Bandhakavi S, Van Riper SK, Tawfik P, et al. A dynamic range compression and three-dimensional peptide fractionation analysis platform expands proteome coverage and the diagnostic potential of whole saliva. *J Proteome Res.* 2009;8:5590–5600.

Bandow JE. Comparison of protein enrichment strategies for proteome analysis of plasma. *Proteomics.* 2010;10:1416–1425.

Beseme O, Fertin M, Drobecq H, Amouyel P, Pinet F. Combinatorial peptide ligand library plasma treatment: Advantages for accessing low-abundance proteins. *Electrophoresis.* 2010;31:2697–2704.

Boschetti E. The use of Thiophilic chromatography for antibody purification: A review. *J Biochem Biophys Methods.* 2001;49:361–389.

Boschetti E, Bindschedler LV, Tang C, Fasoli E, Righetti PG. Combinatorial peptide ligand libraries and plant proteomics: a winning strategy at a price. *J Chromatogr A.* 2009;1216:1215–1222.

Boschetti E, Righetti PG. Plant proteomics methods to reach low-abundance proteins. In: Jorrín Novo JV, Weckerth W, Komatsu S, eds. *Plant Proteomics Methods and Protocols.* New York: Humana Press; 2013 in press.

Bull HB, Breese K. Protein solubility and the lyotropic series of ions. *Arch Biochem Biophys.* 1980;202:116–120.

Burgess RR. Protein precipitation techniques. *Methods Enzymol.* 2009;463:331–342.

Candiano G, Santucci L, Petretto A, et al. 2D-electrophoresis and the urine proteome map: Where do we stand? *J Proteomics.* 2009a;73:829–844.

Candiano G, Dimuccio V, Bruschi M, et al. Combinatorial peptide ligand libraries for urine proteome analysis: investigation of different elution systems. *Electrophoresis.* 2009b;30:2405–2411.

Candiano G, Santucci L, Bruschi M, et al. "Cheek-to-cheek" urinary proteome profiling via combinatorialpeptide ligand libraries: A novel, unexpected elution system. *J Proteomics.* 2012;4:796–805.

Carpentier SB, Witters E, Laukens K, Deckers P, Swennen R, Panis B. Preparation of protein extracts from recalcitrant plant tissues: An evaluation of different methods for two-dimensional gel electrophoresis analysis. *Proteomics.* 2005;5:2497–2507.

Castagna A, Cecconi D, Sennels L, et al. Exploring the hidden human urinary proteome via ligand library beads. *J Proteome Res.* 2005;4:1917–1930.

Chang YC. Efficient precipitation and accurate quantitation of detergent-solubilized membrane proteins. *Anal Biochem.* 1992;205:22–26.

Colzani M, Waridel P, Laurent J, Faes E, Ruegg C, Quadroni M. Metabolic labeling and protein linearization technology allow the study of proteins secreted by cultured cells in serum-containing media. *J Proteome Res.* 2009;8:4779–4788.

Cong WT, Hwang SY, Jin LT, Choi JK. Sensitive fluorescent staining for proteomic analysis of proteins in 1-D and 2-D SDS-PAGE and its comparison with SYPRO Ruby by PMF. *Electrophoresis.* 2008;29:4304–4315.

D'Amato A, Bachi A, Fasoli E, et al. In-depth exploration of cow's whey proteome via combinatorial peptide ligand libraries. *J Proteome Res.* 2009;8:3925–3936.

D'Amato A, Cereda A, Bachi A, Pierce JC, Righetti PG. In depth exploration of the haemolymph of *Limulus polyphemus* via combinatorial peptide ligand libraries. *J Proteome Res.* 2010;9:3260–3269.

D'Ambrosio C, Arena S, Scaloni A, et al. Exploring the chicken egg white proteome with combinatorial peptide ligand libraries. *J Proteome Res.* 2008;7:3461–3474.

De Bock M, de Seny D, Meuwis MA, et al. Comparison of three methods for fractionation and enrichment of low molecular weight proteins for SELDI-TOF-MS differential analysis. *Talanta.* 2010;82:245–254.

Decramer S, Gonzalez de Peredo A, Breuil B, et al. Urine in clinical proteomics. *Mol Cell Proteomics.* 2008;7:1850–1862.

Demeulemester C, Peltre G, Laurent M, Panheleux D, David B. Cyanogen bromide-activated nitrocellulose membranes: A new tool for immunoprint techniques. *Electrophoresis.* 1987;8:71–73.

Di Girolamo F, Boschetti E, Chung MC, Guadagni F, Righetti PG. "Proteomineering" or not? The debate on biomarker discovery in sera continues. *J Proteomics.* 2011;74:589–594.

Disni MK, Dayarathna R, Hancock MS, Hincapie M. A two step fractionation approach for plasma proteomics using immunodepletion of abundant proteins and multi-lectin affinity chromatography: Application to the analysis of obesity, diabetes, and hypertension diseases. *J Sep Sci.* 2008;31:1156–1166.

Dwivedi RC, Krokhin OV, Cortens JP, Wilkins JA. Assessment of the reproducibility of random hexapeptide peptide library-based protein normalization. *J Proteome Res.* 2010;9:1144–1149.

Ernoult E, Bourreau A, Gamelin E, Guette C. A proteomic approach for plasma biomarker discovery with iTRAQ labelling and OFFGEL fractionation. *J Biomed Biotech.* 2010;2010:927917.

Esteve C, D'Amato A, Marina ML, García MC, Righetti PG. In-depth proteomic analysis of banana (*Musa spp.*) fruit with combinatorial peptide ligand libraries. *Electrophoresis.* 2013;34:207–214.

Fakelman F, Felix K, Buechler MW, Werner J. New pre-analytical approach for the deep proteome analysis of sera from pancreatitis and pancreas cancer patients. *Arch Physiol Biochem.* 2010;116:208–217.

Fasoli E, Pastorello EA, Farioli L, et al. Searchning for allergens in maize kernels via proteomic tools. *J Proteomics.* 2009;72:501–510.

Fasoli E, Farinazzo A, Sunb CJ, et al. Interaction among proteins and peptide libraries in proteome analysis: pH involvement for a larger capture of species. *J Proteomics.* 2010;73:733–742.

Fasoli E, D'Amato A, Kravchuk AV, Boschetti E, Bachi A, Righetti PG. Popeye strikes again: The deep proteome of spinach leaves. *J Proteomics.* 2011;74:127–136.

Fic E, Kedracka-Krok S, Jankowska U, Pirog A, Dziedzicka-Wasylewska M. Comparison of protein precipitation methods for various rat brain structures prior to proteomic analysis. *Electrophoresis.* 2010;31:3573–3579.

Fonslow BR, Carvalho PC, Academia K, et al. Improvements in proteomic metrics of low abundance proteins through proteome equalization using ProteoMiner prior to MudPIT. *J Proteome Res.* 2011;10:3690–3700.

Fortis F, Guerrier L, Righetti PG, Antonioli P, Boschetti E. A new approach for the removal of protein impurities from purified biologicals using combinatorial solid phase ligand libraries. *Electrophoresis.* 2006a;27:3018–3027.

Fortis F, Guerrier L, Areces LB, et al. A new approach for the detection and identification of protein impurities using combinatorial solid phase ligand libraries. *J Proteome Res.* 2006b;5:2577–2585.

Froebel J, Hartwig S, Passlack W, et al. ProteoMiner™ and SELDI-TOF-MS: A robust and highly reproducible combination for biomarker discovery from whole blood serum. *Arch Physiol Biochem.* 2010;116:174–180.

Fröhlich A, Lindermayr C. Deep insights into the plant proteome by pretreatment with combinatorial hexapeptide ligand libraries. *J Proteomics.* 2011;74:1182–1189.

Furka A. The "Portioning-Mixing" (spit-mix) synthesis. In: *2nd Int Electr Conf Synth Org Chem*; Sept 1998.

Gengenheimer P. Preparation of extracts from plants. *Methods Enzymol.* 1990;182:174–193.

Guerrier L, Claverol S, Finzi L, et al. Contribution of solid-phase hexapeptide ligand libraries to the repertoire of human bile proteins. *J Chromatogr.* 2007a;1176:192–205.

Guerrier G, Claverol S, Fortis F, et al. Exploring the platelets proteome via combinatorial hexapeptide ligand libraries. *J Proteome Res.* 2007b;6:4290–4303.

Guerrier L, Thulasiraman V, Castagna A, et al. Reducing protein concentration range of biological samples using solid-phase ligand libraries. *J Chromatogr B.* 2006;833:33–40.

Hagiwara T, Saito Y, Nakamura Y, Tomonaga T, Murakami Y, Kondo T. Combined use of a solid-phase hexapeptide ligand library with liquid chromatography and two-dimensional difference gel electrophoresis for intact plasma proteomics. *Int J Proteomics.* 2011;2011:739615.

Hanash SM, Tubergen DG, Heyn RM, et al. Two-dimensional gel electrophoresis of cell proteins in childhood leukemia, with silver staining: A preliminary report. *Clin Chem.* 1982;28:1026–1030.

Hashimoto F, Horigome T, Kanbayashi M, Yoshida K, Sugano H. An improved method for separation of low-molecular-weight polypeptides by electrophoresis in sodium dodecyl sulfate-polyacrylamide gel. *Anal Biochem.* 1983;129:192–199.

Hofmeister F. Zur Lehre von der Wirkung der Salze. *Arch Exp Pathol Pharmacol.* 1888;24:247–260.

Hu BH, Jones MR, Messersmith PB. Method for screening and MALDI-TOF MS sequencing of encoded combinatorial libraries. *Anal Chem.* 2007;79:7275–7285.

Huhn C, Ruhaak LR, Wuhrer M, Deelder AM. Hexapeptide library as a universal tool for sample preparation in protein glycosylation analysis. *J Proteomics.* 2012;75:1515–1528.

Hwang SH, Lehman A, Cong X. OBOC small molecule combinatorial library encoded by halogenated mass-tags. *Org Lett.* 2004;6:3830–3832.

Kim ST, Cho KS, Jang YS, Kang KY. Two-dimensional electrophoretic analysis of rice proteins by polyethylene glycol fractionation for protein arrays. *Electrophoresis.* 2001;22:2103–2109.

Koroleva OA, Bindschedler LV. Efficient strategies for analysis of low abundance proteins in plant proteomics. In: Ivanov AR, Lazarev AV, ed. *Sample Preparation in Biological Mass Spectrometry.* London: Springer; 2011:363–380.

Kristiansen TZ, Bunkenborg J, Gronborg M, et al. A proteomics analysis of human bile. *Mol Cell Proteomics.* 2004;3:715–728.

Kullolli M, Hancock MS, Hincapie M. Automated platform for fractionation of human plasma glycoproteome in clinical proteomics. *Anal Chem.* 2010;82:115–120.

Kurien BT, Scofield RH. Protein blotting: A review. *J Immunol Methods.* 2003;274:1–15.

Laemmli UK. Cleavage of structural proteins during the assembly of the head of bacteriophage T4. *Nature.* 1970;227:680–685.

Lam KS, Lebl M, Krchnak V. The "one-bead-one-compound" combinatorial library method. *Chem Rev.* 1997;97:411–448.

Lam KS, Salmon SE, Hersh EM, Hruby VJ, Kazmierski WM, Knapp RJ. A new type of synthetic peptide library for identifying ligand-binding activity. *Nature.* 1991;354:82–84.

Lathrop JT, Fijalkowska I, Hammond D. The bead blot: a method for identifying ligand–protein and protein–protein interactions using combinatorial libraries of peptide ligands. *Anal Biochem.* 2007;361:65–76.

Leger T, Lavigne D, Lecaer JP, et al. Solid-phase hexapeptide ligand libraries open up new perspectives in the discovery of biomarkers in human plasma. *Clin Chim Acta.* 2011;412:740–747.

Lilley KS, Friedman DB. All about DIGE: quantification technology for differential-display 2D-gel proteomics. *Expert Rev Proteomics.* 2004;1:401–409.

Liu L, Marik J, Lam KS. A novel peptide based encoding system for "one-bead–one-compound" peptidomimetic and small molecule combinatorial library. *J Am Chem Soc.* 2002;124:7678–7680.

Lloyd-Williams P, Albericio F, Giralt E. *Chemical Approaches to the Synthesis of Peptides and Proteins.* Boca Raton: CRC Press; 1997.

Maillard N, Clouet A, Darbre T, Reymond J-K. Combinatorial libraries of peptide dendrimers. *Nature Protocols.* 2009;4:132–142.

McCabe MS, Garratt LC, Schepers F, et al. Effects of P(SAG12)-IPT gene expression on development and senescence in transgenic lettuce. *Plant Physiol.* 2001;127:505–616.

Mechin V, Damerval C, Zivy M. Total protein extraction with TCA-acetone. *Methods Mol Biol.* 2007;355:1–8.

Meng R, Gormleya M, Bhatb VB, Rosenbergc A, Quong AA. Low abundance protein enrichment for discovery of candidate plasma protein biomarkers for early detection of breast cancer. *J Proteomics.* 2011;75:366–374.

Mihr C, Braun HP. Proteomics in plant biology. In: Conn P, ed. *Handbook of Proteomics*. Totowa, NJ: Humana Press; 2003:409–416.

Mijnsbrugge KV, Meyermans H, Van Montagu M, Bauw G, Boerjan W. Wood formation in poplar: Identification, characterization, and seasonal variation of xylem proteins. *Planta*. 2000;210:589–598.

Mouton-Barbosa E, Roux-Dalvai F, Bouyssié D, et al. In-depth exploration of cerebrospinal fluid combining peptide ligand library treatment and label-free protein quantification. *Mol Cell Proteomics*. 2010;9:1006–1021.

Redman JE, Wilcoxen KM, Ghadiri MR. Automated mass spectrometric sequence determination of cyclic peptide library members. *J Comb Chem*. 2003;5:33–40.

Righetti PG, Boschetti E, Kravchuk A, Fasoli E. The proteome buccaneers: How to unearth your treasure chest via combinatorial peptide ligand libraries. *Expert Rev Proteomics*. 2010;7:373–385.

Righetti PG, Boschetti E. Breakfast at Tiffany's? Only with a low-abundance deep proteomic signature! *Electrophoresis*. 2012;33:2228–2239.

Righetti PG, Boschetti E. Plant proteomics methods to reach low-abundance proteins. In: Jorrín Novo JV, Weckerth W, Komatsu S, eds. *Plant Proteomics Methods And Protocols*. Humana Press; 2013, in press.

Rivers J, Hughes C, McKenna T, et al. Asymmetric proteome equalization of the skeletal muscle proteome using a combinatorial hexapeptide library. *PLoS ONE*. 2011;6:e28902.

Rose JK, Bashir S, Giovannoni JJ, Jahn MM, Saravanan RS. Tackling the plant proteome: practical approaches, hurdles and experimental tools. *Plant J*. 2004;39:715–733.

Roux-Dalvai F. Gonzalez de Peredo A, Simó C, et al. Extensive analysis of the cytoplasmic proteome of human erythrocytes using the peptide ligand library technology and advanced spectrometry. *Mol Cell Proteomics*. 2008;7:2254–2269.

Saez V, Fasoli E, D'Amato A, Simo-Alfonso E, Righetti PG. Artichoke and Cynar liqueur: Two (not quite) entangled proteomes. *Biochim Biophys Acta*. 2013;1834:119–126.

Sakuma R, Nishina T, Kitamura M. Deproteinizing Methods evaluated for determination of unc acid in serum by reversed-phase liquid chromatography with ultraviolet detection. *Clin Chem*. 1987;33:1427–1430.

Santucci L, Candiano G, Petretto A, et al. Combinatorial ligand libraries as a two-dimensional method for proteome analysis. *J Chromatogr A*. 2013; in press.

Saravanan RS, Rose JKC. A critical evaluation of sample extraction techniques for enhanced proteomic analysis of recalcitrant plant tissues. *Proteomics*. 2004;4:2522–2532.

Shahali Y, Sutra JP, Fasoli E, et al. Allergomic study of cypress pollen via combinatorial peptide ligand libraries. *J Proteomics*. 2012;77:101–110.

Sihlbom C, Kanmert I, Bahr H, Davidsson P. Evaluation of the combination of bead technology with SELDI-TOF-MS and 2-D DIGE for detection of plasma proteins. *J Proteome Res*. 2008;7:4191–4198.

Staby A, Johansen N, Wahlstrøm H, Mollerup I. Comparison of loading capacities of various proteins and peptides in culture medium and in pure state. *J Chromatogr A*. 1998;827:311–318.

Steinberg TH, Haugland RP, Singer VL. Applications of SYPRO orange and SYPRO red protein gel stains. *Anal Biochem*. 1996;239:238–245.

Surawski PP, Battersby BJ, Lawrie GA, et al. Flow cytometric detection of proteolysis in peptide libraries synthesised on optically encoded supports. *Mol Biosyst*. 2008;4:774–778.

Sussulini A, Garcia JS, Mesko MF, et al. Evaluation of soybean seed protein extraction focusing on metalloprotein analysis. *Microchimica Acta*. 2007;158:173–180.

Unlu M, Morgan ME, Minden JS. Difference gel electrophoresis: A single gel method for detecting changes in protein extracts. *Electrophoresis*. 1997;18:2071–2077.

Valdes I, Pitarch A, Gil C, et al. Novel procedure for the identification of proteins by mass fingerprinting combining two-dimensional electrophoresis with fluorescent SYPRO red staining. *J Mass Spectrom*. 2000;35:672–682.

Wang W, Scali M, Vignani R, et al. Protein extraction for two-dimensional electrophoresis from olive leaf, a plant tissue containing high levels of interfering compounds. *Electrophoresis*. 2003;24:2369–2375.

Wigand P, Tenzer S, Schild H, Decker H. Analysis of protein composition of red wine in comparison with rosé and white wines by electrophoresis and high pressure liquid chromatography-mass spectrometry. *J Agric Food Chem*. 2009;57:4328–4333.

Wu XY, Xiao ZQ, Chen ZC, et al. Improvement of protein preparation methods for bronchial epithelial tissues and establishment of 2-DE profiles from carcinogenic process of human bronchial epithelial tissues. *Zhong Nan Da Xue Xue Bao Yi Xue Ban*. 2004;29:376–381.

Xi JH, Wang X, Li SY, et al. Polyethylene glycol fractionation improved detection of low-abundant proteins by two-dimensional electrophoresis analysis of plant proteome. *Phytochemistry*. 2006;67:2341–2348.

Xu X, Fan R, Zheng R, Li C, Yu D. Proteomic analysis of seed germination under salt stress in soybeans. *J Zhejiang Univ Sci B*. 2011;12:507–517.

Yeang HY, Yusof F, Abdullah L. Precipitation of *Hevea brasiliensis* latex proteins with trichloroacetic acid and phosphotungstic acid in preparation for the Lowry protein assay. *Anal Biochem*. 1995;226:35–43.

Zellner M, Winkler W, Hayden H, et al. Quantitative validation of different protein precipitation methods in proteome analysis of blood platelets. *Electrophoresis*. 2005;26:2481–2489.

Zhang Y, Cremer PS. Interactions between macromolecules and ions: The Hofmeister series. *Curr Opinion Chem Biol*. 2006;10:658–663.

Just as we opened this book with an Antiphony, we will now close it with a Polyphony or, if you prefer, "A Chorus Line." It just so happened that a few, pirated copies of the book circulated around, unbeknownst to us and to the publisher. We managed to obtain, surreptitiously, the reports of the various readers, which are listed below. As these reviews were sent to us in Italian, we give the original text, followed by the English translation (in parentheses)*.

Don Profondo
Medaglie incomparabili / camei rari, impagabili …
In aurea carta pecora / Dell'accademie I titoli ….
(Incomparable [Proteomics] medals / rare, priceless cameos,
[Written] in golden parchment / by titled academicians …)

Lo Spagnolo (The Spaniard)
Gran piante genealogiche / Degli avoli e bisavoli …
Colle notizie storiche / Di quel che ognuno fu ….
(A great genealogic survey / of recent and past [literature] …
With precise historical details / of what everyone did …)

La Polacca (The Polish lady)
L'opere più squisite / D'autori prelibati …
Che vanto sono e Gloria / della moderna età …
(Exquisite work / of superb authors …
Who are the pride and glory / of modern age …)

La Francese (The French Lady)
Scatole e scatoline /Con scrigni e cassettine …
Che I bei tesor nascondono / Sacri alla Dea d'amor …
(Chapters and subchapters / with caskets and secret drawers …
Hidden immeasurable treasures / to fall in love with …)

Il Tedesco (The German gentleman)
Dissertazione classica / Sui nuovi effetti armonici …
De' primi Orfei Teutonici / Le rare produzioni …
(A classical dissertation / of the novel [proteomics] effects …
With the rare early production / of German scientists …)

L'Inglese (The English gentleman)
Viaggi d'intorno al globo / Trattati di marina …
Oriundo della China / Sottil perlato thè …
([Proteomic] odyssey around the globe / including marine biology …
Touching down even on China / aromatic like jasmine tea ….)

*By courtesy and permission of the late Gioachino Rossini, Il Viaggio a Reims (to know more, look for the CD by Claudio Abbado, conductor, with the Berliner Philharmoniker, Sony, 1993. Here the role of Don Profondo is played by the basso-baritono Ruggero Raimondi).

Il Francese (The French Baron)
Varie del Franco Orazio / Litografie squisite …
Pennelli con matite /Conchiglie coi colour …
(Exquisite lithographies / of French top scientists …
Colored just like / French impressionist tableaus …)

Il Russo (The Russian Prince)
Notizia tipografica / Di tutta la Siberia …
Di zibellini e martore / preziosa collezione …
(A topographic survey / of all Russian and Siberian [literature] …
A collection (of papers) as precious / as furs of sable and mink …)

Don Profondo (Gran Finale) (Don Profondo, closing symphony)
Sta tutto all'ordine / Non v'è che dire …
Nè più a partire / Si può tardar …
De' snelli e rapidi / Destrier frementi …

Già parmi udire / lo scalpitar …
(Everything is under control / no questions about that …
We can't wait any longer / to get started …
Of racing horses / fast and slim …
I seem to hear / the soaring gallop …

[IMPRIMATUR, IMPRIMATUR!!]

SELECTED CPLL BIBLIOGRAPHY

Relevant Chronological Publications on Combinatorial Peptide Ligand Libraries Applied to Proteomics Investigations

Thulasiraman V, Lin S, Gheorghiu L, et al. Reduction of the concentration difference of proteins in biological liquids using a library of combinatorial ligands. *Electrophoresis.* 2005;26:3561–3571.

Castagna A, Cecconi D, Sennels L, et al. Exploring the hidden human urinary proteome via ligand library beads. *J Proteome Res.* 2005;4:1917–1930.

Righetti PG, Castagna A, Antonucci F, et al. Proteome analysis in the clinical chemistry laboratory: myth or reality? *Clin Chim Acta.* 2005;357:123–139.

Righetti PG, Castagna A, Antonioli P, Boschetti E. Prefractionation techniques in proteome analysis: The mining tools of the third millennium. *Electrophoresis.* 2005;26:297–319.

Righetti PG, Boschetti E, Lomas L, Citterio A. Protein Equalizer Technology: The quest for a "democratic proteome". *Proteomics.* 2006;6:3980–3992.

Fortis F, Guerrier L, Righetti PG, Antonioli P, Boschetti E. A new approach for the removal of protein impurities from purified biologicals using combinatorial solid-phase ligand libraries. *Electrophoresis.* 2006;27:3018–3027.

Fortis F, Guerrier L, Areces LB, et al. A new approach for the detection and identification of protein impurities using combinatorial solid phase ligand libraries. *J Proteome Res.* 2006;5:2577–2585.

Guerrier L, Thulasiraman V, Castagna A, et al. Reducing protein concentration range of biological samples using solid-phase ligand libraries. *J Chromatogr B.* 2006;833:33–40.

Guerrier L, Claverol S, Finzi L, et al. Contribution of solid-phase hexapeptide ligand libraries to the repertoire of human bile proteins. *J Chromatogr A.* 2007;1176:192–205.

Boschetti E, Monsarrat B, Righetti PG. The "Invisible Proteome": How to capture the low-abundance proteins via combinatorial ligand libraries. *Current Proteomics.* 2007;4:198–208.

Boschetti E, Righetti PG. "Omics" combinatorial libraries for capturing the "Hidden" proteome. In: Poole CF, Wilson ID, eds. *Encyclopedia of Separation Science*, online update. Oxford: Elsevier Science Ltd; 2007:1–18.

Lescuyer P, Hochstrasser D, Rabilloud T. How shall we use the proteomics toolbox for biomarker discovery? *J Proteome Res.* 2007;6:3371–3376.

Poon TCW. Opportunities and limitations of SELDI-TOF-MS in biomedical research: Practical advice. *Expert Rev Proteomics.* 2007;4:51–65.

Au JS, Cho WC, Yip TT, et al. Deep proteome profiling of sera from never-smoked lung cancer patients. *Biomed Pharmacother.* 2007;61:570–577.

Cho WCS. Proteomics technologies and challenges. *Genomics Proteomics Bioinformatics.* 2007;5:77–85.

Sarkar J, Soukharev S, Lathrop JT, Hammond DJ. Functional identification of novel activities: Activity-based selection of proteins from complete proteomes. *Anal Biochem.* 2007;365:91–102.

Antonioli P, Fortis F, Guerrier L, et al. Capturing and amplifying impurities from purified recombinant monoclonal antibodies via peptide library beads: A proteomic study. *Proteomics.* 2007;7:1624–1633.

Guerrier L, Claverol S, Fortis F, et al. Exploring the platelet proteome via combinatorial, hexapeptide ligand libraries. *J Proteome Res.* 2007;6:4290–4303.

Boschetti E, Lomas L, Citterio A, Righetti PG. Romancing the "hidden proteome", Anno Domini two zero zero seven. *J Chromatogr A.* 2007;1153:277–290.

Sennels L, Salek M, Lomas L, Boschetti E, Righetti PG, Rappsilber J. Proteomic analysis of human blood serum using peptide library beads. *J Proteome Res.* 2007;6:4055–4062.

Righetti PG, Boschetti E. Sherlock Holmes and the proteome: A detective story. *FEBS J.* 2007;274:897–905.

Decramer S, Gonzalez de Peredo A, Breuil B, et al. Urine in clinical proteomics. *Mol Cell Proteomics.* 2008;7:1850–1862.

Simó C, Bachi A, Cattaneo A, et al. Performance of combinatorial peptide libraries in capturing the low-abundance proteome of red blood cells. 1. Behavior of mono- to hexapeptides. *Anal Chem.* 2008;80:3547–3556.

Bachi A, Simó C, Restuccia U, et al. Performance of combinatorial peptide libraries in capturing the low-abundance proteome of red blood cells. 2. Behavior of resins containing individual amino acids. *Anal Chem.* 2008;80:3557–3565.

D'Ambrosio C, Arena S, Scaloni A, et al. Exploring the chicken egg white proteome with combinatorial peptide ligand libraries. *J Proteome Res.* 2008;7:3461–3474.

Fang X, Zhang WW. Affinity separation and enrichment methods in proteomic analysis. *J Proteomics.* 2008;71:284–303.

Boschetti E, Righetti PG. Hexapeptide combinatorial ligand libraries: The march for the detection of the low-abundance proteome continues. *Biotechniques.* 2008;44:663–665.

Righetti PG, Boschetti E. The ProteoMiner and the FortyNiners: Searching for gold nuggets in the proteomic arena. *Mass Spectrom Rev.* 2008;27:596–608.

Guerrier L, D'Autreaux B, Atanassov C, Khoder G, Boschetti E. Evaluation of a standardized method of protein purification and identification after discovery by mass spectrometry. *J Proteomics.* 2008;71:368–378.

Guerrier L, Righetti PG, Boschetti E. Reduction of dynamic protein concentration range of biological extracts for the discovery of low-abundance proteins by means of hexapeptide ligand library. *Nature Protocols.* 2008;3:883–890.

Chen ST, Tsai HY, Huang TY. Enhance proteomic detection limitation by combinatorial peptide and nucleotide library. *J Prot Bioinform.* 2008;S2:76.

Sihlbom C, Kanmert I, Bahr H, Davidsson P. Evaluation of the combination of bead technology with SELDI-TOF-MS and 2-D DIGE for detection of plasma proteins. *J Proteome Res.* 2008;7:4191–4198.

Devarajan P, Ross GF. SELDI technology for identification of protein biomarkers. *Meth Pharmacol Toxicol.* 2008;251–271.

Liumbruno G, D'Amici GM, Grazzini G, Zolla L. Transfusion medicine in the era of proteomics. *J Proteomics.* 2008;71:34–45.

Boschetti E, Righetti PG. The ProteoMiner in the proteomic arena: A non-depleting tool for discovering low-abundance species. *J Proteomics.* 2008;71:255–264.

Devarajan P, Ross GF. SELDI technology for identification of protein biomarkers. In: Wanf F, Kang YJ, eds. *Biomarker Methods in Drug Discovery and Development.* Humana Press; 2008:251–289.

Shores KS, Udugamasooriva DG, Kodadek T, Knapp DR. Use of peptide analogue diversity library beads for increased depth of proteomic analysis: Application to cerebrospinal fluid. *J Proteome Res.* 2008;7:1922–1931.

Freeby S, Academia K, Wehr T, Liu N, Paulus A. Enrichment of interleukins and low abundance proteins from tissue leakage in serum proteome studies using ProteoMiner beads. *J Proteomics & Bioinformatics.* 2008;S2:171.

Zolla L. Proteomics studies reveal important information on small molecule therapeutics: A case study on plasma proteins. *Drug Discov Today.* 2008;1042–1051.

Pernemalm M, Orre LM, Lengqvist J, Wikström P, Lewensohn R, Lehtiö J. Evaluation of three principally different intact protein prefractionation methods for plasma biomarker discovery. *J Proteome Res.* 2008;7:2712–2722.

Roux-Dalvai F, Gonzalez de Peredo A, Simó C, et al. Extensive analysis of the cytoplasmic proteome of human erythrocytes using the peptide ligand library technology and advanced mass spectrometry. *Mol Cell Proteomics* 2008;7:2254–2269.

Bandhakavi S, Stone MD, Onsongo G, Van Riper SK, Griffin TJ. A dynamic range compression and three-dimensional peptide fractionation analysis platform expands proteome coverage and the diagnostic potential of whole saliva. *J Proteome Res.* 2009;8:5590–5600.

Chi KR. Troubleshooting discovery and validation of protein biomarkers for cancer. *The Scientist.* 2009;23:63–66.

Farinazzo A, Restuccia U, Bachi A, et al. Chicken egg yolk cytoplasmic proteome, mined via combinatorial peptide ligand libraries. *J Chromatogr A.* 2009;1216:1241–1252.

Colzani M, Bienvenut WV, Faes E, Quadroni M. Precursor ion scans for the targeted detection of stable-isotope-labeled peptides. *Rapid Commun Mass Spectrom.* 2009;23:3570–3578.

Sela-Abramovich S, Chitlaru T, Gat, Grosfeld, Cohen O, Shafferman A. Novel and unique diagnostic biomarkers for *Bacillus anthracis* infection. *Appl Environ Microbiol.* 2009;75:6157–6167.

Dihazi H, Dihazi GH, Nolte J, et al. Multipotent adult germline stem cells and embryonic stem cells: Comparative proteomic approach. *J Proteome Res.* 2009;8:5497–5510.

Restuccia U, Boschetti E, Fasoli E, et al. pI-based fractionation of serum proteomes versus anion exchange after enhancement of low-abundance proteins by means of peptide libraries. *J Proteomics.* 2009;72:1061–1070.

Martinkova J, Jelinkova L, Jarkovska K, et al. Protein profiling of human follicular fluid: Quest for biomarkers of ovarian hyperstimulation syndrome. *Cancer Genom Proteom.* 2009;6:58–65.

Bianchi P, Fermo E, Vercellati C, et al. Congenital dyserythropoietic anemia type II (CDAII) is caused by mutations in the SEC23B gene. *Hum Mutat.* 2009;30:1292–1298.

Hartwig S, Czibere A, Kotzka J, et al. Combinatorial hexapeptide ligand libraries (ProteoMiner): An innovative fractionation tool for differential quantitative clinical proteomics. *Arch Physiol Biochem.* 2009;115:155–160.

Callesen AK, Madsen JS, Vach W, Kruse TA, Mogensen O, Jensen ON. Serum protein profiling by solid phase extraction and mass spectrometry: A future diagnostics tool? *Proteomics.* 2009;9:1428–1441.

Jmeian Y, El Rassi Z. Liquid-phase-based separation systems for depletion, prefractionation and enrichment of proteins in biological fluids for in-depth proteomics analysis. *Electrophoresis.* 2009;30:249–261.

Fasoli E, Pastorello EA, Farioli L, et al. Searching for allergens in maize kernels via proteomic tools. *J Proteomics.* 2009;72:501–510.

D'Amato A, Bachi A, Fasoli E, et al. In-depth exploration of cow's whey proteome via combinatorial peptide ligand libraries. *J Proteome Res.* 2009;8:3925–3936.

Petri AL, Simonsen AH, Yip TT, et al. Three new potential ovarian cancer biomarkers detected in human urine with equalizer bead technology. *Acta Obstet Gynecol Scand.* 2009;88:18–26.

Boschetti E, Bindschedler LV, Tang C, Fasoli E, Righetti PG. Combinatorial peptide ligand libraries and plant proteomics: A winning strategy at a price. *J Chromatogr A.* 2009;1216:1215–1222.

Boschetti E, Righetti PG. The art of observing rare protein species in proteomes with peptide ligand libraries. *Proteomics.* 2009;9:1492–1510.

Li L, Sun CJ, Freeby S, et al. Protein sample treatment with peptide ligand library: Coverage and consistency. *J Proteomics Bioinform.* 2009;2:485–494.

Polkinghorne VR, Standeven KF, Schroeder V, Carter AM. Role of proteomic technologies in understanding risk of arterial thrombosis. *Expert Rev Proteomics.* 2009;6:539–550.

Candiano G, Dimuccio V, Bruschi M, et al. Combinatorial peptide ligand libraries for urine proteome analysis: investigation of different elution systems. *Electrophoresis.* 2009;30:2405–2411.

Farinazzo A, Fasoli E, Kravchuk AV, et al. En bloc elution of proteomes from combinatorial peptide ligand libraries. *J Proteomics.* 2009;72:725–730.

Tonack S, Neoptolemos J, Costello E. Principle and applications of proteomics in pancreatic cancer. In: Neoptolemos JP, ed. *Pancreatic Cancer.* Springer Science; 2009:509–534.

Calvete JJ, Fasoli E, Sanz L, Boschetti E, Righetti PG. Exploring the venom proteome of the western diamondback rattlesnake, *Crotalus atrox,* via snake venomics and combinatorial peptide ligand library approaches. *J Proteome Res.* 2009;8:3055–3067.

Colzani M, Waridel P, Laurent J, Faes E, Rüegg C, Quadroni M. Metabolic labeling and protein linearization technology allow the study of proteins secreted by cultured cells in serum-containing media. *J Proteome Res.* 2009;8:4779–4788.

Masseroli M, Bachi A, Boschetti E, Righetti PG. Searching for specific motifs in affinity capture in proteome analysis. *J Proteomics.* 2009;72:791–802.

Pernemalm M, Lewensohn R, Lehtiö J. Affinity prefractionation for MS-based plasma proteomics. *Proteomics.* 2009;9:1420–1427.

Candiano G, Santucci L, Petretto A, et al. 2D-electrophoresis and the urine proteome map: Where do we stand? *J Proteomics.* 2010;73:829–844.

Fasoli E, Sanz L, Wagstaff S, Harrison RA, Righetti PG, Calvete JJ. Exploring the venom proteome of the African puff adder, *Bitis arietans,* using a combinatorial peptide ligand library approach at different pHs. *J Proteomics.* 2010;73:932–942.

Fasoli E, Farinazzo A, Sun CJ, et al. Interaction among proteins and peptide libraries in proteome analysis: pH involvement for a larger capture of species. *J Proteomics.* 2010;73:733–742.

Dwivedi RC, Krokhin OV, Cortens JP, Wilkins JA. An assessment of the reproducibility of random hexapeptide peptide library based protein normalization. *J Proteome Res.* 2010;9:1144–1149.

Ernoult E, Bourreau A, Gamelin E, Guette C. A proteomic approach for plasma biomarker discovery with iTRAQ labelling and OFFGEL fractionation. *J Biomed Biotechnol.* 2010;2010:927917.

Makridakis M, Vlahou A. Secretome proteomics for discovery of cancer biomarkers. *J Proteomics.* 2010;73:2291–2305.

Righetti PG, Boschetti E, Zanella A, Fasoli E, Citterio A. Plucking, pillaging and plundering proteomes with combinatorial peptide ligand libraries. *J Chromatogr A.* 2010;1217:893–900.

Castagna, Cecconi D, Boschetti E, Righetti PG. Prefractionation of urinary proteins. In: Tongboonkerd V, ed. *Renal and Urinary Proteomics.* Wiley-Blackwell; 2010:201–217.

Guerrier L, Righetti PG, Boschetti E. Wider protein detection from biological extracts by the reduction of dynamic concentration range. In: Shah H, ed. *Mass Spectrometry for Microbial Proteins.* Wiley-Blackwell; 2010:175–204.

Boehmer JL, DeGrasse JA, McFarland MA, et al. The proteomic advantage: Label-free quantification of proteins expressed in bovine milk during experimentally induced coliform mastitis. *Vet Immunol Immunopathol.* 2010;138:252–266.

Ortega F. Estrategias analíticas en la investigación de nuevos biomarcadores. *Monografías Real Acad Nacional de Farmacia.* 2010.

D'Alessandro A, Zolla L. Proteomics for quality-control processes in transfusion medicine. *Anal Bioanal Chem.* 2010;398:111–124.

Stasyk T, Holzmann J, Stumberger S, et al. Proteomic analysis of endosomes from genetically modified p14/MP1 mouse embryonic fibroblasts. *Proteomics.* 2010;10:4117–4127.

Mouton-Barbosa E, Roux-Dalvai F, Bouyssié D, et al. In-depth exploration of cerebrospinal fluid by combining peptide ligand library treatment and label free protein quantification. *Mol Cell Proteomics.* 2010;9:1006–1021.

D'Alessandro A, Righetti PG, Fasoli E, Zolla L. The egg white and yolk interactomes as gleaned from extensive proteomic data. *J Proteomics.* 2010;73:1028–1042.

D'Alessandro A, Righetti PG, Zolla L. The red blood cell proteome and interactome: An update. *J Proteome Res.* 2010;9:144–163.

Nordon IM, Brar R, Hinchliffe RJ, Cockerill G, Thompson MM. Proteomics and pitfalls in the search for potential biomarkers of abdominal aortic aneurysms. *Vascular.* 2010;18:264–268.

Bandow JE. Comparison of protein enrichment strategies for proteome analysis of plasma. *Proteomics.* 2010;10:1416–1425.

Drabovich AP, Diamandis EP. Combinatorial peptide libraries facilitate development of multiple reaction monitoring assays for low-abundance proteins. *J Proteome Res.* 2010;9:1236–1245.

D'Amato A, Bachi A, Fasoli E, et al. In-depth exploration of *Hevea brasiliensis* latex proteome and "hidden allergens" via combinatorial peptide ligand libraries. *J Proteomics.* 2010;73:1368–1380.

Chevalier F. Highlights on the capacities of "Gel-based" proteomics. *Proteome Sci.* 2010;8:23–33.

Kolla V, Jenö P, Moes S, et al. Quantitative proteomics analysis of maternal plasma in down syndrome pregnancies using isobaric tagging reagent (iTRAQ). *J Biomed Biotechnol.* 2010;2010:952047.

Selected CPLL Bibliography

Fertin M, Beseme O, Duban S, Amouyel P, Bauters C, Pinet F. Deep plasma proteomic analysis of patients with left ventricular remodeling after a first myocardial infarction. *Proteomics Clin Appl.* 2010;4:654–673.

Beseme O, Fertin M, Drobecq H, Amouyel P, Pinet F. Combinatorial peptide ligand library plasma treatment: Advantages for accessing low-abundance proteins. *Electrophoresis.* 2010;31:2697–2704.

Keidel EM, Ribitsch D, Lottspeich F. Equalizer technology-equal rights for disparate beads. *Proteomics.* 2010;10:2089–2098.

Righetti PG, Boschetti E, Kravchuk AV, Fasoli A. The proteome buccaneers: How to unearth your treasure chest via combinatorial peptide ligand libraries. *Expert Rev Proteomics.* 2010;7:373–385.

D'Amato A, Cereda A, Bachi A, Pierce JC, Righetti PG. In depth exploration of the haemolymph of Limulus polyphemus via combinatorial peptide ligand libraries. *J Proteome Res.* 2010;9:3260–3269.

Marco-Ramell A, Bassols A. Enrichment of low-abundance proteins from bovine and porcine serum samples for proteomic studies. *Res Vet Sci.* 2010;89:340–343.

Overgaard AJ, Hansen HG, Lajer M, et al. Plasma proteome analysis of patients with type 1 diabetes with diabetic nephropathy. *Proteome Sci.* 2010;8:4.

Cereda A, Kravchuk AV, D'Amato A, Bachi A, Righetti PG. Proteomics of wine additives: Mining for the invisible via combinatorial peptide ligand libraries. *J Proteomics.* 2010;73:1732–1739.

Sjodin MO, Bergquist J, Wetterhall M. Mining ventricular cerebrospinal fluid from patients with traumatic brain injury using hexapeptide ligand libraries to search for trauma biomarkers. *J Chromatogr B.* 2010;878:2003–2012.

Marrocco C, Rinalducci S, Mohamadkhani A, D'Amici GM, Zolla L. Plasma gelsolin protein: A candidate biomarker for hepatitis B-associated liver cirrhosis identified by proteomic approach. *Blood Transfus.* 2010;8:s105–s112.

Ye H, Sun L, Huang X, Zhang P, Zhao X. A proteomic approach for plasma biomarker discovery with 8-plex iTRAQ labeling and SCX-LC-MS/MS. *Mol Cell Biochem.* 2010;343:91–99.

Fröbel J, Hartwig S, Paßlack W, et al. ProteoMiner() and SELDI-TOF-MS: A robust and highly reproducible combination for biomarker discovery from whole blood serum. *Arch Physiol Biochem.* 2010;116:174–180.

Zhi W, Purohit S, Carey C, Wang M, She JX. Proteomic technologies for the discovery of type 1 diabetes biomarkers. *J Diabetes Sci Technol.* 2010;4:993–1002.

Righetti PG, Fasoli E, Aldini G, Regazzoni L, Kravchuk A, Citterio A. Les maîtres de l'orge: The proteome content of your beer mug. *J Proteome Res.* 2010;9:5262–5269.

Arena S, Renzone G, Novi G, et al. Modern proteomic methodologies for the characterization of lactosylation protein targets in milk. *Proteomics.* 2010;10:3414–3434.

Fakelman F, Felix K, Büchler MW, Werner J. New pre-analytical approach for the deep proteome analysis of sera from pancreatitis and pancreas cancer patients. *Arch Physiol Biochem.* 2010;116:208–217.

De Bock M, de Seny D, Meuwis M-A, et al. Comparison of three methods for fractionation and enrichment of low molecular weight proteins for SELDI-TOF-MS differential analysis. *Talanta.* 2010;82:245–254.

Stasyk T, Holzmann J, Strumberger S, et al. Proteomic analysis of endosomes from genetically modified p14/MP1 mouse embryonic fibroblasts. *Proteomics.* 2010;10:4117–4127.

Yang Y, Cheng G, Zhao H, Jiang X, Chen S. Differential proteomics analysis of plasma protein from *Escherichia coli* infected and clinical healthy dairy cows. *Xumu Shouyi Xuebao.* 2010;41:1191–1197.

D'Amato A, Kravchuk AV, Bachi A, Righetti PG. Noah's nectar: The proteome content of a glass of red wine. *J Proteomics.* 2010;73:2370–2377.

Dowling P, Meleady P, Henry M, Clynes M. Recent advances in clinical proteomics using mass spectrometry. *Bioanalysis.* 2010;2:1609–1615.

Choi Y, Kim E, Lee Y, Han MH, Kang I-C. Site-specific inhibition of integrin $\alpha v \beta 3$-vitronectin association by a ser-asp-val sequence through an Arg-Gly-Asp-binding site of the integrin. *Proteomics.* 2010;10:72–80.

Zhang Y, Li Y, Qiu F, Qiu Z. Comprehensive analysis of low-abundance proteins in human urinary exosomes using peptide ligand library technology, peptide OFFGEL fractionation and nanoHPLC-chip-MS/MS. *Electrophoresis.* 2010;31:3797–3807.

Brewis IA, Brennan P. Proteomics technologies for the global identification and quantification of proteins. *Adv Protein Chem Struct Biol.* 2010;80C:1–44.

Jeong SK, Lee EY, Cho JY, et al. Data management and functional annotation of the Korean reference plasma proteome. *Proteomics.* 2010;10:1250–1255.

Fang Y, Robinson DP, Foster LG. Quantitative analysis of proteome coverage and recovery rates for upstream fractionation methods in proteomics. *J Proteome Res.* 2010;9:1902–1912.

Braoudaki M, Tzortzatou-Stathopoulou F, Anagnostopoulos AK, et al. Proteomic analysis of childhood de novo acute myeloid leukemia and myelodysplastic syndrome/AML: Correlation to molecular and cytogenetic analyses. *Amino Acids.* 2010;40:943–951.

Overgaard AJ, Thingholm TE, Larsen MR, et al. Quantitative iTRAQ-based proteomic identification of candidate biomarkers for diabetic nephropathy in plasma of type 1 diabetic patients. *Clin Proteomics.* 2010;6:105–114.

Fasoli E, D'Amato A, Kravchuk V, Boschetti E, Bachi A, Righetti PG. Popeye strikes again: The deep proteome of spinach leaves. *J Proteomics.* 2010;74:127–136.

Farina A, Dumonceau JM, Lecuyer P. Proteomics analysis of human bile and potential applications in cancer diagnosis. *Expert Rev Proteomics.* 2010;6:285–301.

Larkin SET, Zeidan B, Taylor MG, et al. Proteomics in prostate cancer biomarker discovery. *Expert Rev Proteomics.* 2010;7:93–102.

Liumbruno G, D'Alessandro A, Grazzini G, Zolla L. Blood-related proteomics. *J Proteomics.* 2010;73:483–507.

Meng Z, Veenstra TD. Targeted mass spectrometry approaches for protein biomarker verification. *J Proteomics.* 2011;74:2650–2659.

Fahiminiya S, Labas V, Roche S, Dacheux J-L, Gérard N. Proteomic analysis of mare follicular fluid during late follicle development. *Proteome Sci.* 2011;9:54–73.

Gaso-Sokac D, Kovac S, Clifton J, Josic D. Therapeutic plasma proteins—application of proteomics in process optimization, validation, and analysis of the final product. *Electrophoresis.* 2011;32:1104–1117.

Meyer B, Papasoturiou DG, Karas M. 100% protein sequence coverage: A modern form of surrealism in proteomics. *Amino Acids.* 2011;41:291–310.

Fahiminiya S, Labas V, Dacheux J, Gérard N. Improvement of 2D-PAGE resolution of human, porcine and equine follicular fluid by means of hexapeptide ligand library. *Reprod Domest Anim.* 2011;46:561–563.

Righetti PG, Boschetti E, Fasoli E. Capturing and amplifying impurities from recombinant therapeutic proteins via combinatorial peptide libraries: A proteomic approach. *Curr Pharm Biotechnol.* 2011;12:1537–1547.

Righetti PG, Fasoli E, Boschetti E. Combinatorial peptide ligand libraries: The conquest of the "hidden proteome" advances at great strides. *Electrophoresis.* 2011;32:960–966.

Felix K, Fakelman F, Hartmann D, et al. Identification of serum proteins involved in pancreatic cancer cachexia. *Life Sci.* 2011;88:218–225.

Liu C, Zhang N, Yu H, et al. Proteomic analysis of human serum for finding pathogenic factors and potential biomarkers in preeclampsia. *Placenta.* 2011;32:168–174.

Léger T, Lavigne D, Le Caër JP, et al. Solid-phase hexapeptide ligand libraries open up new perspectives in the discovery of biomarkers in human plasma. *Clin Chim Acta.* 2011;412:740–747.

Fertin M, Burdese J, Beseme O, Amouyel P, Bauters C, Pinet F. Strategy for purification and mass spectrometry identification of SELDI peaks corresponding to low-abundance plasma and serum proteins. *J Proteomics.* 2011;74:420–430.

Bandhakavi S, Van Riper SK, Tawfik PN, et al. Hexapeptide libraries for enhanced protein PTM identification and relative abundance profiling in whole human saliva. *J Proteome Res.* 2011;10:1052–1061.

Fröhlich A, Lindermayr C. Deep insights into the plant proteome by pretreatment 1 with combinatorial hexapeptide ligand 2 libraries. *J Proteomics.* 2011;74:1182–1189.

Juhasz P, Lynch M, Sethuraman M, et al. Semi-targeted plasma proteomics discovery workflow utilizing two-stage protein depletion and off-line LC-MALDI MS/MS. *J Proteome Res.* 2011;10:34–45.

Di Girolamo F, Boschetti E, Chung MCM, Guadagni F, Righetti PG. "Proteomineering" or not? The debate on biomarkers discovery in sera continues. *J Proteomics.* 2011;74:589–594.

Boschetti E, Righetti PG. Mixed-bed chromatography as a way to resolve peculiar protein fractionation situations. *J Chromatogr B.* 2011;879:827–835.

Griffoni C, Di Molfetta S, Fantozzi L, et al. Modification of proteins secreted by endothelial cells during modeled low gravity exposure. *J Cell Biochem.* 2011;112:265–272.

Stone MD, Chen XB, McGowan T, et al. Large-scale phosphoproteomics analysis of whole saliva reveals a distinct phosphorylation pattern. *J Proteome Res.* 2011;10:1728–1736.

Zhi W, Sharma A, Purohit S, et al. Discovery and validation of serum protein changes in Type 1 diabetes patients using high throughput two dimensional liquid chromatography-mass spectrometry and immunoassays. *Mol Cell Proteomics.* 2011;10:M111.012203.

Tu C, Li J, Young R, et al. A combinatorial peptide ligand libraries treatment followed by a dual-enzyme, dual-activation approach on a nano-flow LC/Orbitrap/ETD for comprehensive analysis of swine plasma proteome. *Anal Chem.* 2011;83:4802–4813.

Liao Y, Alvarado R, Phinney B, Lönnerdal B. Proteomic characterization of human milk whey proteins during a twelve-month lactation period. *J Proteome Res.* 2011;10:1746–1754.

Selvaraju S, El Rassi Z. Reduction of protein concentration range difference followed by multicolumn fractionation prior to 2-DE and LC-MS/MS profiling of serum proteins. *Electrophoresis.* 2011;32:674–685.

Egidi MG, Rinalducci S, Marrocco C, Vaglio S, Zolla L. Proteomic analysis of plasma derived from platelet buffy coats during storage at room temperature. An application of ProteoMiner technology. *Platelets.* 2011;22:252–269.

Chen G, Zhang Y, Jin X, et al. Urinary proteomics analysis for renal injury in hypertensive disorders of pregnancy with iTRAQ labeling and LC-MS/MS. *Proteomics Clin Applic.* 2011;5:300–310.

Sussulini A, Dihazi H, Muller Banzato CE, et al. Apolipoprotein A-I as a candidate serum marker for the response to lithium treatment in bipolar disorder. *Proteomics.* 2011;11:261–269.

Luque-Garcia JL, Cabezas-Sanchez P, Camara C. Proteomics as a tool for examining the toxicity of heavy metals. *Trends Anal Chem.* 2011;30:703–716.

Simonato B, Mainente F, Tolin S, Pasini G. Immunochemical and mass spectrometry detection of residual proteins in gluten fined red wine. *J Agric Food Chem.* 2011;59:3101–3110.

Dowling P, Clynes M. Conditioned media from cell lines: A complementary model to clinical specimens for the discovery of disease-specific biomarkers. *Proteomics.* 2011;11:794–804.

Di Girolamo F, Bala K, Chung MC, Righetti PG. "Proteomineering" serum biomarkers. A study in scarlet. *Electrophoresis.* 2011;32:976–980.

Selected CPLL Bibliography

D'Amato A, Fasoli E, Kravchuk AV, Righetti PG. Going nuts for nuts? The trace proteome of a cola drink, as detected via combinatorial peptide ligand libraries. *J Proteome Res.* 2011;10:2684–2686.

Albrethsen J. The first decade of MALDI protein profiling: A lesson in translational biomarker research. *J Proteomics.* 2011;74:765–773.

Kentsis A. Challenges and opportunities for discovery of disease biomarkers using urine proteomics. *Pediatrics International.* 2011;53:1–6.

D'Amici GM, Rinalducci S, Zolla L. An easy preparative gel electrophoretic method for targeted depletion of hemoglobin in erythrocyte cytosolic samples. *Electrophoresis.* 2011;32:1319–1322.

Zanusso G, Fiorini M, Righetti PG, Monaco S. Specific and surrogate cerebrospinal fluid markers in Creutzfeldt–Jakob Disease. *Adv Neurobiol.* 2011;2:455–467.

Ziganshin R, Arapidi G, Azarkin I, et al. New method for peptide desorption from abundant blood proteins for plasma/serum peptidome analyses by mass spectrometry. *J Proteomics.* 2011;74:595–606.

Bellei E, Monari E, Bergamini S, Ozben T, Tomasi A. Optimizing protein recovery yield from serum samples treated with beads technology. *Electrophoresis.* 2011;32:1414–1421.

Millioni, Tolin S, Puricelli L, et al. High abundance proteins depletion vs low abundance proteins enrichment: Comparison of methods to reduce the plasma proteome complexity. *PLoS One* 2011;6:e19603.

Fasoli E, D'Amato A, Kravchuk AV, Citterio A, Righetti PG. In-depth proteomic analysis of non-alcoholic beverages with peptide ligand libraries. I: Almond milk and orgeat syrup. *J Proteomics.* 2011;74:1080–1090.

Raposo RAS, Trudgian DC, Thomas B, van Wilgenburg B, Cowley SA, James W. Protein kinase C and NF-κB–dependent CD4 downregulation in macrophages induced by T cell-derived soluble factors: Consequences for HIV-1 infection. *J Immunol.* 2011;187:748–759.

Fonslow BR, Carvalho PC, Academia K, et al. Improvements in proteomic metrics of low abundance proteins through proteome equalization using ProteoMiner prior to MudPIT. *J Proteome Res.* 2011;10:3690–3700.

D'Amato A, Fasoli E, Kravchuk AV, Righetti PG. Mehercules, adhuc Bacchus! The debate on wine proteomics continues. *J Proteome Res.* 2011;10:3789–3801.

Panfoli I, Bruschi M, Ravera S, Candiano G. "Proteomineering": Has the mine been excavated? *Expert Rev Proteomics.* 2011;8:443–445.

Di Girolamo F, Righetti PG, D'Amato A, Chung MC. Cibacron Blue and proteomics: The mystery of the platoon missing in action. *J Proteomics.* 2011;74:2856–2865.

Biosa G, Addis MF, Tanca A, et al. Comparison of blood serum peptide enrichment methods by Tricine SDS-PAGE and mass spectrometry. *J Proteomics.* 2011;75:93–99.

Bousquet-Dubouch MP, Fabre B, Monsarrat B, Burlet-Schiltz O. Proteomics to study the diversity and dynamics of proteasome complexes: From fundamentals to the clinic. *Expert Rev Proteomics.* 2011;8:459–481.

Cunsolo V, Muccilli V, Fasoli E, Saletti R, Righetti PG, Foti S. Poppea's bath liquor: The secret proteome of she-donkey's milk. *J Proteomics.* 2011;74:2083–2099.

Li T, Huang Z, Zheng B. Influence of different pre-treatments for serum on the effects of two-dimensional electrophoresis. *China Med Engineering.* 2011;01.

Wang Y-S, Cao R, Jin H, et al. Altered protein expression in serum from endometrial hyperplasia and carcinoma patients. *J Hematol Oncol.* 2011;4:15–23.

Koroleva OA, Bindschedler LV. Efficient strategies for analysis of low abundance proteins in plant proteomics. In: Ivanov AR, Lazarev AV, eds. *Sample Preparation in Biological Mass Spectrometry.* London: Springer; 2011:363–380.

Tu C, Li J, Young R, et al. Combinatorial peptide ligand library treatment followed by a dual-enzyme, dual-activation approach on a nanoflow liquid chromatography/orbitrap/electron transfer dissociation system for comprehensive analysis of swine plasma proteome. *Anal Chem.* 2011;83:4802–4813.

Skrzypczak WF, Ozgo M, Lepczynski A, Herosimczyk A. Defining the blood plasma protein repertoire of seven day old Dairy calves—A preliminary study. *J Physiol Pharmacol.* 2011;62:313–319.

Zhi W, Wang M, She J-X. Selected reaction monitoring (SRM) mass spectrometry without isotope labeling can be used for rapid protein quantification. *Rapid Commun Mass Spectrom.* 2011;25:1583–1588.

Di Girolamo F, D'Amato A, Righetti PG. Horam nonam exclamavit: sitio. The trace proteome of your daily vinegar. *J Proteomics* 2011;75:718–724.

Meng R, Gormley M, Bhat VB, Rosenberg A, Quong AA. Low abundance protein enrichment for discovery of candidate plasma protein biomarkers for early detection of breast cancer. *J Proteomics.* 2011;75:366–374.

Genoux A, Pons V, Radojkovic C, et al. Mitochondrial inhibitory factor 1 (IF1) is present in human serum and is positively correlated with HDL-cholesterol. *PLoS One.* 2011;6:e23949.

Fahiminiya S, Labas V, Roche S, Dacheux J-J, Gérard N. Proteomic analysis of mare follicular fluid during late follicle development. *Proteome Sci.* 2011;9:54–73.

Monari E, Casali C, Cuoghi A, et al. Enriched sera protein profiling for detection of non-small cell lung cancer biomarkers. *Proteome Sci.* 2011;9:55–65.

Hagiwara T, Saito Y, Nakamura Y, Tomonaga T, Murakami Y, Kondo T. Combined use of a solid-phase hexapeptide ligand library with liquid chromatography and two-dimensional difference gel electrophoresis for intact plasma proteomics. *Int J Proteomics.* 2011;2011:1–11.

Gonzalez-Begne G, Lu B, Liao L, et al. Characterization of the human submandibular/sublingual saliva glycoproteome using lectin affinity chromatography coupled to Multidimensional Protein Identification. *J Proteome Res.* 2011;10:5031–5046.

Moxon JV, Padula MP, Clancy P, et al. Proteomic analysis of intra-arterial thrombus secretions reveals a negative association of clusterin and thrombospondin-1 with abdominal aortic aneurysm. *Atherosclerosis*. 2011;219:432–439.

Halfinger B, Sarg B, Amann A, Hammerer-Lercher A, Lindner HH. Unmasking low-abundance peptides from human blood plasma and serum samples by a simple and robust two-step precipitation/immunoaffinity enrichment method. *Electrophoresis*. 2011;132:1706–1714.

Galata Z, Moschonis G, Makridakis M, et al. Plasma proteomic analysis in obese and overweight prepubertal children. *Eur J Clin Invest*. 2011;41:1275–1283.

Ly L, Wasinger VC. Protein and peptide fractionation, enrichment and depletion: Tools for the complex proteome. *Proteomics*. 2011;11:513–534.

Griffin TJ, Bandhakavi S. Dynamic range compression: A solution for proteomic biomarker discovery? *Bioanalysis*. 2011;3:2053–2056.

Mihai DM, Deng H, Kawamura A. Reproducible enrichment of extracellular heat shock proteins from blood serum using monomeric avidin. *Bioorg Med Chem Lett*. 2011;21:4134–4137.

Chakrabarti A, Bhattacharya D, Basu A, Basu, Saha S, Halder S. Differential expression of red cell proteins in hemoglobinopathy. *Proteomics Clin Appl*. 2011;5:98–108.

Miller I. Protein separation strategies. In: Eckersall D, Whitfield PD, eds. *Methods in Animal Proteomics*. Wiley-Blackwell; 2011:41–76.

Shetty V, Jain P, Nickens Z, Sinnathamby G, Mehta A, Philip R. Investigation of plasma biomarkers in HIV-1/HCV mono- and coinfected individuals by multiplex iTRAQ quantitative proteomics. *OMICS*. 2011;15:705–717.

Fernández C, Santos HM, Ruíz-Romero C, Blanco FJ, Capelo-Martínez J-L. A comparison of depletion versus equalization for reducing high-abundance proteins in human serum. *Electrophoresis*. 2011;32:2966–2974.

Coombs KM. Quantitative proteomics of complex mixtures. *Expert Rev Proteomics*. 2011;8:659–677.

Rivers J, Hughes C, McKenna T, et al. Asymmetric proteome equalization of the skeletal muscle proteome using a combinatorial hexapeptide library. *PLoS ONE*. 2011;6:e28902.

Prudent M, Tissot JD, Lion N. Proteomics of blood and derived products: What's next. *Expert Rev Proteomics*. 2011;8:717–737.

Doucette AA, Tran JC, Wall MJ, Fitzsimmons S. Intact proteome fractionation strategies compatible with mass spectrometry. *Expert Rev Proteomics*. 2011;8:787–800.

Calvete JJ. Proteomic tools against the neglected pathology of snake bite envenoming. *Expert Rev Proteomics*. 2011;8:739–758.

Selvaraju S, El Rassi Z. Liquid-phase-based separation systems for depletion, prefractionation and enrichment of proteins in biological fluids and matrices for in-depth proteomics analysis—An update covering the period 2008-2011. *Electrophoresis*. 2012;33:74–88.

Huhn C, Ruhaak LR, Wuhrer M, Deelder AM. Hexapeptide library as a universal tool for sample preparation in protein glycosylation analysis. *J Proteomics*. 2012;75:1515–1528.

Bhonsle HS, Korwar AM, Kote SS, et al. Low plasma albumin levels are associated with increased plasma protein glycation and HbA1c in diabetes. *J Proteome Res*. 2012;11:1391–1396.

Boschetti E, Righetti PG. Breakfast at Tiffany's? Only with a deep proteomic signature!. *Electrophoresis*. 2012;33:2228–2239.

Fasoli E, D'Amato A, Citterio A, Righetti PG. Ginger Rogers? No, Ginger Ale and its invisible proteome. *J Proteomics*. 2012;75:1960–1965.

Lehr S, Hartwig S, Sell H. Adipokines: A treasure trove for the discovery of biomarkers for metabolic disorders. *Proteomics Clin Appl*. 2012;6:91–101.

Wallinder J, Bergström J, Henriksson AE. Discovery of a novel circulating biomarker in patients with abdominal aortic aneurysm: A pilot study using a proteomic approach. *Clin Tran Sci*. 2012;5:56–59.

Burgener A. Profiling cervical lavage fluid by SELDI-TOF mass spectrometry. *Methods Mol Biol*. 2012;818:143–152.

Boschetti E, Righetti PG. Mixed-beds: Beyond the frontiers of classical chromatography for proteins. In: Grinberg N, ed. *Advances in Chromatography*. vol. 50. CRC Press; 2012:1–46.

Woodbury RL, McCarthy DL, Bulman AL. Profiling of urine using ProteinChip technology. *Methods Mol Biol*. 2012;818:97–107.

Boschetti E, Chung MCM, Righetti PG. "The quest for biomarkers": Are we on the right technical track? *Proteomics Clin Appl*. 2012;6:22–41.

Bulman A, Dalmasso EA. Purification and identification of candidate biomarkers discovered using SELDI-TOF MS. *Methods Mol Biol*. 2012;818:49–66.

Santucci L, Candiano G, Bruschi M, et al. Combinatorial peptide ligand libraries for the analysis of low-expression proteins. Validation for normal urine and definition of a first protein map. *Proteomics*. 2012;12:509–515.

Medvedev A, Kopylov A, Buneeva O, Zgoda V, Archakov A. Affinity-based proteomic profiling: Problems and achievements. *Proteomics*. 2012;12:621–637.

Fassbender A, Waelkens E, Verbeeck N, et al. Proteomics analysis of plasma for early diagnosis of endometriosis. *Obstet Gynecol*. 2012;119:276–285.

Molinari CE, Casadio YS, Hartmann BT, et al. Proteome mapping of human skim milk proteins in term and preterm milk. *J Proteome Res*. 2012;11:1696–1714.

Liang C, Tan GS, Chung MC. 2D DIGE Analysis of serum after fractionation by ProteoMiner™ beads. *Methods Mol Biol*. 2012;854:181–194.

Tan SH, Kapur A, Baker MS. Chicken immune responses to variations in human plasma protein ratios: A rationale for polyclonal IgY ultraimmunodepletion. *J Proteome Res*. 2012;11:6291–6294.

Hartwing S, Lehr S. Combination of highly efficient hexapeptide ligand library-based sample preparation with 2D DIGE for the analysis of the hidden human serum/plasma proteome. *Methods Mol Biol*. 2012;854:169–180.

Cumová J, Jedličková L, Potěšil D, et al. Comparative plasma proteomic analysis of patients with multiple myeloma treated with Bortezomib-based regimens. *Klin Onkol.* 2012;25:17–25.

Palmblad M, Deelder AM. Molecular phylogenetics by direct comparison of tandem mass spectra. *Rapid Commun Mass Spectrom.* 2012;26:72–732.

Pedreschi R, Norgaard J, Maquet A. Current challenges in detecting food allergens by shotgun and targeted proteomic approaches: A case study on traces of peanut allergens in baked cookies. *Nutrients.* 2012;4:132–150.

Konecna H, Muller L, Dosoudilova H, et al. Exploration of beer proteome using Off-Gel prefractionation in combination with two-dimensional gel electrophoresis with narrow-pH-range gradients. *J Agric Food Chem.* 2012;60:2418–2426.

Fekkar A, Pionneau C, Brossas JY, et al. DIGE enables the detection of a putative serum biomarker of fungal origin in a mouse model of invasive aspergillosis. *J Proteomics.* 2012;75:2536–2549.

Gil-Donesa F, Dardeb VM, Alonso-Orgaza S, et al. Inside human aortic stenosis: A proteomic analysis of plasma. *J Proteomics.* 2012;75:1639–1653.

Esteve C, D'Amato A, Marina ML, Garcia MC, Citterio A, Righetti PG. Identification of olive (*Olea europaea*) seed and pulp proteins by nLC-MS/MS via combinatorial peptide ligand libraries. *J Proteomics.* 2012;75:2396–2403.

Millioni R, Tolin S, Fadini JP, et al. High confidence and sensitivity four-dimensional fractionation for human plasma proteome analysis. *Amino Acids.* 2012;43:2199–2202.

Koene MGJ, Mudler HA, Stockhofe-Zurwieden N, Kruijt L, Smits MA. Serum protein profiles as potential biomarkers for infectious disease status in pigs. *MBC Vet Disease.* 2012;8:32–46.

Guo X, Abliz G, Reyimu H, et al. The association of a distinct plasma proteomic profile with the cervical high-grade squamous intraepithelial lesion of Uyghur women: A 2D liquid-phase chromatography/mass spectrometry study. *Biomarkers.* 2012;17:352–361.

Fröhlich A, Gaupels F, Sarioglu H, et al. Looking deep inside: Detection of low-abundant proteins in leave extracts of *Arabidopsis thaliana* and phloem exudates of *Cucurbita maxima. Plant Physiol.* 2012;159:902–1114.

Kolla V, Jen P, Moes S, Lapaire O, Hoesli I, Hahn S. Quantitative Proteomic (iTRAQ) Analysis of 1st trimester maternal plasma samples in pregnancies at risk for preeclampsia. *J Biomed Biotechnol.* 2012;2012:ID 305964, 8 pages.

Marco-Ramell A, Arroyo L, Saco Y, et al. Proteomic analysis reveals oxidative stress response as the main adaptive physiological mechanism in cows under different production systems. *J Proteomics.* 2012;75:4399–4411.

Toniolo L, D'Amato A, Saccenti R, Gulotta D, Righetti PG. The Silk Road, Marco Polo, a bible and its proteome: A detective story. *J Proteomics.* 2012;75:3365–3373.

Soares R, Franco C, Pires E, et al. Mass spectrometry and animal science: Protein identification strategies and particularities of farm animal species. *J Proteomics.* 2012;75:4190–4206.

Fasoli E, D'Amato A, Citterio A, Righetti PG. Anyone for an aperitif? Yes, but only a Braulio DOC with its certified proteome. *J Proteomics.* 2012;75:3374–3379.

Jagtap P, McGowan T, Bandhakavi S, et al. Deep metaproteomic analysis of human salivary supernatant. *Proteomics.* 2012;12:992–1001.

Couttas TA, Raftery MJ, Padula MP, Herbert BR, Wilkins MR. Methylation of translation-associated proteins in *Saccharomyces cerevisiae*: Identification of methylated lysines and their methyltransferases. *Proteomics.* 2012;12:960–972.

Shahali Y, Sutra JP, Fasoli E, et al. Allergomic study of cypress pollen via combinatorial peptide ligand libraries. *J Proteomics.* 2012;77:101–110.

Guerrier L, Fortis F, Boschetti E. Solid-phase fractionation strategies applied to proteomics investigations. *Methods Mol Biol.* 2012;818:11–33.

Righetti PG, Boschetti E, Candiano G. Mark Twain: How to fathom the depth of your pet proteome. *J Proteomics.* 2012;75:4783–4791.

Di Girolamo F, D'Amato A, Righetti PG. Assessment of the floral origin of honey via proteomic tools. *J Proteomics.* 2012;75:3688–3693.

Fröhlich A, Gaupels F, Sarioglu H, et al. Looking deep inside: Detection of low-abundant proteins in leave extracts of Arabidopsis thaliana and phloem exudates of Cucurbita maxima. *Plant Physiol.* 2012;159:902–914.

You Q, Verschoor CP, Pant SD, Macri J, Kirby GM, Karrow NA. Proteomic analysis of plasma from Holstein cows testing positive for mycobacterium avium subsp. Paratuberculosis (map). *Vet Immunol Immunopathol.* 2012;148:243–251.

Fasoli E, D'amato A, Righetti PG, Barbieri R, Bellavia D. Exploration of the sea urchin coelomic fluid via combinatorial peptide ligand libraries. *Biol Bull.* 2012;222:93–104.

Krief G, Deutsch O, Zaks B, Wong DT, Aframian DJ, Palmon A. Comparison of diverse affinity based high-abundance protein depletion strategies for improved bio-marker discovery in oral fluids. *J Proteomics.* 2012;75:4165–4175.

Jagtap P, Bandhakavi S, Higgins LA, et al. Workflow for analysis of high mass accuracy salivary dataset using MaxQuant and ProteinPilot search algorithm. *Proteomics.* 2012;12:1726–1730.

Wu MJ, Rogers PJ, Clarke FM. 125th Anniversary Review: The role of proteins in beer redox stability. *J Instit Brewing.* 2012;118:1–11.

Hernández-Castellano LR, Martinho de Almeida A, Ventosa M, et al. The effect of colostrum intake on blood plasma proteomic profiles of newborn lambs. *Farm Animal Proteomics.* 2012;Part III:148–151.

Rasaputra KS, Liyanage R, Lay JO, Erf GF, Okimoto R, Rath NC. Changes in serum protein profiles of chickens with tibial dyschondroplasia. *The Open Proteomics J.* 2012;5:1–7.

Pedreschi R, Nørgaard J, Maquet A. Current challenges in detecting food allergens by shotgun and targeted proteomic approaches: A case study on traces of peanut allergens in baked cookies. *Nutrients.* 2012;4:132–150.

Candiano G, Santucci L, Bruschi M, et al. "Cheek-to-cheek" urinary proteome profiling via combinatorial peptide ligand libraries: A novel, unexpected elution system. *J Proteomics.* 2012;75:796–805.

Hartwig S, Lehr S. Combination of highly efficient hexapeptide ligand library-based sample preparation with 2D DIGE for the analysis of the hidden human serum/plasma proteome. *Methods Mol Biol.* 2012;854:169–180.

D'Amato A, Fasoli E, Righetti PG. Harry Belafonte and the secret proteome of coconut milk. *J Proteomics.* 2012;75:914–920.

Righetti PG, D'Amato A, Fasoli E, Boschetti E. In taberna quando sumus: A drunkard's cakewalk through wine proteomics. *Food Technol Biotechnol.* 2012;50:253–260.

Righetti PG, Boschetti E. Plant proteomics methods to reach low-abundance proteins. In: Jorrín Novo JV, Weckerth W, Komatsu S, eds. *Plant Proteomics Methods and Protocols.* Humana Press; 2012. In press.

Fassbender A, Verbeeck N, Börnigen D, et al. Combined mRNA microarray and proteomic analysis of eutopic endometrium of women with and without endometriosis. *Hum Reprod.* 2012;7:2020–2029.

Sergeant K, Renaut J, Hausman J-F. Proteomics as a toolbox to study the metabolic adjustment of trees during exposure to metal trace elements. In: *Metal toxicity in plants: Perception, Signaling & Remediation.* 2012:143–164.

Capriotti AL, Caruso G, Cavaliere C, Piovesana S, Samperi R, Laganà A. Comparison of three different enrichment strategies for serum low molecular weight protein identification using shotgun proteomics approach. *Anal Chim Acta.* 2012;740:58–65.

Esteve C, D'Amato A, Marina ML, García MC, Righetti PG. Identification of avocado (*Persea americana*) pulp proteins by nanoLC-MS/MS via combinatorial peptide ligand libraries. *Electrophoresis.* 2012;33:2799–2805.

Milan E, Lazzari C, Anand S, et al. SAA1 is over-expressed in plasma of non small cell lung cancer patients with poor outcome after treatment with epidermal growth factor receptor tyrosine-kinase inhibitors. *J Proteomics.* 2012;76:91–101.

Kramer G, Moerland PD, Jeeninga RE, et al. Proteomic analysis of HIV–T cell interaction: an update. *Frontiers Microbiol.* 2012;3:1–6.

Brown KJ, Formolo CA, Seol H, et al. Advances in the proteomic investigation of the cell secretome. *Expert Rev Proteomics.* 2012;9:337–345.

Picariello G, Mamone G, Nitride C, et al. Shotgun proteome analysis of beer and the immunogenic potential of beer polypeptides. *J Proteomics.* 2012;75:5872–5882.

Malaud E, Piquer D, Merle D, et al. Carotid atherosclerotic plaques: Proteomics study after a low-abundance protein enrichment step. *Electrophoresis.* 2012;33:470–482.

Rebulla P. From pH to MALDI-TOF: Hundreds of spotted opportunities? *J Proteomics.* 2012;76:270–274.

Tølbøll TH, Danscher AM, Andersen PH, Codrea MC, Bendixen E. Proteomics: A new tool in bovine claw disease research. *Vet J.* 2012;193:694–700.

D'Amici GM, Timperio AM, Rinalducci S, Zolla L. Combinatorial peptide ligand libraries to discover liver disease biomarkers in plasma samples. *Methods Mol Biol.* 2012;909:311–319.

Coscia A, Orru S, Di Nicola P, et al. Detection of cow's milk proteins and minor components in human milk using proteomics techniques. *J Matern Fetal Med.* 2012;25:4–51.

Lorkova L, Pospisilova J, Lacheta J, et al. Decreased concentrations of retinol-binding protein 4 in sera of epithelial ovarian cancer patients: a potential biomarker identified by proteomics. *Oncol Rep.* 2012;27:318–324.

D'Alessandro A, Zolla L. Food safety and quality control: Hints from proteomics. *Food Technol Biotechnol.* 2012;50:275–285.

Callesen AK, Mogensen O, Jensen AK, et al. Reproducibility of mass spectrometry based protein profiles for diagnosis of ovarian cancer across clinical studies: A systematic review. *J Proteomics.* 2012;75:2758–2772.

Jiang J, Opanubi KJ, Coombs KM. Non-biased enrichment does not improve quantitative proteomic delineation of reovirus T3D-infected HeLa cell protein alterations. *Frontiers Microbiology / Virology.* 2012;3:1–16.

Yang F, Shen Y, Camp DG, Smith RD. High-pH reversed-phase chromatography with fraction concatenation for 2D proteomic analysis. *Expert Rev Proteomics.* 2012;9:129–134.

D'Amici GM, Rinalducci S, Zolla L. Depletion of hemoglobin and carbonic anhydrase from erythrocyte cytosolic samples by preparative clear native electrophoresis. *Nature Protocols.* 2012;7:36–44.

Kuruvillaet J, Cristobal S. Saliva proteomics: Tool for novel diagnosis for farm animal diseases from body fluids. *Farm Animal Proteomics.* 2012;Part II:84–86.

Roncada P, Piras C, Soggiu A, Turk R, Urbani A, Bonizzi L. Farm animal milk proteins. *J Proteomics.* 2012;75:4259–4274.

Otto M, Bowser R, Turner M, et al. Roadmap and standard operating procedures for biobanking and discovery of neurochemical markers in ALS. *Amyotroph Lateral Scler.* 2012;13:1–10.

Martinho de Almeida A, Bendixen E. Pig proteomics: A review of a species in the crossroad between biomedical and food sciences. *J Proteomics.* 2012;75:4296–4314.

Zerefos PG, Aivaliotis M, Baumann M, Vlahou A. Analysis of the urine proteome via a combination of multi-dimensional approaches. *Proteomics.* 2012;12:391–400.

Walpurgis K, Kohler M, Thomas A, et al. Validated hemoglobin-depletion approach for red blood cell lysate proteome analysis by means of 2D PAGE and Orbitrap MS. *Electrophoresis.* 2012;33:2537–2545.

Castagnola M, Cabras T, Iavarone F, et al. The human salivary proteome: A critical overview of the results obtained by different proteomic platforms. *Expert Rev Proteomics.* 2012;9:33–46.

van Gool AJ, Hendrickson RC. The proteomic toolbox for studying cerebrospinal fluid. *Expert Rev Proteomics.* 2012;9:165–179.

Minton O, Stone PC. The identification of plasma proteins associated with cancer-related Fatigue syndrome (CRFS) in disease-free breast cancer patients using proteomic analysis. *J Pain Symptom Manage* 2012. In press.

Selected CPLL Bibliography

Yang YX, Wang JQ, Bu DP, et al. Comparative proteomics analysis of plasma proteins during the transition period in dairy cows with or without subclinical mastitis after calving. *Czech J Anim Sci.* 2012;57:481–489.

Wang W, Fonslow BR, Wong CCL, Nakorchevsky A, Yates JR. Improving the comprehensiveness and sensitivity of sheathless CE-MS/MS for proteomic analysis. *Anal Chem.* 2012;84:8505–8513.

Gao X, Lee J, Malladi S, Melendez L, Lascelles BDX, Al-Murrani S. Feline degenerative joint disease: a genomic and proteomic approach. *J Feline Med Surg.* 2012; in press.

Senapati S, Barnhart KT. Biomarkers for ectopic pregnancy and pregnancy of unknown location. *Fertil Steril.* 2012; in press.

Rubin O, Delobel J, Prudent M, et al. Red blood cell–derived microparticles isolated from blood units initiate and propagate thrombin generation. *Transfusion.* 2012; in press.

Di Girolamo F, Del Chierico F, Caenaro G, Lante I, Muraca M, Putignani L. Human serum proteome analysis: new source of markers in metabolic disorders. *Biomarkers in Medicine.* 2012;6:759–773.

Imure T, Sato K. Beer proteomics analysis for beer quality control and malting barley breeding. *Food Res Intern.* 2012; in press.

Siciliano RA, Mazzeo MF, Arena S, Renzone G, Scaloni A. Mass spectrometry for the analysis of protein lactosylation in milk products. *Food Res Intern.* 2012; in press.

Amado FML, Ferreira RP, Vitorino R. One decade of salivary proteomics: Current approaches and outstanding challenges. *Clin Biochem.* 2012; in press.

Badrealam KF, Upadhyay RC, Singh AK, Mohanty AK, Chaudhary N, Choudhary J. Prefractionation strategies for bovine plasma/serum proteomics. *Wayamba J Animal Sci.* 2012;4:259–265.

Gaupels F, Sarioglu H, Beckmann M, et al. Deciphering systemic wound responses of the pumpkin extrafascicular phloem by metabolomics and stable isotope-coded protein labeling. *Plant Physiol.* 2012;160:2285–2299.

Coscia A, Orrù S, Di Nicola P, et al. Detection of cow's milk proteins and minor components in human milk using proteomics techniques. *J Matern Fetal Neonatal Med.* 2012;25:54–56.

Lista S, Faltraco F, Hampel F. Biological and methodical challenges of blood-based proteomics in the field of neurological research. *Prog Neurobiol.* 2013;101–102:18–34.

Afroz A, Zahur M, Zeeshan N, Komatsu S. Plant-bacterium interactions analyzed by proteomics. *Front Plant Sci.* 2013;24:21.

Boschetti E, Candiano G, Righetti PG. Two-dimensional electrophoresis associated with dynamic range reduction: A winning combination to discover low-abundance protein markers. In: Veenstra TD, Issaq HJ, eds. *Proteomic and Metabolomic Approaches to Biomarker Discovery.* Amsterdam: Elsevier; 2013. In press.

Lee CP, Taylor NL, Millar AH. Recent advances in the composition and heterogeneity of the arabidopsis mitochondrial proteome. *Front Plant Sci.* 2013;4:1–8.

Schulz BL, Cooper-White J, Punyadeera CK. Saliva proteome research: Current status and future outlook. *Crit Rev Biotechnol.* 2013. In press.

Rodríguez-Suárez E, Whetton AD. The application of quantification techniques in proteomics for biomedical research. *Mass Spectrom Rev.* 2013;32:1–26.

Saez V, Fasoli E, D'Amato A, Simo-Alfonso E, Righetti PG. Artichoke and Cynar liqueur: Two (not quite) entangled proteomes. *Biochim Biophys Acta.* 2013;1834:119–126.

Napoli C, Zullo A, Picascia A, Infante T, Mancini FP. Recent advances in proteomic technologies applied to cardiovascular disease. *J Cell Biochem.* 2013;114:7–20.

Rolland AQ, Lavigne R, Dauly C, et al. Identification of genital tract markers in the human seminal plasma using an integrative genomics approach. *Hum Reprod.* 2013;28:199–209.

Wright B, Stanley RG, Kaiser WJ, Gibbings JM. The integration of proteomics and systems approaches to map regulatory mechanisms underpinning platelet function. *Proteomics Clin Appl.* 2013;7:144–154.

Santucci L, Candiano G, Petretto A, et al. Combinatorial ligand libraries as a two-dimensional method for proteome analysis. *J Chromatogr A.* 2013; in press.

Su L, Cao L, Zhou R, et al. Identification of Novel biomarkers for sepsis prognosis via urinary proteomic analysis using iTRAQ labeling and 2D-LC-MS/MS. *PLoS One.* 2013;8:e54237.

Gomes C, Almeida A, Ferreira JA, et al. Glycoproteomic analysis of serum from patients with gastric precancerous lesions. *J Proteome Res.* 2013; in press.

Kruger T, Lehmann T, Rhode H. Effect of quality characteristics of single sample preparation steps in the precision and coverage of proteomic studies - A review. *Anal Clin Acta.* 2013; in press.

Marrocco C, D'Allessandro A, Girelli G, Zolla L. Proteomic analysis of platelets treated with gamma irradiation versus a commercial photochemical pathogen reduction technology. *Transfusion.* 2013; in press.

Yates JR. The Revolution and Evolution of Shotgun Proteomics for Large-Scale Proteome Analysis. *J Am Chem Soc.* 2013;135:1629–1640.

Martos G, López-Fandiño R, Molina E. Immunoreactivity of hen egg allergens: Influence on in-vitro gastrointestinal digestion of the presence of other egg white proteins and of egg yolk. *Food Chem.* 2013;136:775–781.

von Toerne C, Kahle M, Schäfer A, et al. Apoe, Mbl2 and Psp plasma protein levels correlate with diabetic phenotype in NZO mice - an optimized rapid workflow for SRM-based quantification. *J Proteome Res.* 2013; in press.

Tomatis VM, Papadopulos A, Malintan NT, et al. Myosin VI small insert isoform maintains exocytosis by tethering secretory granules to the cortical actin. *J Cell Biol.* 2013; in press.

Righetti PG. Bioanalysis: *heri, hodie, cras. Electrophoresis.* 2013; in press.

Note: Page numbers followed by *f* indicate figures, and *t* indicate tables.

Printed and bound by CPI Group (UK) Ltd, Croydon, CR0 4YY

03/10/2024

01040325-0018